Macromolecules · 1

Structure and Properties

Second Edition, Revised and Expanded

Macromolecules

Macromolecules · 1

Structure and Properties

Second Edition, Revised and Expanded

Hans-Georg Elias

Michigan Molecular Institute
Midland, Michigan

Translated from German by **John W. Stafford**

PLENUM PRESS • *NEW YORK AND LONDON*

Library of Congress Cataloging in Publication Data

Elias, Hans-Georg, 1928–
  Macromolecules: structure and properties.

  Translation of: Makromoleküle.
  Bibliography: p.
  Includes index.
  1. Macromolecules. I. Title.
QD381.E4413   1983   547.8   83-19294
ISBN 0-306-41077-X (v. 1)

© 1984 Plenum Press, New York
A Division of Plenum Publishing Corporation
233 Spring Street, New York, N.Y. 10013

Printed in the United States of America

Acknowledgments

The permission of the following publishers to reproduce tables and figures is gratefully acknowledged:

Academic Press, London/New York, D. Lang, H. Bujard, B. Wolff and D. Russell, J. Mol. Biol. 23 (1967) 163, (Fig. 4-15); R. S. Baer, *Adv. Prot. Chem.* 7 (1952) 69, (Fig. 30-5); C. D. Han, *Rheology in Polymer Processing*, 1976, (Fig. 35-14)

Akademic-Verlag, Berlin, H. Dautzenberg, Faserforschg. Textiltechn. 21 (1970) 117, (Fig. 4–19); K. Edelmann, *Faserforschg. Textiltechn.* 3 (1952) 344, (Fig. 7-6)

Akademische Verlagsgesellschaft, Leipzig, G. V. Schulz, A. Dinglinger and E. Husemann, Z. Physik. Chem. B 43 (1939) 385 (Fig. 20-6)

American Chemical Society, Washington, D.C.; S. I. Mizushima and T. Shimanouchi, *J. Amer. Chem. Soc.* 86 (1964) 3521, (Fig. 4-7); M. Goodman and E. E. Schmidt, *J. Amer. Chem. Soc.* 81 (1959) 5507, (Fig. 4-22); K. G. Siow and G. Delmas, *Macromolecules* 5 (1972) 29, (Fig. 6-13); P. J. Flory, *J. Amer. Chem. Soc.* 63 (1941) 3083, (Fig. 17-9); G. V. Schulz, *Chem. Tech.* 3/4 (1973) 224, (Fig. 18-5); G. V. Schulz, *Chem. Tech.* 3/4 (1973) 221, (Fig. 20-2); M. Litt, Macromolecules 4 (1971) 312, (Fig. 22-11); N. Ise and F. Matsui, *J. Am. Chem. Soc.* 90 (1968) 4242, (Fig. 23-1); H. P. Gregor, L. B. Luttinger and E. M. Loebl, *J. Phys. Chem.* 59 (1955) 34, (Fig. 23-8); J. S. Noland, N. N.-C. Hsu, R. Saxon and D. M. Schmitt, in N. A. J. Platzer, Ed., Multicomponent Polymer Systems, *ACS Adv. Chem. Ser.* 99. (Fig. 35-5)

American Institute of Physics, New York, W. D. Niegisch and P. R. Swan, *J. Appl. Phys.* 31 (1960) 1906, (Fig. 5-16); M. Shen, D. A. McQuarrie and J. L. Jackson, *J. Appl. Phys.* 38 (1967) 791, (Fig. 11-4); H. D. Keith and F. J. Padden, jr., *J. Appl. Phys.* 30 (1959) 1479, (Fig. 11-22); R. S. Spencer and R. F. Boyer, *J. Appl. Phys.* 16 (1945) 594, (Fig. 11-17)

Applied Science Publishers, London, C. B. Bucknall, Toughened Plastics (1977), (Fig. 35-12 and Fig. 35-13)

Badische Anilin- & Soda-Fabrik AG, Ludwigshafen/Rh., -, Kunststoff-Physik im Gespräch, 2 Aufl., (1968), S. 103 and 107, (Figs. 11-8 and 11-10)

The Biochemical Journal, London, P. Andrews, *Biochem. J.* 91 (1964) 222, (Fig. 9-18)

Butterworths, London, H. P. Schreiber, E. B. Bagley and D. C. West, *Polymer* 4 (1963) 355, (Fig. 7-8); A. Sharples, *Polymer* 3 (1962) 250, (Fig. 10-8); A. Nakajima and F. Hameda, IUPAC, *Macromol. Microsymp.* VIII and IX (1972) 1, (Fig. 10-9); A. Gandica and J. H. Magill, *Polymer* 13 (1972) 595, (Fig. 10-10); C. E. H. Bawn and M. B. Huglin, *Polymer* 3 (1962) 257, (Fig. 17-4); D. R. Burfield and P. J. T. Tait, *Polymer* 13 (1972) 307, (Figs. 19-2 and 19-3); I. D. McKenzie, P. J. T. Tait, D. R. Burfield, *Polymer* 13 (1972) 307, (Fig. 19-4)

Chemie-Verlag Vogt-Schild AG, Solothurn, G. Henrici-Olive and S. Olive, *Kunststoffe-Plastics* 5 (1958) 315 (Fig. 20-3 and 20-4)

M. Dekker, New York, H.-G. Elias, S. K. Bhatteya and D. Pae, *J. Macromol. Sci. [Phys.]* B-12 (1976) 599 (Fig. 11-1); R. L. McCullough, *Concepts of Fibers-Resins Composites*, (1971) (Fig. 35-15 and 35-16)

Engineering, Chemical & Marine Press, London, R. A. Hudson, *British Plastics* 26 (1953) 6, (Fig. 11-18)

The Faraday Society, London, R. M. Barrer, *Trans. Faraday Soc.* 35 (1939) 628, (Tab. 7-5); R. B. Richards, *Trans. Faraday Soc.* 42 (1946) 10, (Fig. 6-23); L. R. G. Treloar, *Trans. Faraday Soc.* 40 (1944) 59, (Fig. 11-5); F. S. Dainton and K. J. Ivin, *Trans. Faraday Soc.* 46 (1950) 331, (Tab. 16-7)

W. H. Freeman and Co., San Francisco, M. F. Perutz, *Sci. American* (Nov. 1964) 71, (Fig. 4-14); H. Neurath, *Sci American* (Dec. 1964) 69, (Fig. 30-2)

Gordon and Breach, New York, G. R. Snelling, *Polymer News* 3/1 (1976) 36, (Fig. 33-4)

Gazetta Chimica Italiana, Rom, G. Natta, P. Corradini and I. W. Bassi, *Gazz. Chim. Ital.* 89 (1959) 784, (Fig. 4-8)

General Electric Co., Schenectady, N. Y., A. R. Schultz, *GE-Report* 67-C-072, (Fig. 6-18); F. A. Karasz, H. E. Bair and J. M. O'Reilly, *GE-Report* 68-C-001, (Fig. 10-3)

C. Hanser, München, G. Rehage, *Kunststoffe* 53 (1963) 605, (Fig. 6-14); H.-G. Elias, *Neue polymere Werkstoffe* 1969-1974 (1975), (Fig. 33-2 and 33-5); H.-G. Elias, *Kunststoffe* 66 (1976) 641, (Fig. 33-6)

Interscience Publ., New York, T. M. Birshtein and O. B. Ptitsyn, *Conformation of Macromolecules*, (1966), (Fig. 4-4); P. J. Flory, *Statistical Mechanics of Chain Molecules*, 1969, (Fig. 4-18); P. Pino, F. Ciardelli, G. Montagnoli and O. Pieroni, *Polymer Letters* 5 (1967) 307, (Fig. 4-23); A Jeziorny and S. Kepka, *J. Polymer Sci. B* 10 (1972) 257, (Fig. 5-4); P. H. Lindenmeyer, V. F. Holland and F. R. Anderson, *J. Polymer Sci. C* 1 (1963) 5, (Fig. 5-17, 5-18, 5-19); H. D. Keith, F. J. Padden and R. G. Vadimsky, *J. Polymer Sci. [A-2]* 4 (1966) 267, (Fig. 5-22); G. Rehage and D. Moller, *J. Polymer Sci. C* 16 (1967) 1787, (Fig. 6-17); T. G. Fox, *J. Polymer Sci. C* 9 (1965) 35, (Fig. 7-9); Z. Grubisic, P. Rempp and H. Benoit, *J. Polymer Sci. B* 5 (1967) 753, (Fig. 9-19); G. Rehage and W. Borchard, in R. N. Haward, Hrsg., *The Physics of the Glassy State* (1973), (Fig. 10-2); N. Overbergh, H. Bergmans and G. Smets, *J. Polymer Sci. C* 38 (1972) 237, (Fig. 10-14); O. B. Edgar and R. Hill, *J. Polymer Sci.* 8 (1952) 1, (Fig. 10-18); P. I. Vincent, *Encycl. Polymer Sci. Technol.* 7 (1967) 292, (Fig. 11-15); P. J. Berry, *J. Polymer*

Sci. **50** (1961) 313, (Fig. 11-20); H. W. McCormick, F. M. Brower and L. Kin, *J. Polymer Sci.* **39** (1959) 87, (Fig. 11-23); N. Berendjick, in B. Ke, Hrsg., *Newer Methods of Polymer Characterization*, (1964), (Fig. 12-1); E. J. Lawton, W. T. Grubb and J. S. Balwit, *J. Polymer Sci.* **19** (1956) 455, (Fig. 21-2); G. Molau and H. Keskkula, *J. Polymer Sci.* [A-1] **4** (1966) 1595, (Fig. 35-8); A. Ziabicki, in H. Mark, S. M. Atlas and E. Cernia, *Man-Made Fibers*, Vol. 1 (1967), (Fig. 38-6, 38-7)

IPC Business Press, G. Allen, G. Gee and J. P. Nicholson, *Polymer* **2** (1961) **8**, (Fig. 35-6)

Journal of the Royal Netherlands Chemical Society, s' Gravenhage, D. T. F. Paals and J. J. Hermans, *Rec. Trav.* **71** (1952) 433, (Fig. 9-25)

Kodansha, Tokio, S. Iwatsuki and Y. Yamashita, Progr. *Polymer Sci. Japan* **2** (1971) 1, (Fig. 22–9)

Kogyo Chosakai Publ. Co., Tokyo, M. Matsuo, *Japan Plastics* (July 1968), (Fig. 5-33)

McGraw-Hill Book Co., New York, A. X. Schmidt and C. A. Marlies, *Principles of High Polymer Theory and Practice* (1948), (Fig. 10-5)

Pergamon Press, New York, J. T. Yang, *Tetrahedron* **13** (1961) 143, (Fig. 4-26); H. Hadjichristidis, M. Devaleriola and V. Desreux, *Europ. Polym. J.* **8** (1972) 1193, (Fig. 9-27); J. M. G. Cowie, *Europ. Polym. J.* **11** (1975) 295, (Fig. 10-22)

Plenum Publ., New York, J. A. Manson and H. Sperling, *Polymer Blends and Composites* (1976), (Fig. 35-10)

The Royal Society, London, N. Grassie and H. W. Melville, *Proc. Royal Soc.* *[London] A* **199** (1949) 14, (Fig. 23-6)

Societa Italiana di Fisica, Bologna, G. Natta and P. Corradini, *Nuovo Cimento Suppl.* **15** (1960) 111 (Fig. 5-9)

Society of Plastics Engineers, Greenwich, Conn., J. D. Hoffman, *SPE Trans.* **4** (1964) 315, (Fig. 10-13); S. L. Aggarwal and R. L. Livigni, *Polymer Engng. Sci.* **17** (1977) 498, (Fig. 35-9)

Springer-Verlag, New York, H.-G. Elias, R. Bareiss and J. G. Watterson, *Adv. Polymer Sci.* **11** (1973) 111, (Fig. 8-5)

D. Steinkopff-Verlag, Darmstadt, A. J. Pennings, J. M. M. A. van der Mark and A. M. Keil, Kolloid-Z. **237** (1970) 336, (Fig. 5-28); G. Kanig, *Kolloid-Z.* **190** (1963) 1, (Fig. 35-1)

Textile Research Institute, Princeton, NJ, H. M. Morgan, *Textile Res. J.* **32** (1962) 866, (Fig. 5-36)

Van Nostrand Reinhold Co., New York, R. C. Bowers and W. A. Zisman, in E. Baer, *Hrsg., Engng. Design for Plastics*, (Fig. 13-4)

Verlag Chemie, Weinheim/Bergstrasse, G. V. Schulz, *Ber. Dtsch. Chem. Ges.* **80** (1947) 232, (Fig. 9-1); H. Benoit, *Ber. Bunsenges.* **70** (1966) 286, (Fig. 9-5); G. Rehage, *Ber. Bunsenges.* **74** (1970) 796, (Fig. 10-1); K.-H. Illers, *Ber. Bunsenges.* **70** (1966) 353, (Fig. 10-25); J. Smid, *Angew. Chem.* **84** (1972) 127, (Fig. 18-1); F. Patat and Hj. Sinn, *Angew. Chem.* **70** (1958) 496, (Eq. 19-12); G. V. Schulz, *Ber. Dtsch. Chem. Ges.* **80** (1947) 232 (Fig. 20-5); E. Thilo, *Angew. Chem.* **77** (1965) 1057, (Fig. 32-4)

Didici in mathematicis ingenio, in natura experimentis, in legibus divinis humanisque auctoritate, in historia testimoniis nitendum esse.

G.W. Leibniz

(I learned that in mathematics one depends on inspiration, in science on experimental evidence, in the study of divine and human law on authority, and in historical research on authentic sources.)

Preface

The second edition of this textbook is identical with its fourth German edition and it thus has the same goals: precise definition of basic phenomena, a broad survey of the whole field, integrated representation of chemistry, physics, and technology, and a balanced treatment of facts and comprehension. The book thus intends to bridge the gap between the often oversimplified introductory textbooks and the highly specialized texts and monographs that cover only parts of macromolecular science.

The text intends to survey the whole field of macromolecular science. Its organization results from the following considerations.

The chemical structure of macromolecular compounds should be independent of the method of synthesis, at least in the ideal case. Part I is thus concerned with the chemical and physical structure of polymers.

Properties depend on structure. Solution properties are thus discussed in Part II, solid state properties in Part III. There are other reasons for discussing properties before synthesis: For example, it is difficult to understand equilibrium polymerization without knowledge of solution thermodynamics, the gel effect without knowledge of the glass transition temperature, etc.

Part IV treats the principles of macromolecular syntheses and reactions. The emphasis is on general considerations, not on special mechanisms, which are discussed in Part V. The latter part is a surveylike description of important polymers, especially the industrially and biologically important ones. Part V also contains a new chapter on raw materials and energy. It no longer surveys monomer syntheses because this information can be found in several recent good books.

Part VI is totally new. It is an introduction into polymer technology and thus discusses thermoplasts, thermosets, elastomers, fibers, coatings, and adhesives with respect to their end-use properties. It also contains chapters on additives, blends and composites.

About 70% of the text has been rewritten. Outdated sections have been replaced, newly available information has been added. But even a book of this size cannot treat everything and so I decided to give biochemical, biophysical, and biotechnological problems only cursory treatment.

Nomenclature and symbols follow in general the recommendations of the Systéme International and the International Union of Pure and Applied Chemistry, although sometimes deviations had to be chosen for the sake of clarity.

Undergraduate-level knowledge of inorganic, organic, and physical chemistry is assumed for the study of certain chapters. Whenever possible, all treatments and derivations were developed step-by-step from basic phenomena and concepts. In certain cases, I found it necessary to replace rigorous and mathematically complex derivations by simpler ones. I very much hope that this makes the book suitable for self-study.

A textbook must, of necessity, rely heavily on secondary literature available as review articles and monographs. Although I have consulted more than 5000 original papers before, during, and after the compilation of the individual chapters, I have (with the exception of one area) not cited the original literature. The exception is in the historical development of the subject, and this exception has been made because I was unable to find an accessible, balanced account treating macromolecular science in terms of the development of its ideas and concepts. I believe also that reading these old original works rewards the student with an insight into how a better understanding of the observed phenomena developed from the difficulties, prejudices, and ill-defined concepts of the times. However, because of the width and diversity of the field, a fully comprehensive and historically sound treatment of the development of its ideas and discoveries is beyond the scope of this work. Thus, since I have not been able to give due recognition to the work of individual chemists and physicists, I have only used names in the text when they have become *termini technici* in relation to methodology, phenomena, and reactions (for example: Ziegler catalysis, Flory-Huggins constant, Smith-Harkins theory, etc.). The occasional use of trade names cannot be taken to mean that these are free for general use.

In writing this book, I have tried to follow the practice of Dr. Andreas Libavius,* who had

> principally taken, from the most far-flung sources, individual data from the best authors, old and new, and also from some general texts, and these were then, according to theoretical considerations and the widest possible experience, carefully interpreted and painstakingly molded into a homogeneous treatise.

It is a pleasure for me to thank all of my colleagues who supported me through advice and reprints. My special thanks go to the translator, Dr. John W. Stafford, Basel, and his benevolent attitude toward my attempts to convert his good Queen's English into my bad American.

Midland   *Hans-Georg Elias*

Alchemia, chemistry textbook from the year 1597; new edition in German, Gmelin Institute, 1964.

Contents

Vol. 1. Structure and Properties
(A selected literature list is given at the end of each chapter)

Part I. Structure

Part II. Solution Properties

Part III. Solid State Properties

Contents

Vol. 2. Syntheses and Materials
(A selected literature list is given at the end of each chapter)

Part IV. Syntheses and Reactions

Part V. Materials

Chapter 24. Raw Materials...........................

Part VII. Appendix

Notation

As far as possible, the abbreviations have been taken from the "Manual of Symbols and Terminology for Physicochemical Quantities and Units," *Pure and Applied Chemistry* **21**(1) (1970). However, for clarity, some of the symbols used there had to be replaced by others.

The ISO (International Standardization Organization) has suggested that all extensive quantities should be described by capital letters and all intensive quantities by lower-case letters. IUPAC does not follow this recommendation, however, but uses lower-case letters for specific quantities.

The following symbols are used above or after a letter:

Symbols Above Letters

—   signifies an average, e.g., $\overline{M}$ is the average molecular weight; more complicated averages are often indicated by $\langle \; \rangle$, e.g., $\langle R_G^2 \rangle_z$ is another way of writing $(\overline{R_G^2})_z$

~   stands for a partial quantity, e.g., $\tilde{v}_A$ is the partial specific volume of the compound A; V_A is the volume of A, whereas $\tilde{V}_A^m$ is the partial molar volume of A

Superscripts

$^{\circ}$ pure substance or standard state

∞ infinite dilution or infinitely high molecular weight

m molar quantity (in cases where subscript letters are impractical)

(q) the q order of a moment (always in parentheses)

$\ddagger$ activated complex

Subscripts

0   initial state

1   solvent

2   solute

3   additional components (e.g., precipitant, salt, etc.)

am	amorphous
B	brittleness
bd	bond
bp	boiling process
cr	crystalline
crit	critical
cryst	crystallization
e	equilibrium
E	end group
G	glassy state
i	run number
i	initiation
i	isotactic diads
ii	isotactic triads
is	heterotactic triads
j	run number
k	run number
m	molar
M	melting process
mon	monomer
n	number average
p	polymerization, especially propagation
pol	polymer
r	general for average
s	syndiotactic diads
ss	syndiotactic triads
st	start reaction
t	termination
tr	transfer
u	conversion
U	monomeric unit
w	mass average
z	z average

Prefixes

at	atactic
ct	*cis*-tactic
eit	erythrodiisotactic
it	isotactic
st	syndiotactic
tit	threodiisotactic
tt	*trans*-tactic

Square brackets around a letter signify molar concentrations. (IUPAC prescribes the symbol c for molar concentrations, but to date this has consistently been used for the mass/volume unit).

Angles are always given by °.

Apart from some exceptions, the meter is not used as a unit of length; the units cm and mm derived from it are used. Use of the meter in macromolecular science leads to very impractical units.

Symbols

A absorption (formerly extinction) ($= \log \tau_i^{-1}$)

A surface

A Helmholtz energy ($A = U - TS$)

A^m molar Helmholtz energy

A preexponential constant [in $k = A \exp(-E^{\ddagger}/RT)$]

A_2 second virial coefficient

a activity

a exponent in the property/molecular weight relationship ($E^{\ddagger} = KM^a$); always with an index, e.g., a_η, a_s, etc.

a linear absorption coefficient, $a = L^{-1} \log (I_0/I)$

a_0 constant in the Moffit-Yang equation

b bond length

b_0 constant in the Moffit-Yang equation

C cycle, axis of rotation

C heat capacity

C^m molar heat capacity

C_N characteristic ratio

C_{tr} transfer constant ($C_{tr} = k_{tr}/k_p$)

c specific heat capacity (formerly: specific heat); c_p = specific isobaric heat capacity, c_v = specific isochore heat capacity

c "weight" concentration(= mass of solute divided by volume of solution); IUPAC suggests the symbol ρ for this quantity, which could lead to confusion with the same IUPAC symbol for density

$\hat{c}$ speed of light in a vacuum, speed of sound

D digyric, twofold axis

D diffusion coefficient

D_{rot} rotational diffusion coefficient

E energy (E_k = kinetic energy, E_p = potential energy, $E^{\ddagger}$ = energy of activation)

E electronegativity

E modulus of elasticity, Young's modulus ($E = \sigma_{11}/\epsilon$)

E general property

$\boldsymbol{E}$ electrical field strength

e elementary charge

e parameter in the Q–e copolymerization theory

e cohesive energy density (always with an index)

e partial electric charge

F force

f fraction (excluding molar fraction, mass fraction, volume fraction)

f molecular coefficient of friction (e.g., f_s, f_D, f_{rot})

f functionality

G *gauche* conformation

G Gibbs energy (formerly free energy or free enthalpy) ($G = H - TS$)

G^m molar Gibbs energy

G shear modulus ($G = \sigma_{21}/$ angle of shear)

G statistical weight fraction ($G_i = g_i/\Sigma_i g_i$)

g gravitational acceleration

g statistical weight

g parameter for the dimensions of branched macromolecules

H height

H	enthalpy
H^m	molar enthalpy
h	height
h	Planck's constant
I	electrical current strength
I	intensity
i	radiation intensity of a molecule
J	flow (of mass, volume, energy, etc.), always with a corresponding index
K	general constant
K	equilibrium constant
K	compression modulus ($p = -K\Delta V/V_0$)
k	Boltzmann constant
k	rate constant for chemical reactions (always with an index)
L	length
L	chain end-to-end distance
L	phenomenological coefficient
l	length
M	molar mass (previously, molecular weight)
m	mass
N	number of elementary particles (e.g., molecules, groups, atoms, electrons)
N_L	Avogadro number (Loschmidt's number)
n	amount of a substance (mole)
n	refractive index
P	permeability coefficient
Pr	production
p	probability
p	dipole moment
$\mathbf{p}_i$	induced dipolar moment
p	pressure
p	extent of reaction
p	number of conformational structural elements per turn
Q	quantity of electricity, charge
Q	heat
Q	partition function (system)
Q	parameter in the Q-e copolymerization equation
Q	polymolecularity index ($Q = \overline{M_w}/\overline{M_n}$)
Q	price
q	partition function (particles)
R	molar gas constant
R	electrical resistance
R	dichroitic ratio
R_G	radius of gyration
R_n	run number
R_ϑ	Rayleigh ratio
r	radius
r	copolymerization parameter
r_0	initial molar ratio of reactive groups in polycondensations
S	sphenoidal or alternating axis of symmetry
S	entropy
S^m	molar entropy
S	solubility coefficient
s	sedimentation coefficient

s	selectivity coefficient (in osmotic measurements)
T	temperature (both in K and in °C)
T	*trans* conformation
T	tetrahedral axis of symmetry
t	time
U	voltage
U	internal energy
U^m	molar internal energy
u	excluded volume
V	volume
V	electrical potential
v	rate, rate of reaction
v	specific volume (always with an index)
W	work
w	mass fraction
X	degree of polymerization
X	electrical resistance
x	mole fraction
y	yield
Z	collision number
Z	z fraction
z	ionic charge
z	coordination number
z	dissymmetry (light scattering)
z	parameter in excluded volume theory
z	number of nearest neighbors
α	angle, especially angle of rotation in optical activity
α	cubic expansion coefficient $[\alpha = V^{-1}(\partial V/\partial T)_p]$
α	expansion coefficient (as reduced length, e.g., α_L in the chain end-to-end distance or α_R for the radius of gyration)
α	degree of crystallinity (always with an index for method, i.e., ir, V, etc.)
α	electric polarizability of a molecule
$[\alpha]$	"specific" optical rotation
β	angle
β	coefficient of pressure
β	excluded volume cluster integral
Γ	preferential solvation
γ	angle
γ	surface tension
γ	linear expansion coefficient
γ	interfacial energy
γ	cross-linking index
γ	velocity gradient
δ	loss angle
δ	solubility parameter
δ	chemical shift
ϵ	linear expansion ($\epsilon = \Delta l/l_0$)
ϵ	expectation
ϵ	energy per molecule
ϵ_r	relative permittivity (dielectric number)
η	dynamic viscosity
$[\eta]$	intrinsic viscosity (called J_0 in DIN 1342)

Θ characteristic temperature, especially theta temperature
θ angle, especially torsion angle (conformation angle)
ϑ angle
κ isothermal compressibility $[\kappa = V^{-1}(\partial V/\partial p)_T]$
κ electrical conductivity (formerly specific conductivity)
κ enthalpic interaction parameter in solution theory
Λ axial ratio of rods
λ wavelength
λ heat conductivity
λ degree of coupling
μ chemical potential
μ moment
μ permanent dipole moment
ν moment, with respect to a reference value
ν frequency
ν kinetic chain length
ν effective network chain molar concentration
ξ shielding ratio in the theory of random coils
Ξ partition function
Π osmotic pressure
π mathematical constant
ρ density
σ mirror image, mirror image plane
σ mechanical stress (σ_{11} = normal stress, σ_{21} = shear stress)
σ standard deviation
σ hindrance parameter
σ cooperativity
σ electrical conductivity
τ bond angle
τ relaxation time
τ_i internal transmittance (transmission factor) (represents the ratio of transmitted to absorbed light)
ϕ volume fraction (volume content)
ϕ angle
$\varphi(r)$ potential between two segments separated by a distance r
Φ constant in the viscosity–molecular-weight relationship
$[\Phi]$ "molar" optical rotation
χ interaction parameter in solution theory
ψ entropic interaction parameter in solution theory
ω angular frequency, angular velocity
Ω angle
Ω probability
Ω skewness of a distribution

Part I
Structure

Chapter 1

Introduction

1.1. Basic Concepts

Macromolecules are molecules built from a large number of atoms. They can be of natural origin, like cellulose, proteins, and natural rubber, or they may be synthetically produced, like poly(ethylene), nylon, and silicones. All macromolecules consist of at least one chain of atoms bonded together and running through the whole molecule. This "backbone" can consist of, for example, carbon atoms, as in

$$R—(CH_2)_N—R'$$
poly(methylene)

or of carbon and oxygen atoms, as with

$$R—(OCH_2CH_2)_N—R'$$
poly(oxyethylene);
poly(ethylene oxide);
poly(ethylene glycol)

or of carbon and nitrogen atoms, as in

$$R—(NH—CHR''—CO)_N—R'$$
polypeptides

or contain no carbon atoms, as with

$$R—(O—Si(CH_3)_2)_N—R'$$
poly(dimethyl siloxane)

The bonds between atoms in the "backbone," or main chain, do not necessarily have to be covalent. Macromolecules may have coordinate bonds

3

or electron-deficient bonds in the main chain (see Chapter 2). By definition, however, neither ionic bonds nor metallic bonds give macromolecular chains.

Each main chain consists of a series of *constitutional units* and of two *end groups*. With poly(oxyethylene), such end groups may, for example, be R=H and R'=OH, so that the molecule has two hydroxyl end groups. The constitutional units in poly(oxyethylene) are the groups

$$CH_2, \quad CH_2CH_2, \quad O, \quad OCH_2, \quad OCH_2CH_2, \quad OCH_2CH_2O, \quad CH_2OCH_2, \text{ etc.}$$

The constitutional unit OCH_2CH_2 is the smallest constitutional unit whose repetition completely describes the main chain structure. Consequently, it is called the *constitutional repeating unit*.

The repeating unit concept always refers to the structure of the macromolecular chain. The concept of the *monomeric unit* (or *mer*, also known as the base unit or monomeric base unit), on the other hand, is based on the polymerization process. It refers to the largest constitutional unit contributed by a single monomer during a polymerization process. Consequently, the constitutional repeating unit may, according to the structure or method of synthesis of the macromolecule, be the same size as or smaller or larger than the monomer unit (see Table 1-1).

Originally, a *polymer* was described as a *molecule* consisting of many (from the Greek, πολυ) monomeric units (from the Greek, μεροσ). Today, however, according to IUPAC (International Union for Pure and Applied Chemistry) a polymer is defined as "a *substance* composed of molecules characterized by the multiple repetition of one or more species of atoms or groups of atoms (constitutional units) linked to each other in amounts sufficient to provide a set of properties that do not vary markedly with the addition or removal of one or a few of the constitutional units."

If the macromolecules have identical constitutional repeating units but differ only in the number N of such units per macromolecule, then the macromolecules are described as belonging to a *polymer homologous series*.

Table 1-1. Monomeric Units and Constitutional Repeating Units of Various Macromolecules

Initial monomers	Monomeric units	Repeating units
$CH_2=CH_2$	$-CH_2-CH_2-$	$-CH_2-$
CH_2N_2	$-CH_2-$	$-CH_2-$
$Cl(CH_2)_2Cl +$ $Na(CH_2)_2Na$	$-CH_2-CH_2$	$-CH_2-$
$Cl(CH_2)_2Cl +$ $Na(CH_2)_3Na$	$-CH_2-CH_2-$ $-CH_2-CH_2-CH_2-$	$-CH_2-$
$NH_2(CH_2)_6NH_2 +$ $HOOC(CH_2)_4COOH$	$-NH(CH_2)_6NH-$ $-CO(CH_2)_4CO-$	$-NH(CH_2)_6NH-CO(CH_2)_4CO-$

The homologous series concept of low-molecular-mass chemistry is thereby extended. In the latter case, for example, one refers to aliphatic unbranched alcohols as homologous:

$$CH_3OH, \ CH_3CH_2OH, \ CH_3CH_2CH_2OH, \ \text{etc.}$$

The property "alcohol" is produced in this case by the hydroxyl group, one of the two end groups when considered in terms of macromolecular chemistry. With the polymer homologous series concept, however, it is the central methylene groups which are of decisive importance: the chemical nature of the end groups is only of secondary interest.

The number N of monomeric units joined together in a macromolecular chain is defined as the *degree of polymerization* X of a molecule. The theoretically important degree of polymerization cannot be directly measured. It can, however, be calculated from the experimentally measurable *molar masses* (molecular weights), M of the polymer, M_m of the monomeric units, and M_e of the end groups:

$$X \equiv (M - M_e)/M_m \approx M/M_m \qquad (1\text{-}1)$$

Thus, the degree of polymerization derives neither from the number of constitutional repeating units nor from the number of constitutional units, but from the number of monomeric units. Only in exceptional cases is the degree of polymerization identical with the *chain link number*. The chain link number is the number of atoms joined sequentially together in the chain. The chain link number is equal to N for poly(methylenes), to $2N$ for poly(dimethyl siloxanes), and to $3N$ for poly(oxyethylenes).

Most polymers consist of mixtures of polymer homologous molecules of different degrees of polymerization: they are polymolecular or polydisperse.† Consequently, the degree of polymerization of such polymers is always an average whose value depends on the statistical weights of the data used to obtain it. Of special importance is the number average (weighted with respect to the amount of material, n, or the equivalent number N of molecules, since $n_i = N_i/N_L$)

$$\langle X \rangle_n \equiv \sum_i n_i X_i \bigg/ \sum_i n_i, \qquad \langle M \rangle_n \equiv \sum_i n_i M_i \bigg/ \sum_i n_i \qquad (1\text{-}2)$$

and the weight average (weighted with respect to the weight or mass, m, or the equivalent mass fraction w)

$$\langle X \rangle_w \equiv \sum_i m_i X_i \bigg/ \sum_i m_i, \qquad \langle M \rangle_w \equiv \sum_i m_i M_i \bigg/ \sum_i m_i$$

---

† Most authors use the word *polydisperse* instead of *polymolecular*. Polydispersity refers, however, to the dispersion of properties of any particles, whereas polymolecularity refers to the distribution of molecular masses only. A distinction is necessary because a given polymolecularity may lead to different polydispersities in aggregating systems.

The term *weight average* has now been superceded by the term *mass average*. Often, *average* is replaced by *mean*. Not only degrees of polymerization and molar masses, but also other properties occur mostly as averages.

The properties of a representative of a polymer homologous series alter systematically with the degree of polymerization. Some properties such as melt viscosity or boiling point increase continuously with increasing degree of polymerization (see Figure 1-1), whereas other properties such as melting temperature or tensile strength are virtually independent of molar mass at very high degrees of polymerization. Consequently, no sharp borderline exists between the low- and high-molecular-mass representatives of a polymer homologous series.

For convenience, however, polymers with a small number of constitutional units and unspecified end groups are called *oligomers*. "Small" in this case is relative: it may refer to 2–20 constitutional units in synthetic polymer chemistry, or several hundred units in nucleotide chemistry. *Telomers* are oligomers with end groups consisting of fragments of chain transfer agents (see Part IV). *Telechelic polymers*, however, are oligomers with known functional end groups.

Constitutional repeating units (or monomeric units), end groups and degrees of polymerization describe the *constitution* of an unbranched ("linear") chain (Chapter 2). With copolymers, consisting of two or more types of monomeric units, the monomeric unit sequence must also be given with the mean composition of the chain. Chains may also be joined together

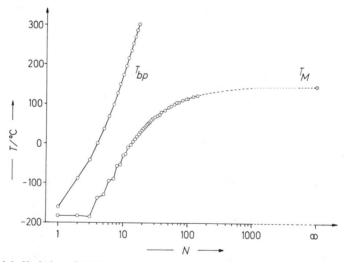

Figure 1-1. Variation of boiling temperature T_{bp} and melting temperature T_M with molar mass for a homologous series of alkanes.

by branching or by cross link networks. Differences in molar mass as well as molar mass distribution (molecular weight distribution) differences also count among the kinds of constitutional isomerism, but these will be discussed in Part II, since practically the only ways to determine molar masses are via solution properties.

The *configuration* (Chapter 3) describes the spatial arrangement of substituents about a given atom and the sequence of such microconfigurations within the main chain. The configuration of substituents corresponds to that of low-molecular-mass chemistry and so will not be discussed.

The *conformation* (called constellation in the older German literature; or configuration, as it is known by the physicist) describes the preferred positions taken up by groups of atoms during rotation about single bonds (Chapter 4). In contrast to configurations, conformations can interchange without destruction and reformation of individual bonds. The sequence of microconformations about individual bonds determines the macroconformation, or shape, of the whole macromolecule. The macroconformations of polymers in solution and in the solid state can be very different from one another.

Orientation (Chapter 5) refers to the preferred positioning of parts or groups of molecules in the bulk phase without the establishment of long-range order. *Crystallinity* (Chapter 5) presupposes not only a three-dimensional preferential arrangement of the chains, but also definite interrelationships between the lattice points. The chain atoms, with their substituents, can be considered as lattice points. Thus, with polymers, in contrast to low-molecular-mass chemistry, it is not the mutual arrangement of the lattice points of various molecules which must be considered, but also the arrangement of the lattice points of an individual macromolecule relative to the other lattice points of the same molecule.

Constitution and configuration embody the chemical structure; orientation and crystallinity describe the physical structure. Conformation can be classified within the chemical structure as well as within the physical structure. Thus, the chemical structure concept deals essentially with the isolated molecule, whereas the physical structure concept considers the structure of groups of molecules together. Thus, crystallinity and orientation are a consequence of the conformation, and the conformation itself is a consequence of the constitution and the configuration.

Chemical and physical structure, together with mobility or flexibility of chain segments and molecules, determine the properties and applications of synthetic and natural macromolecules. The chemical structure of the macromolecule influences its reactivity; the physical structure, however, determines its material properties. Nucleic acids, for example, carry genetic information and/or act as matrices for protein synthesis. Enzymes are very specific catalysts. With synthetic polymers, on the other hand, the chemical properties are of subsidiary importance: synthetic polymers should, as materials, be as chemically stable as possible. The sharp increase in polymer production in recent decades (see Part VI) is understandable when one considers that polymer materials often show significantly better properties and process-ability with respect to materials such as metals, glass, and ceramics. In terms of volume and value, plastics have almost reached the production levels of steel, and in terms of quantity, synthetic fibers have almost reached the production levels of natural fibers. The production of synthetic elastomers is already double that of natural rubber.

1.2. Historical Development

There is no remembrance of former things; neither shall there be any remembrance of things that are to come with those that shall come after.

Ecclesiastes 1:11

Since ancient times, naturally occurring polymers have been used by mankind for various purposes. Proteins from meat and polysaccharides from grain are important sources of food. Wool and silk, both proteins, serve as clothing. Wood, the main component of which is cellulose, a polysaccharide, is used for building and fire-making. Amber, a high-molar mass resin, was worn by the Greeks as a jewel. The use of asphalt as an adhesive is mentioned in the Bible.

In 1839, Simon[1] observed that styrene changed on heating from a clear liquid to a solid, transparent mass—to a poly(styrene) in the modern sense. Since the overall composition of carbon and hydrogen remained constant, Berthelot[2] called this event a polymerization. So the word *polymerization* originally meant simply that several molecules formed one larger association without alteration of the overall composition. This left open to question whether one was dealing with physical molecules (associations) or with genuine macromolecules, in the present sense of the word. Addition polymerization is now defined as the repeated addition of monomers onto a growing activated chain, for example, as with styrene:

$$R{\sim}CH_2{-}\underset{\underset{C_6H_5}{|}}{C}H^* \xrightarrow{\ +\ CH_2{=}CH(C_6H_5)\ } R{\sim}CH_2{-}\underset{\underset{C_6H_5}{|}}{C}H{-}CH_2{-}\underset{\underset{C_6H_5}{|}}{C}H^* \quad \text{etc.} \qquad (1\text{-}4)$$

Berthelot also noticed that depolymerization of the solid mass back to styrene occurred at even higher temperatures. This simple conversion of styrene ⇌ poly(styrene) ⇌ styrene, obtainable only through temperature change, later formed an apparently reliable basis for the micelle theory of such substances.

Before Berthelot, Wurtz[3] had already converted ethylene oxide into low-molar mass poly(ethylene oxides) (using the modern terminology):

$$n\,CH_2\!-\!CH_2 \longrightarrow \;+CH_2\!-\!CH_2\!-\!O+_n \tag{1-5}$$

Lourenço[4] carried out this transformation in the presence of ethylene halides and isolated substances up to $n = 6$ from the reaction medium. He established that the overall structure for compounds of this kind approached that of pure ethylene oxide, although the properties of these substances were different from those of ethylene oxide. Lourenço also noticed that this compound, which was a liquid at room temperature, showed increased viscosity with an increasing degree of polymerization n and he proposed a chain formula for the products.

Shortly after this study, Graham[5] discovered that certain substances, such as lime, diffuse much more slowly than, say, common salt, and do not readily permeate a membrane. Since this behavior was characteristic of limelike, noncrystalline substances, whereas the crystalline substances known at that time all diffused and permeated rapidly, Graham differentiated between crystalloids and colloids (from the Greek $\kappa o \lambda \lambda \alpha$ = lime). He thus attributed the colloidal behavior to the structure of the colloids and not to their state.

Further subdivision of the colloids concerned many researchers. For example, studies of coagulation processes lead Müller[6] to connect suspensions with physical disintegration processes and large molecules with chemical precipitation methods. He designated as "high molecular" such substances as albumin and colloidal silica. The later classification by Staudinger[7] into colloidal dispersions, micellar colloids (assocation colloids), and colloidal molecules (macromolecules) proved to be very suitable and forms the foundation of modern textbooks on colloid science.[8]

It was later found that inorganic substances can also behave as colloids. Examples include the basic oxides of iron and aluminum. The reactivity of these colloids, however, is not very different from that of their cyrstalloid forms. It appears that all substances can be converted under suitable conditions into the colloid state and then reconverted to the noncolloid state. Consequently, colloids are general possible *states* of material, and not specific *materials* (W. Ostwald,[9] P.P. von Weimarn[10].

The correct conclusion, that all low-molar mass substances can be transferred into the colloidal state, led to a false antithesis to this principle, namely, that all colloidal particles or aggregates are therefore composed of smaller molecules, i.e., physical polymers.

In the years between Graham's discovery and Ostwald's postulation, the modern idea of true macromolecules was very much alive. For example, in 1871 Hlasiwetz and Habermann[11] assumed that proteins and polysaccharides were macromolecules, but with the methods available at that time they could not prove the high molar masses which they suggested.

The possibility of such proof became available from laws governing the relation between vapor pressure and mole fraction, or between osmotic pressure, concentration, temperature, and molar mass, which were discovered by Raoult[12] (1882–1885) and van't Hoff[13] (1887–1888). With these methods, very high molar masses (between 10 000 and 40 000 g/mol) were subsequently obtained from rubber, starch, and cellulose nitrate. Other authors found similarly high values of the same materials: e.g., Gladstone and Hibbert[14] found 6000–12 000 g/mol for rubber, and Brown and Morris[15] obtained cryoscopically about 30 000 g/mol for a product of starch obtained by degradation hydrolysis.

To most research workers at that time, however, these high molar masses appeared untrustworthy. That is to say the same methods used on covalently structured crystalloids gave molar masses which agreed satisfactorily with the chemical formula mass. But since formula masses for colloids could not be found unequivocally, the molar masses obtained by physical methods also appeared suspect.

In addition, the proportionality between vapor pressure and concentration (Raoult's law) and that between osmotic pressure and concentration (van't Hoff's law) had to be satisfied. Both requirements were adequately fulfilled within the limits of experimental error by the covalent crystalloids then studied, but not by the colloids. This "error" concerning the two laws made the high molar masses of the colloids also seem suspect. However, we know today that both laws are only limiting laws for infinite dilution. A molar mass apparently dependent on concentrations, i.e., calculated from the limiting laws, is also the rule rather than the exception for low-molar-mass substances. This effect, dependent on the interaction between the molecules in the solution, was known as early as 1900 from ebullioscopic measurements by Nastukoff,[16] who also proposed an extrapolation to zero solute concentration. Caspari[17] obtained a molar mass of 100 000 g/mol for rubber using osmotic measurements by a similar extrapolation procedure.

At that time, however, the formulation of the laws of Raoult and van't Hoff as limiting laws seemed unacceptable. The apparent unlimited validity of the laws for covalent crystalloids and the retention of identity of these low-molar-mass compounds in the colloid contradicted this theorem. In addition, colloids were not the only class of compound which showed a marked deviation from Raoult's law. Similar discrepancies were found in electrolytes. Since the electrolytes known at that time were solely inorganic compounds and could theoretically be formed into colloids, this was coming close to the notion that some peculiar forces were involved.

The results of the first great advances in full-fledged modern organic chemistry also spoke against the assumption that macromolecules are held together by covalent bonds. The great success of classical organic chemistry was based mainly on three principles: the formulation of a reaction in terms of the smallest constitutional change occurring; the use of elemental analysis as a basic test of a proposed constitutional formula and the possibility of crystallizing pure substances. At that time, however, colloids could not be crystallized. Indeed, even in low-molar mass organic chemistry there were substances that were difficult to crystallize, such as alcohol or sugar, but these were considered primarily as inexplicable exceptions. In addition, colloids lacked a broad purity criterion as used in organic chemistry. A substance was considered "pure" when it showed a definite structural formula with a definite molar mass. For one class of colloids known at that time, however, there were obviously similar structural formulas with varying molar masses.

The nature of the peculiar binding forces exhibited by such colloids was therefore important. The existence of intermolecular forces was known from the study of gases.[18] Similar forces could exist in solution. For organic molecules the law of partial valence seemed appropriate and could be effective, according to Thiele,[19] in substances with conjugated double bonds. This thesis seemed established since molecular compounds such as quinhydrones existed.[20]

The hypothesis of partial valence gave a convenient explanation for the behavior of natural rubber. The overall formula C_5H_8 already proposed by Faraday in 1826[21] pointed to one double bond per unit. Harries[22] confirmed this conclusion by ozonolysis of natural rubber and subsequent hydrolysis of the ozonides. Since he also found C_5H_8 to be the overall formula, he did not think that he would have to consider any "end groups." From the observed low molar masses he concluded initially that there were rings of two isoprene units (Figure 1-2). Later, he concluded that there were five to seven isoprene units per cyclic molecule.

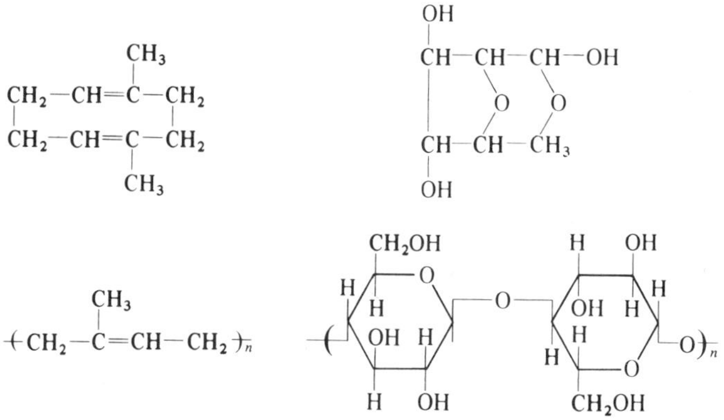

Figure 1-2. Superceded ring formulas (above) for natural rubber (left) and cellulose (right) in comparison with their modern representations (below).

The fact that rubber cannot be distilled also seems to suggest low-molar-mass cyclic compounds held together by partial valence. It was known that associated substances have a much higher boiling point than nonassociated ones. Pickles,[23] on the other hand, suggested the even now accepted chain structure for rubber. As proof of constitution he carried out the first relevant polymer modification, namely, the addition of bromine across the double bonds of rubber. Since the bromine addition did not alter the molecular size, Pickles considered natural rubber to be a true molecule. However, his conclusions were not generally accepted.

Rings similar to those of rubber were then proposed for numerous organic colloids. For example, the structural formula for cellulose was written as a ring (Figure 1-2). With its three hydroxyl groups and the hemiacetal group, the formula satisfactorily reproduced the chemical nature of cellulose. The colloidal character was explained to a large extent by an association of many cyclic compounds of this kind. The fact that no end groups were found was also in agreement with the assumption of cyclic compounds. We know today that the contribution made by end groups to the high molar mass was much too small to be detected by the methods then in use.

Association was affirmed by a further observation: The rotation of optically active diamyl itaconate was, in fact, more or less equal for monomer and "polymer."[24] Otherwise, differences were found for substances which varied constitutionally.

Around 1910–1920, all facts strongly supported the theory that organic colloids were physical groups of particles and not true covalent macromolecules. Organic colloids had the same overall composition and reactivity as their known noncolloidal basic components. Furthermore, they could often be converted back to the noncolloidal material and were not crystallizable. Anomalies appeared during molar mass determinations.

All these phenomena were already known in inorganic colloids. These, too, could only be crystallized, if at all, with the loss of colloidal character. In addition, since no end groups were found all observations seemed to confirm the assumption of low-molar mass rings. Theoretically, the colloidal character was easily explained: It was held together by special forces, e.g., by Thiele's partial valence.

Staudinger, however, disputed the assumption of molecular complexes in organic colloids. In his studies on ketenes,[25] he obtained "polymeric" products which he considered to be cyclobutane derivatives. Since another author[26] considered these dimers to be molecular complexes, Staudinger compiled all the arguments for covalent bonds in a work which has since become classic.[27] End groups did not seem to contradict this idea, since it was generally accepted at that time that the reactivity of a group decreased with increasing molar mass.

In his later work, Staudinger tried to establish experimentally his concept of organic colloids as true macromolecules. For this, it was first necessary to refute the idea of the so-called first micellar theory, which states that in organic colloids small rings were held together by partial valence. In 1922, Staudinger and Fritschi[28] hydrogenated rubber. Since the hydrogenated rubber no longer contained any double bonds, it should according to the micellar theory no longer show any colloidal properties. Yet in actual fact the colloidal properties were retained, as Pickles had already found in his bromination study. The hydrogenation of poly(styrene) to poly(vinyl cyclohexane) also excluded colloids based on Thiele's partial valence. Staudinger therefore concluded that these organic colloids consisted of many atoms joined together by covalent bonds, i.e., true "macromolecules."[29] Since bonds consisting of van der Waals forces are much weaker than covalent bonds, covalently bonded molecular colloids, unlike associated colloids, should retain their colloidal character in all solvents.[30]

However, this discovery was not accepted as proof by most scientists. Cryoscopic molar mass studies on natural rubber in camphor, for example, gave molar masses of 1400–2000 g/mol,[31] whereas Staudinger had found values of 3000–5000 g/mol for hydrogenated rubber. The X-ray studies carried out on such compounds seemed to contradict Staudinger's idea of molecular colloids. A large section of the organic colloids gave X-ray diagrams more like those of liquids than those of low-molar mass crystalloids. Further, the more crystal-like X-ray pictures for organic colloids revealed a relatively small unit cell. Yet is was known from measurements on homologous series of low-molar mass substances that the size of the unit cell is directly proportional to the molar mass. Since it was impossible to imagine that such a proportionality should not be true for all molar mass ranges it was concluded that colloids were composed of low-molar-mass compounds.

The X-ray measurements also contradicted the existence of small rings and supported the idea of chain structures,[32] since rings did not correspond with the elementary cell structure that had been found. An analysis of the X-ray diffraction pattern for rubber revealed crystal lengths of about 30–60 nm. Assuming that the crystal length is identical to the molecular length, K. H. Meyer and H. Mark thus found molar masses of about 5000–10 000 g/mol for natural rubber. The much greater molar mass of about 150 000–380 000 g/mol found for rubber in solution was interpreted as the mass of the solvated chain and, later, also by the supposition that the micelles found by X-ray crystallography were present in solution as associated structures. The second micellar theory, contrary to the first, assumed chains instead of rings, as well as higher molar masses. To explain the very high molar masses obtained experimentally, it was assumed that these chains associated to form colloidal aggregates.

In contrast, H. Staudinger and R. Signer† stressed that the crystal length need not have anything to do with the molecular length. Since the crystal structure was essentially dependent on the constitution of the compound, Staudinger tried to prove his theory by polymer-analogous reactions. In reactions of this type, the side groups of a compound are replaced without attacking the bonds of the main chain. Unbranched poly(vinyl acetate) could be converted by saponification to poly(vinyl alcohol), and then by esterification back to poly(vinyl acetate) again:

$$
\begin{array}{c}
\text{-}(CH_2\text{--}CH)_n \xrightarrow[-CH_3COOH]{+H_2O} \text{-}(CH_2\text{--}CH)_n \xrightarrow[H_2O]{+CH_3COOH} \text{-}(CH_2\text{--}CH)_n \qquad (1\text{-}6) \\
| \qquad\qquad\qquad\qquad\qquad\qquad | \qquad\qquad\qquad\qquad\qquad\qquad | \\
O \qquad\qquad\qquad\qquad\qquad\qquad OH \qquad\qquad\qquad\qquad\qquad\qquad O \\
| \qquad\qquad\qquad\qquad\qquad\qquad\qquad\qquad\qquad\qquad\qquad\qquad | \\
CO \qquad\qquad\qquad\qquad\qquad\qquad\qquad\qquad\qquad\qquad\qquad\qquad CO \\
| \qquad\qquad\qquad\qquad\qquad\qquad\qquad\qquad\qquad\qquad\qquad\qquad | \\
CH_3 \qquad\qquad\qquad\qquad\qquad\qquad\qquad\qquad\qquad\qquad\qquad CH_3
\end{array}
$$

If the same degree of polymerization is obtained for the polymer analogues in different solvents, then it is quite improbable that association colloids are present because of the differing polymer–solvent interactions (see Reference 34).

Staudinger formulated his ideas from studies of natural products, such as cellulose and amylose, or of addition polymerization products, such as poly(styrene). The reaction mechanisms producing these natural and synthetic products were, however, not known at this time. A stepwise addition to an active center of undefined chemical nature was indeed suspected[35] for the polymerization of styrene, and the active center was later identified as a free radical.[36,37] The actual starting step, on the other hand, remained undetermined. Mechanisms involving the addition of monomer to free radicals[38] or via activated complexes between styrene and, for example, dibenzoyl peroxide[39] were discussed. The problem was finally solved in 1941–1943 by labeling the initiators.[40−42] It could be shown that labeled initiator fragments were incorporated into the polymer as end groups.

At that time the addition polymerization mechanism leading to compounds used by Staudinger as model substances was anything but clear. For the second group of compounds that he used—natural macromolecular products—even less was known about the mechanism of formation. Carothers[43] therefore decided to build macromolecular compounds, stepwise, using condensation reactions familiar in low-molar mass organic chemistry, e.g., reacting diols with dicarboxylic acids:

---

†See, e.g., the description of the historical development given by Mark.[33]

$$HO-R-OH + HOOC-R'-COOH \xrightarrow{-H_2O} HO-R-OCO-R'-COOH \qquad (1\text{-}7)$$

$$HO-R-OCO-R'-COOH + HO-R-OH \xrightarrow{-H_2O}$$

$$HO-R-OCO-R'-COO-R-OH \quad \text{etc.}$$

In contrast to addition polymerization, low-molar mass compounds are evolved in polycondensation (condensation polymerization).

Carothers was able to show that many macromolecules could be built, not only through some mysterious process, but also with the known methods of organic chemistry. His work yielded further proof for the formation of organic molecular colloids by covalent bonds, and led to the first synthetic fiber produced industrially on a large scale, nylon 6,6 [poly(hexamethylene-adipamide)]. This polymer is obtained from hexamethylenediamine and adipic acid:

$$n\,H_2N-(CH_2)_6-NH_2 + n\,HOOC\{CH_2\}_4COOH \longrightarrow \qquad (1\text{-}8)$$

$$H\{NH-(CH_2)_6-NHCO\{CH_2\}_4CO\}_n OH + (2n-1)H_2O$$

Further questioning of the micellar theory came from biochemistry. Sumner[44] succeeded in crystallizing the enzyme urease in 1926, and Northrop[45] crystallized the enzyme pepsin in 1930. Thus, the theory that colloids could only be crystallized by losing their colloidal properties was refuted. Then, over a period 1927–1940, Svedberg,[46] using the ultracentrifuge that he had invented, showed that a colloidal solution containing proteins, when ultracentrifuged at various temperatures and in different salt solutions, proved to be uniform with regard to molar mass. This discovery also contradicted the idea of colloidal assocation. Using the electrophoresis which he developed,[47] Tiselius showed conclusively that a specific protein always had the same charge per mass, and this, too, was contradictory to inorganic colloidal association. Thus, at the beginning of the 1930s, it was realized that true macromolecules were being dealt with in organic colloids.

The inability of many synthetic organic macromolecules to crystallize was traced back in the 1940s to the irregularity in configuration of connecting monomeric units. In 1948, Schildknecht and co-workers[48] found that vinyl ether gave polymers with different physical properties according to the catalyst used. Poly(vinylmethylethers) proposed via free radical polymerizations were amorphous, whereas cationic polymerizations at low temperatures yielded crystalline material. This discovery was correctly interpreted as being dependent on differences in the steric makeup of the polymer. However, these results were not much heeded because no other stereoregular polymers can be generated in this way.

The key to the desired synthesis of stereoregular polymers was first provided by the discovery of Ziegler catalysts. Ziegler found that catalyst systems of aluminum alkyls and titanium tetrachloride can polymerize ethylene to poly(ethylene) even at room temperature and normal pressure.[49] Up to this time, poly(ethylene) had been made exclusively by radical polymerization of ethylene at high pressure. Natta and his co-workers[50] observed that these catalysts enabled α-olefins to be converted into stereoregular polymers that could often be crystallized. Later alterations to the Ziegler catalysts also made the polymerization of other monomers possible, so that today a large number of stereoregular polymers is known.

Chain-forming macromolecules contain many bonds in the main chain. The individual chain atoms can therefore assume many different arrangements in space relative to each other. Kuhn[51] already knew in the 1930s that the problem of the spatial arrangement of chain macromolecules can be solved particularly well by statistical analysis and statistical calculation methods. Generally speaking, statistics plays a large role in macromolecular chemistry, as was shown especially by P. J. Flory in a masterly manner.[52]

Literature

Section 1.1. Basic Concepts

Definitions

International Union of Pure and Applied Chemistry, Macromolecular Division, Commission on Macromolecular Nomenclature, Basic Definitions of Terms Relating to Polymers 1974, Pure Appl. Chem. **40**(3), 479 (1974).

Handbooks and Series

Houben-Weyl, *Methoden der organischen Chemie* (ed. E. Müller), G. Thieme, Stuttgart, Vol. XIV, *Makromolekulare Stoffe*, Part 1, *Polymerisation*, 1961; Part 2, *Polykondensate, Reaktionen an Polymeren*, 1963; Vol. XV, *Makromolekulare Naturstoffe* (planned).

R. Vieweg (ed.), *Kunststoff-Handbuch*, C. Hanser, Munich 1963–1975 (12 volumes).

H. Mark, N. G. Gaylord, and N. M. Bikales (eds.), *Encyclopedia of Polymer Science and Technology*, Wiley, New York, 15 Volumes, 1966–1972, Additional volumes (Supplements) since 1976.

J. Brandrup and E. H. Immergut (eds.), *Polymer Handbook*, second ed., Wiley, New York, 1975.

Specialist Periodical Reports, *Macromolecular Chemistry*, The Royal Society of Chemistry, Vol. 1ff (1980ff).

Bibliography

P. Eyerer, *Informationsführer Kunststoffe*, VDI-Verlag, Düsseldorf, 1976.

J. Schrade, *Kunststoffe (Hochpolymere), Bibliographie aus dem deutschen Sprachgebiet*, Erste Folge 1911–1969, Dr. J. Schrade, Swiss Aluminium Ltd., Zürich, 1976.

O. A. Battista, *The Polymer Index*, McGraw-Hill, New York, 1976.
E. R. Yescombe, *Plastics and Rubbers: World Sources of Information*, second ed., Appl. Sci. Publ. Barking, Essex, England, 1976.
A. Anthony, *Guide to Basic Information Sources in Chemistry*, Halsted, New York, 1979.

Section 1.2. Historical Development

1. E. Simon, *Ann. Chim. Phys.* **31**, 265 (1839).
2. M. Berthelot, *Bull. Soc. Chim. France* **6**(2), 294 (1866).
3. A. Wurtz, *Compt. Rend.* **49**, 813 (1859); **50**, 1195 (1860).
4. A.-V. Lourenco, *Compt. Rend.* **49**, 619 (1859); **51**, 365 (1860); *Ann. Chim. Phys.* **67**(3), 273 (1863).
5. T. Graham, *Philos. Trans. R. Soc. London* **151**, 183 (1861); *J. Chem. Soc. (London)* **1864**, 318.
6. A. Müller, *Z. Anorg. Chem.* **36**, 340 (1903).
7. H. Staudinger, *Organische Kolloidchemie*, Vieweg, Braunschweig, Germany, first ed. 1940, third ed. 1950.
8. J. Stauff, *Kolloidchemie*, Springer, Berlin, 1960.
9. W. Ostwald, *Kolloid-Z.* **1**, 291, 331 (1907).
10. P. P. v. Weimarn, *Kolloid-Z.* **2**, 76 (1907/1908).
11. H. Hlasiwetz and J. Habermann, *Ann. Chem. Pharmacol.* **159**, 304 (1871).
12. F. M. Raoult, *Compt. Rend.* **95**, 1030 (1882); *Ann. Chim. Phys.* **2**(6), 66, (1884): *Compt. Rend.* **101**, 1056 (1885).
13. J. H. van't Hoff, *Z. Phys. Chem.* **1**, 481 (1887); *Philos. Mag.* **26** (5), 81 (1888).
14. J. H. Gladstone and W. Hibbert, *J. Chem. Soc. (London)* **53**, 679 (1888): *Philos. Mag.* **28**(5), 38 (1889).
15. H. T. Brown and G. H. Morris, *J. Chem. Soc. (London)* **55**, 462 (1889).
16. A. Nastukoff, *Ber. Dtsch. Chem. Ges.* **33**, 2237 (1900).
17. W. A. Caspari, *J. Chem. Soc. (London)* **105**, 2139 (1914).
18. J. D. van der Waals, *Die Kontinuität des gasförmigen und flüssigen Zustands*, thesis, Leiden, 1873: second ed., J. A. Barter, Leipzig, 1895 and 1900; *Die Zustandsgleichung*, Nobelpreisrede. Akad. Verlagsgesellschaft, Leipzig, 1911.
19. J. Thiele, *Liebigs Ann. Chem.* **306**, 87 (1899).
20. P. Pfeiffer, *Liebigs Ann. Chem.* **404**, 1 (1914); **412**, 253 (1917).
21. M. Faraday, *Q. J. Sci.* **21**, 19 (1826).
22. C. Harries, *Ber. Dtsch. Chem. Ges.* **37**, 2708 (1904); **38**, 1195, 3985 (1909).
23. S. S. Pickles, *J. Chem. Soc. (London)* **97**, 1085 (1910).
24. P. Walden, *Z. Phys. Chem.* **20**, 383 (1896).
25. H. Staudinger, *Die Ketene*, F. Enke, Stuttgart, Germany, 1912, p. 46.
26. G. Schroeter, *Ber. Dtsch. Chem. Ges.* **49**, 2697 (1916).
27. H. Staudinger, *Ber. Dtsch. Chem. Ges.*, **53**, 1073 (1920).
28. H. Staudinger and J. Fritschi, *Helv. Chim. Acta* **5**, 785 (1922).
29. H. Staudinger, *Ber. Dtsch. Chem. Ges.* **57**, 1203 (1924).
30. H. Staudinger, *Ber. Dtsch. Chem. Ges.* **59**, 3019 (1926): H. Staudinger, K. Frey, and W. Starck, *Ber. Dtsch. Chem. Ges.* **60**, 1782 (1927).
31. R. Pummerer, H. Nielsen, and W. Gündel, *Ber. Dtsch. Chem. Ges.* **60**, 2167 (1927).
32. K. H. Meyer and H. Mark, *Ber. Dtsch. Chem. Ges.* **61**, 593, 1939 (1928).
33. H. Mark, *Physical Chemistry of High Polymeric Systems*, Interscience, New York, 1940.
34. H. Staudinger and E. Husemann, *Liebigs Ann. Chem.* **527**, 195 (1937).
35. H. Staudinger and E. Urech, *Helv. Chim. Acta* **12**, 1107 (1929).

36. W. Chalmers, *J. Am. Chem. Soc.* **56,** 912 (1934).

37. H. Staudinger and W. Frost, *Ber. dtsch. chem. Ges.* **68,** 2351 (1935).

39. G. V. Schulz and E. Husemann, *Z. Phys, Chem.* **B39,** 246 (1938).

40. C. C. Price, R. W. Kell, and E. Kred, *J. Am. Chem. Soc.* **63,** 2708 (1941); **64,** 1103 (1942).

41. W. Kern and H. Kämmerer, *J. Prakt. Chem.* **161,** 81, 289 (1942).

42. P. D. Bartlett and S. C. Cohen, *J. Am. Chem. Soc.* **65,** 543 (1943).

43. W. H. Carothers, *Chem. Rev.* **8,** 353 (1931); H. Mark and G. S. Whitby (eds.), *Collected Papers of W. H. Carothers,* Interscience, New York, 1940.

44. J. B. Sumner, *J. Biol. Chem.* **69,** 435 (1926).

45. J. H. Northrop, *J. Gen. Physiol.* **13,** 739 (1930).

46. T. Svedberg and K. O. Pedersen, *The Ultracentrifuge,* Oxford University Press, London and New York, 1940.

47. A. Tiselius, *Kolloid-Z.* **85,** 129 (1938).

48. C. E. Schildknecht, S. T. Gross, H. R. Davidson, J. M. Lambert, and A. O. Zoss, *Ind. Eng. Chem.* **40,** 2104 (1948).

49. K. Ziegler, *Angew. Chem.* **76,** 545 (1964).

50. G. Natta, *Angew. Chem.* **76,** 553 (1964).

51. W. Kuhn, *Ber. Dtsch. Chem. Ges.* **65,** 1503 (1930).

52. P. J. Flory, *Principles of Polymer Chemistry,* Cornell University Press, Ithaca, New York, 1953; P. J. Flory, *Statistical Mechanics of Chain Molecules,* Interscience, New York, 1969.

Chapter 2
Constitution

A great many more parameters are required to describe the constitution of macromolecular compounds than are necessary with low-molecular-mass compounds: the constitution of the main chain itself, the constitution of substituents to the main chain, and also, the molecular architecture. The constitution of the main chain is given by the kind and sequence of main chain atoms, monomeric units, and repeating units, that of the substituents by the nature of the side and end groups, and the molecular architecture by the type and order of the branching or cross-linking.

2.1. Nomenclature

Naturally occurring polymers generally carry trivial names. These trivial names describe the origin (e.g., cellulose), the nature (e.g., nucleic acids), or the function (e.g., catalase) of the polymer.

In the early days of macromolecular chemistry, synthetic polymers were simply labeled according to the monomer from which they were prepared. Thus, ethylene polymers became poly(ethylenes), styrene polymers became poly(styrenes), and those from lactams became poly(lactams). In other cases, the choice of name was provided by a characteristic group occurring in the final polymer. Thus, polymers from diamines and dicarboxylic acids were called polyamides, and those from diols and dicarboxylic acids were called polyesters. This phenomenological nomenclature fails, of necessity, when more than one kind of monomeric unit can be formed from a given monomer.

The IUPAC nomenclature for macromolecular substances, on the other hand, is based on the constitution. This nomenclature derives predominantly from the nomenclature of low-molar-mass inorganic and organic molecules. The nomenclature of low- and high-molar-mass inorganic molecules follows the principle of additivity, that of low-molar-mass organic molecules follows the substitution principle. The nomenclature of organic macromolecules is based on a hybrid form of these two principles: the smallest constitutional repeating units, considered as diradicals, are named, and then, however, the names are additively combined.

Thus, the names of macromolecules consist of the names of the constitutional repeating units and a prefix which is the number, in Greek, of constitutional repeating units per molecule. The prefix *poly* is used to describe an indefinite, but large number of constitutional repeating units. End groups are generally not specified. If they are known, then their names, in the form of radicals, preceded by the Greek letters α and ω, are written before that of the polymer. For example:

$Cl(CH_2)_nCCl_3$:	α-chloro-ω-trichloromethyl-poly(methylene)
$Cl(S)_7SCl$:	α-chloro-ω-chlorosulfur-*catena*-hepta(sulfur)

2.1.1. Inorganic Macromolecules

The name of the constitutional repeating unit consists of the names of the central atoms and the names of bridging and side groups. The central atom is chosen as the one which occurs latest in the sequence prescribed for the nomenclature of inorganic compounds:

$\longrightarrow$ F Cl Br I At O S Se Te Po N P As Sb Bi C Si Ge Sn Pb
B Al Ga In Tl Zn Cd Hg Cu Ag Au Ni Pd Pt Co Rh Ir Fe
Ru Os Mn Tc Re Cr Mo W V Nb Ta Ti Zr Hf Sc Y La Lu Ac
Lr Be Mg Ca Sr Ba Ra Li Na K Rb Cs Fr He Ne Ar Kr Xe
Rn $\longrightarrow$

(see, for example, Example 5 in Table 2-1). The oxidation state of the central atom is given in parentheses immediately after it.

All groups joined to the central atom are called ligands. All ligands are listed in alphabetical order, irrespective of the number of ligands of a given type present or of whether they are bridging or side group ligands. Bridging groups are provided with a μ directly preceding their name; they are also separated from the other ligands by a dash. When the same group is present as both bridging and side group, then the bridging group is named first (see Example 9 in Table 2-1). If a bridging group is joined to a central atom by more than one bond, then this group is also a chelating group: the symbols of the coordinating atoms are also written in italics into the name of the bridging groups (see Example 11, Table 2-1).

In addition to the names of the end groups and the prefix for the number

Table 2-1. *Trivial and Structural Names of Inorganic Macromolecules*

No.	Constitution	Trivial name	Structural name
1	S_8		*Cyclo*-octa(sulfur)
2	$(S)_n$	Polymeric sulfur	*Catena*-poly(sulfur)
3	$(SiF_2)_n$	Silicon difluoride	*Catena*-poly(difluorosilicon)
4	$(O{-}Si(C_6H_5)_2)_n$	Poly(diphenyl siloxane)	*Catena*-poly[μ-oxy-diphenyl silicon(IV)]
5	$(N{=}PCl_2)_n$	Poly(phosphoronitrilochloride); poly(dichlorophosphazene)	*Catena*-poly[dichloro-μ-nitrido-phosphorus(V)]
6	$(NC{-}Ag)_n$	Silver cyanide	*Catena*-poly[μ-cyano-C:N-silver(I)]
7	$\left[\begin{smallmatrix} F \\ Au \\ F \end{smallmatrix}\right]_n$	Gold trifluoride	*Catena*-poly[*cis*-μ-fluoro-difluoro-gold(III)]
8	$2nK^+\left[\begin{smallmatrix} F & F \\ Al & \\ F & F \end{smallmatrix}\right]_n^{2-}$		*Catena*-poly[*trans*-μ-fluoro-tetrafluoro-aluminate(III)]
9	$\left[\begin{smallmatrix} NH_3 \\ Cl{-}Zn \\ Cl \end{smallmatrix}\right]_n$		*Catena*-poly[ammino-μ-chloro-chloro-zinc(II)]
10	$\left[\begin{smallmatrix} Cl \\ Cl{-}Pd \end{smallmatrix}\right]_n$	Palladium chloride	*Catena*-poly[di-μ-chloropalladium(II)]
11			*Catena*-poly{μ-[2,5-dihydroxy-*p*-benzoquinonate(2-)-$O^1,O^2:O^4,O^5$]-zinc (II)}

Table 2-1. (Cont.)

No.	Constitution	Trivial name	Structural name
12	$\begin{bmatrix} & C_6H_5 & & C_6H_5 & \\ -Si & -O-Si- & O- & \\ & | & | & | & | \\ -Si & -O-Si- & O- & \\ & C_6H_5 & & C_6H_5 & \end{bmatrix}_n$	Poly(phenyl sesquisiloxane)	μ'-Oxy-bis{*catena*-poly[μ-oxy-phenyl-silicon(IV)]}
13	$N\equiv C-CH_3$ $\begin{bmatrix} & Cl & \\ -Cu & -Cu- \\ & | & \\ & Cl & \end{bmatrix}_n$ $N\equiv C-CH_3$		Bis(*Cu-Cl'*, *Cl-Cu*){*catena*-poly[aceto-nitrilo-chloro-copper(I)]}

of constitutional repeating unit per molecule in inorganic polymers, a symbol, written in italics, is also given for the dimensionality of the macromolecule. The prefixes, in italics, *cyclo, catena, phyllo,* and *tecto* describe ring-formed, single-strand ("unidimensional"), parquet ("two-dimensional"), and network ("three-dimensional") types of polymers, respectively. The polymerization of *cyclo*-octasulfur thus produces *catena*-poly(sulfur) (see Examples 1 and 2 in Table 2-1).

In multistrand polymers, therefore, each strand is first named as if it were an individual polymer chain. The joining groups between the individual strands obtain the symbol μ' before their ligand names. The two central atoms joined by the bridge are written in italics (see Example 13, Table 2-1).

2.1.2. *Organic Macromolecules*

The smallest constitutional repeating unit in an unbranched organic polymer is a divalent free radical. The name of this biradical is formed in the same way as for low-molar-mass organic chemistry. Thus, the group $-CH_2-$ is called "methylene" and the corresponding polymer is called poly(methylene). Examples of names for other diradicals are as follows:

$-O-$	$-S-$	$-NH-$	$-CO-$	
oxy	thio	imino	carbonyl	(for when joined to heteroatoms)
			oxo	(in all other cases)

$-CH=CH-$			
vinylene	1,4-phenylene	1,4-cyclohexylene	4,6-quinolinediyl

The trivial names are retained for diradicals occurring often. Consequently, $-CH_2CH_2-$ is called "ethylene" and not "dimethylene," the unit $-CH_2CH(CH_3)-$ is "propylene" and not "1-methyl ethylene," and the unit $-OC-C_6H_4-CO-$ is known as "terephthaloyl" and not "carbonyl-1,4-phenylene-carbonyl" or "oxo-1,4-phenylene-oxo" (Example 7 in Table 2-2).

The structural name of complicated compounds is made up of the names of simple divalent radicals ordered according to sequence rules. The component with the highest seniority is listed first, other components following in decreasing order of priority. Heterocyclic rings have highest seniority. Linear groups containing heteroatoms follow. Next come carbocyclic rings, and then, finally, linear groups containing only carbon atoms. Substituents also follow this seniority sequence. Examples are: poly(oxymethylene) (Example 3, Table 2-2), poly(iminoethylene) (Example 4), and poly(1-oxotrimethylene) (Example 5). The names of individual divalent radicals are not separated by a dash here. The main chain direction implied by

Table 2-2.   Trivial and Structural Names of Organic Macromolecules

Structure number and formula	Trivial name (based on the structure of the monomers)	Structural name
1 $+CH_2+_n$	Poly(methylene)	Poly(methylene)
2 $+CH=CHCH_2CH_2+_n$	1,4-Poly(butadiene)	Poly(1-butenylene)
3 $+OCH_2+_n$	Poly(formaldehyde)	Poly(oxymethylene)
4 $+NHCH_2CH_2+$	Poly(ethylene imine)	Poly(iminoethylene)
5 $+CO—CH_2CH_2+$	Poly(ethylene-co-carbon monoxide)	Poly(1-oxotrimethylene)
6 $+NHC(CH_2)_4CNH(CH_2)_6+_n$ O O	Poly(hexamethylene adipamide); nylon 6,6	Poly(iminoadipoyliminohexamethylene)
7 $+OCH_2CH_2OC$ ... $C=O$	Poly(ethylene terephthalate)	Poly(oxyethyleneoxy-terephthaloyl)
8 $+CH—CH—CH—CH_2+_n$ O=C C=O C₆H₅ O	Poly(maleic anhydride-co-styrene)	Poly[(tetrahydro-2,5-dioxo-3,4-furandiyl)(1-phenylethylene)]
9 $+CHOHCH_2+_n$	Poly(vinyl alcohol)	Poly(1-hydroxyethylene)
10 $+C(CH_3)—CH_2+_n$ COOCH₃	Poly(methyl methacrylate)	Poly[1-(methoxycarbonyl)-1-methylethylene]

the structural names is generally not identical to the direction of chain growth.

The following special rules are observed for each group of divalent radicals:

1. Hetero atoms follow the sequence O, S, Se, Te, N, P, As, Sb, Bi, Si, Ge, Sn, Pb, B, and Hg. If, for example, aliphatic groups occur between two heteroatoms, or two rings, or a ring and a heteroatom, then the repeating units are given such that the shortest distance occurs between components with the highest seniority. The compound $-O-CH_2-NH-CHCl-CH_2-SO_2-(CH_2)_6-$ is written in the order given, since the shortest distance between O (highest seniority) and S (second highest seniority) is via the element $-CH_2-NH-CHCl-CH_2-$, with four chain atoms. If the sequence $-O-(CH_2)_6-SO_2-CH_2-CHCl-NH-CH_2-$ would be chosen, then the correct sequence of seniorities for O, S, and N is indeed obtained, but the distance between O and S is now six chain atoms, and consequently longer.

2. In contrast to heteroatoms, nitrogen has the highest seniority for heterocyclics, followed then by O, S, Se, Te, P, As, etc. (see point 1). Heterocyclic rings are given in the following sequence: (a) the largest nitrogen-containing ring, and, indeed, irrespective of the number of nitrogen atoms in the ring, (b) with rings of the same size, the ring with the largest number of nitrogen atoms, and (c) the ring system with the largest number of other heteroatoms of highest seniority. Examples for decreasing order of seniority are:

3. With different rings, the sequence is (a) systems with the highest number of rings, (b) the largest individual ring, and (c) the least hydrogenated ring. Examples are

4. When two rings of the same kind are present, the following rules apply: rings with most substituents have highest seniority. With the number of substituents per ring being the same, the ring with substituents in the lowest position numbers comes first, and if both the number and position numbers of substituents are the same, then

the ring with substituents having names occurring earlier in the alphabet has highest seniority. Examples are

5.   In each ring, the shortest path is always taken.

Substituents are given before the name of the diradical. Poly(vinyl alcohol) is thus correctly called "poly(hydroxyethylene)" (Example 9) and poly(methyl methacrylate) is called, according to IUPAC, "poly[(1-methoxycarbonyl)-1-methyl ethylene]" (Example 10).

Double-strand polymers possess constitutional repeating units bonded at four positions. The relationships of these positions to each other is given by two pairs of numbers, written in italics and separated by a colon. Examples are

poly(*1,4:2,3*-butanetetrayl)   poly(*2,3:6,7*-naphthalenetetrayl-6-methylene)   poly(*2,3:6,7*-naphthalenetetrayl-6,7-dimethylene)

When there are tetravalent radicals as well as diradicals in the structural element (e.g., in spiro polymers), the tetravalent radicals take a higher seniority than the diradical components:

poly[2,4,8,10-tetraoxaspiro(5,5)-   poly[1,3-dioxa-2-silacyclohexane-5,2-
undecane-3,9-diylidene-9,9-bis(octamethylene)]   diylidene-2,2-bis(oxymethylene)]

According to IUPAC nomenclature rules, the trivial names of common polymers do not necessarily have to be replaced by structural names. Therefore, both trivial and structural names of polymers will be used in this book. In general, the standard abbreviations of trivial names will only be used in diagrammatic illustrations (see, for example, Table VII-6). Trade names of thermoplasts, thermosets, elastomers, and fibers will not be used in the text.

2.2. *Atomic Structure and Polymer Chain Bonds*

2.2.1. *Overview*

A molecule is seen as a solid unit with a spatial arrangement stable in time in low-molar-mass chemistry. Such a definition is very extensive and includes, in principle, a rock salt crystal as well as a piece of iron. It is conveniently further limited in that bonding relationships are also drawn on.

In this book, a compound will be considered to be a macromolecule when the atoms in the main chain are joined by directional valencies and the electrons of a bond joining two atoms being shared by both bonded atoms. This kind of definition limits the types of bond to covalent bonds and intermediate types through to ionic or metallic bonds, i.e., coordinate bonds and electron-deficient bonds. Atomic groups joined by metallic bonds are not counted as macromolecules, since, although the electrons are shared between the bound atoms, the bonds are not directional. Ionic crystals are also not considered to be macromolecules, since, with ideal ionic bonds, the electrons are not shared between bonded atoms, nor are the bonds directional.

Macromolecular chains defined in this way may be classified according to the nature of the atoms joined in the main chain, the constitutional repeating units and monomeric units, as well as by the molecular architecture.

Isochains consist of identical main chain atoms. *Heterochains* have two or more different kinds of atoms in the main chain. Isochains are very readily formed by carbon atoms, but less readily, however, by other elements. Chains containing no carbon atoms in the main chain are also known as inorganic chains.

Isochains and heterochains may be substituted or unsubstituted. Here, the different significances of the word *substitution* in inorganic and organic chemistry must be noted. In the strictest sense, unsubstituted isochains occur in *catena*-poly(sulfur). On the other hand, silanes, $H(SiH_2)_nH$, should be considered as substituted in the sense of inorganic nomenclature. By contrast, the poly(alkanes), $H(CH_2)_nH$, are considered to be unsubstituted chains in organic chemistry, since the basic structure is here considered to be the alkanes, and not diamond.

Poly(methylene), poly(oxyethylene), and poly(dimethyl siloxane) possess only one constitutional repeating unit, which also always occurs arranged in the same way. Such chains are called *regular* chains. *Irregular* chains, on the other hand, have the constitutional repeating unit arranged irregularly. The irregularity may result from irregular bonding of a constitutional repeating unit, or from irregularities on the sequence of successive different constitutional repeating units.

Polymers consisting of macromolecules with only one kind of monomeric unit are called *homopolymers*. *Copolymers* are produced from two, three, four, etc., different monomers and are consequently also called bipolymers, terpolymers, quaterpolymers, etc. Previously, copolymers were also known as heteropolymers, interpolymers, or mixed polymers.

2.2.2. Isochains

According to definition, isochains can be formed from all elements for which isolatable compounds containing at least three successive identical atoms in a chain exist. Isolatable compounds are considered to be those that can be obtained in any fluid form, for example, as gas, melt, or solution.

The elements of the first period (see Table 2-3) exhibit the largest chain link number. Within each group, the chain link number decreases with increasing period number.

The whole of organic chemistry depends, finally, on the ability of *carbon* to form isochains. Diamond, as three-dimensional carbon, can, for example, be systematically named as *tecto*-poly(carbon); as is well known, it is the basic structure from which the alkanes may be derived. Alkanes, or poly(alkanes), with the main chain $R(CH_2)_N R$, can be obtained with practically "infinitely high" chain link numbers.

Silicon also occurs as a polymer in the solid state. However, silanes, $H(SiH_2)_N H$, can only be isolated with a chain link number up to $N = 45$. On the other hand, silanes as $(SiH)_N$, exist as high-molar-mass polymers, probably as *phyllo*-poly(silanes) with six-membered, flat annellated rings.

The tendency to form isochains decreases still further with the germanes, $H(GeH_2)_N H$, and stannanes, $H(SnH_2)_N H$. Elemental *germanium* occurs in polymeric form, while *tin* exists as a polymer in one modification, whereas in another it exhibits metallic bonding.

Table 2-3. Periodic Position of the Elements and Their Ability to Form Isochains[a]

IIIB 2s, 1p	IVB 2s, 2p	VB 2s, 3p	VIB 2s, 4p	VIIB 2s, 5p
[5] B, ~5	[6] C, ∞	[7] N, ∞?	[8] O, ∞?	[9] F, 2
[13] Al, 1	[14] Si, 45	[15] P, <10	[16] S, 30,000	[17] Cl, 2
[31] Ga, 1	[32] Ge, 6	[33] As, 5	[34] Se, ?	[35] Br, 2
[49] In, 1	[50] Sn, 5	[51] Sb, 3	[52] Te, ?	[53] I, 2
[81] Tl, 1	[82] Pb, 2	[83] Bi, ?	[84] Po, ?	[85] At, 2

[a]The figures at the right are the highest chain-link numbers so far observed for isochains after isolation.

Only *boron* in the third periodic group, and then, only as the element, occurs in the solid state as a polymer. With the boranes (borohydrides), main chain bonds are only partly boron–boron bonds; they also possess boron–hydrogen–boron bond character.

With *nitrogen* in the fifth group, a blue substance with the composition $(NH)_N$ is obtained by decomposing hydrogen azide, HN_3, at $1000°C$ and shock cooling the reaction product with liquid nitrogen. Isochains presumably occur in this product. The product converts to ammonium azide, NH_4N_3, at $-125°C$. On the other hand, nitrogen is not polymeric in elemental form.

The elements *phosphorus, arsenic,* and *antimony,* on the other hand, exist polymerically in what are known as their allotropic modifications. The best known example is black phosphorus (see Section 33.4.1). The corresponding hydrogen compounds, that is, the phosphanes, arsanes, and stilbanes, are only known in the form of short isochains or small rings. In the sixth group, an ozone modification occurring at low temperatures probably consists of *oxygen* isochains. *Selenium* and *tellurium* form chain polymers in the solid state, and *sulfur* does so in the melt. In contrast to the elements of group V, that is, phosphorus, antimony, and arsenic, the tendency to form hydrogen compounds in chain form, i.e., sulfanes, selanes, and telluranes, is distinctly decreased. For sulfur, for example, chain form "polysulfanes," $HS_nH, n = 2–8$, "halogenosulfanes," XS_nX, with $X = Cl, Br,$ or $I,$ and sulfur cations, S_n^{2+} (with $n = 4, 8, 16$), and sulfur anions, S_n^{2-} (with $n = 4, 9$), are known.

Isochains from elements of other groups are unknown. The experimental observation that only a quite distinct and limited number of elements close together in the periodic table can form such isochains can be explained as follows: Elements of the first row do not possess available *d* orbitals. They can therefore form no more than four σ bonds per atom, which corresponds to sp^3 hybridization. Only carbon and the elements to the right of it have enough electrons to contribute at least one electron to each complete bond (σ, π_x, π_y). The elements to the right of carbon thus show a lower bonding energy than carbon (Table 2-4), and acting as electron donors with the corresponding electron acceptors, they form heterochains particularly readily. Nitrogen, oxygen, and fluorine, when compared with their successors in the same group, have too low a bonding energy, and this, according to K. S. Pitzer, results from the strong mutual repulsion of the free electron pairs.

The elements to the left of carbon, on the other hand, have less available electrons than unoccupied orbitals. Since the atoms tend to fill their energetically more accessible outer orbitals, multicenter bridge bonds, involving hydrogen, methyl groups, etc., will occur with boron and beryllium.

Elements of the second row possess *d* orbitals of sufficiently low energy to

Table 2-4. Bond Energies $(10^{-5}\ J/mol\ bond)$ *for Bonds between Like Elements*[a]

Bond	Energy	Bond	Energy	Bond	Energy	Bond	Energy
						H—H	4.3
C—C	3.5	N—N	1.6	O—O	1.6	F—F	1.6
Si—Si	1.8	P—P	2.1	S—S	2.1	Cl—Cl	2.4
Ge—Ge	1.6	As—As	1.3	Se—Se	1.8	Br—Br	1.5
Sn—Sn	1.4	Sb—Sb	1.2	Te—Te	1.4	I—I	1.5

[a]After L. Pauling.

participate in bonding. In general, however, d orbitals are not used for σ bonding, but, via hybridization, for π bonding. This hybridization increases the stability of the molecule. Hybridization is most marked in silicon, phosphorus, and sulfur. These elements are therefore polymeric in the solid state, and even less condensed states show to some extent a high chainlink number.

In the fourth row the d orbitals are used more for σ bonding than for π bonding. As would be expected, the bonding capacity of the third-row elements lies between that of the second- and fourth-row elements.

It is therefore to be expected that the bond energy within each period falls with increasing atomic number, and likewise within each row (cf. Table 2-4). The higher the atomic number of the element, the less marked the macro-molecular character of the inorganic polymer will be.

The bond energies reproduced in Table 2-4 are largely what one would expect. Admittedly they are not entirely unambiguous, as, on the whole, they are not values for chain structures, but were measured on low-molecular-weight compounds containing only one isobond. The bond strengths of such bonds, however, are influenced by the nature of the substituents. The bond energies of Table 2-4 are thus "average" bond energies from a numerous set of compounds, and are not genuine bond dissociation energies. Complications can be expected even in macromolecules themselves, since their one-, two-, or three-dimensional structures are strongly affected to a varying extent by polarization effects, and therefore exhibit different bond stabilities.

Thus, only a few elements form *isolatable* isochains. However, for all elements for which crystallographic data are available, only 11 do not have at least one macromolecular form in the crystalline state. In a study involving about 1200 crystallographically investigated compounds of two elements, only 5% were nonmacromolecular, and of the others, 1.5% were linear, 7.5% were parquet, and 86% were layer polymers.

2.2.3. Heterochains

Heterochains are known in larger numbers. Carbon forms a whole series of different heterochains with elements such as oxygen, nitrogen, and sulfur, whereby the heteroatom is sometimes arranged alternately with, and sometimes at larger distances from the carbon atom. Examples are

+CRR'—O+   +CRR'—CR"R"'—O+   +R—CO—O+   +R—SO₂+

polyacetals   polyethers   polyesters   polysulfones

+CRR'—CR"R"'—NR+   +R—NH—CO+

polyimines   polyamides

+R—NH—CO—NH+   +R—CO—NH—CO+

polyureas   polyimides

Inorganic heterochains occur in much more varied form than organic heterochains. Inorganic heterochains can be classified as linear chains without (I) or with (II) rings within the chain, or as "spiro" compounds with small (III) or with large (IV) rings:

I	II	III	IV
A = B, Al, Si, Ge, Sn, Pb, P, Sb, Ti, V, Cr, Fe, Co	Si, P, B	Be, Si, Pd, Nb	Be, Zn, Co, Cr
D = O, S, Se, Na	O, N		
X, Y = O, Hal, org. residue		H, F, Cl, Me (with Be), O, S (with Si), Cl (with Pd), I (with Nb)	OH, H_2O
R = Ar =	Me, Ph Phenylene, Borazine, Phosphazine, etc.		Me, Ph

The bonding and stability of heterochains depends basically on the electronegativity of the participating atoms. The electronegativity E is a measure of the capacity of the element to compete at any instant with the other elements for the larger proportion of the electronic charge. Electronegativity cannot be measured directly, but is estimated from ionization potentials, atomic radii, bond force constants, or bond energies. The Pauling electronegativity series based on bond energies is the best known.

In Pauling's electronegativity series, the most electronegative element, fluorine, with $E = 4.0$, acts as a standard or reference element. In this scheme, carbon has a value of $E = 2.5$ and hydrogen, a value of 2.1. Every combination of elements with electronegativity higher than 2.5 with those of electronegativity below 2.5 leads, within the observance of certain selection rules, to heterochains. Oxygen (3.5), nitrogen (3.0), and sulfur (2.5) thus form heterochains with boron (2.0), aluminum (1.5), silicon (1.8), germanium (1.8), tin (1.8), lead (1.8), titanium (1.5), zirconium (1.4), phosphorus (2.1), arsenic (2.0), antimony (1.9), bismuth (1.9), and vanadium (1.6).

On the basis of the electronegativity value, it is to be expected that sulfur and selenium are very similar to carbon in their tendency to form chains (cf. Table 2-3). Equally, $(C—S)_x$ chains are also relatively stable.

The tendency to form multiple bonds competes with the tendency to form heterochains. In multiple bonds σ bonds are also present with π bonds. A π bond possesses a bond order of two. Bond orders can be calculated from force constants, which, in turn, can be obtained from vibrational spectra (infrared and Raman spectroscopy).

From data of this kind, the following conditions were drawn up empirically for π-bond occurrence:

1. Both atoms between which the bond is formed must be electron deficient.
2. The sum of the Pauling electronegativities for both bond partners must be at least five.
3. The difference in the Pauling electronegativities for both bond partners must be as low as possible, i.e., less than 1.5.

The significance of these rules for polymer chemistry can be demonstrated by the following example:

The nitrogen atom has an electronegativity of 3.0. Thus for the nitrogen–nitrogen bond, the sum of the electronegativities ΣE is 6.0, whereas the difference ΔE is zero. Two nitrogen atoms therefore form a very stable multiple (triple) bond and polymeric nitrogen is not stable under normal conditions.

In hydrogen cyanide, $H—C \equiv N$, the values for the carbon–nitrogen bond are $\Sigma E = 5.5$ and $\Delta E = 0.5$. The triple bond is thus stable, but weaker than the nitrogen–nitrogen triple bond. Correspondingly, hydrogen cyanide is also known in polymeric form. In the boron–nitrogen compounds, ΣE falls to 5.0, while ΔE rises to 1.0. HBNH no longer exists as a monomer but is exclusively polymeric.

Too great a difference in electronegativity leads to ionic bonding, and therefore no macromolecular character. Correspondingly, it is found that bond energies between elements of the first and second row, which have fully occupied orbitals, increase to a first approximation (Table 2-5) with the electronegativity difference between the bonded atoms. In contrast, the bond energies between boron and carbon (440 kJ/mol bond) and boron and

Table 2-5. *Bond Energies and Differences in Electronegativity*

Bond	Bond energy in 10^{-5} J/mol bond	Difference in electronegativity
C—S	2.6	0
C—N	2.9	0.5
C—Si	2.9	0.7
C—O	3.5	1.0
Si—O	3.7	1.7

nitrogen (830 kJ/mol bond) can be traced back to the electronic structure of boron. In comparisons of this kind, therefore, allowance must always be made for the position of the element in the periodic table, i.e., for the "ionic bond contribution."

Bond energies (actually bond dissociation energies) primarily indicate the susceptibility of the bond to thermal scission, and therefore give evidence of the thermal stability of the macromolecule. The vulnerability of a bond to other reagents, on the other hand, depends mainly on the ionic character of the bond and whether there remain unoccupied orbitals or electron pairs in the molecule, as these will lower the activation energy of reaction with the reagent. The resistance to reduction, oxidation, hydrolysis, etc., decreases with increasing atomic number in every period. Thus hydrocarbons, $C_n H_{2n+2}$ are not hydrolyzed, but silanes, $Si_n H_{2n+2}$, certainly are. In the latter case there are only four completely filled orbitals of a possible maximum coordination number of six.

In bonds between carbon and nitrogen, phosphorus, oxygen, sulfur, selenium, or the halogens, the carbon is made more positive ($C^{\delta^+}-E^{\delta^-}$) according to the position of the element in the periodic table, and is thus more easily attacked by nucleophilic reagents. If, on the other hand, the bond partner of carbon is a metal atom, then the now negative carbon ($C^{\delta^-}-Mt^{\delta^+}$) can only be attacked by electrophilic reagents. Thus all macromolecules with heteroatoms in the main chain react more readily than pure carbon chains. Under most conditions, they undergo exchange equilibrium, and are readily attacked chemically. The vulnerability of carbon is further dependent on its substituents. These act as electron donors (e.g., methyl groups) or as electron acceptors (e.g., halogens) and can either stabilize or activate the main-chain bonds once they have been formed.

Similar considerations apply to carbon bonded to noncarbon atoms in the main chain. The —Si—CH$_3$ is not so polar as —SiH$_3$, and so the dimethyl derivative is sufficiently stable to enable poly(dimethylsiloxanes), $+Si(CH_3)_2-O+_n$, to be produced commercially. By contrast, Ti—C and Al—C bonds are sensitive to oxygen and hydrogen (electronegativity of metals!). Thus organic groups can only be bonded with these metal atoms via ether or carboxyl groups.

In boron and beryllium there are fewer electrons than available orbitals. The outer orbitals can be filled, for example, in dimethyl beryllium, by a mechanism whereby one carbon atom orbital overlaps with an orbital from each of two neighboring beryllium atoms. The resulting high-molecular-weight chain of dimethyl beryllium thus has three center bonds, which leads to an absurd structural formula when valences are represented in the normal way. Boron hydrides and some aluminum compounds have similar structures:

$$\left(\underset{\underset{CH_3}{\diagdown}}{\overset{\diagup CH_3}{Be}}\right)\qquad -\!\!\left(\underset{\underset{F}{F}}{\overset{F}{\diagdown}}Al\underset{\underset{F}{\diagdown}}{\overset{\diagup F}{\diagdown}}\right)^{-2}\!\!-\qquad \left(\underset{\underset{Cl}{\diagdown}}{\overset{\diagup Cl}{Pd}}\right)\qquad \left(Nb\underset{\underset{I}{\diagdown}}{\overset{\diagup I}{\diagdown}}\right)$$

$$\text{I}\qquad\qquad\qquad\text{II}\qquad\qquad\qquad\text{III}\qquad\qquad\text{IV}$$

Main and subgroup elements of the earlier rows often exhibit a dynamic equilibrium between different coordination numbers. Elements such as fluorine, chlorine, oxygen, etc., can donate one or two free electron pairs to vacant lower energy orbitals of these metal atoms. Fluorine can therefore act as a bifunctional bridging atom, and oxygen can even be mono- to tetrafunctional, according to its bond partner. In all these cases the coordination number is increased above normal. In the case of fluorine, fluorine bridge bonds exist, for example, in the anion of the complex between thallium fluoride and aluminum fluoride, where every chain unit carries a double-negative charge. Chlorine bridging occurs, for example, in palladium (II) chloride, and iodine bridging atoms in niobium (II) iodide.

All of these compounds have many vacant orbitals at their disposal. They are therefore readily attacked, and in all of the common solvents they decompose into smaller units. For this reason they were previously not even considered to be macromolecules in the solid state.

2.3. Homopolymers

2.3.1. Monomeric Unit Bonding

According to the IUPAC definition, a homopolymer is a polymer formed from a single monomer species. Such a nomenclature based on the process will only give an insight into the chemical structure of the initial monomer, it will provide no insight into the structure of the polymer itself.

In many cases, the chemical structure of the polymer can be "intuitively" deduced from the structure of the monomer. For example, polymers with the same monomeric unit can be obtained by the ring-opening polymerization of lactones and by the polycondensation of ω-hydroxycarboxylic acids:

$$(CH_2)_x \overset{CO}{\underset{O}{<}} \qquad \longrightarrow \quad {\left[O {-} (CH_2)_x {-} CO \right]} \qquad (2\text{-}1)$$

$$HO {-} (CH_2)_x {-} COOH \qquad {}_{-H_2O}$$

The structure of the monomeric units are not altered by conversion from monomer to polymer in these polymerizations. In addition, bonding of the monomeric units in the polymer always occurs in the same direction; peroxide and diketo structures are not formed.

These two characteristics are not always encountered, especially in the cases of addition polymerization of monomers with double bonds. Isomerization of the monomers may occur, or "false" bonding of monomeric units into the chain may take place during the actual step of adding monomer onto the polymer chain. Consequently the assumed chemical structure of the polymer must always be carefully verified by analytical methods. Analysis is especially important with industrial polymer production, since the preparation history is not often known exactly. The chemical names of industrial or commercial polymers are often nothing more than a kind of generic name. Commercial poly(ethylenes), for example, despite the ascribed name, are often not homopolymers, but copolymers of ethylene and propylene. As well as that, commercial polymers practically always contain additives such as antioxidants, uv absorbers, fillers, etc. In the addition polymerization of monomers with multiple bonds, head-to-head and tail-to-tail structures are always to be expected together with the normal head-to-tail bonding, as can be seen, for example, with vinyl compounds such as $CH_2{=}CHR$:

$$CH_2{=}\underset{R}{CH} \quad
\begin{array}{l}
\xrightarrow{\text{1,2 addition}} \quad {-}CH_2{-}\underset{R}{CH}{-}CH_2{-}\underset{R}{CH}{-}CH_2{-}\underset{R}{CH}{-} \quad \text{head-to-tail} \\[2em]
\xrightarrow{\text{1,1 addition}} \quad {-}CH_2{-}\underset{R}{CH}{-}\underset{R}{CH}{-}CH_2{-}CH_2{-}\underset{R}{CH}{-} \quad \begin{array}{l}\text{head-to-head}\\ \text{or}\\ \text{tail-to-tail}\end{array}
\end{array} \qquad (2\text{-}2)$$

The end with the largest substituent is generally considered to be the head.

Larger proportions of head-to-head and tail-to-tail linkages are to be expected when steric effects are small and the reasonance stabilization of the growing macroradical or macroion is slight. For example, in the free radical polymerization of vinyl fluoride, I, and vinylidene fluoride, II, $\sim 6\%\text{--}10\%$ and $10\%\text{--}12\%$ head-to-head structures, respectively, are formed; the increase being attributable to the fact that the atomic radius of fluorine, with 0.067 nm, is smaller than that of hydrogen, with 0.077 nm. Similar effects are also responsible for the 40% head-to-head structures formed during the polymerization of propylene oxide, III, with diethyl zinc/water as initiator:

$$CH_2{=}CHF \qquad CH_2{=}CF_2 \qquad H_2C\underset{\displaystyle O}{\diagdown\diagup}CH(CH_3)$$

$$\textbf{I} \qquad\qquad\qquad \textbf{II} \qquad\qquad\qquad \textbf{III}$$

The head-to-head structure contents of poly(vinyl fluoride) and poly(vinylidene fluoride) can be determined by ^{19}F-nuclear magnetic resonance measurements. The physical procedure is, however, not sufficiently sensitive for determining smaller "false" linkage contents, and chemical methods must therefore be used.

Poly(vinyl alcohol) contains $\sim 1\%\text{--}2\%$ head-to-head linkages, as was found by hydroxyl group oxidation. The head-to-head structures are, namely, oxidized to oxalic and succinic acids by periodic acid, H_5IO_6:

$$-CH_2-\underset{\underset{\displaystyle HO}{|}}{CH}-\underset{\underset{\displaystyle OH}{|}}{CH}-CH_2-CH_2-\underset{\underset{\displaystyle HO}{|}}{CH}-\underset{\underset{\displaystyle OH}{|}}{CH}-CH_2- \;\longrightarrow\; \begin{array}{c} HOOC-COOH \\ + \\ HOOC-CH_2-CH_2-COOH \end{array}$$

$$(2\text{-}3)$$

On the other hand, head-to-tail structures are oxidized to acetic acid by CrO_3, while head-to-head linkages are not attacked:

$$-CH_2-\underset{\underset{\displaystyle OH}{|}}{CH}-CH_2-\underset{\underset{\displaystyle OH}{|}}{CH}- \;\longrightarrow\; -CH_2-\underset{\underset{\displaystyle O}{\|}}{C}-CH_2-\underset{\underset{\displaystyle O}{\|}}{C}- \;\longrightarrow\; CH_3COOH \qquad (2\text{-}4)$$

In all these cases, different modes of monomeric linkage are obtained, but the structure of the monomeric unit itself remains the same. Some other monomeric units, however, can be abnormally incorporated into the polymer even under normal polymerization conditions. In the free radical polymerization of methacrylonitrile, for example, a limited amount of polymerization occurs via the nitrile group, as can be shown spectroscopically:

$$\underset{\underset{\displaystyle C{\equiv}N}{|}}{\overset{\overset{\displaystyle CH_3}{|}}{CH_2{=}C}} \;\longrightarrow\; +\!\!\Big(CH_2-\overset{\overset{\displaystyle CH_3}{|}}{C}{=}C{=}N\Big)\!\!\frac{}{x} \qquad (2\text{-}6)$$

Under certain conditions, such "false" monomeric units can be the major portion of the polymerized linkages in ionic polymerizations. For example, acrylamide is polymerized free radically to poly(acrylamide), but, anionically, by proton shift, to poly(β-alanine):

$$+ CH_2 - CH \xleftarrow{\text{free radical}} CH_2 = CH \xrightarrow{\text{anionic}} + CH_2CH_2CONH + \qquad (2\text{-}7)$$
$$\qquad\quad | \qquad\qquad\qquad\qquad\quad | $$
$$\qquad\quad CONH_2 \qquad\qquad\qquad\quad CONH_2$$

Styrene *p*-sulfonamide reacts similarly:

$$+ CH_2CH + \xleftarrow[\text{radical}]{\text{free}} CH_2 = CH \xrightarrow{\text{anionic}} + (CH_2CH_2 - \bigcirc - SO_2NH_2) + \qquad (2\text{-}8)$$

Proton shifts occur occasionally in cationic polymerizations. 4,4-Dimethylpentene-1 polymerizes quite normally via the carbon–carbon double bond at low temperatures, whereas additional monomeric units with three carbon main chain atoms per monomeric unit are produced at higher temperatures:

$$+ CH_2 - CH + \xleftarrow{-130°C} CH_2 = CH \xrightarrow{0°C} + CH_2CH_2CH + $$
$$\qquad\quad | \qquad\qquad\qquad\qquad | \qquad\qquad\qquad\qquad | \qquad\qquad (2\text{-}9)$$
$$\quad CH_2C(CH_3)_3 \qquad\qquad CH_2C(CH_3)_3 \qquad\qquad C(CH_3)_3$$

The "false" monomeric unit content can often be determined by spectroscopic procedures, such as infrared or nuclear magnetic resonance spectroscopy. In the special cases of acrylamide or styrene *p*-sulfonamide polymers, it can be determined by hydrolysis as well. Polymers for which no monomer corresponding to the actual monomeric unit exists are called phantom or exotic polymers.

The "false" monomeric unit linkages produced in small amounts are often less stable than the "normal" linkages. They are often consequently recognized as "weak links" during degradation studies. On the other hand, their presence is often intimated solely from degradation studies, which, of course, does not constitute proof. In addition, weak links may also be produced through incorporation of foreign material, for example, through incorporation of small amounts of oxygen.

2.3.2. Substituents

Neutral substituents on macromolecular chains require no special treatment in terms of constitution or nomenclature to differentiate them from low-molar-mass substances. Practically all the groupings and classifications of low-molar-mass organic, organometallic, and inorganic chemistry can be used with substituents on macromolecular chains.

Special considerations are required for polymers with substituents possessing ionically dissociating bonds, and these polymers can be quite generally seen to be a subclass of polymers with ionically dissociating bonds.

Such polymers are known as *polyelectrolytes* if the ionic concentration is high, and the polymer water soluble. By contrast, water-insoluble polymers with small ionically dissociating bond concentrations are known as *ionomers.*

Polyelectrolytes may be polyacids, polybases, or polyampholytes. They dissociate into polyions and oppositely charged gegenions. Polyacids release protons on dissociation and become, then, polyanions. Poly(phosphoric acid), I, is an example of a polyacid with a dissociating group in the main chain; examples of polyacids with the dissociating group as substituent to the main chain include poly(vinyl phosphonic acid), II; poly(vinyl sulfuric acid), III; poly(vinyl sulfonic acid), IV; and poly(vinyl carboxylic acid) = poly-(acrylic acid), V, with the monomeric units

$$
\begin{array}{ccccc}
\overset{\displaystyle O}{\underset{\displaystyle OH}{-O-\overset{\|}{\underset{|}{P}}-}} & -CH_2CH- & -CH_2CH- & -CH_2CH- & -CH_2CH- \\
& \underset{\displaystyle PO_3H_2}{|} & \underset{\displaystyle SO_3H}{|} & \underset{\displaystyle SO_2H}{|} & \underset{\displaystyle COOH}{|} \\
\textbf{I} & \textbf{II} & \textbf{III} & \textbf{IV} & \textbf{V}
\end{array}
$$

Polybases correspondingly take up protons or even methyl groups on ionizing, and become polycations by such "quaternization." Poly(ethylene imine), VI, is an example of a polybase with a proionic group in the main chain. Examples of polybases with the proionic group as substituent to the main chain are poly(vinyl amine), VII, and poly(4-vinyl pyridine), VIII, with the monomeric units

$$
\begin{array}{ccc}
-CH_2CH_2NH- & -CH_2CH- & -CH_2CH- \\
& \underset{\displaystyle NH_2}{|} & \\
\textbf{VI} & \textbf{VII} & \textbf{VIII}
\end{array}
$$

Polyampholytes are polymers which can carry both positive and negative charges.

Polyions, polyanions, polycations, etc., must be differentiated from macroions, macroanions, macrocations, etc. In contrast to polyions, macroions carry only one or a few ionic groups. For example, the growing chain end in a cationic polymerization has a positive charge, and the polymer chain attached to this growing end is thus a macrocation, but not a polycation. Corresponding definitions apply to macroanions and macroradicals. In conventional free radical polymerization, macroradicals occur, but polyradicals are present during free radical graft polymerization.

2.3.3. End Groups

By definition, end groups are the groups that occur at the ends of a macromolecular chain. An unbranched linear chain has two end groups, a star-shaped polymer with four arms has four end groups.

The identification of the end groups of a polymer yields information on the mechanism of synthesis and also, in favorable circumstances, allows the determination of the molar mass and/or the degree of branching. With otherwise identical chain architecture, the end group concentration diminishes, naturally, with increase in molar mass. Consequently, if the molecular architecture is known, the molar mass can be calculated from the end group concentration. Since the end group fraction is inversely proportional to the molar mass, this method of molar mass determination must yield the number average molar mass (see also Section 8). In general, the number average molar mass is calculated from the amount of substance n_i, of the i kinds of end groups, the number N_{end}, of all end groups per macromolecule present, and the mass, m, of the sample used in the analysis:

$$\langle M \rangle_{n,end} = N_{end} \, m \Big/ \sum_{i=1}^{i=i} n_i \tag{2-10}$$

Thus, the number average molar mass does *not* result from the arithmetic mean of the equivalent masses calculated for every single kind of end group. It is the reciprocal of the mean divided by the reciprocal of the equivalent masses.

The sensitivity of such molar mass determinations depends on the analytical method as well as on the nature of the end groups. Molar masses up to $\sim 40,000$ can be obtained by titration, up to 100,000 by microanalytical determination of iodine, up to 200,000 by radioactively labeled groups, and up to 1,000,000 g/mol with intensely colored groups.

In every case, the number of ends per macromolecule and the structure of the end groups must be known with certainty. On the other hand, the number of branches per macromolecule can be determined by end group analysis if the molar mass is determined by another method.

2.4. Copolymers

2.4.1. Definitions

According to IUPAC, copolymers are produced from more than one kind of monomer. They are classified as bi-, ter-, quater-, quinterpolymers, etc., according to the number of different kinds of monomers. They are further classified into alternating, gradient, and random copolymers according to the sequence the various kinds of monomeric units follow in the macromolecular chain. In alternating copolymers, both types of monomeric unit, A and B, of a bipolymer alternate with each other along the macromolecular chain. With random copolymers, on the other hand, there is an irregular sequence of both monomeric units. In gradient, or tapered, copolymers, there is a gradient of copolymer composition along the chain, such that, for example, one end is rich in A-type monomeric unit, and the other chain end is rich in B-type monomeric unit. According to IUPAC, block polymers consist of blocks of homosequences joined together; graft polymers, in contrast, consist of B side chains grafted onto an A chain. Multiblock polymers with short blocks are also called segment or segmented polymers. Block and graft polymers are to be differentiated from block and graft *co*polymers which contain at least one chain or side branch that is an alternating, random, or gradient copolymer. The following names and types apply to bipolymers:

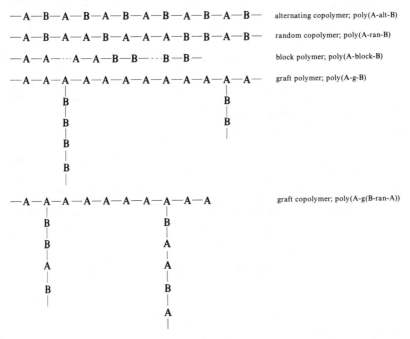

Bipolymers with unknown monomeric unit sequence are written as poly(A/B). Copolymers with more than two monomeric units are classified and named analogously to bipolymers.

Copolymers occur widely in nature and industry. For example, proteins are copolymers with approximately 20 different kinds of amino acids in irregular order, but with the same sequence in each and every chain. Block copolymers also occur in nature with α-amino acid and sugar residue monomeric units.

Synthetic copolymers are produced to improve certain properties, or even as the only means of attaining these properties. Here, alternating and random polymers are synthesized in one step from a mixture of A and B monomers, or are obtained by transformation of an already synthesized polymer. In contrast, block and graft polymers are generally produced by two successive and different polyreactions, and so are often called "multistep" polymers.

The analysis of copolymers is significantly more complicated than that of homopolymers, since the sequence must also be determined in addition to the constitutional composition. In many cases, the determination of the distribution of monomeric units in terms of the proportions of di-, tri-, ter-, and quatersequences is ignored and one is content with classifying the polymer as "constitutionally heterogeneous."

2.4.2. *Constitutional Composition*

The average composition of copolymers can be determined most readily for the case where the monomeric units can be isolated and identified by suitable scission or degradation reactions. This is the usual method for the elucidation of protein structure. The proteins are hydrolyzed by acids and/or bases in an automated apparatus. The resulting amino acids are chromatographically separated and assayed quantitatively via the color reaction with ninhydrine in so-called amino acid analyzers.

Such mild degradative reactions are usually not successful with synthetic polymers and are completely inapplicable for carbon chains. When synthetic polymers are pyrolyzed under controlled conditions, a gas-chromatographic analysis of the degradation products provides a kind of fingerprint for the composition and sequence of the polymer in question. Since the method is rapid but not absolute, it is preferentially used for industrial quality control.

The composition of carbon-chain polymers with monomeric units having widely differing analytical composition, characteristic elements or groups, or radioactive labels can be readily determined. Chemical (microanalysis, functional group determination, etc.) and spectroscopic methods (infrared, ultraviolet, nuclear magnetic resonance, etc.), as well as the determination of radioactivity, yield the average composition of the polymer. The mean composition can also be determined from the refractive indices of solid samples. The composition can be calculated from the principle that the copolymer is considered to be a solution of one unipolymer (from one of the monomeric units) in the other. The composition can also be found by means of the refractive index increment dn/dc in solution, which gives the variation in refractive index with concentration. The mass fraction w_A of the monomeric unit A can be calculated from

$$\left(\frac{dn}{dc}\right)_{copolymer} = \left(\frac{dn}{dc}\right)_A w_A + \left(\frac{dn}{dc}\right)_B w_B \qquad (2\text{-}11)$$

where $w_A + w_B = 1$. The temperature, solvent, and wavelength, as well as the refractive indices of both homopolymers A and B, must be known for this

Table 2-6. *Results for a Styrene–Methyl Methacrylate Copolymer from Different Analytical Procedures*[a]

Sample number	% Methyl methacrylate in the polymers				
	C,H,O	ir	uv	nmr	dn/dc
C12	74.4	74.0	78.5	73.5	72.8
C14	58.1	53.0	57.7	—	57.0
C16	42.2	41.0	48.5	40.2	41.5
C18	23.0	23.5	28.7	24.1	21.5

[a]Data from H.-G. Elias and U. Gruber.

method of determination. Table 2-6 gives a summary of the results of various analytical methods for styrene–methyl methacrylate copolymers. The results from uv (ultraviolet) analysis differ markedly from those of other methods because the position of the uv bands is strongly dependent on the styrene sequence lengths.

The cloud-point titration method (Section 6.6.5) can also be applied to copolymers whose monomeric units are chemically not very different. In this method, solutions of various concentrations are titrated with nonsolvent to the first cloud point. By extrapolation to 100% polymer, a critical volume fraction ϕ_3^θ of the nonsolvent is obtained, which normally depends linearly on the copolymer composition.

Given certain assumptions, the cloud-point method allows a homo-polymer to be distinguished from copolymers, and this enables the results of graft copolymerization to be monitored. All the other methods so far described do not allow any differentiation between copolymers and polymer mixtures (polymer blends). Polymer mixtures can also be characterized by ultracentrifugation in a density gradient (Section 9.7.5) and, in certain cases, also by fractionation (Section 6.6.4).

2.4.3. Constitutional Heterogeneity

A copolymer with two or more constitutionally different monomeric units is not sufficiently characterized by its average composition alone. A product with, for example, a 50% proportion of component A and 50% of component B can be a true copolymer with a composition that is constant for all the molecules present, a true copolymer with different A : B ratios among the component molecules, a polyblend from two homopolymers, or a corresponding mixture from two homo- or bipolymers. Consequently, copolymers have to be characterized according to their compositional distribution, that is, the distribution of amounts of monomeric units and of molar masses must be determined with respect to composition.

There are two especially suitable procedures for this: fractional precipitation or dissolution (fractionation according to solubility) and equilibrium sedimentation or centrifugation in a density gradient. Equilibrium sedimentation is only suitable for polymers of very high molar mass (see Section 9), and, so, will not be discussed here. In contrast, fractionation according to solubility is universally applicable. Precipitant is added gradually to a solution of a copolymer. Fractions with lowest solubility precipitate first, followed by those of medium solubility, and then by those of greatest solubility. The precipitated fractions usually have a gel-like consistency, hence the name, *gel phase*. The gel consists of polymer fractions swollen by solvent/precipitant. Since the solubility depends on the constitutional composition, a series of fractions of different mean constitutional compositions is obtained. Table 2-7 shows the results of such a fractionation carried out on a copolymer of vinyl acetate and vinyl chloride. The proportions of vinyl chloride and vinyl acetate do not conform to the fraction number (order in which the fractions are removed), since in such fractionations, the solubility depends not only on the constitutional composition, but also on the molar mass. Thus, through appropriate choice of solvent or precipitant, the

Table 2-7. Results from the Fractionation of a Copolymer of Vinyl Acetate and Vinyl Chloride^a

Fraction number	Mass in mg	Mass fraction in %	E_{vc}	Σw_i^*
2	41.0	5.32	0.363	2.660
1	56.0	7.27	0.364	8.955
5	78.5	10.19	0.412	17.683
3	43.5	5.65	0.414	25.600
4	61.5	7.98	0.510	32.414
6	64.5	8.37	0.577	40.591
15	26.5	3.44	0.587	46.495
11	38.0	4.93	0.595	50.681
13	38.0	4.93	0.595	55.613
7	72.5	9.41	0.625	62.784
9	51.0	6.62	0.636	70.798
8	63.5	8.24	0.638	78.228
10	32.0	4.15	0.642	84.425
14	56.0	7.27	0.665	90.136
12	48.0	6.23	0.676	96.885
	$\Sigma W_i = 770.5$	$\Sigma w_i = 100$	$\bar{E}_{vc} = 0.550$	
			$= (\bar{E}_w)_{vc} = \bar{E}_w$	

^{a}The results are given in terms of increasing content of vinyl chloride monomeric unit E_{vc} in the fractions (from H.-J. Cantow and O. Fuchs).

fractionation may proceed predominantly according to molar mass or composition (see also Section 6). On the other hand, unsuitable choice of solvent and precipitant can cause an erroneous imitation of homogeneous polymer behavior.

The composition distribution of bipolymers can be displayed on a two-axis graph, in which the abscissa gives the measured property, as, for example, the vinyl chloride monomeric unit content, and the ordinate gives the corresponding sum of mass fractions. Such an integral mass distribution of the properties can be, as for example, for the copolymer in Table 2-7, obtained as follows from the fractionation data:

Each fraction is not homogeneous, per se, with respect to the measured property, but possesses a distribution. The property measured for this fraction is also a mean value. To a first approximation, one half of the fraction will have a composition above, the other fraction half will have a composition below, the mean fraction value. For fraction 2 with $E_{vc} = 0.363$, the fraction value, $w_2 = 0.0532$ should not be taken, only half this value, that is, $w_2^* = 0.0266$, should be used. For the next fraction, fraction 1, the fraction value to be used for the ordinate corresponds to the whole fraction of fraction 2, and half the fraction for fraction 1, that is $w_1^* = 0.0532 + (0.0727/2) = 0.08955$, etc.

The plot shows that the integral vinyl chloride monomeric unit content is certainly not "smooth," that is, it does not have only one single inflexion

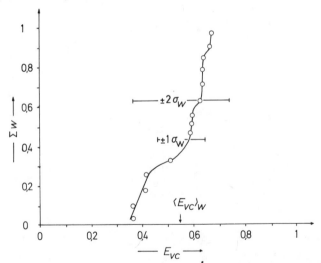

Figure 2-1. Integral mass distribution of the vinyl chloride content of a poly(vinyl chloride-co-vinyl acetate) with the mass average composition $(\bar{E}_w)_{vc}$ and the mass average standard deviation, σ_w, of the distribution.

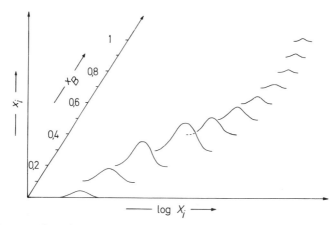

Figure 2-2. Three-dimensional plot of the differential distribution of the mole fraction, x_i, of copolymer molecules of mole fraction, x_B, of monomeric units, B, and of degree of polymerization X_i (schematic).

point. On the contrary, there are marked gradations, which means, of course, that a second fractionation is necessary to show that the gradations are real and not an experimental artifact. It can also be seen that the distribution curve is distinctly "skewed" since the mean composition corresponds to a mass fraction of 0.35, and not one of 0.5. Additionally, the composition distribution does not extend over the whole composition range of $E_{vc} = 0$ to $E_{vc} = 1.0$, but only from about 0.33 to 0.73 (Figure 2-1).

A three-dimensional plot is always preferable to a two-dimensional plot because of the influence of molar mass on solubility, which is invariably encountered. In such a plot, the molar mass distribution for each mean monomeric unit composition is drawn in, such that a graph in relief (Figure 2-2) is obtained. The effort involved in producing such a graph is quite large, however.

2.4.4. Sequences

The monomeric units of a bipolymer prepared from monomers A and B recur in a given chain according to a sequence which depends on the polyreaction mechanism, i.e.,

$$-\underline{A}-\underline{B}-\underline{A}-\underline{A}-\underline{B}-\underline{B}-\underline{A}-\underline{B}-\underline{A}-\underline{A}-\underline{B}-\underline{A}-\underline{B}-\underline{B}-\underline{B}-\underline{A}-\underline{A}-\underline{A}-\underline{B}-\underline{B}-$$

The underlined sequences consist of 1,2,3, etc., identical monomeric units. What are called homomonads A and B, homodiads AA and BB, homotriads AAA and BBB, etc., are to be found in the chain. Heterodiads AB and BA, heterotriads AAB, ABA, BAA, BBA, BAB, and ABB, and heterotetrads, etc. can be analogously defined. Each monomeric unit belongs to a monad, two diads, three triads, four tetrads, etc. Consequently, with linear polymers of degree of polymerization X, the number of monomeric units belonging to each j-ad is equal to $X - j + 1$. With ring-shaped polymers, the number of monomeric units in each j-ad is always equal to the degree of polymerization, whereby, of course, j may not exceed X.

Of necessity, certain relationships exist between j-ads which are independent of the mechanism of formation. A-monads occur, for example, in the diads AA, AB and BA. The last two diad types are not, however, experimentally distinguishable, and are measured together as the fraction f_{AB}. In addition, the j-ads must be systematically counted, such that each of the j-ads is only counted once among the $(j + 1)$-ads. The following is obtained for the relationships between monads and diads, whereby the j-ads counted are underlined:

$$f_{\underline{A}} = f_{\underline{A}A} + f_{\underline{A}B} = f_{AA} + (1/2)f_{\overline{AB}} \tag{2-11}$$

$$f_{\underline{B}} = f_{\underline{B}B} + f_{\underline{B}A} = f_{BB} + (1/2)f_{\overline{AB}} \tag{2-12}$$

The relationships between diads and triads are:

$$f_{\underline{AA}} = f_{\underline{AA}A} + f_{\underline{AA}B} = f_{AAA} + (1/2)f_{\overline{AAB}} \tag{2-13}$$

$$f_{\underline{AB}} = f_{\underline{AB}A} + f_{\underline{AB}B} = f_{ABA} + (1/2)f_{\overline{ABB}} \tag{2-14}$$

$$f_{\underline{BA}} = f_{B\underline{AA}} + f_{\underline{BA}B} = (1/2)f_{\overline{BAA}} + f_{BAB} \tag{2-15}$$

$$f_{\underline{BB}} = f_{\underline{BB}B} + f_{\underline{BB}A} = f_{BBB} + (1/2)f_{\overline{BBA}} \tag{2-16}$$

$$f_{\underline{AB}} = f_{BA} \quad \text{(indistinguishable or equivalent)} \tag{2-17}$$

$$f_{ABA} + (1/2)f_{\overline{ABB}} = (1/2)f_{\overline{BAA}} + f_{BAB} \tag{2-18}$$

A bipolymer can be characterized by the fractions, f_{AA}, $f_{\overline{AB}}$, and f_{BB} of AA, (AB and BA) and BB bonds. A triangular plot is especially useful here, since the dependence of the three diad type contents on, for example, reaction conditions can be shown. In such a case, a curve is obtained instead of a point (see Figure 2-3).

2.4.5. Sequence Lengths

The sequence length can be found from a series of physical and chemical methods. Generally, all the methods are strongly dependent on the polymer constitution; thus they can often only be applied to a specific polymer.

The chemical methods depend almost exclusively on two principles: chain scission or neighboring side group reactions. The methods that work on

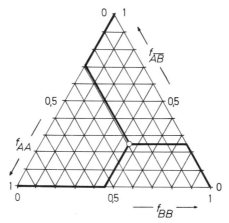

Figure 2-3. Description of the triad composition of a bipolymer by representation on a triangular graph. For the example given, $f_{AA} = 0.30$, $f_{\overline{AB}} = 0.25$, and $f_{BB} = 0.44$.

the chain scission principle use the fact that one of the two components of a bipolymer will be attacked by a specific reaction process, while the other component is stable. With the copolymer of isobutylene and isoprene, for example, the main chain double bond of the isoprene monomeric unit can be split by ozonolysis. The mean sequence length of the isobutylene sequences can be calculated from the molar mass of the remaining isobutylene oligomer.

With proteins and peptides, on the other hand, monomeric units are successively removed from one chain end through the use of suitable enzymes. An example of this is given by copolymers of L- and D-leucine. Carboxypeptidase A enzyme removes L-leucine residues from the carbon end of the peptide chain until it encounters a L/D bond.

In reactions involving neighboring groups, the fact that in kinetically controlled reactions not all the groups can react is used (cf. Section 23.4.5). For example, the hydroxyl groups of poly(vinyl alcohol) cannot be completely acetylated with butyraldehyde, since isolated OH groups remain because of the random nature of the reaction. Of course, reactions must be carried out as a sufficiently high dilution so that possible intermolecular reactions do not occur. In addition, the solvent must be as "good" as possible, because then the polymer coil is greatly expanded and reaction between distant OH groups on the same polymer chain is reduced.

Two types of procedure can be similarly distinguished among the physical methods. One group deals with relatively short, the other with relatively long sequences. Nuclear magnetic resonance (nmr), uv, and ir spectroscopy belong to the first group, while X-ray analysis and differential thermal analysis belong to the second. In the ir spectrum, for example, the intensity of the $-(CH_2-)_n$ rocking frequency shifts from 815 ($n = 1$) to 752

($n = 2$), to 733 ($n = 3$), to 726 ($n = 4$), and 722 cm^{-1} ($n \geq 5$), so that it is possible to assay short methylene sequences. In the far-ir, an isolated styrene unit ($n = 1$) in $+CH_2-CH(C_6H_5)+_{\overline{n}}$ shows a broad band at 560 cm^{-1}, while a well-defined band is found at 540 cm^{-1} for $n \geq 6$. This band results from deformation of the aromatic ring, which stems from and is coupled with a deformation of the polymer chain; it can therefore be used for the analysis of styrene–butadiene copolymers. In favorable circumstances, nmr enables the pentad sequences to be assayed, while uv studies are suitable up to triad sequences. X-Ray and differential thermal analysis can be employed for sequence analysis because longer stereoregular sequences can crystallize more readily than shorter ones, and longer atactic sequences show a distinctly different glass-transition temperature than shorter ones. In order to obtain anything conclusive from such a procedure, there must, in general, be a minimum of 15–20 units per sequence. Therefore this procedure does not distinguish between block copolymers and polymer blends. However, both X-ray analysis and differential thermal analysis are less direct than the other methods mentioned, since "false" groups can, in certain circumstances, be built into the chain without altering the ability to crystallize or shift the glass-transition temperature.

The number average sequence length of A homosequences, $\langle X_a \rangle_n$, is given by

$$\langle X_a \rangle_n = \sum_i (N_a)_i (X_a)_i \bigg/ \sum_i (N_a)_i \qquad (2\text{-}19)$$

where $(N_a)_i$ is the number of A homosequences of "length" $(X_a)_i$. As a rule, this parameter is not experimentally directly accessible. It can, however, be obtained from other data in the following way:

The numerator is none other than the total number of A-monomeric units in the polymer:

$$N_A = \sum_i (N_a)_i (X_a)_i \qquad (2\text{-}20)$$

The denominator is obtained through the following considerations: each A homosequence begins with a BA bond and ends with an AB bond. Consequently, the number of A sequences must be equal to half the number of all AB bonds:

$$\sum_i (N_a)_i = (1/2) N_{AB} \qquad (2\text{-}21)$$

This relationship is exact for ring-shaped macromolecules and is a good approximation for linear macromolecules of high degrees of polymerization.

Additionally, the number N_{AB} of AB bonds can be expressed in terms of the fraction f_{AB}:

$$f_{AB} = \frac{N_{AB}}{N_{AA} + N_{\overline{AB}} + N_{BB}} = \frac{N_{AB}}{(\langle X \rangle_n - 1) N_{cop}} \tag{2-22}$$

It is to be remembered here that the number of bonds in *one* molecule, $(N_{AA} + N_{\overline{AB}} + N_{BB})$, is always one less than the degree of polymerization, and the total number of *all* bonds, of course, still depends on the total number of copolymer molecules, N_{cop}. The number, N_A, of A monomeric units, and the number, N_{cop}, of copolymer molecules are given by the masses, m, and molar masses, M:

$$N_A = m N_L / M_A \tag{2-23}$$

$$N_{cop} = (m_A + m_B) N_L / \langle M \rangle_n \tag{2-24}$$

The number average molar mass, $\langle M \rangle_n$, of the copolymer is related to the corresponding degree of polymerization by

$$\langle X \rangle_n = \frac{N_A + N_B}{N_{cop}} = \langle M \rangle_n \left(\frac{w_A}{M_A} + \frac{w_B}{M_B} \right) \tag{2-25}$$

Setting Equations (2-20)–(2-24) into Equation (2-19) and transforming leads to

$$1/\langle X_a \rangle_n = 0.5 f_{\overline{AB}} [1 - \langle M \rangle_n^{-1} (M_A/w_A) + (w_B M_A/w_A M_B)] \tag{2-26}$$

In this way, the number average sequence length of A sequences is calculated from the mass fractions, w_A and w_B, and the molar masses, M_A and M_B, of the monomeric units, together with the fraction $f_{\overline{AB}}$, of all AB and BA bonds as well as the number average molar mass of the copolymer. Very often, the sequence number, or run number, is used. This number $\langle R \rangle_n$ gives the total number of all blocks or homosequences per 100 monomeric units:

$$\langle R \rangle_n = (2 \times 10^2)(\langle X_a \rangle_n + \langle X_b \rangle_n)^{-1} \tag{2-27}$$

2.5. Molecular Architecture

Macromolecules differ not only in their constitutional composition and bonding or sequence of monomeric units, but also in terms of their molecular architecture. The molecular architecture relates to the way individual macromolecular chains are bonded to each other. Unbranched chains, branched chains, ordered networks, and disordered or irregular networks can be distinguished. Interpenetrating networks, (IPNs), consist of two interpenetrating networks that are not bonded to each other.

2.5.1. Branching

The simplest molecular architecture is shown by the unbranched or "one-dimensional" chain, for example, the sulfur, I, or poly(methylene), II, chain:

$$-S-S-S-S-S-S- \qquad -CH_2-CH_2-CH_2-CH_2- \qquad -CH_2-CH-$$
$$\qquad\qquad\qquad\qquad\qquad\qquad\qquad\qquad\qquad\qquad\qquad\qquad\qquad\qquad\qquad | $$
$$\qquad\qquad\qquad\qquad\qquad\qquad\qquad\qquad\qquad\qquad\qquad\qquad\qquad\qquad (CH_2)_5 H$$

$$\text{I} \qquad\qquad\qquad\qquad\qquad\qquad \text{II} \qquad\qquad\qquad\qquad\qquad\qquad \text{III}$$

For historical reasons, unbranched chains are called "linear" chains, since it was originally thought that such chains would occur fully extended in space. In fact the random distribution of microconformations dictates that an isolated chain of this type adopts a random coil shape, which also occurs in the amorphous state. Regularly recurring substituents such as in poly(heptene-1), III, are not regarded as branching; consequently, III is also an unbranched polymer.

Branching, by definition, initially occurs through a polyreaction; branches are not found in the initial monomer. For this reason, branched macromolecules produced in one-step reactions have branches with the same constitution as the main chain. The main chain is the longest of the chains joined together in such a branched macromolecule. The other chains joined to the main chain are called long-chain or short-chain branches according to their length. Short chains can be regarded as oligomers joined to the main chain, but long chains represent joined polymers. If the branches to the main chain are themselves branched, then this is known as series branching. With very extensive series branching, the polymer has a kind of fir-tree-shaped structure. An example is provided by the short- and long-chain branching of poly(ethylenes):

$$CH_3 \qquad\qquad\qquad\qquad\qquad\qquad\qquad\qquad\qquad CH_3$$
$$| \qquad\qquad\qquad\qquad\qquad\qquad\qquad\qquad\qquad\qquad | $$
$$(CH_2)_3 \qquad\qquad\qquad\qquad\qquad\qquad\qquad\qquad (CH_2)_x$$
$$| \qquad\qquad\qquad\qquad\qquad\qquad\qquad\qquad\qquad\qquad | $$
$$\sim\!CH_2-CH-(CH_2)_x-CH-(CH_2)_y\!\sim \qquad \sim\!CH_2-CH-(CH_2)_y-CH\!\sim$$
$$\qquad\qquad\qquad\qquad\qquad | \qquad\qquad\qquad\qquad\qquad\qquad\qquad\qquad | $$
$$\qquad\qquad\qquad\qquad\qquad CH_2 \qquad\qquad\qquad\qquad\qquad\qquad\qquad (CH_2)_z$$
$$\qquad\qquad\qquad\qquad\qquad | \qquad\qquad\qquad\qquad\qquad\qquad\qquad\qquad | $$
$$\qquad\qquad\qquad\qquad\qquad CH_3 \qquad\qquad\qquad\qquad\qquad\qquad\qquad CH_3$$

short-chain branching $(x,y \gg 1)$ long-chain branching $(x,y,z \gg 1)$

Star-shaped branching radiates out from a single branch point. Comb-shaped molecules, on the other hand, contain branches of generally equal length joined more or less equidistantly along the main chain (see Figure 2-4).

The determination of the number of branched points per macromolecule, the mean branch length, the branch length distribution, and the main chain length between branch points represent one of the difficult analytical problems of macromolecular chemistry. The branch points or end groups of

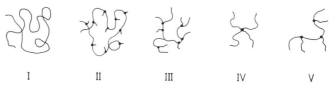

Figure 2-4. Schematic representation of branched and unbranched chains of the same degree of polymerization. I, Unbranched; II, short-chain branched; III, long-chain branched; IV, star-shaped branched; V, comb-shaped branched. Branched points are shown by a small black circle.

strongly short-chain-branched polymers can often be spectroscopically or chemically determined. However, with long-chain branching, the number of branch points is very small compared to the number of chain-links. Consequently, the existence and extent of long chain branching is mostly estimated from the dimensions of the macromolecule in solution. A long-chain-branched macromolecule, namely, exhibits smaller molecular dimensions than the unbranched macromolecule with the same molar mass, as can be clearly seen in Figure 2-4.

2.5.2. Graft Polymers and Copolymers

Graft polymers are branched polymers whereby the branches have a constitution or configuration which differs from that of the primary, or backbone, chain. The primary polymer chain, onto which other chains are grafted, is also known as the graft base or graft substrate. Graft copolymers are a subgroup of graft polymers (see also Section 2.4.1).

Graft polymers are generally characterized by a number of parameters which can be clarified by considering the example of grafting B monomers onto a polymer with A monomeric units. The product of the reaction shall have 40% A monomeric units and 60% B monomeric units according to chemical or spectroscopic analysis; it is not necessarily a graft polymer, since in addition to the desired graft polymer, poly(A-g-B), it may contain still unreacted graft substrate, poly(A), and/or homopolymer, poly(B), formed in addition. Preparative fractionation gives an analysis of 10% poly(A), 70% poly(A-g-B) and 20% poly(B). From these figures, it is found that the graft polymer contains 30% of all A groups and 40% of all B groups of the products of the reaction. Consequently, the graft polymer contains 42.9% A groups and 57.1% B groups.

The *graft success rate* is then defined as the fraction of grafted substrate with respect to the total amount of substrate, and is 75% in the above example. The *graft yield* relates correspondingly to the fraction of B monomers with respect to the total number of B monomeric units produced; it is 66.7% in this example. The *graft degree* gives the amount of grafted B monomer with respect to the total amount of substrate and is 100% in this example. The *graft extent*, on the other hand, gives the grafted amount of B with respect to the amount of A to which it is grafted, and is consequently 133%.

2.5.3. Irregular Networks

Cross-linked polymers contain at least two cross-links per chain joining it to other chains. The cross-links may be intercatenary or intracatenary, also called interchenary or intrachenary. They should not, on the other hand, be called inter- or intramolecular cross-links, since a cross-linked polymer molecule represents a single molecule, and all cross-links, therefore, are intramolecular. Cross-linking reactions, in contrast, are generally inter- molecular. During network formation, intercatenary polyreactions lead to larger molecular size by addition of chains to the cross-linked network without increase in cross-link density, whereas intracatenary polyreactions only involve functional groups actually on the cross-link network itself; when these react, the cross-link density is increased but the network size is not.

Intracatenary cross-linked macromolecules occur very often in nature. With the enzyme, ribonuclease, for example, the single peptide chain is cross-linked to itself via four disulfide bridges (Figure 2-5). Another protein, insulin, has, on the other hand, two chains, A and B, of differing composition and peptide residue sequence, joined together by a total of two disulfide bridges.

Such small and intercatenary cross-linked macromolecules are just like branched macromolecules in generally being soluble in some solvent or other. Thus, they characteristically differ from actual cross-linked networks, which possess many intercatenary cross-links. Such cross-linked molecules, or networks, are "infinitely" large in comparison with the usual branched or unbranched macromolecules. They are not soluble in any solvent. On the other hand, not all insoluble polymers are cross-linked, of course.

If the crosslink density is not too high, such cross-linked networks may

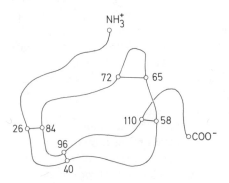

Figure 2-5. The enzyme ribonucle- ase, which consists of 124 amino acid units (—NH—CHR—CO—). The four intramolecular cross-link- ing points (26–84; 40–96; 58–110; 65–72) involve cystine units.

still be swollen to what is known as a gel. Gels with dimensions of about 10–1000 nm are known as microgels. A microgel can generally be dissolved or suspended in a solvent.

Gel formation always occurs in irreversible polyreactions when the functionality of the molecules exceeds 2. Just after the gel point is passed, some of the monomeric units are joined in a gel which spans the whole of the reaction volume. The name partly cross-linked is given to such a reaction product. The concept of partly cross-linked is applied to the whole of the reaction product, whereas cross-linked network only applies to the architecture of the molecule itself. Consequently, a branched product is not partly cross-linked, whereas a partly cross-linked product is most often also branched.

Characterizing cross-linked networks according to molar mass is unrealistic in regard to their "infinite" size. Such cross-linked networks are classified according to the network chain length, kind of cross-link, and cross-link density. Here a cross-link point is defined as a group from which more than two network chains extend. A network chain corresponds to that portion of the chain which joins two cross-link points together.

The degree of cross-linking, x_c, also known as the cross-link or network density, is the mole fraction of monomeric units which are actual cross-link points with respect to the total number of monomeric units. The number average molar mass of a network chain may be calculated from the molar mass, M_u, of monomeric units and the cross-link density:

$$\langle M_c \rangle_n = M_u / x_c \tag{2-28}$$

The network chain molar mass can not only be calculated from the network density, but also from what is known as the cross-link index, γ. The cross-link index gives the number of monomeric units which are cross-link points per primary polymer chain. A primary polymer chain is the linear macromolecule which existed before cross-linking.

$$\gamma = \frac{\langle M_n \rangle_0}{\langle M_c \rangle_n} = \frac{\langle M_n \rangle_0 x_c}{M_u} = x_c \langle X \rangle_n \tag{2-29}$$

All of these considerations refer to ideal networks, that is, networks without free chain ends. The proportion of free chain ends is inversely proportional to the molar mass of the primary polymer chain molecule. For $\langle M_n \rangle_0 > \langle M_c \rangle_n$, the effective network chain concentration, $[M_c]_{eff}$, can be calculated from the total chain concentration, $[M_c]$, via

$$[M_c]_{eff} = [M_c] (1 - 2\langle M_c \rangle_n / \langle M_n \rangle_0) \tag{2-30}$$

This correction formula may not be used for very highly cross-linked networks, since additional free ends occur.

   Descriptions of networks in terms of cross-link index and density provide
no information on the internal architecture, that is, the homogeneity, of the
network. Depending on how they are produced, most networks are more or
less inhomogeneous, that is, the local density has a distribution. With
multifunctional polycondensation, the gel point is most often reached at
relatively high yields, and the network formed is quite homogeneous. In
addition polymerization, the gel point occurs already at relatively low yields.
The polymerization continues around the spatially fixed network structured
centers, and, so, densely cross-linked centers are produced within a less
densely cross-linked matrix.

   Heterogeneities can be produced in networks for quite a number of
reasons, for example, because of unreacted functional groups, loose chain
ends, chain entanglements, or intracatenary ring formation. They can be
caused by preordering of monomer in the un-cross-linked network, dilution
effects, differing monomer functional group reactivities, demixing effects
caused by diffusion controlled propagation steps, or by phase separation
effects.

   Phase separations can be utilized in the synthesis of what are called
macroporous or macroreticular networks. In this process, the polymerization
is carried out in the presence of a solvent which is a precipitant for the
corresponding un-cross-linked polymer. An example is the cross-linking free
radical polymerization of styrene with some divinyl benzene in hexane.
Hexane dissolves both styrene and divinyl benzene but is a precipitant for
poly(styrene). During polymerization the poly(styrene) segments phase
separate with simultaneous cross-linking, even at low conversions. Further
polymerization then occurs in the neighborhood of these precipitated
propagation centers such that, according to conditions, porous continuous
matrix networks or globular structures with pore-like voids are produced (see
Figure 2-6). Macroreticular networks are much more porous to solvents and
dissolved material than are homogeneous networks with the same cross-link
density. In addition, they swell less and resist pressure better. Consequently,

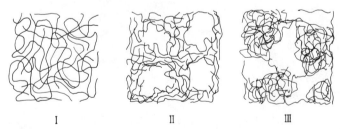

I   II   III

Figure 2-6. Schematic representation of irregular networks. I, Homogeneous network, II and
III, macroreticular networks.

they are preferred for ion exchange and gel permeation chromatography columns.

The chemical nature of the cross-link points is quite unimportant to typical cross-linked network properties such as elasticity and swelling in solvents. Most chemical cross-linking occurs via covalent bonds, but cross-linking can also be achieved with coordinate or electron-deficient bonds. Cross-link-like effects can also be caused by purely physical phenomena, for example, by crystallite regions in partially crystalline polymers, amorphous domains in block polymers, or molecular entanglements in amorphous polymers and polymer melts.

2.5.4. Ordered Networks

Ordered networks, in contrast to irregularly cross-linked networks, possess only structurally equivalent monomeric units. They are classified as 0-, 1-, 2-, and 3-types according to whether the ordered network structure extends to zero, one, two, or three dimensions in space (Figure 2-7). They may also be called zero, one, two, or three dimensional. The IUPAC nomenclature utilizes the prefixes *catena, phyllo,* and *tecto* for the 1-, 2-, and 3-types.

Ordered networks can be obtained by equilibrium and nonequilibrium reactions. Generally, with reversible reactions, the equilibrium must be shifted through use of solubility factors in order that high yields of ordered networks may be obtained. With kinetically controlled reactions, ordered networks are obtained either directly via stereospecific homopolymerization, or by cyclo-polycondensation of rigid monomers.

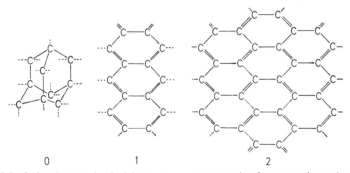

0   1   2

Figure 2-7. Ordered networks. 0, Adamantane as an example of a cage polymer (type 0); 1, Cyclized and dehydrogenated 1,2-poly(butadiene) as an example of a ladder or double-strand polymer (type 1); 2, graphite as a layer or parquet polymer (type 2). (· · ·) Carbon-to-hydrogen bonds; (— and ═) carbon-to-carbon bonds.

0-types form cage structures and are not macromolecular. Examples of this class are adamantane and bullvalene.

With *1-types*, it is possible to distinguish between bridge and spiro structures. Theoretically, spiro chains may be classed as single-chain polymers (see Section 2.2.3). 1-Type structures with bridging links are also known as double-strand polymers, or, because their structure is reminiscent of a ladder, they are also known as ladder polymers. Ladder polymers generally have good thermal stability, since, in contrast to linear chain polymers, the breaking of one main chain bond does not lead to a lower molar mass.

Theoretically, the various types of helices also belong to this class. Since, however, the "network" nature depends on special conformation effects, they will be treated in Section 4 instead of here.

Layer, parquet, or planar polymers of *type 2* are seen in graphite and its derivatives. Diamond is a network polymer of *type 3*. Network polymers exist exclusively, and parquet polymers almost exclusively, in the solid state. They are also known as monoaggregatable materials. Certain cell walls of bacteria consist of baglike macromolecules, which are a special case of parquet polymers.

In carbon compounds, the number of parquet and network polymers is limited by the tetravalence of carbon. In inorganic compounds, however, they exist in large numbers, for example, in quartz $(SiO_2)_x$, in black phosphorus $(P)_x$, etc. Theoretically, the same synthetic problems occurring with ladder polymers arise in the synthesis of parquet polymers. In order to obtain the desired arrangement in one dimension (type 1) or in two dimensions (type 2), it is necessary to avoid reaction paths leading to irregular structures. In the preparation of synthetic graphite, this is achieved by the meticulous exclusion of all centers of crystallization.

Literature

2.1. Nomenclature

IUPAC, *Nomenclature of Inorganic Chemistry, Definitive Rules 1970*, second ed., Butterworths, London, 1971.

IUPAC Commission on Macromolecular Nomenclature, Nomenclature for Regular Single-Strand and Quasi-Single-Strand Inorganic and Coordination Polymers, *Pure and Appl. Chem.* **53** 2283 (1981).

IUPAC, *Nomenclature of Organic Chemistry*, third ed., Butterworths, London, 1971.

IUPAC, Macromolecular Nomenclature Commission, Nomenclature of regular single-strand organic polymers, *Macromolecules*, **6**, 149 (1973); *J. Polym. Sci. (Polym. Lett. Ed.)* **11**, 389 (1973).

Deutscher Zentralausschuss für Chemie, Internationale Regeln für die chemische Nomenklatur und Terminologie, Verlag Chemie, Weinheim, 1977ff.

R. S. Cahn and O. C. Dermer, *Introduction to Chemical Nomenclature*, fifth ed., Butterworths, Woburn, Massachusetts, 1979.

2.2. Atomic Structure and Polymer Chain Formation

M. F. Lappert and G. J. Leigh, *Developments in Inorganic Polymer Chemistry*, Elsevier, Amsterdam, 1962.

F. G. A. Stone and W. A. G. Graham (eds.), *Inorganic Polymers*, Academic Press, New York, 1962.

K. Andrianov, *Metal Organic Polymers*, Interscience, New York, 1965.

F. G. R. Gimblett, *Inorganic Polymer Chemistry*, Butterworths, London, 1973.

A. L. Rheingold, *Homoatomic Rings, Chains and Macromolecules of Main-Group Elements*, Elsevier, Amsterdam, 1977.

2.3. Homopolymers

General Analytic Methods

G. M. Kline (ed.), *Analytical Chemistry of Polymers*, Interscience, New York, 3 parts, 1959–1962.

M. P. Stevens, *Characterization and Analysis of Polymers by Gas Chromatography*, Marcel Dekker, New York, 1969.

J. Haslam, H. A. Willis, and D. C. M. Squirrel, *Identification and Analysis of Plastics*, Iliffe, London, 1972.

E. Schröder, J. Franz, and E. Hagen, *Ausgewählte Methoden zur Plastanalytik*, Akademie-Verlag, Berlin, 1975.

E. G. Brame, Jr., *Applications of Polymer Spectroscopy*, Academic Press, New York, 1978.

D. O. Hummel, F. Scholl (eds.), *Atlas of Polymer and Plastics Analysis*, Verlag Chemie, Weinheim, 1978–1981 (3 vols.).

F. Heatley, Nuclear magnetic resonance. Synthetic macromolecules, *Nucl. Magn. Reson.* **8,** 266 (1979).

H. W. Siesler and K. Holland-Moritz, *Infrared and Raman Spectroscopy of Polymers*, Marcel Dekker, New York, 1980.

H. W. Siesler, Fourier transform infrared (FT IR) spectroscopy in polymer research, *J. Mol. Struct.* **59,** 15 (1980).

J. Urbanski, W. Czerwinski, K. Janicka, F. Majewska, H. Zowall, *Handbook of Analysis of Synthetic Polymers and Plastics*, Wiley, New York 1977.

V. G. Berezkin, V. R. Alishoyev, and I. B. Nemirovskaya, *Gas Chromatography of Polymers*, Elsevier, Amsterdam, 1977.

M. Hoffmann, H. Kromer, and R. Kuhn, *Polymeranalytik*, Thieme-Verlag, Stuttgart, 1977 (2 vols.).

Spectroscopy

S. Krimm, Infrared spectra of high polymers, *Fortschr. Hochpolym. Forschg.* **2,** 51 (1960/61).

R. Zbinden, *Infrared Spectroscopy of High Polymers*, Academic Press, New York, 1964.

J. C. Henniker, *Infrared Spectroscopy of Industrial Polymers*, Academic Press, New York, 1967.

A. Elliot, *Infrared Spectra and Structure of Organic Long-Chain Polymers*, Arnold, London, 1969.

J. L. Koenig, Raman spectroscopy of biological molecules: A review, *J. Polym. Sci. D* [*Macromol. Revs.*] **6**, 59 (1972).

J. Dechant, *Ultrarotspektroskopische Untersuchungen an Polymeren*, Akademie-Verlag, Berlin, 1972.

J. L. Koenig, Raman scattering of synthetic polymers, *Revs. Appl. Spectrosc.* **4**, 233 (1971).

D. O. Hummel (ed.), *Polymer Spectroscopy*, Verlag-Chemie, Weinheim, 1975.

E. G. Brame, Jr., *Application of Polymer Spectroscopy*, Academic Press, New York, 1978.

H. W. Siesler and K. Holland-Moritz, *Infrared and Raman Spectroscopy of Polymers*, Marcel Dekker, New York, 1980.

Structures

S. R. Palit and B. M. Mandal, End-group studies using dye techniques, *J. Macromol. Sci.* [*Revs.*] **C2**, 225 (1968).

M. F. Hoover, Cationic quaternary polyelectrolytes—a literature review, *J. Macromol. Sci.* [*Chem.*] **A4**, 1327 (1970).

F. Oosawa, *Polyelectrolytes*, Marcel Dekker, New York, 1971.

L. Holiday, *Ionic Polymers*, Halsted Press, New York, 1975.

R. G. Garmon, End group determinations, *Techn. Methods Polym. Eval.* **4**(1), 31 (1975).

A. Eisenberg and M. King, Ion-containing polymers; physical properties and structure, in: *Polymer Physics*, Vol. 2, R. S. Stein (ed.), Academic Press, New York, 1977.

N. C. Billingham, *Molar Mass Measurements in Polymer Science*, Wiley, New York, 1977 (includes end group analysis).

2.4. Copolymers (*see also 2.3*)

G. Schnell, Ultrarotspektroskopische Untersuchungen an Copolymerisaten, *Ber. Bunsenges.* **70**, 297 (1966).

U. Johnsen, Die Ermittlung der molekularen Struktur von sterischen und chemischen Co-polymeren durch Kernspinresonanz, *Ber. Bunsenges.* **70**, 320 (1966).

J. C. Randall, *Polymer Sequence Determination—Carbon 13 NMR Method*, Academic Press, New York, 1977.

M. M. Coleman and P. C. Painter, Fourier transform infrared studies of polymeric materials, *J. Macromol. Sci.—Rev. Macromol. Chem.* **C16**, 197 (1977–1978).

2.5. Molecular Architecture

W. Funke, Über die Strukturaufklärung vernetzter Makromoleküle, insbesondere vernetzter Polyesterharze, mit chemischen Methoden, *Adv. Polym. Sci.* **4**, 157 (1965/67).

W. De Winter, Double strand polymers, *Revs. Macromol. Sci.* **1**, 329 (1966).

H.-G. Elias, Die Struktur vernetzter Polymerer, *Chimia* **22**, 101 (1968).

V. A. Grečanovskij, Verzweigungen an Polymerketten (Russ.), *Usp. Khim.* (*Adv. Chem.*) **38**, 2194 (1969); *Rubber Chem. Technol.* **45**, 519 (1972).

C. G. Overberger and J. A. Moore, Ladder polymers, *Adv. Polym. Sci.* **7**, 113 (1970).

G. Delzenne, Recent advances in photo-crosslinkable polymers, *Revs. Polym. Technol.* **1,** 185 (1972).

N. A. Platé and V. P. Shibaev, Comb-like polymers. Structure and properties, *J. Polym. Sci.* [*Macromol. Revs.*] **8,** 117 (1974).

P. A. Small, Long-chain branching in polymers, *Adv. Polym. Sci.* **18,** 1 (1975).

D. Klempner, Polymer networks with mutual penetration, *Angew. Chem. Int. Ed. Eng.* **17,** 97 (1978).

Y. Ikeda, Characterization of graft copolymers, *Adv. Polym. Sci.* **29,** 47 (1978).

S. Bywater, Preparation and properties of star-branched polymers, *Adv. Polym. Sci.* **30,** 89 (1979).

J. Roovers, Model star and comb systems: Characterization and properties, *Polym. News* **5,** 248 (1979).

Chapter 3
Configuration

3.1. Overview

3.1.1. Symmetry

The configuration and conformation of a molecule are partial aspects of its static stereochemistry, that is, the spatial arrangement of the atoms of the molecule. Stereochemical considerations presume knowledge of symmetry properties.

According to group theory, the symmetry properties of an object, and thus, also of a molecule, are described by symmetry elements and symmetry operations. The basic symmetry elements are the symmetry center, the symmetry axes or axes of rotation, and the symmetry planes or mirror image planes. These elements correspond to the symmetry operations such as inversion about a point, rotation about an axis, and the mirror imaging of the object in a plane. A fourth symmetry operation involves the translation and three-dimensional repetition of identity of the "infinitely" extended object. Combinations of symmetry operations are called complexes. The combination of rotation and mirror image reflection is known as rotational mirror imaging, and the combination of rotation and inversion is called rotational inversion. The symmetry elements corresponding to these operations are the alternating axis of symmetry and the rotational inversion axis.

An object changes its position during a symmetry operation, but does not change its appearance. At least one point, however, of the object maintains its position during a symmetry operation. This means that the object may be described by what are known as point groups or symmetry groups, which are groups of symmetry elements. Here, the 32 possible point groups correspond to the 32 crystal classes of crystallography. Whereas the point groups are also known as Bravais points, after the early crystallographer, Bravais, symmetry elements are characterized by Schoenflies symbols.

According to the Schoenflies symbolism, T corresponds to a tetrahedral, O to an octahedral, and S to a sphenoidal or alternating axis of symmetry, C to a cyclical or rotational axis, and D corresponds to a digyric or twofold axis with other twofold axes perpendicular to this axis. A rotational axis is n-fold, with the symbol C_n, if the object can be superimposed on itself n times during a full rotation of $360°$. Mirror images, or the corresponding mirror image planes, are also often given the symbol σ.

The highest-fold number axis is called the main axis or main rotational axis, which is always arranged vertically. All other symmetry elements are arranged relative to this main axis. The symbol σ_h is given to all mirror image planes which lie in a plane perpendicular to the main axis, or, in other words, lie in the horizontal plane. Mirror image planes which are vertical to the main axis also contain the main axis of rotation and are given the symbol σ_v. The symbol σ_d is often used instead of the symbol σ_v since, in this last case, certain angles are bisected by the mirror image plane and are dihedral. The most important symmetry groups are given in Table 3-1 with some molecular examples in Figure 3-1. The further classification of the symmetry groups according to their n-fold value then leads to all the 32 point groups.

3.1.2. Stereoisomerism

Two molecules with the same sequence of atoms but with different spatial arrangement of atoms are called stereoisomers. They thus differ from constitutional isomers which have different sequences of atoms for the same overall chemical formula. Consequently, constitutional and stereo isomers are subclasses of isomers which are molecules with the same overall chemical formula, but differing in structure. An isomer may be either a constitutional or a stereo isomer; it can never be simultaneously both.

Stereoisomers can be classified according to their symmetry properties into enantiomers and diastereomers. Enantiomers are mirror images of each other and, so, are often called antipodes (Figure 3-2); each isomer has the same energy. Diastereoisomers are not, on the other hand, mirror images; they do not have the same internal energy. Consequently, two stereoisomers may be either enantiomers or diastereomers; they may not be both simultaneously.

Configurational and conformational isomers are also stereoisomers. The concepts are not so distinctly defined, and are not completely mutually exclusive. Thus, in low molar mass organic chemistry, the classical concepts of

Table 3-1. *Symmetry Group Definitions. Chiral Symmetry Groups are Marked with an Asterisk*

Symmetry group	Symbol	Examples
Without symmetry axis		
*Onefold axis of rotation	C_1	L-Alanine
An inversion center (center of symmetry)	C_i, S_2, i	2,3-Dibromobutane
A mirror plane (plane of symmetry)	C_s, S_1, σ	Vinyl chloride
With an n-fold symmetry axis		
*n-Fold axis of rotation	C_n	1,3-Dibromoallene (C_2)
n-Fold axis of rotation and symmetry plane perpendicular to this	$C_{nh} = C_n + \sigma_n$	*Trans*-1,2-dibromoethylene (C_{2h})
n-Fold axis of rotation with n symmetry planes	$C_{nv} = C_n + n\sigma_v$	1,1-Dibromoethylene
n-Fold alternating axis of symmetry, n even, $n \neq 2$ (rotation–reflection axes, mirror axes, improper axes, alternating axes)	S_n	Tetramethyl spiro-pyrrolidinium ion (S_4)
An n-fold and n twofold symmetry axes		
*An n-fold axis of rotation and n twofold axes of rotation perpendicular to this	$D_n = C_n + nC_2$	Doubly bridged biphenyl (D_2)
An n-fold axis of rotation, n two-fold axes of rotation, and n vertical mirror planes	$D_{nd} = C_n + nC_2 + n\sigma_d$	Allene
An n-fold axis of rotation, n twofold axes of rotation, n vertical mirror planes, and a horizontal mirror plane	$D_{nh} = C_n + nC_2 + n\sigma_v + \sigma_h$	Ethylene
Several n-fold symmetry axes with $n > 2$		
Tetrahedron	$T_d = 4C_3 + 3C_2 + 6\sigma$	Methane
Octahedron	$O_h = 3C_4 + 4C_3 + 6C_2 + 9\sigma$	Chromium hexacarbonyl
Sphere	$K_h =$ All symmetry elements	

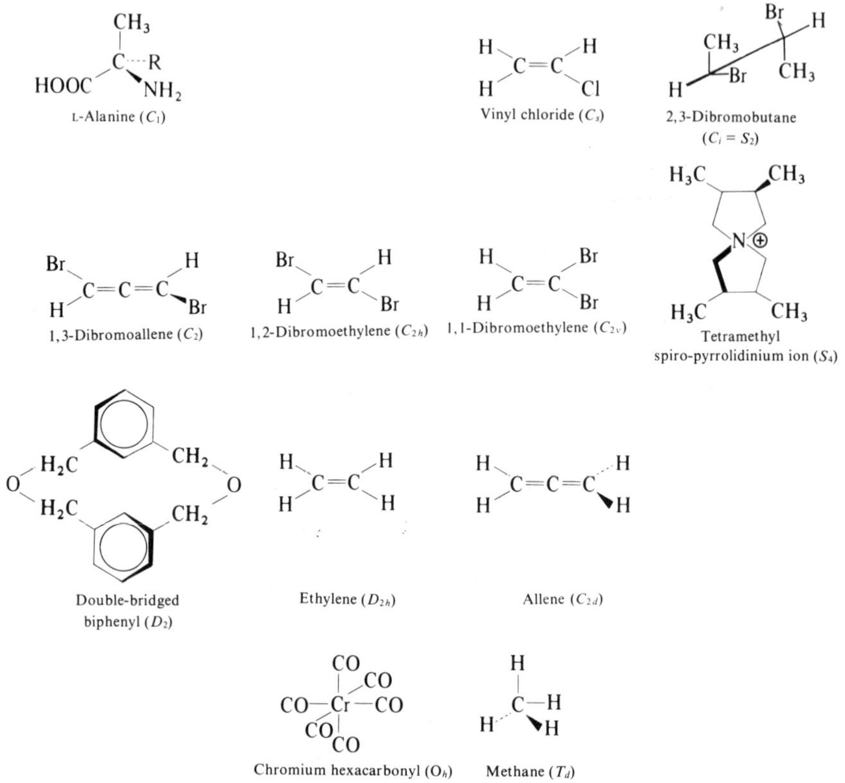

Figure 3-1. Examples of symmetry groups.

configuration and conformation are indeed still used; they are, however, subsidiary to the enantiomer and diastereomer concept. Since, in macro-molecular chemistry, a given configuration may give rise to many more conformations than in low-molar-mass chemistry, it is convenient to treat configuration and conformation separately. Of course, in macromolecular chemistry, a whole series of ordered or disordered sequences of configurations and conformations can occur along the chain in addition to the configurations about a single atom or group and the conformations about a single bond. Translation as a symmetry operation and identity as a symmetry element must also be considered in macromolecular stereochemistry.

Since enantiomers are image/mirror images, they always possess "chiral-ity," that is, they are to each other as the left hand is to the right hand (Greek, $\chi\epsilon\tau\rho$ = hand). Chirality is characterized by the absence of symmetry centers, symmetry planes, and alternating axes of symmetry. n-Fold symmetry axes, on the other hand, may be present. Because of this special property, symmetry axes are called symmetry elements of the first kind, to differentiate them from

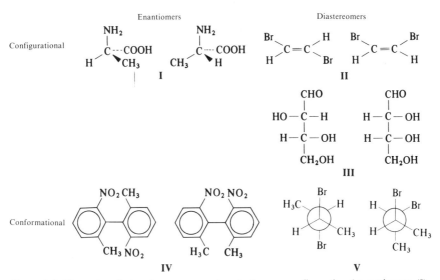

Figure 3-2. Examples of stereoisomers. D- and L-alanine are configurational enantiomers (I), *cis-* and *trans*-dibromoethylene (II), threose (III, left), and erythrose (III, right) are configurative diastereoisomers. Both atropoisomers of 2,8-dinitro-6,12-dimethyl biphenyl (IV) and both conformers of 2,3-dibromobutane shown (V) are conformational enantiomers.

all other symmetry elements, which are called symmetry elements of the second kind.

Diastereomers may be either chiral or achiral. For example, the *cis/trans* 1,2-dibromoethylenes represent an achiral diastereomer pair, whereas threose/erythrose are chiral diastereomers. Epimers are a subclass of diastereomers in that they are molecules with several asymmetric centers with one being different and all the other centers being exactly the same.

A chiral molecule may be asymmetric or disymmetric. Asymmetry occurs in the complete absence of symmetry centers of the first and second kind, including manyfold rotational axes. Of the three chiral symmetry groups shown in Table 3-1, only one is also asymmetric, namely, the symmetry group C_1 with a onefold rotational axis. Asymmetry is always found with molecules having four unlike ligands arranged tetrahedrally about a central atom which may be carbon, Si, P^+, N^+, etc. Conversely, molecules with four ligands on a symmetry plane are not asymmetric. Asymmetric molecules are always optically active.

On the other hand, chiral molecules with manyfold rotational axes are disymmetric, and have the symmetry groups C_n and D_n. Disymmetric molecules are also optically active, since optical activity occurs with the absence of any *n*-fold alternating axis of symmetry. It must be emphasized here that disymmetry as defined here differs from the older definition, whereby "disymmetry," for example, is considered to be interchangeable with the modern "chirality."

Consequently, chirality is the necessary and sufficient condition for optical activity. In the special case of a molecule consisting of several chiral centers, which compensate to the extent that the overall effect is achirality, the molecule is said to be a *meso* compound. Conversely, racemates are mixtures of equal proportions of two enantiomers.

Two further and often used concepts are prochirality and pseudoasymmetry. Pseudoasymmetric molecules possess only a few asymmetric properties. The best-known example is trihydroxy glutaric acid. The central carbon atom can, in this case, be joined to two ligands with the same (I-1 and I-2) or different (I-3 and I-4) configurations. In the first two structures the central carbon atom is achiral, but the molecule is chiral, since no symmetry elements of the second kind are present. With the second two structures, the central carbon atom is, by definition, asymmetric, since they each have four tetrahedrally arranged unlike ligands. However, in contrast to a genuine chiral center, the structures are bisected by a symmetry plane. On exchanging ligands, enantiomers are not formed, but two different *meso* forms are obtained. Thus, such carbon atoms are called pseudoasymmetric. They always occur with carbon compounds of the type CabFꟻ, where a and b are two different achiral ligands and F and ꟻ are two constitutionally equal, but mirror image (enantiomorphous) ligands. Pseudoasymmetric carbon atoms are written with the two lower case letters r and s:

$$
\begin{array}{cccc}
\text{COOH} & \text{COOH} & \text{COOH} & \text{COOH} \\
| & | & | & | \\
\text{H}-\overset{R}{\text{C}}-\text{OH} & \text{HO}-\overset{S}{\text{C}}-\text{H} & \text{H}-\overset{R}{\text{C}}-\text{OH} & \text{H}-\overset{R}{\text{C}}-\text{OH} \\
| & | & | & | \\
\text{H}-\text{C}-\text{OH} & \text{HO}-\text{C}-\text{H} & \text{H}-\overset{r}{\text{C}}-\text{OH} & \text{HO}-\overset{s}{\text{C}}-\text{H} \\
| & | & | & | \\
\text{H}-\overset{R}{\text{C}}-\text{OH} & \text{H}-\overset{s}{\text{C}}-\text{OH} & \text{H}-\overset{S}{\text{C}}-\text{OH} & \text{H}-\overset{s}{\text{C}}-\text{OH} \\
| & | & | & | \\
\text{COOH} & \text{COOH} & \text{COOH} & \text{COOH} \\
\text{I-1} & \text{I-2} & \text{I-3} & \text{I-4} \\
\end{array}
$$

enantiomers   *meso*-form diastereomers

On exchange of one of the two equal achiral ligands, a, of a molecule of the type Caabc by a new ligand, d, the molecule is converted to chirality. Such ligand arrangements or molecules are said to be prochiral or potentially chiral. They have symmetry centers or planes, but do not, however, have any symmetry axis passing through the prochiral carbon atom. Consequently, the two achiral ligands do indeed react equally rapidly with achiral reagents, but not with chiral reagents. If these two ligands are marked as a′ and a″, respectively, then, namely, the triangle cba″ seen from a′ is the mirror image, when in a prochiral arrangement, to the triangle cba′ as seen from a″. One arrangement has the seniority a > b > c arranged anticlockwise, and, so, is labeled Si (from the Latin *sinister* for left), and the other, which corresponds to a mirror image, is labeled Re (*rectus*, or Latin for right). Since both triangles, cba′ and cba″, are planar, the same considerations can be extended to planar trigonal arrangements of the type Cabc, for example, to carbonyl groups. Acetaldehyde, consequently, has a back and front side, it is prochiral, or chiral in two dimensions:

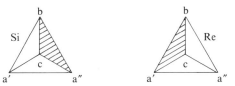

3.1.3. DL and RS Systems

The spatial arrangement of atoms about a chiral or rigid center of a low-molar-mass molecule is called the configuration. The chiral part may be a chiral center, the rigid part may be a double bond or rigid ring.

Configurations may be described by either the DL or the RS system. The older DL system is only useful for chiral centers of the type $R-CHX-R'$. The newer RS system, on the other hand, may be used with any desired chiral center, as well as the chiral axes and chiral planes.

With the DL system, the right-rotating (+)-glycerine aldehyde is assigned the D configuration for no special reason. In the D configuration, according to definition in the Fischer projection, the X substituent is on the right-hand side, where, the (+)-glycerine aldehyde, X is OH. For sugars, with their several chiral centers, the center farthest from the C atom of highest oxidation level determines the classification. Thus, the right-rotating (+)-glucose has the D configuration. On the other hand, the left-rotating L-(−)-serine is the reference substance for α-amino acids:

$$
\begin{array}{ccc}
\text{CHO} & \text{CHO} & \text{COOH} \\
| & | & | \\
\text{H}-\text{C}-\text{OH} & \text{H}-\text{C}-\text{OH} & \text{H}_2\text{N}-\text{C}-\text{H} \\
| & | & | \\
\text{CH}_2\text{OH} & \text{HO}-\text{C}-\text{H} & \text{CH}_2\text{OH} \\
 & | & \\
 & \text{H}-\text{C}-\text{OH} & \\
 & | & \\
 & \text{H}-\text{C}-\text{OH} & \\
 & | & \\
 & \text{CH}_2\text{OH} & \\
\end{array}
$$

D-(+)-glycerine aldehyde   D-(+)-glucose   L-(−)-serine

The RS system is based on purely topological considerations and is consequently independent of reference substances. The individual ligands of a chiral center are ordered according to a sequence in this system whereby the assignment of R- or S- can be derived from the order. The ligands are given different seniorities in the system such that the position in the periodic system of the atom bonded directly to the chirality center determines the seniorities:

I, Br, Cl, HSO_3, HS, F, C_6H_5COO, CH_3COO, HCOO, C_6H_5O,
$C_6H_5CH_2O$, C_2H_5O, CH_3O, HO, NO_2, NO, $(CH_3)_3N^+$, $(C_2H_5)_2N$,
$(CH_3)_2N^+H$, CH_3NH, NH_3^+, NH_2, CCl_3, COCl, CF_3, $COOCH_3$,
COOH, $CONH_2$, C_6H_5CO, CH_3CO, CHO, CR_2OH, CH_2OH, $(C_6H_5)_3C$,
C_6H_5, $(CH_3)_3C$, C_6H_{11}, CH_2=CH, $(CH_3)_2CH$, $C_6H_5CH_2$,
$(CH_3)_2CHCH_2$, C_6H_{13}, C_5H_{11}, C_4H_9, C_3H_7, C_2H_5, CH_3, Li, D,
H, lone electron pair

The chiral center is viewed from the side opposite to and along the bond joining the chiral center to the ligand of lowest seniority or preference in the above sequence list. If the remaining ligands are arranged clockwise in order of decreasing seniority, then the chiral center is given the symbol R (= rectus). If the remaining ligands are arranged anticlockwise in order of decreasing preference, then the configuration is an S configuration (S = sinister). Thus, D-(+)-glycerine aldehyde is S-glycerine aldehyde because of the seniorities $OH > CHO > CH_2OH > H$ and L-(−)-serine is R-serine because of the seniorities $NH_2 > COOH > CH_2OH > H$. The RS system is unambiguous but formal, since very similar compounds can belong to different series because of the differing seniorities or preferences of the ligands. An example of this is (S)-alanine (I) and (R)-trifluoroalanine (II):

$$CH_3-\overset{\overset{\displaystyle NH_2}{|}}{\underset{\underset{\displaystyle COOH}{|}}{C}}-H \qquad CF_3-\overset{\overset{\displaystyle NH_2}{|}}{\underset{\underset{\displaystyle COOH}{|}}{C}}-H$$

$$\text{I} \qquad\qquad\qquad \text{II}$$

3.1.4. Stereo Formulas

The spatial arrangement of ligands about a chiral center can be represented by different stereo formulas. Three-dimensional models can be represented such that a ▶, a —, and a --- represent a bond to a ligand above, in and below the plane of the paper on which it is drawn, respectively. Thus, an R molecule of formula CzyxH with the four ligands having the seniorities $z > y > x > H$ is drawn as I with the asymmetric C atom and the ligands y and z arranged in the plane of the paper, and x being above and H being below this plane.

$$\text{I} \qquad \text{II} \qquad \text{III} \qquad \text{IV}$$

With Fischer projection two-dimensional formulas, the projection plane is so arranged through the asymmetric carbon atom that two ligands are above and two ligands lie below the projection plane (= plane of the paper) (II). The lower ligands assume the vertical, the upper ligands the horizontal positions in the projected axis crossing (III). In the resulting Fischer projection (IV), the positions of the ligands with respect to the plane of the paper are no longer defined by special symbols.

Molecules with two or more chiral centers are analogously drawn, whereby conformational differences are ignored. For this reason, a series of different stereo formulas are prefered, namely, the flying wedge, the perspective, and the Newman projections.

Fischer   flying wedge   perspective (sawhorse)   Newman

Without discussing conformations further at this stage, it can be mentioned that the Fischer projections correspond to a synperiplanar (eclipsed, *cis*) and the other projections to an antiperiplanar (staggered, *trans*) conformation (see also Section 4.1.1).

When representing macromolecular chains, it is useful to rotate the Fischer projections about 90° or to show the main chain, ⌇⌇, as a hypothetical zig-zag structure. Thus, the representation

whereas the structure

In this convention, groups lying above the main chain are considered to be in positions above the paper plane, and groups under the main chain lie below the paper plane. The representation, ⌇⌇CHR—CHR⌇⌇, is reserved for polymers of unknown steric structures. However, deviations from these representations must often be made for didactic reasons when considering chemical reactions.

3.2. Ideal Tacticity

3.2.1. Definitions

Regular macromolecules such as, for example,

$$+CHCH_3\text{-}\!\!\text{)}_{\overline{n}} \qquad +CH(CH_3)-CH_2\text{-}\!\!\text{)}_{\overline{n}} \qquad +NH-CO-CH(CH_3)\text{-}\!\!\text{)}_{\overline{n}}$$

$$\textbf{I} \qquad\qquad\qquad \textbf{II} \qquad\qquad\qquad\qquad \textbf{III}$$

have the repeating units shown. Only with poly(methyl methylene) (I), however, is the constitutional repeating unit also identical with the two possible configurational base units. In contrast, four configurational base units, as well as two constitutional repeating units can already be distinguished with poly(propylene) (II):

$$
\begin{array}{cccc}
\text{H} & \text{CH}_3 & \text{CH}_3 & \text{H} \\
| & | & | & | \\
-\text{C}-\text{CH}_2- & -\text{C}-\text{CH}_2- & -\text{CH}_2-\text{C}- & -\text{CH}_2-\text{C}- \\
| & | & | & | \\
\text{CH}_3 & \text{H} & \text{H} & \text{CH}_3 \\
\textbf{II-1} & \textbf{II-2} & \textbf{II-3} & \textbf{II-4}
\end{array}
$$

The configurational base units, II-1 and II-2, are to each other as image and mirror image; they are enantiomers. II-3 and II-4 are also enantiomers to each other. In contrast, the pairs II-1/II-4 and II-2/II-3 are diastereoisomeric pairs, since they possess different configurational monomeric units.

Thus, not only the kind of configurational monomeric units, but also their mutual bonding to what are called stereorepeating units are important with polymers. The three simplest stereorepeating units in stereoregular poly(propylene) are

$$
\begin{array}{ccc}
\text{H} & \text{CH}_3 \quad \text{H} & \text{CH}_3 \quad \text{CH}_3 \quad \text{H} \\
| & | \qquad | & | \qquad\; | \qquad\; | \\
-\text{C}-\text{CH}_2- & -\text{C}-\text{CH}_2-\text{C}-\text{CH}_2 & -\text{C}-\text{CH}_2-\text{C}-\text{CH}_2-\text{C}-\text{CH}_2- \\
| & | \qquad | & | \qquad\; | \qquad\; | \\
\text{CH}_3 & \text{H} \qquad \text{CH}_3 & \text{H} \qquad \text{H} \qquad \text{CH}_3 \\
\textbf{II-1} & \textbf{II-5} & \textbf{II-6}
\end{array}
$$

Here it is immaterial whether II-1 or II-2 is chosen as the simplest stereorepeating unit. Infinitely long poly(propylene) molecules from II-1, of course, differ from those from II-2 only in the orientation of the repeating units. Thus, chains consisting of II-1 and II-2 are not enantiomeric to each other, which is in contrast to their configurational monomeric units.

The configurational repeating unit sequence defines the tacticity, and the stereorepeating unit sequence determines the stereoregularity of the polymer. According to IUPAC, a stereoregular polymer is defined as a regular polymer that contains a type of stereorepeating unit bonded together in its molecules by a single sequential arrangement. A tactic polymer correspondingly contains a single species of configurational repeating unit bonded by a single sequential arrangement. Consequently, a stereoregular polymer is always also a tactic polymer. Conversely, a tactic polymer is not always a stereoregular polymer, since not *all* stereoisomeric centers need to be defined with a tactic polymer. For example, IV-1 is a tactic polymer with respect to the configuration of the main chain atom carrying the ester group, but it is not a stereoregular polymer since the configuration of the main chain atom carrying the methyl group is not defined. IV-2 is also tactic, but not stereoregular. On the other hand, IV-3 is stereoregular, and consequently, also tactic:

$$
\begin{array}{ccc}
\quad\text{H} & \text{H} & \text{H}\quad\text{CH}_3 \\
\quad| & | & |\qquad| \\
{+}\text{C}{-}\text{CH(CH}_3){+} \quad & {+}\text{CH(COOR)}{-}\text{C}{+} \quad & {+}\text{C}{-\!\!-}\text{C}{-\!+} \\
\quad| & | & |\qquad| \\
\quad\text{COOR} & \text{CH}_3 & \text{COOR}\,\text{H} \\
\quad\textbf{IV-1} & \textbf{IV-2} & \textbf{IV-3}
\end{array}
$$

3.2.2. Monotacticity

Tactic polymers are classified as isotactic, syndiotactic, or heterotactic according to how the configurational monomeric units are bonded to each other. If two defined stereoisomeric positions per configurational monomeric unit occur, then this is known as a ditactic polymer, which may, for example, be diisotactic or disyndiotactic. If the configurational monomeric unit has three defined stereoisomeric positions, then the polymer is a tritactic polymer. Polymers of higher tactic number are similarly defined.

With isotactic polymers, only one kind of configurational monomeric unit is present; therefore, the configurational monomeric unit is identical to the configurational repeating unit. An example of this is poly(propylene) with the monomeric unit II-1. In contrast, both enantiomeric configurational monomeric units alternate in the main chain of syndiotactic polymers; an example of this case is poly(propylene) with the configurational repeating unit II-5. On the other hand, heterotactic polymers possess configurational repeating units consisting of two identical configurational monomeric units and one which is enantiomeric to these. An example of this would be the hypothetical heterotactic polymer consisting of the unit II-6. Figure 3-3 shows other examples of isotactic and syndiotactic polymers. Here, isotactic polymers have all the ligands "on the same side" in, and only in, the Fischer projection. These ligands are only all on the same side of the plane of the paper in the flying wedge projection when the monomeric unit has an even number of main chain atoms.

Isotactic and syndiotactic polymers can be readily distinguished by what is known as the bond rule. With a carbon atom as central atom, the three different substituents, r, R, and ⁓⁓(chain), for example, are arranged so

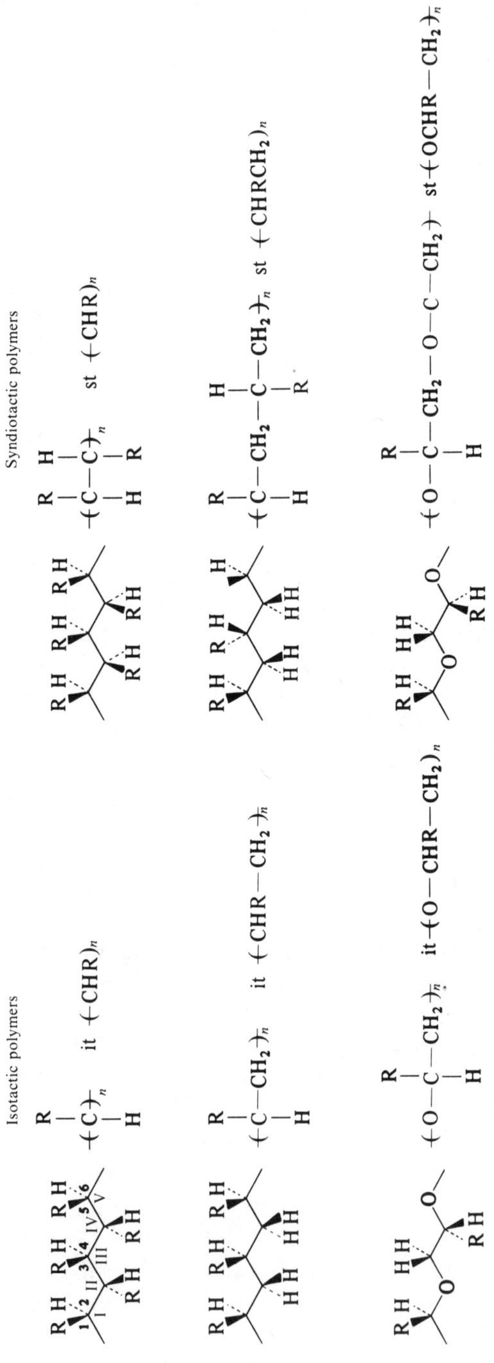

Figure 3-3. Examples of isotactic and syndiotactic polymers.

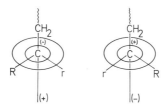

Figure 3-4. Definition of (+) and (−) bonds about a central atom C with the substituents r, R, and
〜〜(main chain).

that the size of the substituents increases anticlockwise relative to the bond
—(Figure 3-4, left). The bond, —, leading to this atom can the be described as
a (+) bond. The bond leading away from the central atom to the chain 〜〜is
then, of necessity, a (−) bond. If, on the other hand, the substituents are
arranged clockwise according to size, then the bond leading to the central
atom will be a (−) bond, and that going away will be a (+) bond (Figure 3-4,
right).

Two central atoms, or the monomeric units containing them, are
configurationally identical if the corresponding bonds are characterized by
the same (+) and (−) sign sequence. Polymers are defined as isotactic when all
their central atoms have the same configuration. In the chain, therefore, (+)
and (−) bonds alternate, i.e., (+) (−) (+) (−) (+) (−), etc. In syndiotactic
polymers, on the other hand, every second central atom has the opposite
configuration, and each central atom has one neighbor with the same, the
other with the opposite, configuration, i.e., the bonds follow in the sequence
(+) (−) (−) (+) (+) (−) (−) (+) (+), etc.

An isotactic carbon chain with the monomeric unit ─(CRH)─ is
considered as an example (Figure 3-3, left). Starting from bond 1, the three
substituents at carbon atom I are arranged counterclockwise in relation to
their size. Bond 1 is referred to as a (+), and bond 2 as a (−) bond in relation to
the carbon atom. If one moves stepwise along the chain, then, according to
definition, bond 2 with respect to carbon atom II of an isotactic polymer must
be (+) and bond 3 (−). The three substituents around carbon atom III must
likewise be arranged counterclockwise. This means that the substituent R in
carbon atom II must lie below the plane of the paper. Thus, in an isotactic
polymer with the base unit ─(CRH)─, substituent R lies alternatively above
and below the plane of the paper, whereas in a corresponding syndiotactic
polymer all like substituents are found on one and the same side relative to the
plane of the paper (Figure 3-3, right).

So, it is seen that stereoisomers are considered differently in macromolecular chemistry to what they are in low-molar-mass chemistry. In
macromolecular chemistry, one begins at one end of the main chain and
considers the configuration of each main chain atom relative to the preceding

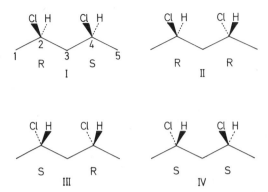

Figure 3-5. The four possible configurations of 2,4-dichloropentane.

one. In low-molar-mass chemistry, on the other hand, the absolute configuration about every single chain atom is determined. This latter mode of
considering configurations leads to all kinds of difficulties because translational symmetry operations are ignored.

In 2,4-dichloropentane, for example, there are two asymmetric carbon
atoms with four ligands having the seniorities $Cl > CH_2—CHCl—CH_3 >
CH_3 > H$ (Figure 3-5). Molecule I has, according to the chirality rules, the
configuration RS. The two asymmetric central atoms have opposing configurations; so, the molecule is a *meso* compound. Since molecules I and III can be
converted into each other by a rotation about 180°, these two molecules must
be identical. Molecules II and IV cannot be transformed into each other by a
symmetry operation, and so, they must be enantiomers; a 1:1 mixture of these
two is therefore racemic.

The situation is similar with longer-chain molecules. With the heptamers
of propylene, having one isopropyl and one isobutyl end group, the
configuration of the molecule with stereo formula I is reversed about the
center by application of the "absolute" configurational analysis (Figure 3-6).
The upper half has an S configuration, and the lower half has an R
configuration. The stereo formulas I and IV belong, correspondingly, to the
same *meso* compound. The configurational reversal, however, is only caused
by the labeling convention, and, so, is only apparent. In actual fact, all the
asymmetric carbon atoms have the same relative configuration with respect to
each other; the molecule is isotactic. The stereo formulas II and III for a
syndiotactic molecule have been called "racemic" in misunderstood analogy
to the above considerations. One is not actually dealing with two enantiomers,
but with a single compound. There is no genuine configurational reversal.

Stereoblock polymers are defined analogously to constitutional block
polymers: each block has a different species of stereorepeating unit to the
block preceding it, but has the same constitutional monomeric units. Tactic

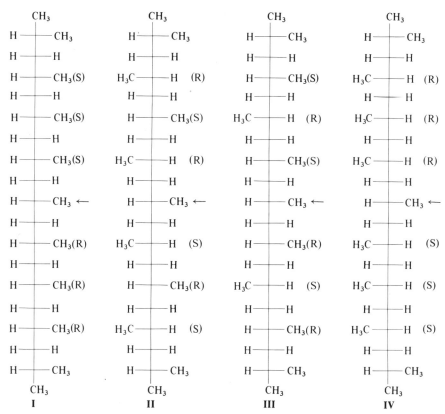

Figure 3-6. The configuration of heptamers of propylene, each having one isopropyl and one isobutyl end group. The molecule has two equal end groups with respect to configurational analysis (structural analysis), but not with respect to the polymerization (process analysis). The configuration is given in the Fischer projection (see below). The arrows point to pseudoasymmetric carbon atoms. R and S are used in the sense as defined by organic chemistry configurational analysis.

block polymers bear the same relationship to stereoblock polymers as tactic polymers to stereoregular polymers: with the former, not all stereoisomeric centers are defined, in the latter, they are. Stereoblock copolymers and tactic block copolymers correspondingly consist of more than one monomer.

Atactic polymers are also regular polymers. They contain, by definition, the possible configurational monomeric units in equal proportions, but with an ideally random distribution from molecule to molecule. Such distributions are caused by symmetric Bernoulli mechanisms during polymerization (see Section 15). They are distinguished by having equal numbers of iso- and syndiotactic diads ($N_i = N_s$), iso-, hetero-, and syndiotactic triads ($N_{ii} = N_{is} =$

$N_{si} = N_{ss}$), etc. This concept of atacticity, as strictly interpreted here, is not often met with in the literature; "atacticity" is more often defined as "not tactic" or "not found to be predominantly tactic under the experimental conditions used" than really "atactic" in the sense used above.

Polymers of different geometric isomerism are also tactic. *Cis*-tactic (ct) and *trans*-tactic (tt) polymers can be distinguished according to the steric arrangement of chain components about double bonds in the main chain. Samples of this are *cis*- and *trans*-1,4-poly(butadiene):

3.2.3. Ditacticity

Ditactic polymers possess two stereoisomeric centers per constitutional monomeric unit, and tritactic polymers possess three. Ditactic polymers may be formed by the polymerization of 1,2-disubstituted ethylene derivatives, as, for example, with pentene-2:

$$n\,CH_3-CH=CH-C_2H_5 \longrightarrow +CH(CH_3)-CH(C_2H_5)\!+_n \qquad (3\text{-}1)$$

In principle, the poly((1-ethyl)(2-methyl)ethylene) formed can occur in four different configurations, since two arrangements are possible for each of the two asymmetric carbon atoms. However, the number of arrangements for the two asymmetric centers is restricted by the fact that the centers in the monomeric unit retain the configuration about the bond that joined them as monomer. Both centers are thus only isotactic or only syndiotactic, so that the polymer is either diisotactic or disyndiotactic. In analogy to the usual nomenclature of low-molar-mass compounds, polymers with the same sequence of monomeric units in the Fischer projection are referred to as erythropolymers, and those with an alternating sequence as threopolymers (Figure 3-7).

In the erythro-diisotactic configuration (eit), the substituents R and R' all lie on the same side in a Fischer projection, whereas in a flying wedge projection, all the R substituents are found on one side, but the R' substituents on the other side of the plane of the paper. In the Newman projection of the eclipsed conformation, R lies above R' and H above H in the eit configuration. The characteristic features of the three other configurations can be seen in Figure 3-7.

By polymerizing the double bonds of unsaturated rings, it is possible to synthesize polymers with rings as stereoisomeric centers. The other ring atoms bonded directly onto the main chain atoms of the ring should be considered as

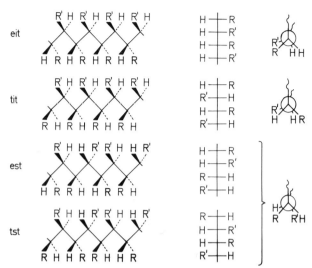

Figure 3-7. The four configurations of ditactic poly(2-pentene). eit = erythrodiisotactic, tit = threodiisotactic, est = erythrodisyndiotactic, tst = threodisyndiotactic; $R = CH_3$, $R' = C_2H_5$ or vice versa.

substituents. Poly(cyclohexene), therefore, forms four different configurations, just as poly(2-pentene) does (Figure 3-8). The only special case occurs for the bonds that represent the entry or exit points of the main chain into the ring. They are *cis* in erythro and *trans* in threo configurations.

3.3. Real Tacticity

3.3.1. J-ads

The structures of stereoregular and tactic polymers described above are ideal. Real polymers are always irregular; they do not have perfect steric or tactic structures. Thus, the mean sequence and composition of steric and configurational diads, triads, etc. must be given by suitable statistical parameters, in analogy to the case for constitutional diads, triads, etc.

Configurational diads must be either isotactic or syndiotactic. Thus the sum of their mole fractions must be unity:

$$x_i + x_s \equiv 1 \qquad (3-2)$$

It must be remembered here that the mole fractions of isotactic diads in chiral monomeric units, as, for example, with α-amino esters, is given by the sum of

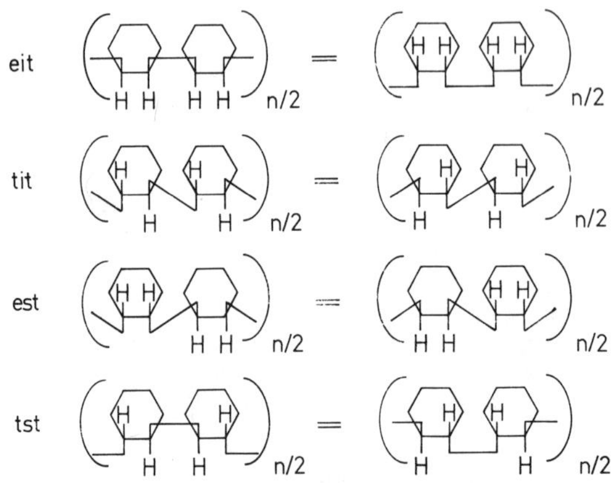

Figure 3-8. The four configurations of ditactic poly(cyclohexene).

mole fractions of the DD and LL links and of syndiotactic diads, by the sum of the mole fractions of DL and LD links:

$$x_i \equiv x_{DD} + x_{LL} \tag{3-3}$$

$$x_s \equiv x_{DL} + x_{LD} \tag{3-4}$$

Each configurational triad consists of two diads. The two diads may be isotactic–isotactic, isotactic–syndiotactic, or syndiotactic–syndiotactic. The mole fraction of what are known as heterotactic triads, $x_h = x_{is} + x_{si} = x_{\overline{is}}$, does not distinguish between directions, both si and is triads are included in $x_{\overline{is}}$. Six different tetrads can be analogously distinguished. Similarly, there are ten different configurational pentads, etc. The number, N_j, of possible j-ads (diads, triads, tetrads, etc.) is given by

$$N_j = 2^{j-2} + 2^{k-1} \tag{3-5}$$

When j is an even number (diads, tetrads, hexads, etc.), then $k = j/2$. When j is uneven (triads, pentads, etc.), $k = (j - 1)/2$. Mole fractions are given by

$$x_{ii} + x_{\overline{is}} + x_{ss} \equiv 1 \quad \text{(tetrads)} \tag{3-6}$$

$$x_{iii} + x_{\overline{iis}} + x_{isi} + x_{\overline{iss}} + x_{sis} + x_{sss} \equiv 1 \quad \text{(pentads)} \tag{3-7}$$

Relationships independent of the polymerization mechanism must exist between the various j-ads since each of the j-ads, from triads onward, consists of two or more diads. The relationships between the fractional compositions of diads and triads are

$$x_i = x_{ii} + 0.5x_{\overline{is}}, \qquad x_s = x_{ss} + 0.5x_{\overline{is}} \tag{3-8}$$

and for tetrads, analogously:

$$x_{ii} = x_{iii} + 0.5x_{\overline{iis}}$$

$$x_{ss} = x_{sss} + 0.5x_{\overline{iss}} \tag{3-9}$$

$$x_{\overline{is}} = 0.5x_{\overline{iis}} + x_{sis} + 0.5x_{\overline{iss}} + x_{isi}$$

The sequence lengths of configurational *j*-ads are defined and treated analogously to constitutional *j*-ad sequence lengths (see Section 2.4.5). Thus, the number average sequence length of isotactic sequences is

$$\langle X_I \rangle_n = 2x_i / x_{\overline{is}} \tag{3-10}$$

Since the junction between a syndiotactic and an isotactic sequence is always a heterotactic triad, the number average sequence length of *all* isotactic and syndiotactic sequences is given by the reciprocal mole fractions of heterotactic triads:

$$\langle X \rangle_n = 1 / x_{\overline{is}} \tag{3-11}$$

3.3.2. Experimental Methods

Methods for determining the presence, kind, and amount of configurational base units can be classified as relative or absolute. Absolute methods do not require calibration with polymers of known tacticity. Relative methods, on the other hand, require comparison with standard substances. X-ray crystallography, nuclear magnetic resonance, infrared spectroscopy, and optical activity measurements are all absolute methods. Relative methods include crystallinity, solubility, glass transition temperature, and melting temperature measurements as well as chemical reactions (Table 3-2).

3.3.2.1. X-Ray Crystallography

X-ray crystallography (Section 5.2.1) is an absolute method. With it, it is possible to determine the distances between the atoms in crystalline regions, and then the configuration from the position and intensity of the diffractions. The method does not depend on knowledge of model compounds. However, it is only applicable to substances that crystallize well and have a high steric purity. X-ray crystallography is used in configurational studies to calibrate relative methods.

3.3.2.2. Nuclear Magnetic Resonance Spectroscopy

Nuclear magnetic resonance (nmr) spectroscopy of polymers in solution is a very important method of studying polymer configuration since noncrystalline as well as crystalline compounds can be studied. The method depends on the fact that the chemical shift of the signals of bonded hydrogen

Table 3-2. Methods of Determining Tacticities

Method	Determination of the			Remarks
	Presence	kind	Quantity	
X-ray crystallography	Only with crystalline polymers	Yes	No	
Nuclear magnetic resonance spectroscopy	Yes	In principle	Diads and triads, often also tetrads and pentads	^1H, ^{13}C, ^{19}F, etc. in solution, ^{13}C also in solid state
Infrared spectroscopy	Only sometimes	Sometimes	Only diads	Via conformation
Optical activity	Only with chiral molecules	Yes	Yes	Mainly via conformation
Crystallinity	Questionable	Mostly not	Questionable	
Solubility	Questionable	No	Questionable	Diad content after calibration via cloud point titrations
Glass transition and melting temperatures	Sometimes	No	Sometimes	
Chemical reactions	Sometimes	No	Yes	Calibration required

atoms (protons), ^{13}C and ^{19}F atoms, etc., in fixed chemical environments depends on the configuration of the main chain. In theory, the technique represents an absolute method, but, on technical grounds, it can often only be used as a relative method. An example of this is the analysis of the spectra of poly(methyl methacrylates) of various tacticities.

In poly(methyl methacrylate), $\{CH_2-C(CH_3)(COOCH_3)\}$, signals can be expected from the methylene protons CH_2, from the α-methyl protons CH_3, and from the methyl ester protons $COOCH_3$. The assignment of the signals from the three types of protons is made possible by comparison with the spectrum of methyl pivalate $(CH_3)_3C-COOCH_3$. In both poly(methyl methacrylate) and methyl pivalate, the α-methyl protons and the methyl ester protons appear at the same position in the nmr spectrum. Information about the tacticity of the polymer can be obtained as described below.

In st-poly(methyl methacrylate), the two methylene protons occur in a chemically equivalent environment, since every proton is flanked by an α-methyl group and a methyl ester group. In it-poly(methyl methacrylate), on the other hand, the two methylene protons are not chemically equivalent, since one proton is surrounded by two α-methyl groups, and the other by two methyl ester groups (Figure 3-9). It is immaterial whether the conformations represented in Figure 3-3 are really the only acceptable ones which occur or not, since only the average conformation is detected. The two equivalent methylene protons of st-PMMA thus lead to a single proton resonance signal, whereas the chemically nonequivalent methylene protons lead to an AB quartet (Figure 3-9).

Since the two hydrogen atoms of a methylene group in an isotactic diad are not nmr-spectroscopically equivalent, they have also been called "*meso*" (and also called heterosteric or diastereotopic). In analogy to this, the methylene group of a syndiotactic diad has been called "racemic" (and also called homosteric or enantiotopic). For these reasons the composition fractions of isotactic and syndiotactic diads are often given in the literature as (m) or (r) instead of as x_i or x_s. The names racemic and meso are not equivalent to those used in classical organic chemistry and are therefore misleading. These terms, racemic and meso, are also superfluous, since the terms isotactic and syndiotactic are unambiguously defined in terms of configuration, and one should not base a structural definition on a phenomenon of a particular method of measurement.

The resonance signals of the α-methyl protons appear at various points in the spectrum according to tacticity. It is not possible to draw any conclusions about the configuration from the position of these signals alone, since it is only with difficulty that any inferences can be drawn about magnetic screening by neighboring groups. However, assignments can easily be made if the signals coming from the methylene protons are known.

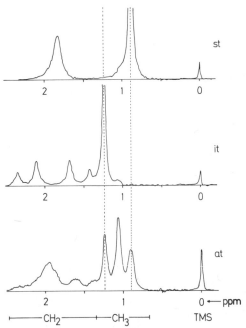

Figure 3-9. Section from the proton resonance spectra of isotactic (it), syndiotactic (st), and atactic (at) poly(methyl methacrylates). The signals of the methyl ester protons are not shown. TMS = reference signal of tetramethyl silane. (According to P. Goeldi and H.-G. Elias.)

Using the assignments thus obtained, it is possible to analyze the spectrum of nonholotactic PMMA. From the spectrum of a so-called atactic PMMA, it can readily be seen that evidence on the proportion of iso- and syndiotactic diads is only accessible with difficulty from the methylene proton signals. The signals of the singlets and the quartets are not very well defined. With the α-methyl proton signals, the situation is better. Here, three different signals are observed, one of which is in the position corresponding to a signal from the iso- and another from the syndiotactic polymer. The third signal lies between these two. One therefore concludes that the outer α-methyl proton signals correspond to the syndiotactic and isotactic triads and that the central signal arises from the heterotactic triads. The area beneath the signals is proportional to the proportion of the corresponding triads.

The methyl ester protons give rise to a single signal, which is independent of tacticity (not shown in Figure 3-9). Therefore, the methyl ester proton signal cannot be used for tacticity determinations. The chemical shift is not affected by the configuration of the main chain since they are too far removed from an asymmetric center.

Generally speaking, the signals obtained from polymer solutions are broader than those of low-molar mass model compounds. In low-molecular-weight compounds the higher the concentration and the lower the temperature, the broader are the signals. The broadening of the signals results from the strong magnetic interactions between different nuclei (see Section 10.2.4). If the concentration is lowered, the nuclei become less oriented and the signals become narrower. The same effect can be achieved by raising the temperature.

In polymers, however, the individual monomeric units are bonded together into a chain. Since the broadness of the signal depends on nearest-neighbor influence, dilution of the solution does not lead to well-resolved signals. The resolution of the signals in random coils is also largely independent of molar mass. Sharper signals can therefore only be achieved by measurements at elevated temperatures. The splitting of the signals from nonholotactic polymers also depends to a certain extent on the nature of the solvent used. At the present time it remains to be shown whether this influence of the solvent results from a conformational shift or from a specific interaction between the solvent and the base units (solvation).

For polymers of the $+CH_2-CHR+$ type, spin–spin coupling of neighboring CH_2 and CH groups can lead to complex proton resonance spectra which are difficult to interpret. This difficulty can be overcome by the double-resonance technique and/or higher magnetic field strengths.

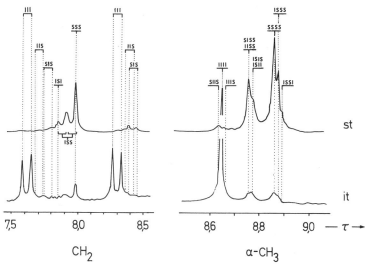

Figure 3-10. Proton resonance spectroscopy of poly(methyl methacrylates) in ~0.1 g/ml chlorobenzene solution at 135°C and 220 MHz. Left: the β-methylene signals given by tetrads. Right: the α-methylene proton signals given by pentads. Above: predominantly syndiotactic PMMA. Below: predominantly isotactic PMMA.

In general, 60-MHz proton resonance spectra can only yield data on diad and triad contents. Higher magnetic field strengths lead to greater chemical shifts, giving better resolution (Figure 3-10). Such measurements can, for example, be carried out at 220 or 300 MHz using superconducting magnets cooled with liquid helium. Higher *j*-ad contents can often be determined from ^{13}C-spectra, since the chemical shift of ^{13}C is much greater than that of protons (up to 250 ppm, in contrast to up to 10 ppm). In addition, tacticities can not only be determined from chemical shifts, but also from spin/spin and spin/lattice relaxation times. Indeed, relaxation times are often more sensitive to tactitities than are chemical shifts.

3.3.2.3. Infrared Spectroscopy

Infrared spectroscopy is also frequently used in the quantitative determination of the proportion of diads. As a rule, the assignment of the different diad signals is made possible from results with polymers or oligomers of known configuration. In certain cases, the calculation of the absorption frequency for the individual types has already been performed. The CH and CH_2 deformational vibrations refer directly to various configurations, as is also often the case with the amide I bands of poly(α-amino acids). Since products of different stereoregularity crystallize to different extents, and since ir spectra are sensitive to crystallinity in the range 670–1000 cm^{-1}, then the diad content can also be determined by means of what are called crystallinity

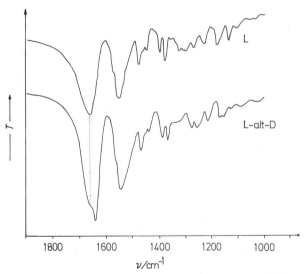

Figure 3-11. Influence of the tacticity on the amide I-band of poly(leucines) at 1643 cm^{-1}. Above: poly(L-leucine); below: poly(L-alt-D-leucine). (After F.-G. Fick, J. Semen, and H.-G. Elias.)

bands. However, the method if often unsuitable, since the crystallinity of the polymer depends on its previous thermal history (Chapter 5).

3.3.2.4. Other Methods

There is a series of other methods which likewise use the different crystallinities of polymers of various degrees of stereoregularity. However, none of these methods is completely unequivocal, for two reasons. On the one hand it is known that extensively "atactic" polymers, such as the poly(vinyl alcohol) obtained by saponification of radically polymerized vinyl acetate, can also crystallize relatively well. On the other hand, large substituents can impede or prevent the crystallization of stereoregular polymers. For example, isotactic poly(styrene), which can be crystallized, can, in a series of polymer analog reactions be converted through noncrystallizable poly(p-iodostyrene) and poly(p-lithiumstyrene) back into crystallizable, isotactic poly(styrene) without any change in the configuration:

$$+CH_2-CH+ \xrightarrow{+I_2/HIO_3} +CH_2-CH+ \xrightarrow{+Li} +CH_2-CH+ \xrightarrow{+H_2O} +CH_2-CH+ \qquad (3\text{-}12)$$
$$\overset{|}{C_6H_5} \qquad\qquad \overset{|}{C_6H_4I} \qquad\qquad \overset{|}{C_6H_4Li} \qquad\qquad \overset{|}{C_6H_5}$$

Stereoregular polymers can be separated from atactic ones in certain circumstances since crystalline polymers dissolve less readily than do amorphous ones. However, in addition to crystallinity, the solubility is also dependent on the degree of stereoregularity and the molar mass. Thus, readily soluble fractions of high-molar-mass, atactic polymers can also contain low-molar-mass fractions of stereoregular material and vice versa. When samples have received identical thermal pretreatment, information on differing crystallinity and stereoregularity can be obtained from the values of the melt and glass transition temperatures (see Chapter 10).

Other methods used for stereoregularity determinations involve the use of dipole moments, streaming birefringence, rate of saponification, and cloud-point titration. However, all these methods are only applicable to special polymers and/or are only indirect methods, and so they have not found general application.

Literature

Nomenclature

M. L. Huggins, C. Natta, V. Desreux, and H. Mark, Report on nomenclature dealing with steric regularity in high polymers, *J. Polym. Sci.* **56**, 153 (1962).

R. S. Cahn, C. Ingold, and V. Prelog, Specification of molecular chirality, *Angew. Chem. Int. Ed. (Eng.)* **5**, 385 (1966).

IUPAC, Commission on Nomenclature of Organic Chemistry, 1974 Recommendations for
Section E, Fundamental Stereochemistry, *Pure Appl. Chem.* **45**, 11 (1976).
IUPAC, Commission on Macromolecular Nomenclature, Stereochemical definitions and
notations relating to polymers, *Pure Appl. Chem.* **51**, 1101 (1979).

General Reviews

K. Mislow, *Introduction to Stereochemistry*, W. A. Benjamin, New York, 1965; *Einführung in
die Stereochemie*, Verlag Chemie, Weinheim, 1967.
G. Natta and M. Farina, *Struktur and Verhalten von Molekülen im Raum*, Verlag Chemie,
Weinheim, 1976; *Stereochemistry*, Longman, New York, 1972.
W. Bähr and H. Theobald, *Organische Stereochemie*, Springer, Berlin, 1973.
E. L. Eliel, *Stereochemistry of Carbon Compounds*, McGraw-Hill, New York, 1962; *Stereo-
chemie der Kohlenstoffverbindungen*, Verlag Chemie, Weinheim, 1966.
E. L. Eliel, *Grundlagen der Stereochemie*, Birkhäuser, Basel, 1972.
B. Testa, *Principles of Organic Stereochemistry*, Marcel Dekker, New York, 1979.

Group Theory

F. A. Cotton, *Chemical Application of Group Theory*, Interscience, New York, 1963.
H. H. Jaffe and M. Orchin, *Symmetrie in der Chemie*, Hüthig, Heidelberg, 1967.
K. Mathiak and P. Stingl, *Gruppentheorie*, second ed., Akad. Verlagsges., Frankfurt am Main,
1969.
J. D. Donaldson and S. D. Ross, *Symmetry in the Stereochemistry*, Intertext Books, London,
1972.
R. L. Flurry, *Symmetry Groups, Theory and Chemical Applications*, Prentice-Hall, Englewood-
Cliffs, New Jersey, 1980.

Configuration and Tacticity in Macromolecules

L. Dulog, Taktizität and Reaktivität, di- und tritaktische Polymere, *Fortschr. Chem. Forschg.* **6**,
427 (1966).
G. Natta and F. Danusso, *Stereoregular Polymers and Stereospecific Polymerizations*,
Pergamon, Oxford, 1967 (two vols., original papers from the Natta school).
A. D. Ketley (ed.), *The Stereochemistry of Macromolecules*, Marcel Dekker, New York, three
vols., 1967–1968.
F. A. Bovey, *Polymer Conformation and Configuration*, Academic Press, New York, 1969.
J. L. Koenig, *Chemical Microstructure of Polymer Chains*, Wiley, New York, 1980.

Infrared Spectroscopy

S. Krimm, Infrared spectra of high polymers, *Fortschr. Hochpolym. Forschg.* **2**, 51 (1960).
G. Schnell, Ultrarotspektroskopische untersuchungen an copolymerisaten, *Ber. Bunsenges.* **70**,
297 (1966).

Nuclear Magnetic Resonance Spectroscopy

P. R. Sewell, The nuclear magnetic resonance spectra of polymers, *Ann. Rev. NMR Spectrosc.* **1,** 165 (1968).

Hung Yu Chen, Application of high resolution NMR spectroscopy to elastomers in solution, *Rubber Chem. Technol.* **41,** 47 (1968) (also contains data on thermoplasts, etc.).

M. E. Cudby and H. A. Willis, Nuclear magnetic resonance spectra of polymers, *Ann. Rev. NMR Spectrosc.* **4,** 363 (1971).

F. A. Bovey, *High Resolution NMR of Macromolecules*, Academic Press, New York, 1972.

F. Heatley, Nuclear magnetic resonance, Synthetic macromolecules, *Nucl. Magn. Res.* **8,** 266 (1979).

Chapter 4

Conformation

The expression "conformation" is always used with respect to a single bond; such conformations may also be called microconformations. There are a great many microconformations of this kind in a macromolecule, such that the macromolecule adopts an overall macroconformation. The macroconformation determines the shape of the molecule.

4.1. Basic Principles

4.1.1. Conformations about Single Bonds

The arrangements of atoms or groups of atoms in space about a single bond of molecules of definite configuration are known as conformations or constellations when these spatial arrangements are not superimposable. Torsion stereoisomers produced by rotation about double bonds or partial double bonds, as, for example, with helicenes or amides, are also sometimes included in this classic definition of conformation. The concepts of conformation and configuration are partially merged by this extension.

Of the infinitely large number of theoretically possible conformations, only some will be energetically favorable. These conformational isomers are called conformers, rotational isomers, or rotamers. However, they can only be isolated as single substances when the rotational barrier exceeds about 65–85 kJ/mol bond. The existence of conformers of lower rotational barriers was first presumed, however, in the 1930s on the basis of differences between calculated and observed entropies.

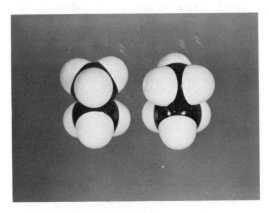

Figure 4-1. Eclipsed or synperiplanar (left) and staggered or antiperiplanar (right) conformations of ethane.

Ethane, for example, has two of these kind of conformers (Figure 4-1). In the staggered, or antiperiplanar (ap), conformation, the H atoms on one carbon atom face the gaps between the H atoms of the other carbon atom, whereas, in the eclipsed, or synperiplanar (sp), conformation, the H atoms are directly opposite each other. The ap conformation has the symmetry group D_{2d} and the sp conformation has the symmetry group D_{3h}; consequently, both conformers are achiral. They convert into each other by rotating one of the methyl groups about 60°. The conformers that lie within these arrangements, on the other hand, belong to the symmetry group D_3; they are chiral.

Molecules of the type A_i–B_j–C_k can, of course, be unambiguously defined geometrically when both bond lengths, $b(A_i, B_j)$ and $b(B_j, C_k)$, and the bond angle, $\tau(B_j) = \tau(A_i, B_j, C_k)$, are given. To completely define molecules of the type A_i–B_j–C_k–D_l geometrically, on the other hand, not only the three bond lengths and two bond angles (valence angles), but also what is known as the torsion angle (also called conformation angle, rotation angle, or dihedral angle) must be given. The torsion angle θ, is the angle between the A_i–B_j–C_k plane and the B_j–C_k–D_l plant (Figure 4-2). It has a value of zero for synperiplanar conformations. A torsion angle is called positive when the A_iB_j bond has to be rotated less than 180° to the right to make it superimposable with the C_kD_l bond. Thus, the torsion angle $\theta(B_j)$ in Figure 4-2 is negative. Torsion angles are measured from $-180°$ to $+180°$ instead of from 0° to 360°.

The configuration of carbon compounds is, of course, determined by the tetrahedral arrangement of the ligands. Consequently there are three energy maxima and three energy minima in the case of ethane, with only two possible conformers: ap and sp. In butane, each carbon atom joined by the central C–C bond also carries three ligands, which are two hydrogen atoms and one methyl group. Thus, only two of the three energy maxima are equal (Figure

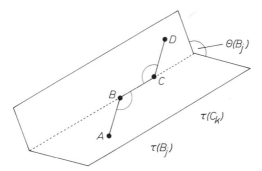

Figure 4-2. The arrangement of four consecutive atoms, A–B–C–D, in space with valence angles $\tau(B_j) = \tau(A_i, B_j, C_k)$ and $\tau(C_k) = \tau(B_j, C_k D_l)$, together with torsion angle $\theta(B_j) = \theta(A_i, B_j, C_k, D_l)$.

4-3). So, instead of three identical synperiplanar positions, there are one synperiplanar and two anticlinal positions, with the anticlinal positions being mirror images of each other. The three identical antiperiplanar positions are also replaced by an antiperiplanar and two synclinal positions. The ap conformation is planar symmetrical and, so, belongs to the symmetry group, C_{2h}. The two sp conformations belong, in contrast, to the C_2 symmetry group, and, so, are chiral and enantiomeric to each other. These terms, used in low-molar-mass organic chemistry, are replaced in macromolecular chemistry by the terms *cis*, *anticlinal*, *gauche*, and *trans*, in accordance with IUPAC proposals. Some older terms are also used:

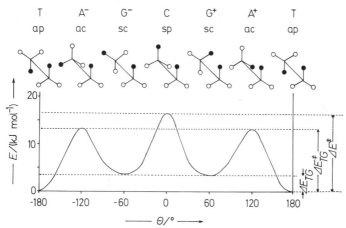

Figure 4-3. Conformations and rotational barrier potentials about the CH_2—CH_2 bond of butane, $CH_3 CH_2 CH_2 CH_3$, as a function of the torsion angle between the two methyl groups (●).

   cis (C) = synperiplanar (sp); eclipsed, planar-syn, cis-eclipsed, ecliptic.

   anti (A) = anticlinal (ac); partially eclipsed, skew-anti, partially ecliptic.

gauche (G) = synclinal (sc); skewed, skew-syn, wind skew, gauche staggered.

   trans (T) = antiperiplanar (ap); staggered, trans staggered, anti, anti-parallel, gapped, atomic gapped.

These conformations correspond to torsion angles of $0°$ (C), $60°$ (G^+), $120°$ (A^+), $\pm 180°$ (T), $-120°$ (A^-), and $-60°$ (G^-). Other conformations are given the same names when they do not deviate from the ideal conformation by more than $\pm 30°$. Thus, a conformation with a torsion angle of $170°$ is also known as *trans*. Enantiomorphic conformations of unknown prefix, $+$ or $-$, are correspondingly called $G/\overline{G}$, $A/\overline{A}$, $C/\overline{C}$, and $T/\overline{T}$; in the latter cases, of course, only when the torsion angle is not exactly $0°$ or $180°$.

Threefold rotational potential energy barriers such as in ethane and butane are not always encountered. Twofold potential rotational energy barriers are produced by 1,4-phenylene groups in the main chain, for example. *Catena*-poly(sulfur) also has a twofold rotational potential energy barrier.

4.1.2. Conformational Analysis

Preferential conformations of a molecule are studied in conformational analysis. This can be done by microwave, uv, ir, nmr, and Raman spectroscopy, X-ray crystallography, kinetic, equilibrium and dipole moment measurements, etc. The existence and stability of conformers can also be determined by energy calculations.

All calculations are based on mechanical molecular models: the atoms are more or less deformable spheres, the bonds are rigid springs. Attraction and repulsion are calculated separately as a function of the torsion angle. A typical exercise, for example, proceeds from the total energy of an ethane molecule, which is made up of five components (Figure 4-4):

    I.   The energy E_{nn} between the nuclei of nonbonded hydrogen atoms.

   II. The energy E_b between the electrons of the carbon–carbon bond.

  III. The energy E_{ee} between the electrons of the carbon–hydrogen bonds.

  IV. The energy E_{ne} between the hydrogen nuclei and the electrons of the bonds to the other hydrogen atoms.

   V. The kinetic energy E_{kin} of the electrons.

The expression "nonbonded" is used here for such atoms which do indeed influence the conformation about a given bond but are not bonded to the

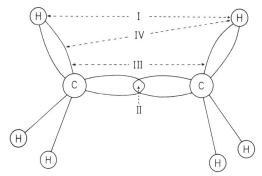

Figure 4-4. Schematic representation of the most important interactions in the ethane molecule (according to T. M. Birshtein and O. B. Ptitsyn). (For explanation of I-IV, see text.)

atom as bond partner which is being considered. So, the three hydrogen atoms bonded to C^2 in ethane, $H_3C^1—C^2H_3$, are "nonbonded" with respect to C^1. The terms *bonded* and *nonbonded* in macromolecules always refer to main chain atoms; they are never used for bonds within substituents to the main chain.

Only such interactions that depend on the torsion angle determine the occurrence of given conformations. Interactions between the electrons of the C—C bond (case II) can only contribute to the conformation energy when there is no cylindrical symmetry about the σ-bond. The symmetry would be broken if the $4f$ state participated in the bonding, since the corresponding regions of overlap of f bonds are not cylindrically symmetrical to the C—C bond. In this case, the electron clouds overlap more strongly in the *cis* than in the *trans* conformation, giving a more stable *cis* form. Experimentally, however, exactly the reverse is observed, namely, a more stable *trans* conformation. Interactions between the electrons of the C—C bond do not or do not significantly contribute to the conformation, i.e., $E_b \approx 0$.

The total attraction is given, thus, by E_{ne}, and the total repulsion consists of $E_{nn} + E_{ee} + E_{kin}$. With the ethane molecule, three equal maxima or minima are given in a curve describing the attraction or repulsion as a function of the torsion angle. The phases of the attraction and repulsion energies are shifted by 120°. The energy difference between minimum and maximum for attraction is 82.5 kJ/mol and for repulsion is 93.8 kJ/mol. The difference between these two energies is called the potential energy barrier, potential energy crest, or rotation energy barrier. It is caused by repulsion in the case of ethane, where it is 11.3 kJ/mol.

The actual influence of the chain on macromolecular conformation is first clearly apparent in *n*-pentane, since here, for the first time, there are two subsequent chain conformations to consider. Since one *trans* and two *gauche* positions are possible in every chain bond of this kind, then for the two subsequent chain conformations there are four different physically distinct combinations or conformational diads (Figure 4-5). Of these, the diad TT possesses the lowest energy, and the combination G^-G^+ the highest. Even with this simplified model, which does not take the *cis* conformations into account, calculations of macroconformation are very difficult. For this reason, the model is further simplified by likewise excluding the combinations G^-G^+ or G^+G^-, and by taking the energy difference between G and T to be a constant (i.e., the energy difference should be independent of neighboring conformations).

The repulsion forces dominating in aliphatic hydrocarbons lead to a *trans* conformation of the chain. The balance between attraction and repulsion forces can be changed for some of the possible conformations by the presence of neighboring atoms with lone electron pairs or by the presence of electronegative substituents. In these cases, the attractions between nuclei and electrons then become sufficiently large. When such atoms are present, the compounds tend to take up the conformations permitting the greatest number of *gauche* interactions between neighboring lone electron pairs and/or electronegative substituents (*gauche* effect). Thus, poly(oxymethylene), with the monomeric unit $+O—CH_2+$, exists in the crystal in the all-*gauche* conformation, the lowest energy state.

The conformation energy of larger molecules is calculated as the weighted sum of the contributions of the individual main chain bonds. Influences extending over more than two main chain bonds, however, are mostly ignored. In such calculations, a given reference conformation is given the energy of zero and a statistical weight of unity. This reference conformation is, for example, the *trans* conformation for poly(ethylene) and the *gauche* conformation for poly(oxymethylene).

Several procedures exist for the calculation of the conformational energy. They differ in the choice of the parameters and potentials. With macromolecules, contributions due to bond length deformation and bond angle distortion are often ignored and the conformational energy E_r is calculated from the contributions of the torsional energy [term (a)], the energy of interaction between nonbonded groups [term (b)], the electrostatic energy [term (c)], and the energy of hydrogen bonding [term (d)]:

$$E_r = \sum_i \underbrace{0.5(E_i^{\ddagger})(1 - \cos N_i\boldsymbol{\theta}_i)}_{(a)} + \sum_{i,j} \underbrace{2\epsilon_{i,j}\left(\frac{d_{i,j}}{r^{12}} - \frac{b_{i,j}}{r^6}\right)}_{(b)}$$

$$+ B\underbrace{\sum_{j,k}(e_je_k/r)}_{(c)} + \underbrace{E_H}_{(d)} \qquad (4\text{-}1)$$

$E_i^{\ddagger}$ is the potential barrier for rotation about the bond i, N_i is the symmetry of the rotation (usually 2, 3, or 6), $\boldsymbol{\theta}_i$ is the torsion angle, r is the distance between

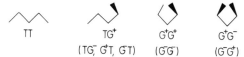

Figure 4-5. Conformational diads in pentane, $CH_3CH_2CH_2CH_2CH_3$.

atomic nuclei, and e_j and e_k are partial charges derived from the dipole moments of the bonds. The factor B contains the Coulomb energy and the apparent relative permittivities (dielectric constants). $\epsilon_{i,j}$, $d_{i,j}$, and $b_{i,j}$ are the parameters describing the potential energy of the contributions from non-bonded atoms. A 9–6 potential is often used instead of the Lennard-Jones 12–6 potential shown in Equation (4-1).

Although the various calculations can differ very strongly from each other according to the choice of bond distance, interaction energy, potentials, etc., the general conclusions from given sets of calculations agree very well with each other. For example, two different groups of workers found that the hydrogen bonding in poly(L-alanine), which occurs in the form of an α helix with $\dashv NH—CH(CH_3)—CO \vdash$ as the constitutional repeat unit, only contributes about 20% of the overall stability of the helix. This is in spite of the fact that the terms used by the different workers for the electrostatic contribution differ even in sign, one being positive, the other negative (Table 4-1).

Lines of equal energy are then plotted against the torsion angles θ_1 and θ_2 of two successive main chain bonds and a contour diagram is obtained (Figure 4-6). The conformation map obtained shows that two energy minima exist for it-poly(propylene), as has been confirmed for crystalline it-PP by X-ray crystallography. One of these minima corresponds to a left-handed, the other to a right-handed 3_1-helix.

Table 4-1. Individual Interaction Energy Contributions to the Stability of the Poly(L-Alanine) α-Helix

	E in J/mol	
Source of contribution	After Ooi, Scott, Van der Kooi, Scheraga	After Kosuge, Fujiwa, Isogai, Saitŏ
Rotation	2,050	2,430
Nonbonded atoms	−25,080	−29,940
Electrostatic interaction	−4,610	10,890
Hydrogen bonding	−7,290	−4,280
Total	−34,930	−20,890
Hydrogen-bonding contribution	20.8%	20.5%

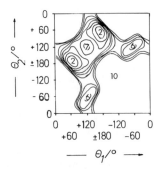
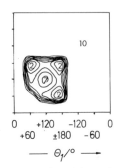

Figure 4-6. Contour diagram of the conformational energy as a function of the torsion angles, θ_1, and θ_2, of two successive main chain bonds for isotactic (left) and syndiotactic (right) poly(propylene). The numbers give the energy in kcal/mol bond (1 kcal = 4.184 kJ). it-PP has two energy minima at $\theta_2 = 180°$ and $\theta_1 = 60°$ (TG$^+$) and at $\theta_2 = 60°$ and $\theta_1 = 180°$ (G$^-$T) which correspond to a left- or a right-handed helix. st-PP possesses three energy minima: one for the TT conformation at $\theta_2 = 180°$ and $\theta_1 = 180°$, the other two are for two helix conformations (TG$^+$ and G$^-$T) (after G. Natta, P. Corradini and P. Ganis).

4.1.3. Constitutional Effects

The energetically preferred *trans* conformations of ethane are separated by only a low potential energy barrier of 11.3 kJ/mol. Thus, the conformers can relatively rapidly interconvert in the fluid phase. The low potential barrier can be overcome by thermal energy supplied by the collision of molecules. On the average, thermal energies of about $0.5RT$ per degree of freedom are transferred during such a collision. Because this is less than needed, the Maxwell–Boltzmann energy distribution dictates that a relatively small fraction of collisions will provide enough energy to cross the energy barrier. The majority of collisions will only deliver energy sufficient for oscillations of up to $\pm 20°$ about the potential minima. The majority of molecules thus remain in conformations associated with a minimum in the potential energy. They can be consequently treated as if they only exist in discrete rotational states. Fluctuations about the minima are not discounted; it is assumed that they compensate each other.

The conformational transitions are generally very fast. The rate constant of such a transition can, of course, be calculated from the potential energy by the following equation:

$$k_{conf} = (kT/h) \exp\left(-E^{\ddagger}/RT\right) \tag{4-2}$$

Conformers can only be preparatively separated when no significant conformational change occurs during the time required for the separation, and this occurs only when k_{conf} is sufficiently low because of high potential energies and/or low temperatures (Table 4-2). Consequently, conformers

Table 4-2. Calculated Equilibrium Constants, k_{conf}, and Times, $t_s \approx 0.05/k_{conf}$, for a conformational transition of 5% (3×10^{12} s $\approx$ 95,000 a)[a]

$\Delta E^{\ddagger}$, kJ/mol	k_{conf}/s^{-1}, for T =			t_s/s, for T =		
	100 K	300 K	500 K	100 K	300 K	500 K
10	1.25×10^7	1.13×10^{11}	9.40×10^{11}	10^{-9}	10^{-13}	10^{-14}
25	0.18	2.77×10^8	2.55×10^{10}	10^{-1}	10^{-10}	10^{-12}
50	1.59×10^{-14}	1.23×10^4	6.23×10^7	10^{12}	10^{-6}	10^{-9}
100	1.21×10^{-40}	2.42×10^{-5}	3.27×10^2	10^{38}	10^3	10^2

[a] a = year.

generally occur in rapid dynamic equilibrium. On the other hand, states with lifetimes of microseconds are observed in electron spin resonance experiments, which corresponds, in comparison to the usual values of $1/k_{conf}$, to a kind of instantaneous registration of the conformer population. Consequently, the conformers appear as definitive species in esr. Thus, it would be difficult to conceptually differentiate configurational from conformational isomers on this time scale.

As expected, the height of the potential energy barrier decreases with increasing bond length, but otherwise equal structure. This is seen with comparison of ethane with methyl silane and disilane (Table 4-3). The potential barrier increases with increasing steric hindrance, as can be seen in the series ethane–propane–isobutane–neopentane, methanol–dimethyl ether, and acetaldehyde–propylene–isobutylene.

The potential barrier also diminishes on going from the threefold CH_3 group to the onefold OH group (compare ethane with methanol). For this reason only, the potential barrier for CH_2/CO and CH_2/O bonds is considerably less than for the CH_2/CH_2 bond.

There are more energetically equivalent conformations for compounds of the type $+CH_2-CH_2+_n$ than there are for compounds of the type $+CH_2-CHR+_n$. Thus, ligands of symmetrical compounds can occur in certain conformations with greater probability than is the case with asymmetrically substituted compounds, i.e., the molecule is more flexible.

Various factors are thus responsible for a high molecular flexibility: (1) a large bond length between chain atoms to give a low potential barrier, (2) many competing positions for like substituents, and (3) a low potential difference between *gauche* and *trans* positions because of the *gauche* effect. All three effects are present in poly(dimethylsiloxane), $+Si(CH_3)_2-O+$, i.e., the Si—O bond length of 0.164 nm is relatively long, the main chain is rotationally symmetric, and there are polar oxygen chain atoms. The high molecular flexibility is mainly responsible for the low glass-transition temperature (see Chapter 10). Linear poly(dimethylsiloxanes) are therefore highly viscous liquids up to molar masses of millions.

Table 4-3. Potential Energy Barriers $\Delta E^{\ddagger}$ and Bond Lengths L for Rotation
about the Single Bonds Shown

Compound	$\Delta E^{\ddagger}$ in kJ/mol bond	L in nm
H_3Si—SiH_3	4.2	0.234
CH_3—SiH_3	7.1	0.193
CH_3—CH_3	12.3	0.154
CH_3—CH_2CH_3	14.9	0.154
CH_3—$CH(CH_3)_2$	16.3	0.154
CH_3—$C(CH_3)_3$	20.1	0.154
CCl_3—CCl_3	42	0.154
CH_3—NH_2	8.3	0.147
CH_3—SH	5.4	0.181
CH_3—OH	4.5	0.144
CH_3—OCH_3	11.3	0.143
CH_3—CHO	4.9	0.154
CH_3—CH=CH_2	8.4	0.154
CH_3—$C(CH_3)$=CH_2	10.0	0.154
—CH_2—CH_2COCH_2—	9.6	0.154
—CH_2—$COCH_2CH_2$—	3.4	0.154
—CH_2—$COOCH_2$—	2.1	0.154
—CH_2—$OOCCH_2$—	5.0	0.143
CH_3—OCH_3	11.3	0.143
—CH_2—$NHCH_2CH_2$—	13.8	0.147
—CH_2—SCH_2CH_2—	8.8	0.181

4.2. Conformation in the Crystal

4.2.1. Inter- and Intracatenary Forces

The macroconformation of chainlike macromolecules in crystalline polymers is principally determined by two factors, which are inter- and intracatenary forces. Calculation of potential barriers of isolated molecules, that is, in a "vacuum," is based exclusively on intracatenarily effective forces (see also Section 4.1.2). Microconformations calculated in this way correspond to an internal energy minimum. According to the equivalence principle, all structural units should adopt geometrically equivalent positions in relation to the crystallographic axes, whereby a monomeric unit, for example, may serve as a structural unit. Thus the regular sequence of microconformations should lead to a regular macroconformation.

The question is now whether intercatenary effects can lead to changes in the microconformation caused by intracatenary forces, and, if so, to what extent. Of course, intercatenary forces quite definitely affect the mutual packing of main chains, and so, give rise to density differences. But the maximum density differences generally correspond to energy differences of only about $1.2 \, kJ/mol$ monomeric unit. Such small differences do not exclude changes in the microconformation relative to that *in vacuo* due to inter-catenary forces; they do, however, ensure that such changes rarely occur. Thus, the macroconformation in a crystal generally corresponds to that *in vacuo*. According to the principle of smallest intracatenary conformation change, a chain in a crystal adopts the conformation with the lowest energy that the equivalence principle allows.

The chains try to pack as densely as possible. The minimum interchain distances produced thereby are determined by the van der Waals radii. So, with the aid of the equivalence principle, knowledge of these radii allows the macroconformation in the crystalline state to be estimated even though knowledge of the nature and effect of individually active forces may be absent.

The distance between H atoms on adjacent carbon atoms in the T conformation of polyethylene can be calculated as 0.25 nm from the bond length of 0.154 nm and valence angle of 109.6°. This distance is greater than the sum of the van der Waals radii of 0.24 nm for the two hydrogen atoms. Consequently, crystalline poly(ethylene) occurs in the T conformation.

With poly(tetrafluoroethylene), on the other hand, the distance between fluorine atoms bonded to adjacent carbon atoms in the T conformation is 0.25 nm, which is smaller than the sum, 0.28 nm, of the van der Waals radii. Thus, the main chain atoms deviate from the ideal T conformation by means of a small change of from 0° to 16° in the torsion angle (see also Figure 4-7). The deviation of carbon chains from the T conformation increases with size of the substituents. For example, with isotactic poly(propylene), having the mono-meric unit $+CH_2-CH(CH_3)+$, every second conformation in the lowest energy state is a G conformation, such that the whole chain adopts the . . . TGTGTG . . . conformation. This sequence of microconformations leads to a helix macroconformation.

4.2.2. Helix Types

Helix conformations occur relatively often in macromolecules. Helices are distinguished by a number, p_q, where p gives the number of con-formational repeating units per q complete turns of the helix. For example, it-poly(propylene) forms a 3_1 helix, that is, three conformational repeating units of the kind $+CH_2-CH(CH_3)+$, for example, are required to give the same spatial arrangement after one turn which differs only by translation from the original arrangement (Figure 4-8). Seven conformational repeating units in two turns are correspondingly required for a 7_2 helix.

Helices are characterized by their direction of turn. A right-handed helix observed along the axis rotates away from the observer in a clockwise

Conformation	Spatial representation		Monomeric units	Helix type	Torsion angle	Example
	Perpendicular to the chain	Along the chain				
T			—CH₂—CH₂— —CH₂— —CH₂—CHCl—	1₁ 2₁ 1₁	0/0 0/0 0/0	Poly(ethylene) Poly(methylene) st-Poly(vinyl chloride)
			—CF₂—CF₂—	13₁	16/16	Poly(tetrafluoroethylene)

TG			—CH₂—CH—
			       R
				3₁	0/120
					0/120
					0/120
						it-Poly(propylene) (R = CH₃)
						it-Poly(styrene) (R = C₆H₅)
						it-Poly(5-methyl-1-heptene)
						(R = CH₂—CH₂—CH—CH—CH₃
						                    |
						                    CH₃

			—CH₂—CH—
			       |
			      CH₂
			CH₃—CH
			      |
			      CH₃
				7₂	−13/110	it-Poly(4-methyl-1-pentene)

			—CH₂—CH— (m-CH₃-phenyl)	11₃	−16/104	it-Poly(m-methyl-styrene)

Spatial representation

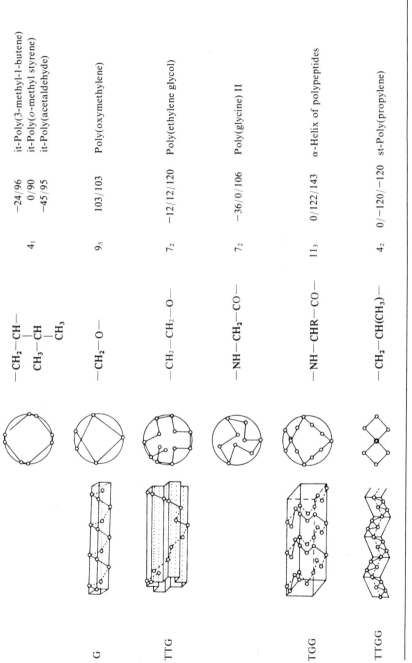

Figure 4-7. Important conformational forms of macromolecules (after S.-L. Mizushima and T. Shimanouchi).

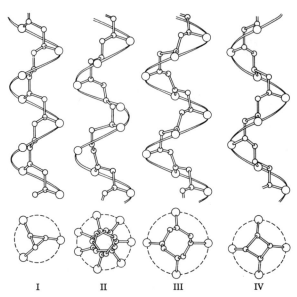

I   II   III   IV

Figure 4-8. Schematic representation of the different kinds of helices occurring in various isotactic polymers $+CH_2-CHR+_{\overline{n}}$. (I) 3_1; (II) 7_2; (III, IV) 4_1 (after G. Natta, P. Corradini, and I. W. Bassi).

direction. With a left-handed helix, the rotation is in the reverse direction. Thus, the helix $..TG^+TG^+TG^+..$ of it-PP is a left-handed helix. The left-handed α-helix of poly(α-amino acids), correspondingly, has the chain conformation $..G^+G^+(trans)G^+G^+(trans)....$ The symbol *(trans)* is often used to designate rigid torsion angles; in this case, about the amide double bond (see Figure 4-7). Right-handed helices are given the symbol P (plus), left-handed helices take the symbol M (minus).

In principle, polymers consisting of a single kind of chiral monomeric unit can produce left- and right-handed helices. But the two kinds of helix are diastereomeric to each other; that is, they are not energetically equal. For this reason, either left or right-handedness is a preferred form for such polymers. For example, polymers of chiral (S)-α-olefins and most poly(D-saccharides) form exclusively left-handed helices. On the other hand, deoxyribonucleic acids and almost all poly(L-α-amino acids) occur as right-handed helices. Polymers of the corresponding monomer antipodes form helices of opposite turn.

In contrast, polymers with prochiral monomeric units form helices which are enantiomeric to each other. In this case, left- and right-handed helices are energetically equivalent, and so, equally probable. Thus isotactic poly(pro-pylene) in the crystalline state forms equal amounts of left- and right-handed helices, with the conformations $(TG^+)_n$ and $(TG^-)_n$, respectively.

Macroconformations consisting of two or three helices intertwined with each other are also sometimes called super helices or super secondary structures. An example is deoxyribonucleic acid, which forms a double helix from two complementary chains, each in the form of a helix (see Section 29). With synthetic polymers, both it-poly(methyl methacrylate) and poly(*p*-hydroxybenzoic acid) appear to form double helices. Triple helices are, for example, formed by the protein, collagen (see Section 30).

4.2.3. Constitutional Effects

In isotactic polyvinyl compounds, $+CH_2-CHR+$, the bulky R substituents on every second chain atom force the chain to change from the all T conformation into the TG conformation. it-Poly(propylene) and it-poly-(styrene) exist in the crystalline state in the form of 3_1 helices with torsion angles of 0° and 120°. The valence angle is also significantly distorted by increasing substituent size: 110° for poly(ethylene), 114° for it-poly(propylene), and 116° for it-poly(styrene). On going from it-poly(5-methyl-1-heptene) to it-poly(4-methyl-1-pentene), the methyl groups come closer to the main chain and the greater steric effect forces the chain atoms to deviate from the *trans* and *gauche* positions (with torsion angles of 0° and +120°) and to adopt torsion angles of −13° and +110°. Poly(4-methyl-1-pentene) exhibits a 7_2 helix, whereas, with poly(3-methyl-1-butene), the methyl groups are in the immediate vicinity of the chain, and the helix formed is expanded further to a 4_1 helix.

Isotactic polymers with two chain atoms per monomeric unit thus tend to occur in more or less ideal TG conformations. In addition, the low energy difference for slight deviations from the ideal torsion angle can lead to various helix types. Rapid crystallization of it-poly(butene-1) produces a 4_1 helix, for example, which, as a high-energy form, changes into a 3_1 helix on annealing (see also Chapter 10).

The substituents of syndiotactic vinyl polymers in all-*trans*-conformations are more widely spaced from each other than in corresponding isotactic compounds. The T conformation is therefore generally the lowest energy conformation for syndiotactic polymers. 1,2-Poly(butadiene), poly(acrylonitrile), and poly(vinyl chloride) belong to this group. In a few cases a series of torsion angles of 0°, 0°, −120°, −120° is more advantageous, and therefore substances such as st-poly(propylene) generally take on a TTGG conformation, but can also crystallize in a T conformation since the energy differences are small.

Poly(vinyl alcohol), with the monomeric unit $+CH_2-CHOH+$, carries a hydroxyl group at every second chain atom. These OH groups can form intramolecular hydrogen bonds. Therefore, in contrast to it-poly(α-

olefins), it-poly(vinyl alcohol) does not form a helix, but an all-*trans* conformation. For the same reason, st-poly(vinyl alcohol) does not exist as a zig-zag chain but as a helix.

The influence of interactions between the electron clouds of the bonds between the main chain atoms is much reduced in heterochain polymers. There are three bonds to be considered in the CH_2 group, but only one in the O linkage. The potential barrier therefore falls to about one third of the value in carbon chains (see also Table 4-3). This means, for example, that molecules with oxygen atoms in the main chain are more flexible than comparable ones with carbon chains. Because of the decreased atomic bond length of 0.144 nm for the C—O bond, as opposed to 0.154 nm in the C—C bond, methyl substituents draw relatively closer together, causing the diameter of the helix to be increased. it-Poly(acetaldehyde) thus exists as a 4_1 helix, but it-poly(propylene) exists as a 3_1 helix. If the influence of the methyl substituents is not present, as in poly(oxymethylene), then the effects of bond orientation make themselves particularly noticeable. Poly(oxymethylene) thus exists as the G conformation, whereas poly(ethylene glycol) is TTG. Like poly(ethylene glycol, poly(glycin) II crystallizes into a 7_2 helix, but this is deformed because of hydrogen bonding. In it-poly(propylene oxide), the repulsion between the methyl groups is increased, and the bond orientation is decreased because of the methyl substituents: This polymer crystallizes in an all-*trans* conformation.

4.3. Conformation in the Melt and in Solution

4.3.1. Low-Molar Mass Compounds

The conformation of molecules is determined in the gaseous state completely, and in the crystalline state almost completely by intracatenary forces. Solvents interact with solutes, and, thus, can alter their conformations. The effect of this interaction, however, is weak when the solute is surrounded by like molecules. Thus, butane has practically the same conformation energy in the gaseous state, 3.35 kJ/mol, as in the liquid state, 3.22 kJ/mol.

The solvent can interact with the chlorine atoms in the case of 1,2-dichloroethane. 1-2-Dichloroethane preferentially adopts the *trans* conformation in most solvents; thus, the conformation energy, E_T-E_G, is negative (Figure 4-9). The conformation energy, however, becomes more positive with increasing solvent polarity, till, finally, the *gauche* conformation predominates for 1,2-dichloroethane in methanol.

The increase in *gauche* conformations in polar solvents caused by the *gauche* effect is also seen with *meso*-2,4-pentane diol (Table 4-4). The proportion of TT diads with this compound decreases with increasing solvent polarity, and the proportions of TG^+ and G^-T diads increase. It is noteworthy that the proportion of pure *gauche* diads is practically zero in all solvents.

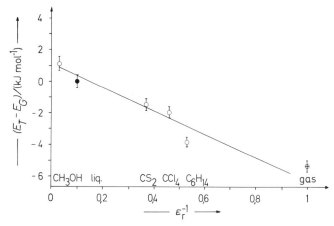

Figure 4-9. Conformation energy of 1,2-dichloroethane as a function of the relative permittivity (dielectric constant) in the liquid phase (●); in various solvents (○); and in the gas phase (⊕).

4.3.2. Macromolecular Compounds

During the melting of crystalline helical polymers, so much energy is given to the crystal lattice, generally, that the stabilizing effect of packing on the macroconformation is eliminated and new, or irregular, conformation sequences are produced. The stability loss of the individual helices is, however, entropically compensated, since the individual macromolecules can now form a great many macroconformers that coexist in rapid equilibria. Thus, only short helical sequences at very low concentrations survive the melting process. Consequently, they can no longer be spectroscopically observed.

An example of this is syndiotactic poly(propylene). The ir spectrum has a pronounced "crystalline" band at 868 cm^{-1} for the stablest crystalline state. According to theoretical considerations, this band, (see Figure 4-10), is practically entirely due to helical (TTGG) conformations. The band disappears on melting.

Table 4-4. Influence of Solvent on Conformation of meso-2,4-Pentanediol at 40° C

Solvent	Relative permittivity	Percentage of conformative diads			
		TT	TG$^+$ and G$^-$T	TG$^-$ and G$^+$T	G$^+$G$^+$ and G$^-$G$^-$
CCl$_4$	2.2	70	10	10	10
CH$_2$Cl$_2$	8.9	90	10	0	0
Pyridine	12.4	45	48	7	0
DMSO	46.7	30	60	10	0
D$_2$O	78.4	5	70	25	0

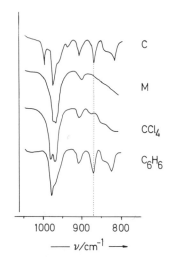

Figure 4-10. Detail from the infrared spectrum of a syndiotactic poly(propylene) at 37° (C) in the crystalline state, at 4% concentration in benzene (C_6H_6) and in carbon tetrachloride (CCl_4) solution, and at 170° C in the melt (M). The bands at 868 cm^{-1} measure the helical sequence (TTGG) concentration (after M. Peraldo and M. Cambini, as well as B. H. Stofer and H.-G. Elias).

In principle, the same effects are observed on dissolving crystalline helical polymers. In addition, the interaction of the solvent with the polymer may be observed, as well as changes in the population and sequence lengths of conformations caused by this interaction. Two limiting cases may be distinguished according to the strength of the interaction:

Strong interactions only occur between polar groups. They can, for example, lead to solvation of macromolecular groups. Alternatively, solvent molecules in the neighborhood of a macromolecular group can induce a *gauche* effect. In such cases, the conformational changes of the macro-molecule in going from the crystalline state to the dissolved state are determined by group-specific effects, i.e., in the last analysis, by enthalpic effects. Only a few of the bonds retain the previous crystal conformation since practically every bond can adopt a new conformation. Thus, the sequence lengths of conformational diads of polar macromolecules are very short in very polar solvents.

In contrast, slight or no group-specific interactions occur in the dissolution of apolar polymers in apolar solvents. Neither solvation nor induced *gauche* effects are driving forces for conformational changes. Thus, conformational changes must be mostly due to entropic effects. For energetic reasons, only a few conformations are transformed. Large sequences are retained in the original conformation (Figure 4-11), whereby, of course, rapid

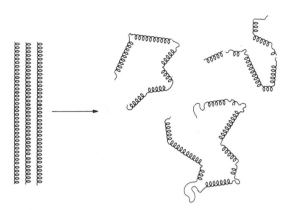

Figure 4-11. Dissolution of a macromolecule occurring in the crystalline state in the form of a helix. Only a few "kinks" are required to produce the macroconformation of a coil, with the helix microconformation being largely retained.

conformational changes of enantiomeric polymers occur, from left-handed to right-handed helices, and vice versa.

Longer conformational homosequences can be stabilized by ordered solvents. For example, benzene forms money-roll-shaped associates, whereas carbon tetrachloride does not have order. st-Poly(propylene) exhibits practically no helical sequences in CCl_4, but does in benzene. The intensity of their band at 868 cm^{-1} due to these sequences in benzene decreases with increasing temperature, until, finally, at about 57° C, all of the helical sequences have melted away (Figure 4-12).

Similar evidence of order in solution can be obtained for other polymers with other experimental methods. Proton magnetic resonance studies on poly(oxyethylene) in benzene solution give a new signal (and a weaker one in CCl_4 solution) for heptamers and higher species which is obviously ascribable to another conformation. Alkanes, $CH_3 \!\!+\!\! CH_2 \!\!\xrightarrow{}_n\!\! CH_3$, in 1-chloronaphthalin give only one methylene proton signal for $n < 14$ in PMR measurements. When $n > 15$, two signals are obtained. This separation into two signals is not observed in CCl_4 solutions or with deuterated alkanes. The signal separation is attributed to an intramolecular chain folding, but could also be due to an intermolecular association of chain segments.

The mean number of monomeric units occurring in helical sequences, N_h, can be estimated from the conformational energy, ΔE. ΔE is given as half the Gibbs free energy for the "reaction" between a left-handed and a right-handed conformational diad:

$$ll + dd \rightleftharpoons ld + dl \qquad (4\text{-}3)$$

and so we have

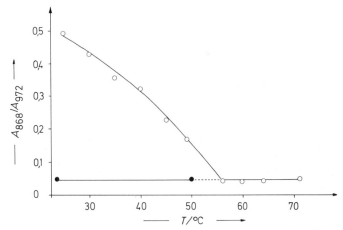

Figure 4-12. Variation with temperature of the ratio of the absorption at 868 cm^{-1} ("crystalline" bands) to that at 972 cm^{-1} (reference band) of a ($\sim 4 \times 10^{-3}$ g/cm^3) solution of st-poly(propylene) in benzene (O) and in carbon tetrachloride (●) solution (After B. Stofer and H.-G. Elias).

$$\Delta E = 0.5 \, \Delta G = -0.5RT \ln \frac{[ld][dl]}{[ll][dd]} = 0.5RT \ln \frac{g_{ll}g_{dd}}{g_{ld}g_{dl}} \qquad (4\text{-}4)$$

The mole fractions here can be replaced by the statistical weights. Each statistical weight is related to the conformational energy of a conformational diad E_{jk} by $E_{jk} = -RT \ln g_{jk}$. Thus from Equation (4-4) we obtain

$$\Delta E = 0.5(E_{ll} + E_{dd}) - 0.5(E_{ld} + E_{dl}) \qquad (4\text{-}5)$$

E_{ll} is the conformational energy of a left-handed monomer unit that follows another left-handed monomer unit; E_{dl} is the energy of a left-handed monomer unit that follows a right-handed monomer unit, etc. With chains containing achiral configurational monomeric units, $E_{ld} = E_{dl}$, but $E_{ld} \neq E_{dl}$ for chains with chiral main chain groups or ligands in the side chain. N_h is then given by

$$N_h = \frac{1 + \exp(-\Delta E/RT)}{\exp(-\Delta E/RT)} \qquad (4\text{-}6)$$

According to these calculations, for example, about 12 monomer units each occur in left-handed and right-handed helical segments in the case of poly(4-methyl-1-pentene), for example. On the other hand, an average of 31 monomeric units occur in a left-handed helix form with poly((S)-4-methyl-1-hexene) but only 2.2 occur as right-handed helical segments.

Since chiral monomeric units containing polymers form helices of preferential handedness, and the helices are themselves chiral, the helicity of such polymers can be studied via their optical activity, or via their optical rotatory dispersion or circular dichroism. Such methods, of course, cannot be used for polymers consisting of prochiral monomeric units. In this case, however, the conformational equilibrium can be "frozen" by chemical reaction. Each configurational diad corresponds, or course, to a definite kind of conformation. The conformations available, then, lead to the corresponding populations of configurational diads, triads, etc., because of the stereochemical equilibrium. Such equilibria can be studied, for example, via the reaction of bases on poly(acrylic compounds), [see Equation (23-10)]. The distributions of configurational diads and triads are already almost the same for constitutional trimers as they are for constitutional dimers (Table 4-5); consequently, only the nearest neighbor, and not more distant groups, have to be considered with conformations. Thus, practically ideal atactic polymers are produced by poly(acrylic compounds) with small substituents. With more bulky substituents, however, higher quantities of syndiotactic diads are formed at the expense of isotactic diads, whereas the heterotactic diad fraction remains practically constant.

Table 4-5. Stereochemical Equilibria for Compounds of the Type $CH_3(CHR-CH_2)_N H$

R	N	$T/°C$	Solvent	Catalyst	x_i	x_{ii}	x_{ss}	$x_{\overline{is}}$
CH_3	4	−75	Pentane	HSO_3Cl	0.46			
CH_3	4	270	Octane	Pd/C	0.49			
CH_3	5	270	Octane	Pd/C	0.49	0.24	0.26	0.50
COOH	3	180	Water	HCl	0.48	0.24	0.27	0.49
$COONH_4$	3	180	Water	NH_4OH	0.48	0.24	0.27	0.49
COONa	3	180	Water	NaOH	0.45	0.20	0.30	0.50
$COOCH_3$	2	25	Methanol	CH_3ONa	0.46			
$COOCH_3$	3	25	Methanol	CH_3ONa	0.45	0.20	0.29	0.51
$COOC_2H_5$	3	25	Ethanol	C_2H_5ONa	0.43	0.18	0.31	0.51
$COO(i-C_3H_7)$	3	25	i-Propanol	$i-C_3H_7ONa$	0.36	0.11	0.40	0.49
C_6H_5	2	25	DMSO	t-BuOK	0.49			
C_6H_5	3	70	DMSO	t-BuOK	0.47	0.22	0.28	0.50
CN	2	25	Methanol	NaOH	0.40			
Cl	2	25	CS_2	LiCl	0.29			
Cl	2	70	DMSO	LiCl	0.36			
Cl	3	70	DMSO	LiCl	0.34	0.11	0.43	0.46

4.4. The Shape of Macromolecules

4.4.1. Overview

The external shape of macromolecules is determined by the number and distribution of the conformations as well as by the interaction between chain segments. Chain segments, or "segments," are defined as portions of the chain of any desired length.

Molecules of perfect helical conformation are rod or cylinder shaped. They are characterized by their external dimensions such as length and diameter.

Two types can be distinguished according to interaction between helical and nonhelical segments in macromolecules containing interrupted helix conformations. The segments pack into more or less dense arrangements determined by the constitution and configuration if the attractive forces are strong. "Compact" bodies with the shapes of spheres or ellipsoid are produced. They are also characterized by their external dimensions, in this case by diameter, axis length, and axis ratio.

If the segments interact predominantly repulsively, then the segments adopt the shape of a coil of more or less randomly intertwined segments, even when there are only a few "kinks" between helical segments (Figure 4-11). Such a coil is characterized by the chain end-to-end distance, L, or by the radius of gyration, R (Figure 4-13). These parameters are often compared with the contour length, L_{cont}, and the maximum possible chain length, L_{max}. L_{cont} is the length of a fully stretched chain of N main chain bonds of length b:

$$L_{cont} = Nb \qquad (4\text{-}7)$$

It is given by measuring the contour length of the chain, and, so, it is independent of the angle between main chain atoms. The physical maximum possible chain length, on the other hand, is also dependent on τ, the

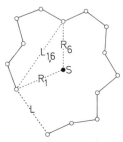

Figure 4-13. Schematic representation of a molecular coil consisting of 14 units of mass. The center of gravity of the molecule is at S. L is the chain end-to-end distance. The radii of gyration about S of the individual point masses are R_i.

main-chain bond angle. It can be calculated from simple geometrical considerations as

$$L_{max} = Nb \sin (0.5 \ \tau) \tag{4-8}$$

The shape and size of macromolecules together with the segment distribution within these forms determine the excluded volume of the polymer. Compact molecular shapes such as helices, ellipsoids, and spheres have only an *external* (intermolecular) excluded volume: the space occupied by a given volume in space cannot be occupied by others, and so is an excluded volume for other molecules. Coils, on the other hand, with their loose internal structure, also have, additionally, an internal (intramolecular) excluded volume, since the space occupied by one segment is not available to another segment of the same molecule.

4.4.2. Compact Molecules

A perfect helical main chain conformation always leads to a rodlike or cylindrical external shape. But each monomeric unit in such a rod contributes a certain flexibility. So, the flexibility of the rod, as a whole, must increase with increasing degree of polymerization, even when the flexibility per monomeric unit remains constant. A macroscopic example of this would be the flexibility of steel wires of equal diameter but different lengths. Thus, even a perfect helix will adopt coil shape if the molecular mass is very high. Because of this, helically occurring macromolecules, and other "stiff" macromolecules, can often be well represented by what is known as the wormlike screw model for macromolecular chains: at low molecular masses, the chains behave like a stiff rod, but for high molecular masses, the behavior is more coil-like. Examples are nucleic acids, many poly(α-amino acids), and highly tactic poly(α-olefins).

Ellipsoidal and spherical shape are especially prevalent among proteins, which are naturally occurring copolymers of α-amino acids. Certain sequences of these α-amino acids are helical, whereas others are not (Figure 4-14). The interaction of these ordered and disordered conformational sequences with each other and with the solvent thus determine the internal structure and external shape. In water, for example, hydrophilic groups will tend to occupy positions on the surface of the protein, and hydrophobic groups will tend towards the interior. With myoglobin, for example, only two amino acid residues with hydrophilic substituents are to be encountered in the protein interior. Such a protein molecule, then, will appear as a quite compact sphere or as an ellipsoid.

Many protein molecules aggregate to form associates consisting of several molecules. Such "quaternary" structures are often so stable that they do not noticeably dissociate into their constituent molecules over the experi-

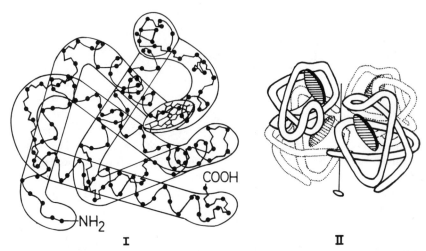

Figure 4-14. Tertiary structure of the protein myoglobin (I), and quaternary structure of the protein hemoglobin (II). The dots indicate amino acid residues. Hemoglobin consists of four myoglobin-type subunits. The heme planes are shown as cross-hatched areas (after M. F. Perutz). Myoglobin is shown on a larger scale.

mentally accessible concentration range. Thus, they appear as molecules, and not as associates. They are also spherical or ellipsoid in external appearance. An example of this is the protein hemoglobin which consists of four "subunits" (molecules) of the myoglobin type (Figure 4-14).

Quaternary structures can, however, often be forced to dissociate through the use of suitable solvents, as, for example, often occurs with proteins in aqueous urea solution. Further, the helical segments themselves may also be dissolved to give coil-like shape when other solvents, such as dichloroacetic acid, are used. In addition, this "denaturing" can also be achieved through careful heating of some aqueous protein solutions; consequently, this process is also known as thermal denaturing. If the protein solutions are more strongly heated, or if chemical reagents, radiation, pressure, or shear stress is applied, the coil-like molecules may finally associate into larger aggregates. This "thermal aggregation" can be seen in the increasing turbidity of the solution and change in the biological activity. In the industrial sector, thermal denaturing and thermal aggregation are not often differentiated, and both process are known simply by the joint term, denaturing.

4.4.3. Coiled Molecules

The coil shape of macromolecules of sufficiently large diameter of the chain can be established by electron microscopy. Since the molecules lie on a

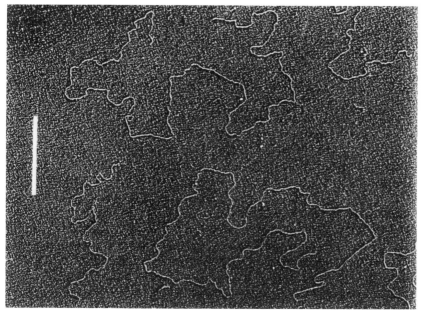

Figure 4-15. An electron microscope picture of the double chain of deoxyribonucleic acid (after D. Lang, H. Bujard, B. Wolff, and D. Russell).

substrate, a projection in two dimensions of the three-dimensional form in solution is always obtained. Deoxyribonucleic acid is seen as a two-dimensional coil, even though the local conformation is that of a double helix (Figure 4-15).

Such pictures are instantaneous representations. The conformations interchange rapidly, producing rapidly changing macroconformations. Thus, even in the case of monodisperse polymers, the experimentally observed dimensions of coils represent means of many macroconformations. With polymers possessing size distributions, the averaging also occurs over the various molecular chain lengths. Here, the number of possible macroconformations is astronomically large. For example, a linear macromolecular chain with N main-chain bonds and three energetically equally probable conformations about each bond can, of course, adopt 3^N different macroconformations. If the degree of polymerization is 1001, then $3^{1000} \approx 10^{477}$ macroconformations are possible. The best electronic data processing machines can only handle $3^{100} \approx 10^{48}$ conformations at present. Happily, many of the properties of chain molecules reach asymptotic values by about 100 chain units.

The shape of coiled molecules is often incorrectly taken to be spherical. In fact, the innumerable macroconformations of a coiled molecule never adopt a simple geometrical form, not even instantaneously. A mean shape may be given to each macroconformation, and this mean should derive from the other possible macroconformations. If this averaging process is not followed correctly, the mean will represent a more symmetrical shape than is actually present.

The instantaneous shape of a macromolecule cannot be experimentally determined from any known method. It may be calculated, however, by the following procedure:

The center of gravity of the molecule is placed at the origin of a Cartesian coordinate system. The molecule is oriented in this coordinate system so that the principal axes of inertia are along the coordinate axes. The vector radius $\mathbf{R}_i$ of each mass point of the chain molecule (see Figure 4-13) can be resolved into the three orthogonal components $(\mathbf{R}_i)_1$, $(\mathbf{R}_i)_2$, and $(\mathbf{R}_i)_3$, where

$$(\mathbf{R}_i)_1^2 + (\mathbf{R}_i)_2^2 + (\mathbf{R}_i)_3^2 = \mathbf{R}_i^2 \qquad (4\text{-}9)$$

Likewise, the radius of gyration can also be resolved into three components, namely, $R_{G,1}^2$, $R_{G,2}^2$, and $R_{G,3}^2$. Since these three components stand in special relationship to the three principal axes of inertia of the molecule, they are often called the main components of the radius of gyration of the chain.

For a fully spherically symmetric coil molecule, we have

$$R_{G,1}^2 = R_{G,2}^2 = R_{G,3}^2 = \tfrac{1}{3}R_G^2 \qquad (4\text{-}10)$$

Calculations have shown, however, that the squares of the principal components are not equal in magnitude, but have about the relationship 11.8:2.7:1 to each other. The instanteous shape of a coil molecule is not spherical; it is more in the form of a kidney. However, increased branching of the main chain will cause the molecule to become more symmetrical.

4.4.4. Excluded Volume of Compact Molecules

The excluded volume of rigid macromolecules is calculated relatively easily, since the excluded volume between molecules need only be considered. The excluded volume within a rigid molecule is, by definition, zero.

The volume of an unsolvated sphere is given by

$$V_{\text{sphere}} = (4\pi/3)\, r_{\text{sphere}}^3 = M_{\text{sphere}} v_{\text{sphere}}/N_L \qquad (4\text{-}11)$$

where r is the radius, M the molar mass, and v the specific volume of the sphere. The closest distance that one sphere can approach to another is $2r_{\text{sphere}}$. The excluded volume u_{sphere} is therefore

$$u_{\text{sphere}} = (4\pi/3)(2r_{\text{sphere}})^3 = 8 M_{\text{sphere}} v_{\text{sphere}}/N_L \qquad (4\text{-}12)$$

and is therefore eight times the volume of the sphere.

In calculating the excluded volume of rods, these are assumed to be

Table 4-6. *Excluded Volume u and Mean Square Radius of Gyration* $\langle R_G^2 \rangle$ *as a Function of Characteristic Dimensions of Various Particle Shapes* (V = *volume*)

Particle	u	$\langle R_G^2 \rangle$
Infinitely thin spherical shell of radius r	$8V$	r^2
Spherical shell of interior radius $r_i = C r_e$ and exterior radius r_e	$8V$	$\dfrac{3}{5}\left(C^2 + \dfrac{C+1}{C^2+C+1}\right)r_e^2$
Sphere of radius r	$8V$	$\frac{3}{5}r^2$
Very thin disk of radius r and thickness h	$\pi(r/h)V$	$0.5r^2$
Rotational ellipsoid	$(3/8)\pi(l/r)V$	$\frac{1}{5}(l^2 + 2r^2)$
a. Prolate with length l and radius $r(l \gg r)$		
b. Oblate with thickness h and radius $r(r \gg h)$	$(3/2)\pi(r/h)V$	$\frac{1}{5}(r^2 + 2h^2)$
Rod of length l and diameter $2r$	$(l/r)V$	$\frac{1}{12}l^2 + r^2$
Coil in the theta state with the chain end-to-end distance $\langle L^2 \rangle_0^{0.5}$	a	$\frac{1}{6}\langle L^2 \rangle_0$
Coil with the hydrodynamically equivalent radius r_h	a	$[8/(3\pi^{0.5})]^2 \langle r_h^2 \rangle$
Coil with the relationship $\langle L^2 \rangle = $ constant $\times (M^{1+\epsilon})$	a	$\langle\frac{1}{6}\rangle L^2 \ (1 + \frac{5}{6}\epsilon + \frac{1}{6}\epsilon^2)^{-1}$

^{a}See Section 4.5.5.

cylindrical with the volume

$$V_{\text{rod}} = \pi r_{\text{rod}}^2 l_{\text{rod}} \qquad (4\text{-}13)$$

with the length l_{rod} and the radius r_{rod}. The problem here is the calculation of the mutual orientation of the rigid rods in space, since not all orientations are allowed with rods separated by a distance of less than l_{rod}. The results of the calculation lead to a substitution of the factor 8 in the excluded volume of a sphere by the factor $l_{\text{rod}}/r_{\text{rod}}$:

$$u_{\text{rod}} = \frac{l_{\text{rod}}}{r_{\text{rod}}} \frac{M_{\text{rod}} v_{\text{rod}}}{N_L} \qquad (4\text{-}14)$$

The excluded volumes of other rigid particles can be analogously calculated. Excluded volumes, radii of gyration, and characteristic ratios of various rigid particles are given in Table 4-6.

4.4.5. Excluded Volume of Coiled Molecules

Real coiled molecules have an external and internal excluded volume. The external excluded volume results from the volume excluded to other molecules; it is intermolecular and its influence disappears at infinite dilution.

The external excluded volume, however, retains a finite value even at infinite dilution.

On the other hand, the internal excluded volume results from the finite thickness of the macromolecular chain, and its influence is retained, even at infinite dilution. The internal excluded volume can be formally separated into two components. Repulsive forces lead to a positive excluded volume. Attractive forces, in contrast, tend to reduce the volume occupied by two contacting parts of the chain from the value obtained by the summation of their individual volumes; the resulting excluded volume is negative.

Positive and negative excluded volumes can compensate each other in special cases. The coil then behaves as if it consisted of an infinitely thin chain. Thus, it adopts the unperturbed state, with unperturbed dimensions; it is an "ideal" coil with a Gaussian distribution of the chain end-to-end distances (see Section A4-4).

The interaction forces responsible for the excluded volume are called "long-range" forces, since they result from the interaction between units of the chain which are separated from each other by many other units of the chain. Restricted rotation is correspondingly caused by "short-range" forces interacting between nonbonded atoms or groups. Consequently, the terms *long range* and *short range* are not used with respect to the actual forces, themselves, or their range of influence, but are used with respect to the distance along the chain between the interacting groups (Figure 4-16).

Because of the internal excluded volume produced, long-range forces cause either an expansion or a contraction of the molecular coil. This change in dimensions can be formally described by an expansion factor, α_R, with respect to the radii of gyration, R_G:

$$\langle R_G^2 \rangle = \alpha_R^2 \langle R_G^2 \rangle_0 \tag{4-15}$$

The expansion factor is unity for an unperturbed coil. The larger α_R, the greater is the expansion and the "better" the solvent is for the polymer.

In Equation (4-15), the expansion factor gives directly the average linear expansion of the coil. The expansion factor calculated in this way is, of course,

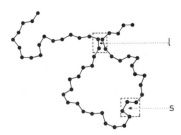

Figure 4-16. Short-range (s) and long-range (l) forces in chainlike macromolecules.

fictitious, since the coil is not spherical in shape, and will probably have various expansion factors for various directions in space. The existence of unequal expansion factors over all directions in space means that the distribution of molecular segments of real coils is no longer the same as for ideal coils. For this reason, the expansion factor defined with respect to the radius of gyration is not the same as that analogously defined with respect to the chain end-to-end distance.

A potential function $\psi(\mathbf{r})$ between two segments $\mathbf{r}$ apart must be assumed before the exluded volume of a coiled molecule can be calculated. The segments in question can belong to the same or to different molecules. Since every molecule possesses many segments, the expressions involved are very complex, and the derivations can only be sketched here.

The excluded volume of a segment u_{seg} is given by what is called the cluster integral, whereby $\psi(\mathbf{r})$ should be very much smaller than $\langle R_G^2 \rangle^{1/2}$:

$$u_{\text{seg}} = 4\pi \int_0^\infty \left[1 - \exp\left(-\frac{\psi(\mathbf{r})}{kT} \right) \right] r^2 \, d\mathbf{r} \qquad (4\text{-}16)$$

All relevant theories agree that u_{seg} can be directly related to the expansion factor α_R. To simplify the mathematics, a parameter z is defined:

$$z \equiv (4\pi)^{-3/2} \frac{u_{\text{seg}}}{M_{\text{seg}}^2} \frac{M^2}{\langle R_G^2 \rangle_0^{3/2}} \qquad (4\text{-}17)$$

The term $u_{\text{seg}}/M_{\text{seg}}^2$ is a constant, independent of the molar mass M but still dependent on the constitution and configuration, since it describes the excluded volume caused by a pair of segments. The parameter z is not directly experimentally accessible; it can only be determined from experimental data with the aid of theories.

When the following assumptions are made, an expression for the expansion factor can be derived from the cluster integral with the aid of z:

1. The probability distribution of bond vectors follows a Gaussian function.
2. The potential for interaction between segments is additive.
3. The pair potential is given by the expression

$$\exp\left[-\frac{\psi(r)}{kT} \right] = 1 - u_{\text{seg}}\delta(r) \approx \exp\left[-u_{\text{seg}}\delta(r) \right]$$

where r is the vector distance between two segments and $\delta(r)$ is the three-dimensional δ function.

Using these assumptions, we obtain for the relationship between the expansion factor α_R (based on the radius of gyration) and z

$$\alpha_R^2 = 1 + (134/105)z - 2.082z^2 + \cdots \qquad (4\text{-}18)$$

and correspondingly, for the expansion factor α_L (based on the chain end-to-end distance):

$$\alpha_L^2 = 1 + (4/3)z - 2.075z^2 + 6.459z^3 - \cdots \qquad (4\text{-}19)$$

These series expansions are exact for coils with pairwise additivity of interactions. But the series only converge very slowly and so are only applicable for $z < 0.1$ (for α_R) or $z < 0.15$ (for α_L). The term α_R^2 should in any case only depend on z, that is, Equation (4-18) should hold universally for all polymer–solvent–temperature systems.

A closed expression for the function $\alpha_R = f(z)$ has been sought without success and it is questionable if such a generally valid expression for $\alpha_R = f(z)$ exists. An expression often used previously is

$$\alpha_R^3 = 1 + 2z \qquad (4\text{-}20)$$

This expression indeed shows the initial trend of the function correctly for $\alpha_R \approx 1$, but leads to strong deviations from experimental data for large expansion factors (Figure 4-17). The experimental data often fit the semiempirical Yamakawa–Tanaka equations quite well:

$$\alpha_R^2 = 0.541 + 0.459(1 + 6.04z)^{0.46} \qquad (4\text{-}21a)$$

$$\alpha_L^2 = 0.572 + 0.428(1 + 6.23z)^{0.50} \qquad (4\text{-}21b)$$

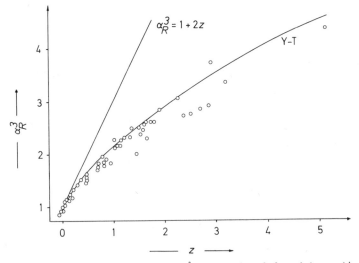

Figure 4-17. Variation of the expansion factor, α_R^3, as a function of z for poly(styrene) in various solvents. Y–T = Yamakawa–Tanaka equation.

4.5. Coiled Molecule Statistics

4.5.1. Unperturbed Coils

Coiled molecules are often characterized by statistical parameters such as the chain end-to-end distance L and the radius of gyration R. There are often simple relationships between L and R as well as between these and the bond length b, the valence angle τ, and the torsion angle θ.

In the simplest model, infinitely thin segments with valence angles of any size desired are assumed. The required relationship between L and the other molecular parameters can be obtained by a vector calculation (see Appendix to Chapter 4). The essentials can be obtained by a simpler and clearer derivation:

Two segments of length b form an angle τ:

The distance L_{00} between the ends of the two segments is given by the cosine rule

$$L_{00}^2 = 2b^2 - 2b^2 \cos \tau \qquad (4\text{-}22)$$

When sufficient freely chosen angles τ are present then L_{00}^2 is replaced by the mean square of all chain end-to-end distances, $\langle L^2 \rangle_{00}$, and $\cos \tau$ is replaced by the mean, $\langle \cos \tau \rangle$. Since, in this freely jointed chain model, all directions are equally probable, then $\langle \cos \tau \rangle = 0$ and Equation (4-22) becomes

$$\langle L^2 \rangle_{00} = 2b^2 \qquad (4\text{-}23)$$

On transferring from two segments to N segments, the following is correspondingly obtained:

$$\langle L^2 \rangle_{00} = Nb^2 \qquad (4\text{-}24)$$

The result is characteristic for all processes based on a random flight model. Such processes also include diffusion. The random flight model is distinguished by the fact that each step (in the above case, bond direction) is independent of the previous one.

However, definite valence angles occur in real macromolecular coils. The relationship between the mean end-to-end distance L_{0f} of this valence angle chain (with implicit assumption of free rotation about single bonds) and the number N of bonds of length b in the chain is, for an infinitely large number N of constant angles τ,

$$\langle L^2 \rangle_{0f} = Nb^2 (1 - \cos \tau)(1 + \cos \tau)^{-1} \qquad (4\text{-}25)$$

Going from the freely jointed chain to one with fixed valence angles produces a coil expansion when the valence angle exceeds 90°.

The constant valence angle chain with free rotation is also unreal, since it overlooks the existence of conformers. Each conformation can be assigned a torsion angle θ (see Figure 4-3). By an analogous mathematical procedure to that used to determine the influence of the valence angle on the coil dimensions, the following is obtained for symmetrically constructed chains [i.e., $+CH_2-CR_2+_{\overline{n}}$] of infinite molecular weight and for nonzero values of θ:

$$\langle L^2 \rangle_0 = Nb^2 \left(\frac{1 - \cos \tau}{1 + \cos \tau} \right) \left(\frac{1 + \cos \theta}{1 - \cos \theta} \right) = Nb^2 \left(\frac{1 - \cos \tau}{1 + \cos \tau} \right) \sigma_{\text{sym}}^2$$

$$(4\text{-}26)$$

Since various microconformations are possible, a mean must be taken of all conformational influences. The term in Equation (4-26) containing the torsion angle is often replaced by a new quantity, σ_{sym}^2, and is inserted as a squared term so that σ, as a measure for a change of dimensions, becomes comparable to the lengths L and b.

For a chain with completely free rotation, $\langle \cos \theta \rangle = 0$ and Equation (4-26) becomes Equation (4-25). An all-*trans* chain is a rigid chain with $\theta = 0$, which is, however, physically absurd with regard to Equation (4-26). Thus, σ is a measure of the hindrance to rotation; it is often called the steric hindrance parameter. With tactic polymers, the relationship between θ and σ is more complicated. One can, however, always use an equation analogous to Equation (4-26):

$$\langle L^2 \rangle_0 = Nb^2 (1 - \cos \tau)(1 + \cos \tau)^{-1} \sigma^2 \qquad (4\text{-}27)$$

The end-to-end chain distance is a readily conceivable but not a directly measurable quantity. Also, it loses all physical significance with branched chains, which possess more than two chain ends. A related quantity is the radius of gyration. It is directly measurable as the mean square radius of gyration, which is defined as the second moment of the mass distribution around the center of mass. The root mean square (rms) radius of gyration, often simply called radius of gyration, is thus

$$\langle R_G^2 \rangle^{1/2} = \left(\frac{\Sigma_i m_i R_i^2}{\Sigma_i m_i} \right)^{1/2} \qquad (4\text{-}28)$$

A definite relationship exists between the end-to-end chain distance and the radius of gyration for linear (nonbranched) chains (with or without fixed valence angles or free rotation). The relationship is derived in the appendix of this chapter for the segment chain. It can be seen from Equations (4-24), (4-25), and (4-26) that on transferring from the segment-chain model to either of the valence-angle-chain models, the end-to-end chain distance is increased for $\tau > 90°$. The radius of gyration likewise increases. Mathematical analysis shows that the same relationship between the chain end-to-end distance and the radius of gyration exists for all three chain models in the limiting case of infinitely high molar mass:

$$\langle L^2 \rangle_0 = 6 \langle R_g^2 \rangle_0 \qquad\qquad (4\text{-}29)$$

4.5.2. Steric Hindrance Parameter and Characteristic Ratio

The steric hindrance parameter σ measures the hindrance to rotation about main chain bonds, and, so, is a measure of the thermodynamic flexibility of the coiled molecule. It can be calculated from the radius of gyration of unperturbed coils via Equations (4-29) and (4-27) if the bond length, valence angle, and number of main chain bonds is known.

The steric hindrance parameter σ is a constant only in the case of apolar polymers in apolar solvents. A distinct dependence of the hindrance parameter on the type of solvent can be observed, however, for polar polymers and/or polar solvent combinations (see Table 4-7). Such effects are to be expected because of changes brought about in the *trans/gauche* ratios of conformers in the chain.

The hindrance parameter increases with the size of the substituents (i.e., with increasing molar mass M_u of the monomeric unit) for polymer–solvent

Table 4.7. Hindrance Parameter of Various "Atactic" Polymers

Polymer	Solvent	Temperature in °C	σ
Poly(ethylene)	Tetralin	100	1.63
Poly(propylene)	Cyclohexanone	92	1.8
Poly(isobutylene)	Benzene	24	1.93
Poly(styrene)	Cyclohexanone	34	2.3
Poly(1-vinyl naphthalene)	Decalin/toluene	25	3.2
Poly(methyl methacrylate)	Benzene	30	2.10
	Toluene	30	2.12
	Benzene/cyclohexane	25	2.14
	Acetone	25	1.86
	Butanone	25	1.89
	Butyl chloride	25	1.87
Cellulose	Copper ethylene diamine	25	2.0
Hydroxyethyl cellulose	Methanol	25	1.9
Poly(potassium vinyl sulphonate)	1 m KCl	45	2.81
Poly(sodium vinyl sulphonate)	1 m NaCl	45	2.97
Poly(acrylic acid)	Dioxan	30	1.85
Poly(sodium acrylate)	1.5 m NaBr/H_2O	15	2.38

pairs having approximately the same interaction. Table 4-7 shows this for the series of polyethylene–poly(propylene)–poly(styrene)–poly(1-vinyl naphthalin).

Cellulose and its derivatives have σ values of about 2, i.e., thermodynamically they are about as flexible as poly(isobutylene). Thus, cellulose chains are not extraordinarily stiff, although they are often assumed to be so on the basis of their high exponents in the intrinsic viscosity–molar mass relationship (see Section 9.9.7). These high exponents are interpreted as arising from the particular (high) draining properties of the cellulose molecule.

In concentrated salt solution, polyelectrolytes exhibit more or less unperturbed dimensions, which, however, still depend to some extent on the nature of the gegenion because of the various binding strengths of different gegenions. The σ values so obtained lie within the usual range. Thus, the large dimensions of polyelectrolytes in dilute salt solutions must result from long-range forces, and not from short-range forces.

In calculating the steric hindrance parameter σ from the square of the unperturbed chain end-to-end distance $\langle L^2 \rangle_0$, it is assumed that the number of bonds N, the bond length b, and the valence angle τ are constants. These assumptions are in order for the number of chain members and the bond length, since the bond energies of main chain bonds are about 40–400 kJ/mol bond. The assumption of a constant valence angle is critical, however. According to spectroscopic measurements and heat of combustion studies on ring molecules, deformation of the C—C—C— bond angle by 5.6° requires only 2 kJ/mol and by 10°, only about 7 kJ/mol bond. Since conformation energies lie within the same range, the assumption of a constant bond angle is not unproblematic.

Consequently, the valence angle component is often not separated from the hindrance parameter in calculations. Instead, a characteristic ratio C_N is defined as a measure of the expansion of a coil in the unperturbed state:

$$C_N \equiv \frac{\langle L^2 \rangle_0}{Nb^2} = (1 - \cos \tau)(1 + \cos \tau)^{-1} \sigma^2 \qquad (4\text{-}30)$$

C_N increases asymptotically to a constant value with increasing number N of bonds (Figure 4-18).

4.5.3. Statistical Chain Element

According to Equation (4-27), the chain end-to-end distance depends on parameters independent of the constitution and configuration, such as the number of bonds, as well as on parameters which are dependent, like bond length, valence angle, and steric hindrance parameter. A certain stiffness of the chain can be caused by increased bond length and valence angle, as well as

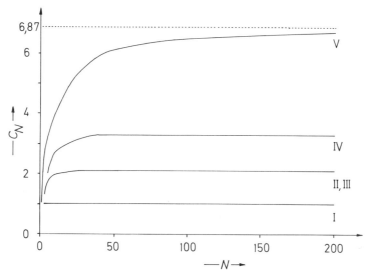

Figure 4-18. Variation of the characteristic ratio C_N with the number N of bonds in the chain molecule. (I) Segment chain; (II) valence angle chain with free rotation; (III) valence angle chain with hindered rotation and three rotational isomers of equal energy; (IV) valence angle chain with hindered rotation and a conformational energy of $E_G - E_T = 2.09$ kJ/mol; and (V) valence angle chain with hindered rotation and neighboring group influence and $\Delta E_{G^\mp} - \Delta E_{G^\pm} = 8.30$ kJ/mol. Valence angle $= 112°$ (after P. J. Flory).

a larger steric hindrance parameter. Formally, all these factors can be included in one unit of several links, thus reducing the number of independent units requiring consideration. One can also write, therefore, in place of Equation (4-27),

$$\langle L^2 \rangle_0 = N_s L_s^2 \qquad (4\text{-}31)$$

where L_s is taken to be the statistical chain element, and there are N_s of these in the chain. The longer L_s, the stiffer will be the chain. L_s can therefore be used as a measure of flexibility in the same way as σ, but it has less physical significance than σ. However, since the calculation of σ is not entirely straightforward and that of L_s is free of these difficulties, L_s and σ can be considered to be of equal worth. Equation (4-31) corresponds to Equation (4-24) for segment chains of unspecified segment length.

In this model (see Section 4.4.1) the contour length is, of necessity, given by the product of the length L_s and the number of statistical chain elements N_s:

$$L_{\text{cont}} = N_s L_s = Nb \qquad (4\text{-}32)$$

so that Equation (4-31) can also be written

$$\langle L^2 \rangle_0 = L_{\text{cont}} L_s = N_s L_s^2 \qquad (4\text{-}33)$$

The length of the statistical chain element can thus be calculated from the contour length L_{cont} and the experimentally determined chain end-to-end distance.

4.5.4. Chains with Persistence

The angle between neighboring segments is not fixed in the segment model. If the bond length is taken as the segment, then the valence angle can still be chosen freely. The valence angle is, however, fixed in a real chain. Thus the segments following the first segment cannot occupy any desired position in space. The chain thus possesses persistence.

The rigidity resulting from this persistence can be described by a persistence length L_{pers}, which is the average projection of the end-to-end distance of an infinite chain in the direction of the first segment. It is defined as

$$L_{pers} \equiv \frac{b}{1 + \cos \tau} \tag{4-34}$$

The chain end-to-end distance of a finite chain with free rotation is, according to Equation (A4-28) in the appendix to this chapter,

$$\langle L^2 \rangle_{0f} = Nb \left\{ b \left(\frac{1 - \cos \tau}{1 + \cos \tau} \right) + \frac{2b \cos \tau}{N} \left[\frac{1 - (-\cos \tau)^N}{(1 + \cos \tau)^2} \right] \right\} \tag{4-35}$$

Combining Equations (4-32)–(4-35), we obtain

$$\langle L^2 \rangle_{0f} = L_{cont} L_{pers}(1 - \cos \tau) + 2 L_{cont} L_{pers} \frac{\cos \tau}{N} \left[\frac{1 - (-\cos \tau)^N}{1 + \cos \tau} \right] \tag{4-36}$$

A chain with an infinite number of segments of zero length and a valence angle approaching 180° is called a wormlike chain (note that the contour length remains constant). This limiting case cannot be derived directly from Equation (4-36), because, although $\tau \to \pi$ (and thus $\cos \tau \to -1$), N simultaneously tends to infinity. Thus, the second term in Equation (4-36) is expressed in terms of the contour length and $(1 + \cos \tau)$ is given in terms of the persistent chain length:

$$\langle L^2 \rangle_{0f} = L_{cont} L_{pers}(1 - \cos \tau) + 2 L_{pers}^2(\cos \tau)[1 - (-\cos \tau)^N] \tag{4-37}$$

Equation (4-37) contains no term including N except for the expression $(-\cos \tau)^N$. In the limiting case of $\tau \to \pi$, this is, $(1 - \cos \tau) \to 2$, Equation (4-37) becomes

$$\langle L^2 \rangle_{0f} = 2 L_{cont} L_{pers} - 2 L_{pers}^2 + 2 L_{pers}^2(-\cos \tau)^N \tag{4-38}$$

In the limiting case $(-\cos \tau)$ is only slightly smaller than 1, but is raised to the power of N. Transformation of the above equations can lead to a more easily handled expression. First, $\cos \tau$ is expressed in terms of Equation (4-34). Then cross-multiplication leads to the introduction of the contour length:

$$\lim_{\substack{N \to \infty \\ \tau \to \pi}} (-\cos \tau)^N = \lim_{N \to \infty} \left(1 - \frac{b}{L_{\text{pers}}} \right)^N = \lim_{N \to \infty} \left(1 - \frac{L_{\text{cont}}}{NL_{\text{pers}}} \right)^N \tag{4-39}$$

and since

$$\lim_{x \to \infty} \left(1 - \frac{1}{x} \right)^x = e^{-1} \tag{4-40}$$

then equation (4-39) can be rearranged and solved like Equation (4-37):

$$\lim_{N \to \infty} \left(1 - \frac{L_{\text{cont}}}{NL_{\text{pers}}} \right)^N = \left[\lim_{N \to \infty} \left(1 - \frac{L_{\text{cont}}}{NL_{\text{pers}}} \right)^{NL_{\text{pers}}/L_{\text{cont}}} \right]^{N_{\text{cont}}/L_{\text{pers}}}$$

$$= \exp \left(\frac{-L_{\text{cont}}}{L_{\text{pers}}} \right) \tag{4-41}$$

Inserting Equation (4-41) into Equation (4-38), we obtain

$$\langle L^2 \rangle_{0f} = 2L_{\text{pers}}^2 [y - 1 + \exp(-y)], \qquad y = L_{\text{cont}}/L_{\text{pers}} \tag{4-42}$$

For the radius of gyration, an analogous derivation gives

$$\langle R_G^2 \rangle_{0f} = L_{\text{pers}}^2 \left[\frac{2}{y^2} [y - 1 + \exp(-y)] - \left(1 + \frac{y}{3} \right) \right] \tag{4-43}$$

For flexible chains, the contour length is much larger than the persistence length. That is, y becomes much larger than one and the expression $\exp(-y)$ tends to zero. Thus, Equation (4-42) becomes

$$\lim_{y \to \infty} \langle L^2 \rangle_{0f} = 2L_{\text{pers}} L_{\text{cont}} \tag{4-44}$$

and Equation (4-43) gives

$$\lim_{y \to \infty} \langle R_G^2 \rangle_{0f} = \lim_{y \to \infty} \frac{L_{\text{pers}} L_{\text{cont}}}{3} \left(1 - \frac{3}{y} + \frac{6}{y^2} - \frac{6}{y^3} \right) = \frac{L_{\text{pers}} L_{\text{cont}}}{3} \tag{4-45}$$

A comparison of Equations (4-33) and (4-44) shows that the persistence length is exactly half the segment length L_s. In this case, the radius of gyration of a wormlike chain bears the same relationship to the chain end-to-end distance as does the radius of gyration of a valence chain with or without free rotation [see Equations (4-44), (4-45), and (4-29)].

On the other hand, for very rigid chains $y \to 0$, and $\exp(-y)$ can then be developed into a series $1 - y + (y^2/2!) - (y^3/3!) + \cdots$, and the relationship obtained for the chain end-to-end distance and the radius of gyration is

$$\langle L^2 \rangle_{0f} = L_{\text{cont}}^2 \left(1 - \frac{y}{3} + \frac{y^2}{12} - \cdots \right) = L_{\text{cont}}^2 \tag{4-46}$$

$$\langle R_G^2 \rangle_{0f} = \frac{L_{\text{cont}}^2}{12} \left(1 - \frac{y}{3} + \frac{y^2}{12} - \cdots \right) = \frac{L_{\text{cont}}^2}{12} \tag{4-47}$$

Thus, a very rigid chain behaves like a rod, since the end-to-end chain distance equals the contour length and the radius of gyration is smaller than the chain end-to-end distances by a factor of $(12)^{1/2}$.

The persistence length model therefore describes the whole spectrum from the more rodlike oligomers (small y) to the well-developed coils (large y). However, the model ignores the finite thickness of the chain, and so it only holds strictly for unperturbed coils. The error due to the finite thickness may be neglected when the persistence length is much greater than the chain thickness.

4.5.5. Dimensions

All of the above considerations were concerned with the radius of gyration of unbranched chain molecules at infinite dilution. With increasing concentration the coils fill the available space more and more. The loose coils tend to become compressed above a certain critical concentration. This critical concentration can be roughly approximated on the basis of hexagonal close packing (about 75% of the total volume) of spheres of radius r:

$$c_{\text{crit}} = \frac{9}{16\pi (5/3)^{3/2} N_L} \frac{M}{\langle R_G^2 \rangle^{3/2}} = 1.38 \times 10^{-25} \frac{M}{\langle R_G^2 \rangle^{3/2}} \frac{\text{g}}{\text{cm}^3} \tag{4-48}$$

A spherically shaped macromolecule of molar mass 1.3×10^6 g/mol and radius of gyration of $(\langle R_G^2 \rangle)^{1/2} = 100$ nm should then have a critical concentration of 1.8×10^{-4} g/cm^3. Above this concentration, the coil will tend to be compressed and the radius of gyration will decrease (Figure 4-19). At still higher concentrations, the radius of gyration begins to increase again, which can indicate association.

The radius of gyration of polymers of given constitution and configuration in the unperturbed state is a function of the square root of the molar mass. From Equations (4-27) and (4-29), the following is, namely, obtained for N_e ($= M/M_e$) chain elements of molar mass M_e per chain element and length L_e:

$$\langle R_G^2 \rangle_0 = L_e^2 (6 M_e)^{-1} (1 - \cos \tau)(1 + \cos \tau)^{-1} \sigma^2 M = K_e M \tag{4-49}$$

A more complicated expression is obtained for states other than the

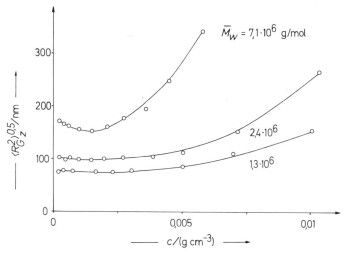

Figure 4-19. Change of the z average root mean square radius of gyration $\langle \bar{R}_G^2 \rangle_z^{0.5}$ with concentration for poly(styrene) in benzene at 20°C and various molar masses (after H. Dautzenberg).

unperturbed state. The radii of gyration are, of course, related to the expansion factor, α_R [Equation (4-15)]. This, in turn, is a complicated function of the z parameter [Equation (4-21)], which, itself, depends on the molar mass [Equation (4-17)]. Thus, on including the dependence of the expansion factor on the molar mass, an empirical molar mass exponent is obtained:

$$\langle R_G^2 \rangle = K_e \alpha_R^2 M = K_R M^{1+\epsilon} = K_R M^{a_R} \quad (\epsilon > 0) \tag{4-50}$$

Such expressions relating the radius of gyration with molar mass hold over surprisingly large molar mass ranges (Figure 4-20).

Measurements on polymers of given constitution and configuration dissolved in randomly chosen solvents generally do not lead to unperturbed dimensions of the polymers, since the dimensions are perturbed by both long-range and short-range forces:

Combining Equations (4-15), (4-17), and (4-20) and rearranging, we obtain

$$\left(\frac{\langle R_G^2 \rangle}{M} \right)^{3/2} = \left(\frac{\langle R_G^2 \rangle_0}{M} \right)^{3/2} + 2(4\pi)^{-3/2} \left(\frac{u_{seg}}{M_{seg}^2} \right) M^{1/2} \tag{4-51}$$

The first term on the right-hand side of Equation (4-51) is a characteristic constant for the unperturbed coil, as can be seen from equation (4-26). Since $\langle R_G^2 \rangle_0 / M$ is a function of the hindrance parameter σ, it is a measure of the short-range interactions. Conversely, the slope contains the constant u_{seg} / M_{seg}^2,

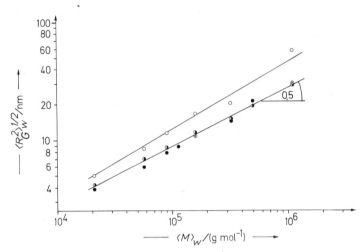

Figure 4-20. Mass average of the radius of gyration, $\langle R_G \rangle_w$, of poly(styrene) as a function of the mass average molar mass $\langle M \rangle_w$ in carbon disulfide (O) cyclohexane (◑) and for deuterated poly(styrene) in the solid state (●). Measurements at 35.4° C (after H. Benoit, D. Decker, and J. S. Higgins, C. Picot, J. P. Cotton, B. Farnoux, and G. Jannink).

and so is a measure of the long-range interactions. Thus, if one plots $(\langle R_G^2 \rangle / M)^{3/2}$ against $M^{1/2}$, long-range and short-range interactions can be separated from each other. The relationship is often valid over relatively wide molar mass ranges (Figure 4-21), but should fail at high molar masses, since the basic function $\alpha_R^3 = 1 + 2z$ then deviates strongly from experimental data (see Fig. 4-16).

It has been shown experimentally that the unperturbed dimensions of iso- and syndiotactic polymers of the same constitution and molar mass can differ by up to 20%, although, in what are known as good solvents, with high α_R, the same radii of gyration are obtained. This effect can be explained in the following way: in the unperturbed state, short-range forces, which are strongly influenced by the microconformation, dominate. But the microconformation, again, depends strongly on the configuration. In good solvents, long-range forces dominate, and since these involve interactions between segments, they are practically independent of the microconformation, and thus, also of the configuration.

A more complicated relationship for coil molecules in good solvents exists between the chain end-to-end distance and the radius of gyration for valence chains with restricted rotation:

$$\langle L^2 \rangle = 6(1 + \tfrac{5}{6}\epsilon + \tfrac{1}{6}\epsilon^2) \langle R_G^2 \rangle \tag{4-52}$$

$\epsilon = 0$ for coils with unperturbed dimensions and Equation (4-52) reduces to Equation (4-29). Conversely, the radius of gyration of infinitely thin rods is

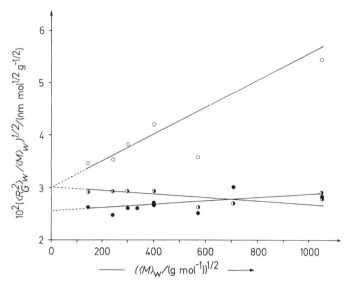

Figure 4-21. Reduced radii of gyration as a function of molar mass for poly(styrenes) (Data from Figure 4-20).

directly proportional to the molar mass, i.e., $\langle R_G^2 \rangle = K_R M^2$. In this case, $\epsilon = 1$. Thus, for rods, Equation (4-52) becomes

$$\langle L^2 \rangle = 12 \langle R_G^2 \rangle \tag{4-53}$$

Branched macromolecules have a higher average segment density than unbranched macromolecules of the same molar mass, and have a lower coil volume. This is easily seen by comparing a star-shaped branched molecule with a linear one. The influence of the branching on the dimensions can be expressed by a g factor

$$\frac{\langle R^2 \rangle^\theta_{\text{branched}}}{\langle R^2 \rangle^\theta_{\text{linear}}} = g \tag{4-54}$$

The magnitude of g depends on the type of branching (star, comb) and on its regularity. The dimensions therefore decrease from linear macromolecules through regular comb branched, irregular comb branched, regular random branched, and irregular random branched to star-shaped branched molecules. The quantitative classification of the dimensions according to type, density, length, and regularity of the branches has not yet found a complete theoretical solution, since the effect also depends on the quality of the solvent. However, it does have some practical significance, since the determination of the dimensions is often the only possible way of detecting long-chain branching in polymers.

4.6. Optical Activity

4.6.1. Overview

Polymers with a single kind of chiral monomeric unit are always optically active, that is, they rotate the plane of polarized light. In general, the effect of end groups disappears for degrees of polymerization above about 10–20; thus, the measured optical activity of high-molar-mass polymers results from that of the chiral monomeric units. An example of this consists of poly(L-α-amino acids) in the coiled state, for example, in dichloroacetic acid.

This chiral influence is superimposed on that of the helix in the case of helical polymers from chiral monomeric units, since the helix is also chiral. Thus, a reinforced effect is observed, which makes the study of polymer dimensions in solution through measurement of optical activity remarkably interesting.

Copolymers consisting of alternatively arranged monomeric units of opposite chirality are optically inactive as a result of the configuration itself. They can, however, produce helices of a specific direction of turn in certain solvents, such that this preferential conformation produces an optical activity. Poly(L-alt-D-leucine), which produces what is known as a π helix in benzene, is an example of this.

A similar effect sometimes occurs with copolymers of chiral and nonchiral monomeric units, whereby their optical activity is higher than the additivity rule allows. The nonchiral monomeric units are obviously drawn into helical conformations by the helical sequences of the chiral monomeric units.

In contrast, polymers from nonchiral monomeric units are not optically active, even when their chains adopt a helical conformation. Since their molecules are enantiomeric to each other, left-handed and right-handed helices occur with equal probability, and the optical activity of the polymer is zero.

4.6.2. Basic Principles

If a linearly polarized electromagnetic wave meets a stereoisomeric center such as an asymmetric carbon atom, the plane of the polarized light is rotated. Since linearly polarized waves can be considered as the superposition of two circularly polarized waves of opposite rotation, the asymmetric electron configuration in the immediate vicinity of a stereoisomeric center causes the two propagation velocities of left-handed and right-handed polarized light to differ, hence resulting in a rotation of the plane of the polarized light.

The rotation of the plane of polarized light is measured as an optical rotation α. What is termed the "specific" optical rotation $[\alpha]$ in organic chemistry is a function of the mass fraction w_2 of the solute, the density of the solution ρ, and the length l of the sample cell, as well as the optical rotation:

$$[\alpha] = \frac{\alpha}{l w_2 \rho} \tag{4-55}$$

Traditionally, α is measured in degrees, l in dm, and ρ in g/cm³.

The "molar" optical rotation $[\Phi]$,

$$[\Phi] = 10^{-2}[\alpha]M_u \tag{4-56}$$

relates the "specific" rotation to the formula molar mass M_u of a monomeric unit or the molar mass of low-molar-mass compounds.

The "effective" molar rotation is also sometimes used. In this case, the refractive index is taken into account with a factor $(n^2 + 2)/3$:

$$[\Phi]_{\mathrm{eff}} = \frac{3}{n^2 + 2}[\Phi] \tag{4-57}$$

$[\alpha]$, $[\Phi]_{\mathrm{eff}}$, and $[\Phi]$ also depend on the temperature and the wavelength of the light used and also often on the concentration.

Generally, the wavelength dependence can be adequately represented by one of the following empirical equations:

$$[\Phi] = a_0 \left(\frac{\lambda_0^2}{\lambda^2 - \lambda_0^2}\right) \quad \text{(one-term Drude equation)} \tag{4-58}$$

$$[\Phi] = a_0 \left(\frac{\lambda_0^2}{\lambda^2 - \lambda_0^2}\right) + b_0 \left(\frac{\lambda_0^2}{\lambda^2 - \lambda_0^2}\right)^2 \quad \text{(Moffit–Yang equation)} \tag{4-59}$$

a_0, b_0, and λ_0 are system-specific constants; λ_0 gives the wavelength of the subsequent absorption maximum. The dependence of optical rotation or quantities dependent on optical rotation on the wavelength is called optical rotatory dispersion (ORD). ORD derives from the differences in refractive index of right-hand and left-hand rotating components of polarized light.

Generally, the Drude equation describes the optical activity of coil molecules; the Moffit–Yang equation is more suitable for helices.

The left-hand and right-hand rotating components of polarized light are absorbed to different extents by optically active compounds. The dependence of the difference in absorption of left- and right-hand rotating components of polarized light on wavelength is called circular dichroism (CD).

Complex behavior is observed (Cotton effect) when measurements of the optical rotatory dispersion are carried out near an absorption band. On one side of the band it passes through a minimum (trough), on the other side a maximum (peak), and at the inflection point, the optical activity is zero. The Cotton effect is termed *positive* when the peak occurs at a higher wavelength than the trough. The inflection point of the curve $[\alpha] = f(\lambda)$ occurs at $[\alpha] = 0$ for nonoverlapping Cotton effects; it corresponds to the maximum in the uv absorption spectrum.

A Cotton effect always occurs when an absorbing group resides in an asymmetric environment. One component of the circularly polarized light is then absorbed more strongly than the other. The more weakly absorbed component has a greater velocity, and therefore a lower refractive index on the lower frequency side of the band. Since the Cotton effect is caused by the asymmetric environment of an absorbing group, its magnitude depends strongly on the helical content of the molecule.

4.6.3. Structural Effects

The molar rotation $[\Phi]$ caused by a given asymmetric carbon atom in compounds of the general type $R-*CH(CH_3)-(CH_2)_y-CH_3$ is only slightly influenced by distant neighboring groups (Table 4-8). The measurable optical rotation depends on the sensitivity of the polarimeter and on any special experimental conditions used. For example, L-malic acid rotates to the left in dilute aqueous solution and to the right in concentrated solution. The optical rotation is zero at a certain concentration in water, although L-malic acid is chiral. All optically active systems are therefore chiral. Whether a chiral system is optically active or not depends on the conditions.

Table 4-8. *Molar Optical Rotation* $[\Phi]_{25}^D$ *of Various Low-Molar-Mass Compounds*
$$R-*CH(CH_3)-(CH_2)_y-CH_3$$
in the Liquid State at 589 nm and 25° C

R	$y = 1$	$y = 2$	$y = 3$	$y = 4$	$y = \infty^a$
			$[\Phi]_{25}^D$ in 10^{-2} deg dem^{-1} cm^3 mol^{-1}		
$(CH_2)_2H$	0	10	11.4	12.5	16.0
$(CH_2)_3H$	−10	0	1.5	2.4	6.0
$(CH_2)_4H$	−11.4	−1.7	0	0.8	5.0
$(CH_2)_5H$	−12.5	−2.4	−0.8	0	4.0
$(CH_2)_2Br$	−38.8	−21.3	−16.8	−14.7	−7.0
$(CH_2)_3Br$	−21.9	−14.5	−8.3	−6.2	−1.0
$(CH_2)_4Br$	−14.9	−7.8	−5.3	−4.0	−0.5
$(CH_2)_2OH$	−9.0	2.1	4.0	6.1	10.5
$(CH_2)_3OH$	−11.9	0	0.7	2.6	7.0
$(CH_2)_4OH$	−12.0	−1.7	0	0.8	5.5

^{a}The data given for $y = \infty$ were obtained by extrapolating $[\Phi]_{25}^D = f(y^{-1})$ to $y^{-1} = 0$.

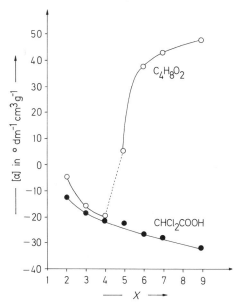

Figure 4-22. The dependence of the specific rotation $[\alpha]_\lambda$ of oligomers of poly(γ-methyl-L-glutamate) of different degrees of polymerization X in dichloroacetic acid (coil) and in dioxane (helix) (after M. Goodman and E. E. Schmitt).

Since the contribution of the end groups to the molecular properties decreases with increasing molecular weight, end groups only influence the optical activity at low degrees of polymerization. Figure 4-22 shows that the specific rotation $[\alpha]$ of the polymer homologous series of poly(γ-methyl-L-glutamate) in the hydrogen-bond-destroying solvent dichloroacetic acid continuously decreases with rising degree of polymerization X, as the influence of the end groups on the optical activity becomes less and less. In the helicogenic (helix-producing) solvent dioxane, the optical activity falls from the monomer through the dimer, trimer, and tetramer, to rise sharply again in pentamers because of helix formation. At higher degrees of polymerization, the contribution per monomeric unit, which is affected by helix formation, becomes less and less, until finally the optical activity becomes independent of the degree of polymerization. Practically constant values for the optical activity are already achieved by a degree of polymerization of $\sim$10–15. The renewed increase in optical activity with degree of polymerization for the pentamers can be explained on the basis of the helical structure. Helices of poly(γ-methyl-L-glutamate) have 3.7 monomeric units per helix turn. Thus, before a helix can be formed, at least four amino acid residues must be joined together. But even with four amino acid residues, the helix is still not sufficiently stabilized by the contribution of nonbonded atomic interaction.

4.6.4. Poly(α-Amino Acids)

Poly(L-α-amino acids) generally form right-handed helices, whereas poly(D-α-amino acids) form left-handed helices. An exception is poly(β-benzyl-L-aspartate), which occurs as left-handed helices. The helical structure is retained in solvents such as dioxan or dimethyl formamide. But in dichloroacetic acid or hydrazine, coiled molecules are present.

The dependence of the molar optical activity of coils on the wavelength of polarized light can be described by a one-term Drude equation. In the case of helices, the Moffitt–Yang equation is better. The constants λ_0 are completely unaffected by the solvent, while a_0 and b_0 assume different values (Table 4-9). For a given polymer in different helicogenic solvents, b_0 is found to be more or less constant, while a_0 still depends on the solvent. For various poly(α-aminocarboxylic acids), b_0 has approximately the same value so long as these are in the helix conformation. b_0 is therefore a typical constant for the helix conformation of poly(α-aminocarboxylic acids), while a_0 contains contributions from the helix and the asymmetric carbon atoms.

4.6.5. Proteins

In proteins of natural origin, copolymers of α-aminocarboxylic acids of uniform sequence (see Chapter 30), the constant b_0 has different values according to the protein. Since proteins usually contain L-amino acids, the helix conformation does not depend a great deal on the size of the substituents, and proteins take up a very compact structure in aqueous solutions (see Section 4.4.2), the constant b_0 has been used as a measure of the helix content of proteins. $b_0 = -650$ has been fixed for a 100% helix conformation. The figures given in Table 4-10 were obtained for different proteins.

Table 4-9. Influence of Overall Conformation of Poly(γ-benzyl-L-glutamate) on Parameters λ_0, a_0, and b_0[a]

Solvent	Overall conformation	Equation used in evaluation	λ_0	a_0	b_0
Dichloroacetic acid	Coil	Drude	190	—	—
Dichloroacetic acid	Coil	Moffitt–Yang	212	—	0
Hydrazine	Coil	Drude	212	—	0
Dimethylformamide	Helix	Moffitt–Yang	212	200	−600
Dioxane	Helix	Moffitt–Yang	212	220	−670
Dioxane	Helix	Moffitt–Yang	212	198	−682
Chloroform	Helix	Moffitt–Yang	212	250	−625
1,2-Dichloroethane	Helix	Moffitt–Yang	212	205	−635

[a]λ_0 in nm; a_0 and b_0 in 10^{-2} deg dm^{-1} cm^3 mol^{-1}

Table 4-10. Helix Content of Various Proteins

Protein	b_0 in 10^{-2} deg dm^{-1} cm^3 mol^{-1}	f_h in %
Tropomyosin	−650	100
Serum albumin	−290	46
Ovalbumin	−195	31
Chymotrypsin	−95	15

The evaluation of the helix content of proteins is important because it allows the influence of the solvent on the conformation to be estimated with regard to the conformation determined by X-ray measurements on the crystalline state. It presupposes a "two-phase" model, i.e., the independent and distinctly separate existence of helix and coil portions. This assumption is confirmed by observations of the helix/coil transition (see below). The determination of the helix content of proteins using b_0 is not completely straightforward, however, since helix portions that are too short do not make a full contribution to b_0 (see Figure 4-22), L-amino acids can occur in left-handed helices as well as right-handed helices with a chance in sign for b_0, and, finally, there can be mixtures of left- and right-handed helices.

4.6.6. Poly(α-Olefins)

The molar optical rotation of optically active it-poly(α-olefins) depends not only on the wavelength and temperature, but also on the optical purity of the monomers (and thus, also, of the polymers) (Figure 4-23). The molar optical rotation of these polymers remains constant when the optical purity of the monomer is high.

The molar optical rotation of a polymer obtained by polymerizing a monomer of given optical purity under various conditions depends on the tacticity of the resulting polymer (Figure 4-24). Extrapolation of the reciprocal optical rotation to that of the tactic polymer allows the molar optical rotation of the latter to be obtained.

A value of $\Phi = 292$ was found for poly[(S)-4-methyl hexene-1], whereas the value for the hydrogenated monomer chosen as model compound is only 9.9 (in each case, expressed in units of 10^{-2} deg dm^{-1} cm^3 mol^{-1}). The increased value for the polymer undoubtedly results from the contribution of the helical structure.

The dependence of the molar optical rotation of poly(α-olefins) on the wavelength of the polarized light used can be well represented by the one-term Drude equation. The constant λ_0 is of approximately the same magnitude for polymers and their hydrogenated monomers (Table 4-11). For polymers,

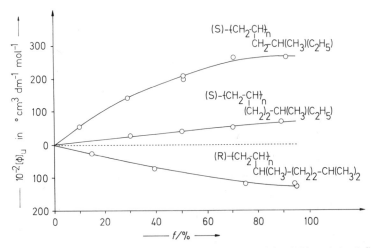

Figure 4-23. Molar optical rotation $[\Phi]_u$ of various methanol insoluble poly(α-olefins) in hydrocarbon solutions as a function of the optical purity, f, of the initial monomer (P. Pino, F. Ciardelli, G. Montagnoli, and O. Pierono).

however, the a_0 values are sometimes considerably larger than they are for the hydrogenated monomers. The values are only slightly influenced by the nature of the solvent used, i.e., the length of the helical segments is independent of solvent.

The molar optical rotation of poly(α-olefins) decreases with increasing temperature. This is interpreted as the "melting" of relatively long, left-

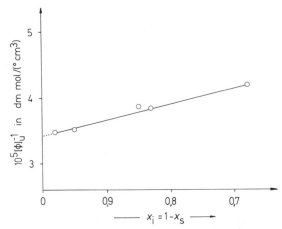

Figure 4-24. Dependence of the reciprocal molar optical rotaton, $[\Phi]_u^{-1}$, with tacticity content of isotactic, x_i, and syndiotactic, x_s, diads for poly (S)-4-methyl-1-hexene. Optical purity of initial monomer was 93% (according to measurements by P. Pino *et al.*).

Table 4-11. Constants a_0 and λ_0 of the One-Term Drude Equation for Various Synthetic Polymers and Their Hydrogenated Monomers as Model Low-Molar-Mass Compounds[a]

	Monomer		λ_0 in nm		a_0 in 10^{-2} deg dm^{-1} cm^3 mol^{-1}	
	Name	Constitution	Model	Polymer	Model	Polymer
I	(S)-3-Methyl-pentene-1	$CH_2=CH-CH(CH_3)(C_2H_5)$	176	167	−113	1143
II	(S)-4-Methyl-1-hexene	$CH_2=CH-CH_2-CH(CH_3)-C_2H_5$	170	165	3078	104
III	(1R, 3R, 4S)-1-Methyl-4-isopropyl-cyclohex-3-yl-vinyl ether	$CH_2=CH-O-$ (cyclohexyl bearing $CH(CH_3)_2$ and CH_3)	155	165	−1144	−2169
IV	[(−)-N-Propyl-N-α-phenylethyl]-acrylamide	$CH_2=CH-CO-N(CH_3)-CH(C_3H_7)-C_6H_5$	280	272	−1518	−1188
V	[(1S, 2R, 4S)-1,7,7-Trimethyl-norborn-2-yl]-acrylate	$CH_2=CH-CO-O-$ (trimethylnorbornyl, H_3C, CH_3)	190	191	−485	−401
VI	[(S)-Methylbutyl]-methacrylate	$CH_2=C(CH_3)-CO-O-CH(CH_3)-C_2H_5$	191	188	59	53

[a] The measurements refer to room temperature, and the same solvent was used for each polymer–model-compound pair. The monomers were polymerized with Ziegler catalysts (I, II), cationically (III), anionically (IV), and radically (V, VI).

handed helical segments. Calculations based on the same model say that the length of the relatively short, right-handed helical segments should not change very much with temperature.

The molar optical rotation of configurational copolymers of (S) and (R) isomers of the same monomer is generally, in the case of poly(α-olefins), a hyperbolic and not a linear function of the optical purity of the monomers. Thus, the molar optical rotation of the copolymers is always greater than that obtained by additivity rules. Whether this is caused by tactic blocks in the polymers or by mixtures of (S) and (R) unipolymers has not been established yet.

4.7. Conformational Transitions

4.7.1. Phenomena

The monomeric units of a macromolecule may lie in helical (h) or nonhelical (c) conformations. Such a polymer chain of helical and non-helical sequences may, for example, have a time-average sequence of . . . hhchhhhhhcccchhcccccchhhh . . . The proportions of helical and non-helical conformations, here, will change with solvent, pressure, and tempera-ture. These changes may be monitored by group-specific methods (ir, uv, nmr, ORD, CD) or by molecule specific methods (radii of gyration, viscosities).

Measurements of what is called the intrinsic viscosity (see Section 9) of poly(methyl methacrylates) as a function of temperature show, for example, a minimum at about 47°C which is interpreted as a conformational transition (Figure 4-25).

The specific rotation of poly(γ-benzyl-L-glutamate) in a mixture of, for example, ethylene dichloride and dichloroacetic acid first increases slightly as more $CHCl_2COOH$ is added, then remains constant over a wide range of mixture compositions, and finally falls sharply to negative values at a $\sim 75\%$ $CHCl_2COOH$ content (Figure 4-26). Since ethylene dichloride is a helicogenic solvent, the initial increase is considered to be a change in the helix structure (an expansion?), but the decrease results from a helix/coil transition.

4.7.2. Thermodynamics

Conformational changes in macromolecules must be cooperative pro-cesses since, at least for regular conformational sequences, each conformation must be influenced by the conformation of neighboring bonds. An equilib-rium constant can be defined for each individual conformational transition.

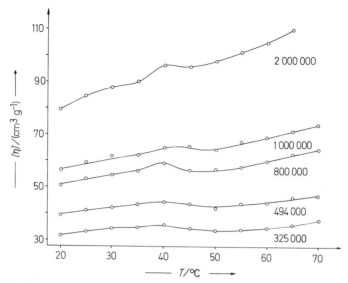

Figure 4-25. Change in the Staudinger index [η] with temperature for poly(methyl meth-acrylates) of various mass average molar masses (after I. Katime, C. Ramiro Vera, and J. E. Figueruelo).

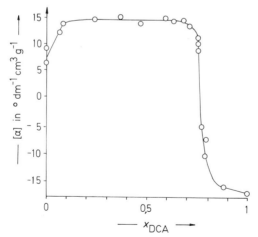

Figure 4-26. Specific rotation $[\alpha]_D$ of a poly(γ-benzyl-L-glutamate) ($\overline{M}_w = 350,000$ g mol) in ethylene dichloride–dichloroacetic acid mixtures at 20° C (after J. T. Yang).

The different conformations can be distinguished from each other by the symbols A and B. A and B can, for example, be *trans* or *gauche* conformations or even the *cis* and *trans* positions of the peptide groups in poly(proline), etc.

The process

$$AAB \rightleftharpoons ABB \qquad (4\text{-}60)$$

(or BAA $\rightleftharpoons$ BBA) describes the *propagation* of already existing sequences of A or B conformations, and an equilibrium constant $K_p = K$ is assigned to this process. In the process

$$AAA \rightleftharpoons ABA \qquad (4\text{-}61)$$

however, a B sequence is begun or destroyed. An equilibrium constant $K_n = \sigma K_p = \sigma K$ can be assigned to this *nucleation* process. σ is a measure of the cooperativeness of the transition. The segments preferentially take on the conformation of neighboring groups for $\sigma < 1$. In this case, conformational diads AA or BB are preferred to AB or BA diads (positive cooperativity). When $\sigma = 1$, $K_p = K_n$; no cooperativity exists. There is no known case of a negative cooperativity or anticooperativity with $\sigma > 1$.

The nucleation process in the interior of a chain must be microscopically reversible. Thus, the equilibrium constant for the process

$$BBB \rightleftharpoons BAB \qquad (4\text{-}62)$$

is $K^{-1}\sigma$. At each end of the chain, however, the conformations have only one neighboring conformation. The σ values for nucleation from the ends of the chain must therefore be different from those for the chain interior as well as depending on the type of conformation (A or B) of the chain ends. To a first approximation, however, the formation of a nucleus at the chain ends can also be described by σ when many transitions are involved.

In a chain consisting of $N = 4$ conformations, the all-A conformation can transform into the all-B conformation in four steps:

$$AAAA \rightleftharpoons BAAA \rightleftharpoons BBAA \rightleftharpoons BBBA \rightleftharpoons BBBB \qquad (4\text{-}63)$$

Three propagation steps follow a nucleation step. The equilibrium concentration is thus given by

$$c_{BBBB} = \sigma K \cdot K \cdot K \cdot K \cdot c_{AAAA} = \sigma K^4 c_{AAAA} \qquad (4\text{-}64)$$

Thus, when $\sigma K^4 = 1$, $c_{BBBB} = c_{AAAA}$.

If, in this case, the equilibrium constant $K \gg 1$, then $1/\sigma^{1/4} \gg 1$ since $\sigma K^4 = 1$. Thus, here, all intermediates must occur at low concentrations with respect to the extreme conformations (i.e., for $\sigma = 10^{-4}$: $c_{BBBB} = 10 c_{BBBA} = 10^2 c_{BBAA} = 10^3 c_{BAAA}$). That is, the conformational transition occurs essentially completely or not at all.

The transition of a four-conformational chain is therefore described by the product σK^4. The expression for a chain with N conformations is consequently σK^N. The fraction f_B of B states formed is then

$$f_B = \frac{\sigma K^N}{1 + \sigma K^N} \tag{4-65}$$

However, f_B can only be calculated from Equation (4-65) when N is small. For an all-or-none process, $1/\sigma^{1/N} \ll 1$ or $\sigma^{1/N} \gg 1$ (see above). If the probabilities for the conformational transitions are the same for all conformations, then the probability per chain must increase with increasing N. Thus, for the chain transition probability, the product of N and $\sigma^{1/N}$ must be considered. The condition for Equation (4-65) is thus

$$N\sigma^{1/N} \ll 1 \tag{4-66}$$

On the other hand, the chain transition is independent of N for large chain lengths. The degree of transition f_B in this case is calculated as

$$f_B = 0.5\left(1 + \frac{K - 1}{[(K - 1)^2 + 4\sigma K]^{0.5}}\right) \tag{4-67}$$

Thus, according to Equation (4-67), K will always be unity no matter what the value of σ may be at the midpoint of the transition ($f_B = 0.5$). However, the sharpness of the transition increases with decreasing σ.

The expressions are more complex for transitions of chains of moderate length. In this region, the transitions depend significantly on the chain length (see also Figure 4-27).

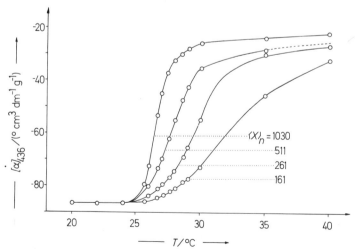

Figure 4-27. Temperature dependence of the specific rotation, $[\alpha]_{436}$, of poly(ϵ-carbobenzyloxy-L-lysines) of various number average degrees of polymerization in *m*-cresol (after M. Matsuoka, T. Norisuye, A. Teramoto, and H. Fujita).

Table 4-12. Thermodynamic Parameters, K and σ for the Helix–Coil
Transition of Poly(α-amino acids) or poly(nucleotides)

Monomeric unit	$T/°C$	K	$10^5 \sigma$
Glycine	60	0.63	1.0
L-Serine	60	0.74	7.5
Hydroxypropyl-L-glutamine	60	0.96	22
L-Alanine	0	1.08	
L-Alanine	60	1.01	80
L-Alanine	80	0.99	
L-Phenyl alanine		1.00	180
L-Leucine	0	1.28	
L-Leucine	60	1.09	330
L-Leucine	80	1.33	
Adenine/thymine (1:1)		0.5	10
Guanine/cytosine (1:1)		2.0	10

According to these considerations, σ is a measure of the effects produced by the helix sequence ends. Of course, the monomeric units situated at these ends experience a different environment because of the nearness of the nonhelical sequences as they would in the center of the helix. It has been found that σ is very small for proteins and poly(α-amino acids) (Table 4-12). So, end effects are not favored by these polymers. Thus, if a helical state of four monomeric units is separated from another helical state of three monomeric units, a helical state of seven monomeric units will tend to form.

On the other hand, the equilibrium constant K indicates the tendency to form helical or nonhelical states. K values in excess of unity denote helix formers; K values much less than unity, conversely, indicate coil-forming sequences. With proteins, proline, serine, glycine, and aspartine, for example, are typical helix breakers. Lysine, thyrosine, aspartic acid, threonine, arginine, cysteine, and phenyl alanine act as neither helical breakers or formers, whereas all other α-amino acids are typical helix formers.

4.7.3. Kinetics

With the exception of helix/coil transitions of polypeptides and polynucleotides, the kinetics of conformational transitions has only been investigated to a slight extent. Relatively large rate constants of 10^6–10^7 s^{-1} have been obtained for these helix/coil transitions. However, rate constants of 10^{-6} to 1 s^{-1} have been obtained for denaturing processes where helix/coil transitions also play a role (see also Chapter 4.4.2). The high rates of helix/coil transitions are undoubtedly due to the cooperativity of the process. The low rates for denaturing must therefore result from nonhelical regions.

Figure 4-28. Kinds of rotational transitions in a chain molecule (after H. Morawetz).

A large part of the molecule must be set in motion in the rotation about a single bond in a long-chain molecule (Figure 4-28a). This should be very difficult in viscous media. Conversely, an increase in the activation energy for coupled rotation about two bonds (Figure 4-25b) with respect to similar low-molar-mass compounds is to be expected.

This subject has been investigated for the case of piperazine polymers and diacyl piperazines as model substances:

$$CH_3-\underset{\underset{O}{\|}}{C}-N\overbrace{}N-\underset{\underset{O}{\|}}{C}-CH_3 \qquad \left(\!\!R-\underset{\underset{O}{\|}}{C}-N\overbrace{}N-\underset{\underset{O}{\|}}{C}\!\!\right)_{\!n}$$

I   II

The N—CO bond of these compounds possesses partial double-bond character. Consequently, the rotation about this bond is comparatively slow. Various absorption bands will be observed in the proton resonance spectrum according to whether neighboring groups are in the *cis* or the *trans* position about the N—CO bond. The Gibbs energy of activation $\Delta G^{\ddagger}$ can be determined from the temperature dependence of the band intensities.

Experimentally it was found that the Gibbs activation energy for diacetyl piperazine (I) and poly(succinyl piperazine) [II, $R = (CH_2)_2$], poly(adipyl piperazine) [II, $R = (CH_2)_4$], and poly(sebacyl piperazine) [II, $R = (CH_2)_8$] are practically the same ($\Delta G^{\ddagger} = 7.6 \times 10^4$ J/mol). The reason for this is not clear. The activation energy can, for example, be stored in the polymer molecule and used for the rotation about another bond. Alternatively, the tension set up by rotation can be compensated by distortion of rotational and valence angles.

A4. Appendix to Chapter 4

A4.1. Calculation of the Chain End-to-End Distance

In the segment model the length and direction of each individual bond are given by a vector b_i (Figure 4-13). The vector distance L_{00} between both ends of the chain is then

$$L_{00} = b_1 + b_2 + \cdots + b_{n-1} =$$

$$\sum_{i=i}^{i=n-1} b_i \qquad (A4\text{-}1)$$

For the mean square chain end-to-end distance L_{00}^2 (average over all molecules and all molecular shapes in time of a molecule), we use the products of the vectors:

$$(\overline{L_{00}^2}) = \overline{L \cdot L} = \overline{\left(\sum_i b_i \cdot \sum_j b_j \right)} = b_1 \cdot b_1 + b_2 \cdot b_2$$

$$+ \cdots + b_{n-1} \cdot b_{n-1} + 2 \sum_i \sum_j \overline{b_i \cdot b_j} \qquad (A4\text{-}2)$$

where the index j has the same significance as the index i and simply shows that every member of the first sum has to be multiplied by every member of the second sum. The scalar product $b_i \cdot b_{i+1}$ is equal to $b_i b_{i+1} \cos(180 - \tau)$, where $(180 - \tau)$ is the angle between two neighboring bonds and τ is the valence angle. In the chosen segment model, every angle τ has the same probability of occurring as the angle $(180 + \tau)$. Therefore, $\cos \tau = -\cos(180 + \tau)$. The double sum in equation (A4-2) vanishes, giving

$$(\overline{L_{00}^2}) = (N - 1)b^2 \approx Nb^2 \qquad (A4\text{-}3)$$

If several different bonds of differing lengths b_q are present (for example, in polyamides), then Equation (A4-3) is altered to the corresponding sum:

$$(\overline{L_{00}^2}) = \sum_q (N - 1)_q b_q^2 \qquad (A4\text{-}4)$$

A4.2. Relationship between the Radius of Gyration and the Chain End-to-End Distance for the Segment Model

It is the radius of gyration $(\overline{R_G^2})^{0.5} \equiv \langle R \rangle$, and not the chain end-to-end distance $\langle L^2 \rangle^{0.5} \equiv \langle L \rangle$, that is experimentally accessible. In this model, nevertheless, the chain end-to-end distance is unambiguously related to the radius of gyration. As shown in Figure 4-13, the masses of the chain atoms can be concentrated into point masses connected together by bonds of length l. r_1 is the vector from the center of gravity of the molecule to the first point mass, r_i the corresponding vector to the ith point mass, and L the vector between both point masses. Therefore, for each point mass

$$r_i = r_1 + L_i \qquad (A4\text{-}5)$$

The center of gravity is defined by requiring that the first moment of the segment distribution around the center of mass be zero

$$\sum_i m_i r_i = 0 \qquad\qquad \text{(A4-6)}$$

Since all point masses are identical, one can write for N point masses

$$\sum_{i=i}^{i=n} r_i = N r_i + \sum_{i=i}^{i=n} L_i = 0 \qquad\qquad \text{(A4-7)}$$

and therefore

$$r_1 = -(1/N) \sum_i L_i \qquad\qquad \text{(A4-8)}$$

The radius of gyration $\langle R \rangle$ is now defined as the root mean square over all the radii r (see above), and this in turn as the second moment of the mass distribution

$$r^2 = \frac{\sum_i m_i \overline{r_i^2}}{\sum_i m_i} \qquad\qquad \text{(A4-9)}$$

or for the mean over all squares (with, for this model, the index 00)

$$\overline{R_{00}^2} = \frac{\sum_i m_i \overline{r_i^2}}{\sum_i m_i} = \langle R^2 \rangle_{00} \qquad\qquad \text{(A4-10)}$$

By definition, all the masses m_i are identical. Since it is possible to average over all sums first or alternatively over all products and then over all sums, one can also write for Equation (A4-10)

$$\langle R^2 \rangle_{00} = \frac{m_i \sum_i \overline{r_i^2}}{\sum_i m_i} \qquad\qquad \text{(A4-11)}$$

and, with $m = \sum_i m_i$ and N the number of chain links ($N = m/m_i$), from Equation (A4-11)

$$\langle R^2 \rangle_{00} = \sum_i \overline{r_i^2}/N \qquad\qquad \text{(A4-12)}$$

With poly(methylene), the number of chain units N is identical to the degree of polymerization $\overline{X}_n$. However, for monomeric units of two units, e.g., poly(styrene), $N = 2\overline{X}_n$. The relation between the radius of gyration and the chain end-to-end distance is then

$$\langle R^2 \rangle_{00} = \frac{1}{N} \sum_{i=1}^{i=n} (r_1 + L_i) \cdot (r_1 + L_i)$$

$$\qquad\qquad \text{(A4-13)}$$

$$\langle R^2 \rangle_{00} = r_1^2 + \frac{1}{N} \sum_{i=1}^{i=n} L_i^2 + \frac{2}{N} r_1 \cdot \sum_{i=1}^{i=n} L_i$$

According to Equation (A4-2), however,

$$r_1^2 = \sum_{i=1}^{i=n} \sum_{j=1}^{j=n} L_i \cdot L_j \tag{A4-2a}$$

and from Equations (A4-8) and (A4-2)

$$\frac{2}{N} r_1 \cdot \sum_{i-1}^{i=n} L_i = -\frac{2}{N^2} \sum_{i=1}^{i=n} \sum_{j-1}^{j=n} L_i \cdot L_j \tag{A4-8a}$$

With Equations (A4-2a) and (A4-8a), Equation (A4-13) becomes

$$\langle R^2 \rangle_{00} = \frac{1}{N} \sum_{i=1}^{i=n} L_i^2 - \frac{1}{N^2} \sum_{i=1}^{i=n} \sum_{j=1}^{j=n} L_i \cdot L_j \tag{A4-14}$$

The vector product is solved with the cosine rule already used in Equation (A4-2)

$$L_{ij}^2 = L_i^2 + L_j^2 - 2L_i \cdot L_j \tag{A4-15}$$

According to definition, the indices i and j have the same significance, so that the sums over the squares of the distances L_i^2 and L_j^2 are identical. On inserting Equation (A4-15) into (A4-14), one obtains

$$\langle R^2 \rangle_{00} = \frac{1}{2N^2} \sum_{i=1}^{i=n} \sum_{j=1}^{j=n} \overline{L_{ij}^2} \tag{A4-16}$$

According to Equation (A4-2) or (A4-3), however, the average $\overline{L_{ij}^2}$ is the chain end-to-end distance of a chain of $j - i$ elements of length b:

$$\overline{L_{ij}^2} = |j - i| b^2 = (|j - i| \overline{L^2})/N \tag{A4-17}$$

The absolute difference corresponds to a product sum, where each sum can be evaluated individually. Summing over all j values, one obtains

$$\sum_{j=1}^{j=n} |j - i| = \sum_{j=1}^{i} (i - j) + \sum_{j=i+1}^{n} (j - 1)$$

$$= i^2 - 0.5i(i + 1) + 0.5(N - i)(N + i + 1) - i(N - i) \tag{A4-18}$$

$$= i^2 - iN + 0.5N^2 + 0.5N - i$$

and for the sum over all i values

$$\sum_{i=1}^{i=n} i^2 = 1^2 + 2^2 + \cdots + N^2 = \frac{N(N + 1)(2N + 1)}{N} \tag{A4-19}$$

thus giving

$$\sum_{i=1}^{i=n} \sum_{j=1}^{i=j} |j - i| = \tfrac{1}{3}(N^3 - N) \cong \tfrac{1}{3}N^3 \qquad \text{for} \qquad N \geq 1 \quad \text{(A4-20)}$$

From Equation (A4-16) with Equations (A4-17) and (A4-20), this gives

$$\langle R^2 \rangle_{00} = \frac{1}{2N^2}\frac{N^3}{3}\frac{L^2}{N} = \frac{1}{6}\overline{L_{00}^2} = \overline{R_{00}^2} \qquad \text{(A4-21)}$$

Equation (A4-21) allows the theoretically important end-to-end chain distance to be calculated from the experimentally accessible (e.g., by light scattering measurements) radius of gyration. This calculation remains valid for the linear valence angle chain and the linear valence angle chain with hindered rotation, but is not valid for polymers in good solvents.

A4.3. Calculation of the Chain End-to-End Distance for Valence Angle Chains

In the segment model, any given angle can occur between neighboring segments. Valence angles, however, must be considered with real chains and, to a first approximation (see Section 4.5.1), they can be taken as constant. As with the segment model [see Equation (A4-2)], all vectors for the N chain units (that is, $N - 1$ bonds) must be multiplied by each other:

$$(\overline{L_{0f}^2}) = \overline{L} \cdot \overline{L} = (b_1 \cdot b_1 + b_2 \cdot b_2 + \cdots + b_{n-1} \cdot b_{n-1})$$
$$+ 2(b_1 \cdot b_2 + b_2 \cdot b_3 + \cdots + b_{n-2} \cdot b_{n-1})$$
$$+ 2(b_1 \cdot b_3 + b_2 \cdot b_4 + \cdots + b_{n-3} \cdot b_{n-1})$$
$$+ \cdots + 2(b_1 \cdot b_{n-2} + b_2 \cdot b_{n-1}) + 2(b_1 \cdot b_{n-1}) \qquad \text{(A4-22)}$$

The scalar product of the vectors b_i and b_j that contain the angle $(180 - \tau)$ is defined as $b_i \cdot b_{i+1} = |b_i||b_{i+1}| \cos (180 - \tau)$. Therefore, from Equation (A4-22),

$$(\overline{L_{0f}^2}) = (N - 1)b^2 + 2(N - 2)b^2 \cos(180 - \tau)$$
$$+ 2(N - 3)(b_1 \cdot b_3) + \cdots + 2(b_1 \cdot b_{n-1}) \qquad \text{(A4-23)}$$

For the scalar product average we have

$$b_1 \cdot b_{j+1} = b^2 \cos^j (180 - \tau) \qquad \text{(A4-24)}$$

Equation (A4-23) then becomes

$$\overline{L_{0f}^2} = (N - 1)b^2 + 2b^2[(N - 2) \cos (180 - \tau) + (N - 3) \cos^2 (180 - \tau)$$
$$+ \cdots + \cos^{N-2} (180 - \tau)] \qquad \text{(A4-25)}$$

It is possible to develop Equation (A4-25) into a series. Writing

$a = N - 1$ and $x = \cos(180 - \tau) = \cos\alpha$, and using the following relationships, valid for $x < 1$,

$$1 + x + x^2 + x^3 + \cdots = 1/(1 - x) \tag{A4-26}$$

$$1 + 2x + 3x^2 + 4x^3 + \cdots = 1/(1 - x)^2 \tag{A4-27}$$

one obtains from Equation (A4-25)

$$\overline{L_{0f}^2} = (N - 1)\, b^2\, \frac{1 + \cos\alpha}{1 - \cos\alpha} - 2b^2(\cos\alpha)\, \frac{1 - \cos^{N-1}\alpha}{(1 - \cos\alpha)^2} \tag{A4-28}$$

For many chain units, e.g., for $N - 1 \gg 1$, the last component is negligible compared to the first, and Equation (A4-28) reduces to

$$\overline{L_{0f}^2} = Nb^2\, \frac{1 + \cos\alpha}{1 - \cos\alpha} = Nb^2\, \frac{1 - \cos\tau}{1 + \cos\tau} \tag{A4-29}$$

One obtains, analogously, for the radius of gyration

$$6\overline{R_{0f}^2} = Nb^2\, \frac{1 + \cos\alpha}{1 - \cos\alpha} = Nb^2\, \frac{1 - \cos\tau}{1 + \cos\tau} \tag{A4-30}$$

A4.4. Distribution of Chain End-to-End Distances

At any given instance, a number of chain molecules of equal length will have a random distribution of chain end-to-end distances. This information can be obtained from a derivation analogous to that used to derive the Maxwellian velocity distribution of molecules in an ideal gas.

The term $p(L_x)$ is the distribution function of the chain end-to-end distance along the x axis. Since space has reflective symmetry, $p(L_x) = p(-L_x)$, and, consequently, $p(L_x)$ must be an even-valued function of L_x and for this simplest case $p(L_x) = f(L_x^2)$.

The three distribution functions for the three possible directions in space must be interdependent when the number of bonds N is small. For $N = 1$, for example, $L_x^2 + L_y^2 + L_z^2 = b^2$ must hold, where b is the bond length. The three components of the overall distribution function become less dependent on each other with increasing number of bonds in the chain molecule. If N is very large, and L^2 is simultaneously much smaller than the square of the length of the fully stretched chain molecule, then the components can be considered to be independent of each other. The total probability is simply the product of the individual probabilities, i.e.,

$$p(L_x)p(L_y)p(L_z) = f(L_x^2)\, f(L_y^2)\, f(L_z^2) \tag{A4-31}$$

This probability cannot depend on the direction in space. It must be a function of the square of the end-to-end distance:

$$L^2 = L_x^2 + L_y^2 + L_z^2 \qquad (A4\text{-}32)$$

and thus,

$$f(L_x^2)\, f(L_y^2)\, f(L_z^2) = F(L^2) = f(L_x^2 + L_y^2 = L_z^2) \qquad (A4\text{-}33)$$

must also hold. The condition in Equation (A4-33) is only satisfied by one mathematical function, that is

$$p(L_x) = f(L_x)^2 = a \exp(-kL_x^2) \qquad (A4\text{-}34)$$

The minus sign is used so that $p(L_x)$ tends to zero when L_x tends to infinity. The constants a and k can be determined as follows: The distribution can be normalized, i.e.,

$$p(L_x)\,dL_x = a \int_{-\infty}^{+\infty} \exp(-kL_x^2)\, dL_x = a(\pi/k)^{0.5} = 1 \qquad (A4\text{-}35)$$

The second moment of the distribution function must give the mean square of the L components, i.e., $\langle L_x^2 \rangle = \langle L^2 \rangle /3 = Nb^2/3$, and thus

$$\langle L_x^2 \rangle = \int_{-\infty}^{+\infty} L_x^2 p(L_x)\,dL_x = a \int_{-\infty}^{+\infty} L_x^2 \exp(-kL_x^2)\,dL_x = \frac{a\pi^{0.5}}{2k^{3/2}} = \frac{Nb^2}{3}$$

$$(A4\text{-}36)$$

Dividing (A4-35) by (A4-36) and inserting into Equation (A4-34), we obtain

$$p(L_x) = \left(\frac{3}{2\pi Nb^2} \right)^{0.5} \exp\left(-\frac{3}{2Nb^2} L_x^2 \right) \qquad (A4\text{-}37)$$

Literature

Sections 4.1–4.3. Conformation

M. V. Volkenstein, *Configurational Statistics of Polymeric Chains*, USSR Academy of Science, Moscow, 1959; Interscience, New York, 1963.

T. M. Birshtein and O. B. Ptitsyn, *Conformations of Macromolecules*, Interscience, New York, 1966.

F. A. Bovey, *Polymer Conformation and Configuration*, Academic Press, New York, 1969.

P. J. Flory, *Statistical Mechanics of Chain Molecules*, Wiley–Interscience, New York, 1969.

G. G. Lowry, *Markov Chains and Monte Carlo Calculations in Polymer Science*, Marcel Dekker, New York, 1970.

A. J. Hopfinger, *Conformational Properties of Macromolecules*, Academic Press, New York, 1973.

Sections 4.4 and 4.5.   Macromolecular Shape

H. Sund and K. Weber, The quartenary structure of proteins, *Angew. Chem. Int. Ed.* **5**, 231 (1966).

G. N. Ramachandran, *Conformation of Biopolymers*, 2 vols., Academic Press, London, 1967.

R. E. Dickerson and I. Geis, *The Structure and Action of Proteins*, Harper and Row, New York, 1969.

V. N. Tsvetkov, V. Ye. Eskin, and S. Ya. Frenkel, *Structure of Macromolecules in Solution*, Butterworths, London, 1970.

H. Yamakawa, *Modern Theory of Polymer Solutions*, Harper and Row, New York, 1971.

H. Yamakawa, Polymer statistical mechanics, *Ann. Rev. Phys. Chem.* **25**, 179 (1974).

H. Morawetz, *Macromolecules in Solution*, second ed., Wiley–Interscience, New York, 1975.

D. A. Rees and E. J. Welsh, Secondary and tertiary structure of polysaccharides in solutions and gels, *Angew. Chem. Int. Ed. Eng.* **16**, 214–223 (1977).

K. Solc, Shape of flexible polymer molecules, *Polym. News* **4**, 67–74 (1977).

R. Jenkins and R. S. Porter, Unperturbed dimensions of stereoregular polymers, *Adv. Polym. Sci.* **36**, 1 (1980).

Section 4.6.   Optical Activity

C. Djerassi, *Optical Rotatory Dispersion*, McGraw-Hill, New York, 1960.

L. Velluz, M. Legrand, and M. Grosjean, *Optical Circular Dichroism*, Verlag Chemie, Weinheim, 1965.

B. Jirgensons, *Optical Rotatory Dispersion of Proteins and Other Macromolecules*, Springer, Berlin, 1969.

P. Pino, F. Ciardelli, and N. Zandomeneghi, Optical activity in steroregular synthetic polymers, *Ann. Rev. Phys. Chem.* **21**, 561 (1970).

P. Crabbé, *ORD and CD in Chemistry and Biochemistry*, Academic Press, New York, 1972.

E. Sélégny (ed.), *Optically Active Polymers* (= Vol. 5, *Charged and Reactive Polymers*), Reidel, Dordrecht, 1979.

Section 4.7.   Conformational Transitions

D. Poland and H. A. Scheraga, *Theory of Helix–Coil Transitions in Biopolymers—Statistical Mechanical Theory of Order–Disorder Transitions in Biological Macromolecules*, Academic Press, New York, 1970.

C. Sadron, (ed.), *Dynamic Aspects of Conformation Changes in Biological Macromolecules*, Reidel, Dordrecht, 1973.

R. Cerf, Cooperative conformational kinetics of synthetic and biological chain molecules, *Adv. Chem. Phys.* **33**, 73 (1975).

A. Teramoto and H. Fujita, Conformation-dependent properties of synthetic polypepides in the helix–coil transition region, *Adv. Polym. Sci.* **18**, 65 (1975).

A. Teramoto and H. Fujita, Statistical thermodynamic analysis of helix–coil transitions in polypeptides, *J. Macromol. Sci.—Rev. Macromol. Chem.* **C15**, 165–278 (1976).

D. Poland, *Cooperative Equilibria in Physical Biochemistry*, Oxford University Press, Oxford, 1978.

Chapter 5

Supermolecular Structures

5.1. Overview

5.1.1. Phenomena

The state of order of polymers can extend from the completely random (amorphous) to the completely ordered (ideally crystalline). The physical structures occurring depend not only on the constitution and configuration of the molecules and the micro- and macroconformations caused by these, but also on the experimental conditions. In other words, the observed solid state physical structures do not generally correspond to states at equilibrium. Thus, for example, a distinction must be made between crystallinity and crystallizability. The crystallizability is governed by the constitution and configuration and not by the crystallization conditions. As a thermodynamic equilibrium state it only depends on temperature and pressure. The crystallinity, conversely, is strongly influenced by the crystallization conditions; it includes frozen-in states of inequilibrium, and is always less than the cyrstallizability.

The crystallization conditions, however, not only influence the crystallinity, but also the morphology of the solid polymer. This can be seen, for example, in electron micrographs of solid state samples of a polyamide-6 which have been prepared under various conditions (Figure 5-1). If a solution of this polyamide in glycerine, heated to 260° C, is poured into glycerine at about 25° C, then globular structures occur (Figure 5-1a). If the same solution is cooled at 1–2 K/min, then fibrillar structures form (Figure 5-1b). At a rate of

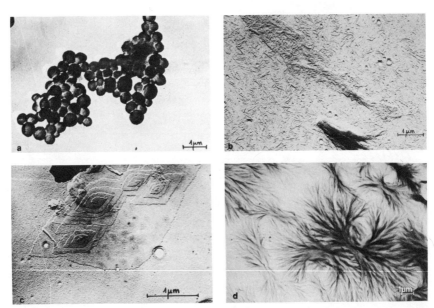

Figure 5-1. Electron micrographs of morphological structures produced in polyamide 6 under various conditions (after Ch. Ruscher and E. Schultz). Above left: a hot (260° C) solution in glycerine is shock cooled by pouring into glycerine at 25° C (part a); above right: a hot (260° C) solution in glycerine is cooled at ~1–2 K/min (part b); below left: a hot (260° C) solution in glycerine is cooled at ~40 K/min (part c); below right: slow evaporation of a solution in formic acid at room temperature (part d).

cooling of about 40 K/min, platelets occur (Figure 5-1c). From the evaporation of a solution of formic acid, on the other hand, sheaflike structures are obtained (Figure 5-1d).

The globular structures show no degree of order that can be recognized electron microscopically. A high degree of order may be presumed in the lamellae; they are reminiscent of crystals. In the fibrillar and dendrite structures, likewise, there are undoubtedly ordered structures. The nature of these states of order cannot be found from electron microscope photographs without additional methods.

Thus, a distinction between at least two states of order must be made with crystalline polymers. Practically ideally crystalline states, for example, can form over short ranges; these are described by the unit and elementary cell. The resulting ordered substructures then arrange themselves with greater or lesser degrees of order into larger structures, whereby the morphology of the solid body is produced.

5.1.2. Crystallinity

In the middle of the 19th century, for example, a crystal was defined as a material with planar surfaces which intersected at definite angles. The electron micrograph in Figure 5-1c shows, for example, planar surfaces, but these are arranged spirally. The density, X-ray scattering, and melting temperature of this material do not correspond to a 100% crystallinity, however (see below). Toward the end of the 19th century, a crystal was redefined as a homogeneous, anisotropic, solid medium. "Homogeneous" means that the physical properties do not alter on transposing along the crystal axes. Crystals are anisotropic because the physical properties vary in different directions, i.e., with rotation. However, this definition also applies to drawn poly(styrene) produced by radical polymerization, which is certainly not crystalline according to all the experimental criteria.

At the beginning of the 20th century, crystals were redefined again on the molecular or atomic basis of the lattice concept. According to this view, highly ordered crystals must give sharp diffraction pictures on being irradiated with X-rays, because the wavelength of the X-ray beam is comparable with atomic distances. A sharp melting temperature also follows from the crystal lattice concept. Macromolecular substances, however, show ill-defined diffraction patterns in structures that had been regarded as ordered from electron microscopic evidence. These data can be interpreted as resulting from crystalline and amorphous regions lying next to one another (two-phase model), or from a crystal with imperfections (one-phase model). If, for example, the X-ray measurements on polyamides are analyzed according to the two-phase model, then the result is a crystallinity of less than 50%. Such a result, however, is barely consistent with the electron microscope photograph in Figure 5-1c. Density measurements lead to the same conclusion: The same density can be interpreted as the effect of a relatively large amorphous region (two-phase model), or a small number of imperfections (one-phase model) (Table 5-1).

It is debatable whether the two-phase model describes the relationships correctly. The concept of a phase has to be used with caution in the interpretation of the physical structure of macromolecular substances because equilibrium states do not usually exist, and the borderlines between "crystalline" and "amorphous" phases are not distinct. As the electron micrographs (particularly Figure 5-1d) show, different states of order can, in fact, occur alongside each other in a particular sample. The individual methods for determining crystallinity thus cover varying degrees of order, and will consequently lead to different degrees of crystallinity (Table 5-1). The concept of "crystallinity" in macromolecules is therefore as ambiguous as that of the molecular weight for polydisperse samples, as long as the method used to

Table 5-1. Comparison of the Crystallinity α_m of a Poly(ethylene) Sample
Calculated from the Density ρ or the Specific Volume $v = 1/\rho$
Using a Two-Phase Model with with the Percentage Number of
Crystal Defects Calculated Using a One-Phase Model

Specification	Density ρ in g/cm^3	Specific volume v in cm^3/g	Crystallinity in %	Defects in %
100% crystalline	1.00	1.000	100	0
—	0.981	1.020	89	1.9
—	0.971	1.030	83	2.9
100% amorphous	0.852	1.174	0	—

determine it is not specified. At the present time it is still not possible to specify to which degree of order the various methods of crystallinity determination refer. The crystallinity is therefore characterized by the method of determination used, and referred to as X-ray crystallinity, density crystallinity, infrared crystallinity, etc. Thus, according to the method used, different crystallinities may be obtained for the same polymer, as is shown in Table 5-2 for cotton and poly(ethylene terephthalate). In poly(ethylene) and crystalline 1,4-*cis*-poly-(isoprene), on the other hand, the different methods yield consistent crystallinity values.

5.2. Crystallinity Determination

5.2.1. X-Ray Crystallography

When fast electrons impinge on matter, electrons are ejected from the inner shell of an atom of the target material, and the atom is ionized. Then electrons undergo transitions from the outer shells to the inner shells. Since the energy levels of the different shells are discrete (quantized), a coherent ray is emitted with energy equal to the difference in energy levels of the two shells. This coherent ray thus possesses a quite specific wavelength, e.g., 0.154 nm for the Cu $K\alpha$ radiation frequently used in X-ray measurements. Electron beams behave analogously, and also possess a specific wavelength, e.g., 0.0213 nm when the electrons are accelerated to 10 000 V.

Because the X-rays are of an electromagnetic nature, they must be diffracted at crystal planes if the distance between the planes is comparable to the wavelength of the X-ray. In crystals, which possess three-dimensional lattice order, the lattice plane arrangement causes diffraction to occur. The rays issuing from the different crystal planes interact systematically with one another, leading to discrete reflections. According to Bragg, the position of the reflection is given by the wavelength λ of the incident X-ray beam, the distance d_{Bragg} between parallel planes, and the angle θ between incident ray

Table 5-2. *Comparison of Degree of Crystallinity as Determined by Various Methods*[a]

	Degree of crystallinity in %		
		Poly(ethylene terephthalate)	
Method	Cellulose (cotton)	Undrawn	Drawn
Hydrolysis	93	—	—
Formylation	87	—	—
Infrared spectroscopy	—	61	59
X-ray analysis	80	29	2
Density	60	20	20
Deuterium exchange	56	—	—

[a] Calculated using the two-phase model. All values are for the same crystallization or draw conditions.

and lattice plane:

$$N\lambda = 2\, d_{\text{Bragg}} \sin \theta \qquad (5\text{-}1)$$

where N is the order of the reflection. In polymers, the reflection with the strongest intensity is frequently found for $N = 1$. If the parallel planes of the different crystallites are randomly skewed relative to one another—as in crystal powders—then a monochromatic primary ray will find enough particles to yield all the reflection positions that fulfill Bragg's conditions (Figure 5-2). Since there are many small crystallites with many different crystal plane orientations, a system of coaxial cones of X-rays with a common apex at the center of the sample is obtained. A vertical section of this system of cones on a photographic plane gives a series of concentric circles, or, using a cylindrical film, sectors of a circle.

X-ray exposures of amorphous polymers on a photographic plate show weak rings upon a strong, diffuse background (Figure 5-3a). These weak maxima are also called halos, and come from the local order in amorphous polymers. Partly crystalline polymers also show these halos, but in addition, they show the relatively strong rings of crystalline reflections (Figure 5-3b). The fact that the diffuse background is always very strong in polymers comes mainly from the scattering by air, with small contributions from the thermal movement in the crystallites, and Compton scattering. Compton scattering is a quantized, incoherent scattering effect which occurs equally in all substances regardless of their physical state.

The intensity of reflections and halos is generally interpreted with regard to the two-phase model as the content of crystalline and amorphous phases. For this, the diffuse background is first separated (see Figure 5-4). To determine the amorphous content, one begins with the lowest scattering angle,

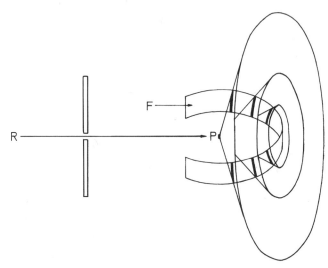

Figure 5-2. The Debye–Scherrer powder method. An X-ray *R* passes through a collimator and then meets a powder preparation *P*. The reflections caused by *P* lie on cones of reflection, which form crescents or arcs on a cylindrical film *F*.

since crystalline reflections are almost always absent here. In addition, crystalline diffraction is normally low in the minima between two maxima if the maxima are more than 3° apart. For further evaluation it is assumed that the diffracted intensity is proportional to the crystalline content and that the intensity of the amorphous halos is proportional to the amorphous content (at a specific angle or specific range of angles.) The proportionality factors also

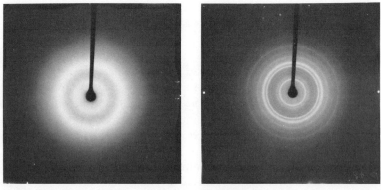

Figure 5-3. X-Ray diagram of amorphous atactic poly(styrene) (left) and drawn partially crystalline isotactic poly(styrene) (right).

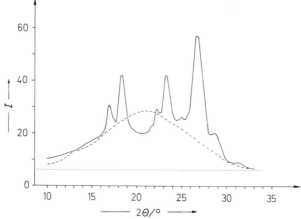

Figure 5-4. X-Ray intensity as a function of Bragg angle of amorphous (---) and crystalline (—) poly(ethylene terephthalate). The amorphous PETP was prepared by precipitation of the polymer from a solution in phenol–tetrachloroethane (1 : 1) with glycerol. Crystalline PETP was prepared by annealing (from A. Jeziorny and S. Kepka).

depend on the observation angle and a specific function. These can be determined, for example, by comparison with completely amorphous or crystalline samples. Amorphous samples can be obtained, for example, by quenching (not always possible) or by a direct X-ray study of the polymer melt. Other methods are sometimes used, e.g., amorphous cellulose can be produced by grinding in a ball mill. The intensity of the reflections is thus a measure of the crystallinity.

The width of the reflections depends on both the size of crystallites and local lattice fluctuations (defects). The smaller the crystallites, the more the diffraction appears to be scattering. Very small crystallites are therefore extremely difficult to detect by X-rays. Also, the crystallites must be present above a certain minimum concentration, otherwise the corresponding reflection intensity is not obtained by this method. In this respect, the qualitative detection of crystallinity by polarization microscopy is more satisfactory in those cases where the crystallite size is greater than the wavelength of light. In order to obtain discrete X-ray diffractions, it is necessary to have a sufficiently high concentration of ordered, three-dimensional regions extending over distances of at least 2–3 nm. This means of calculating crystallite sizes from the width of the reflections is very suspect, however, in chainlike macromolecules. The positions of the monomeric units of the individual chains are, in fact, slightly displaced with regard to one another on crystallization, since, for kinetic reasons, parts of the chain are fixed in states of order before the whole chain attains its ideal lattice position.

Local variations in the lattice constants are caused by this effect, and these similarly enlarge the width of the reflections.

Drawn fibers and films show X-ray diffraction pictures which are similar to the Bragg crystal rotation photographs (Figure 5-5). The crystal rotation method was originally carried out in order to orient favorably as many crystal planes as possible with regard to the X-ray beam. In drawn fibers and films, the molecular axes lie mainly in the direction of elongation (see Section 5.7). A ray that impinges normal to the draw direction will therefore produce reflections of varying sharpness on a photographic plate.

For historical reasons, such X-ray diffraction pictures are called fiber diagrams, although, of course, they are obtained also from drawn films. However, in contrast to crystal-rotation photographs, the fibers do not need to be rotated for fiber photographs, because many crystallites are already oriented. Reflections on the zero line are called "equatorial," and correspond to crystal planes lying parallel to the molecular axis (draw direction). Crystal planes that lie vertical ("normal") to the molecular axis produce what are called meridional reflections. Meridional reflections lie in a plane that bisects the equatorial line. When the crystallites are insufficiently oriented, the reflections (spots) degenerate into crescents (arcs) (see also Section 5.7). Thus, in shape, arcs lie between the spots of the fiber diagram with full crystallite orientation, and the circles produced by randomly oriented crystallites.

The conformation of helices can be elucidated from the observed number and spacing of layer lines in helix-forming macromolecules. In a 3_1 helix every

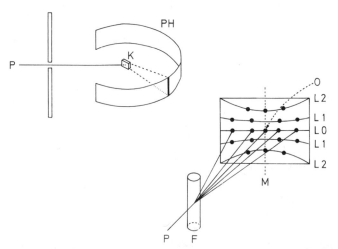

Figure 5-5. Crystal rotation method according to Bragg (upper left) and the production of a fiber diagram (lower right) by X-ray measurements. K, crystal; PH, photographic film; P, primary ray; F, fiber; M, meridian; 0, center; L0, L1, L2: zeroth, first, and second planes. L0, the equator.

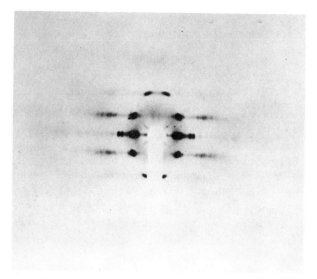

Figure 5-6. Fiber diagram of a drawn film of it-poly(propylene) (R. J. Samuels).

fourth, seventh, etc., chain link is in the same position as the first. For this case, there should be three layer lines, as is shown in Figure 5-6 for a drawn film of it-poly(propylene).

5.2.2. Density Measurements

In general, molecules are more closely packed in the crystalline than in the amorphous state. The density of a crystalline polymer is corespondingly higher ($\rho_{cr} > \rho_{am}$) and its specific volume subsequently lower ($v_{cr} < v_{am}$). A degree of crystallinity α_m by mass can therefore be determined from the observed specific volumes v_{obs}, assuming the two-phase model and additivity for the specific volumes v_{cr} and v_{am}:

$$v_{obs} = \alpha_m v_{cr} + (1 - \alpha_m) v_{am} \tag{5-2}$$

or, solved for α_m,

$$\alpha_m = \frac{v_{am} - v_{obs}}{v_{am} - v_{cr}} \tag{5-3}$$

A degree of crystallinity α_v by volume can be defined analogously:

$$\alpha_v = \frac{\rho_{obs} - \rho_{am}}{\rho_{cr} - \rho_{am}} \tag{5-4}$$

The density ρ_{obs} or the specific volume v_{obs} is determined directly by experiment. The density-gradient column is suitable for this. A density-gradient column contains a liquid whose density increases continuously from the meniscus down to the base. Such liquids can consist, for example, of mixtures of organic solvents or of salt solutions. They should neither dissolve nor swell the macromolecular sample to be studied, but must wet it. With the appropriate mechanical apparatus, for example, it is possible to form density gradients, in which the densities vary, the gradient becoming linear, concave, convex, etc., with the height of the column of liquid. A macromolecular sample then remains suspended at a particular level, according to its density.

The specific volume of an amorphous substance is obtained when the specific volume of the melt is extrapolated past the melting point to lower temperatures. An attempt may also be made to produce completely amorphous standard substances by quenching the melt, etc. The specific volume of a crystalline substance is calculated from the crystal structure. For this, one uses the volume V_e of the unit cell (see Section 5.3.1), containing the number N_i of atoms of type i having atomic weight A_i:

$$\frac{1}{v_{cr}} = \rho_{cr} = \sum \frac{N_i A_i}{V_e} \tag{5-5}$$

The densities of amorphous and crystalline polymers can differ by up to 15% (Table 5-3). Polymers with unsubstituted monomeric units, such as poly(ethylene) and nylon 6,6, for example, show the greatest difference in density. These chains crystallize in an all-*trans* conformation with particularly close packing of molecular chains. In helix-forming macromolecules with large substituents, such as it-poly(styrene), for example, the packing is, by contrast, less efficient.

Table 5-3. *Polymer Densities in Completely Amorphous and Completely Crystalline States*

| | Density in g/cm³ | | |
| | Crystalline | Amorphous | |
Polymer	ρ_{cr}	ρ_{am}	$\rho_{cr} - \rho_{am}$
Poly(ethylene)	1.00	0.852	0.148
it-Poly(propylene)	0.937	0.854	0.083
it-Poly(styrene)	1.111	1.054	0.057
Poly(vinyl alcohol)	1.345	1.269	0.076
Poly(ethylene terephthalate)	1.455	1.335	0.120
Bisphenol A-polycarbonate	1.30	1.20	0.10
Nylon 6,6 (α modification)	1.220	1.069	0.151
trans-1,4-Poly(butadiene)	1.020	0.926	0.094

5.2.3. Calorimetry

Differing states of order give rise to differing specific heats in crystalline and amorphous polymers. Presupposing a two-phase model, a degree of crystallinity can be calculated in partially crystalline polymers by means of the enthalpy, analogous to Equation (5-3):

$$\alpha_{cal} = \frac{H_{am} - H_{obs}}{H_{am} - H_{cr}} = \frac{\Delta H_m}{\Delta H_m^\circ} \tag{5-6}$$

H_{am}, H_{cr}, and H_{obs} are the totally amorphous, totally crystalline, and test sample enthalpies, respectively. Correspondingly, ΔH_m is the heat of fusion of the sample and ΔH_m° is that of a perfectly crystalline material. ΔH_m° is very difficult to determine in macromolecules, since perfectly crystalline substances are never obtained (ΔH_m° is usually found from melting point depression by diluent and occasionally from low-molecular-weight compounds, which naturally leads to some uncertainty). If the samples being studied have very small crystallites, then a quantity $\Delta H_m'$ is measured instead of ΔH_m, and this includes the surface energy γ at the end surfaces of the crystallites, as well as the length L of the crystallite:

$$\Delta H' = \Delta H_m - \frac{2\gamma}{\rho_{cr} L} \tag{5-7}$$

For this, it is assumed that the molecular axes lie parallel to the length L (see Section 10.4.1).

5.2.4. Infrared Spectroscopy

In the ir spectra of crystalline polymers, absorption bands often appear which are completely absent in amorphous polymers (Table 5-4). These bands

Table 5.4. *IR Bands Used in Determining Crystallinity of Polyolefins*

	IR bands in cm^{-1}	
Polymer	Amorphous	Crystalline
Poly(ethylene)	1298	1894; 719
it-Poly(propylene)	4274	975; 894
it-Poly(butene-1) (orthorhombic modification)	4274	815; 922

lie mainly in the range 650–1500 cm^{-1}. Consequently, they originate mainly from bond angle deformations, which are, in turn, affected by the macro-molecular conformation. These bands in the ir spectrum thus relate primarily to the conformation of the individual macromolecules, and not to inter-molecular interactions. However, macromolecules can crystallize into varying conformations, thereby giving rise to differing crystal modifications (see Section 5.3.1). These modifications can, in turn, coexist in one sample. To determine the degree of crystallinity from ir measurements, it is first necessary to ascertain whether all the crystalline contributions are included in the chosen band.

A common method of determining the crystallinity from ir measure-ments relates the measured absorbance (previously:extinction) A_{cr} of a crystalline band to the absorbance A_{cr}° of a 100% crystalline sample via the degree of crystallinity α. For amorphous bands it follows analogously that $A_{am} = (1 - \alpha_{ir}) A_{am}^{\circ}$. This assessment is also based on a two-phase model. Both A_{am} and A_{cr} are naturally also proportional to the total quantity of the sample, but since only the ratio $D = A_{cr}/A_{am}$ is measured, the influence of the amount is eliminated. By inserting the expressions for A_{cr} and A_{am} into that for D, the following is obtained:

$$\alpha_{ir} = \frac{D}{D + (A_{cr}^{\circ}/A_{am}^{\circ})} \tag{5-8}$$

The absorbances A_{cr}° or A_{am}° of the totally crystalline and totally amorphous samples, respectively, must, in turn, be determined independently.

5.2.5. Indirect Methods

Indirect methods of determining the degree of crystallinity start from the fact that a given chemical or physical event proceeds differently in the crystalline phase and in the amorphous phase. Common physical experiments include, for example, the study of water vapor absorption of hydrophilic polymers or the diffusion of a dye into the polymer. Together with a series of chemical reactions (hydrolysis, reaction with HCHO, deuterium exchange), they are used in particular for determining the crystallinity of cellulose.

The degrees of crystallinity obtained by these indirect methods are not, however, very reliable. That is to say, swelling can occur when water and chemical reagents penetrate the solid polymer, but this results in a change in the accessibility of the separate regions, and the degree of crystallinity obtained consequently no longer relates to the original sample.

5.3. Crystal Structure

5.3.1. Molecular Crystals and Superlattices

The concept of a crystal as a three-dimensional lattice consisting of periodically repeating units provides no information, per se, on the structure or size of these units or of the periodicity of their repetition. With substances of low molar mass, these units are often identical to the molecules themselves, and the intermolecular distance determines the periodicity of these molecular crystals.

Similar molecular crystals can also occur with macromolecular substances. Spherically shaped and ellipsoidal proteins produce, for example, protein crystals, with protein molecules occupying the lattice positions. The quite large voids between the lattice points and the vacant spaces within the protein molecules are filled with water or aqueous salt solution. Protein crystals consist of up to 95% water or salt solution. The channels and holes so produced are often so large that low-molar-mass substrates can penetrate and enzymatically react in these spaces. Heavy metals can also diffuse into these spaces. This effect is made use of in X-ray crystallography, since the phases of the X-ray scattering distribution can then be evaluated, and this aids the determination of the internal structure of the protein molecule (see also Section 4.4.2).

Long-chain macromolecules only seldom form large crystals. One of the few examples is poly(oxy-2,6-diphenyl-1,4-phenylene), for which crystals up to centimeter size can be obtained from solutions in tetrachloroethane. These crystals also contain large amounts of solvent, in this case, up to 35%.

Another special case is what are known as superlattices of certain block polymers. In the solid state, like blocks of different molecules arrange themselves into certain shapes—for example, as spherical domains of one kind of block in a continuous matrix of the other kind of block (see also Section 5.6.3). If the blocks are of approximately the same size, the spherical domains recur at regular distances. In this way, they form a lattice, which is called a superlattice because of the magnitude of the interdomain distances in comparison with atomic dimensions.

5.3.2. Elementary and Unit Cells

While the diffracted X-ray intensity is a measure of the crystallinity of the sample, the positions of the diffracted rays give information about the crystal structure, i.e., the unit cell.

In low-molar-mass substances, the unit cell contains at least one whole molecule. The unit cell therefore represents the lowest periodicity in an X-ray spectrum. The largest dimension of this unit cell corresponds, for example, in unbranched low-molecular-weight paraffins, to the length of the extended paraffin molecule in an all-*trans*-conformation. Thus, in *n*-alkanes, $H(CH_2)_nH$, the length L of the unit cell is equal to the chain length up to a chain-link number of $n = 70$ (Figure 5-7). In other molecules, for example, in polyurethanes, the molecular axis is not perpendicular but diagonal to the base plane. In this case L is not equal, but only proportional, to the chain length. Since L increases with increasing n, the reflections resulting from the unit cell lengths are displaced to smaller and smaller angles. The two other dimensions of the unit cell are affected by the elongation of the molecular chain perpendicular to the molecular axis and the distances between the molecules.

Above $n \approx 80$, however, the length L of the alkane unit cell remains constant at room temperature, with $L \approx 10.5$ nm. Since the chain length continues to increase as n increases, the chain must consequently fold back on

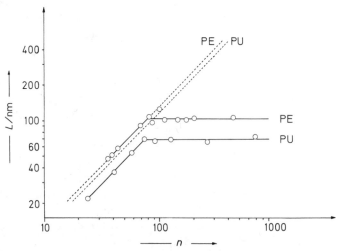

Figure 5-7. Dependence of the length L of the observed long period on the number n of chain links in alkanes (PE) with the constitutional formula $H(CH_2)_nH$ (54°C, c direction), and in polyurethanes (PU) with the constitutional formula $HO\text{-}(CH_2)_2\text{—}O\text{—}(CH_2)_2\text{—}[\text{—}O\text{—}CO\text{—}NH\text{—}(CH_2)_6\text{—}NH\text{—}CO\text{—}O\text{—}(CH_2)_2\text{—}O\text{—}CH_2)_2\text{-}]_x\text{—}OH$ (room temperature). The long periods of the low-molecular-weight polyurethanes are considerably lower than those calculated for an all-*trans* conformation (- - -), the molecular axes must therefore be diagonal to the base plane. [Measurements on alkanes and poly(ethylenes) from various authors, and on urethanes from W. Kern, J. Davidovits, K. J. Rauterkus, and G. F. Schmidt.]

itself into the crystal. An analogous effect is observed in other polymers, for example, with polyurethanes.

In chain macromolecules, however, in addition to this large periodicity coming from the unit cell, a smaller periodicity is also observed. This results from the elementary cell—a subcell of the unit cell. In the unit cell of *n*-alkanes, the CH_2 links recur periodically. Since *n*-alkanes, and also poly-(ethylene), crystallize into the all-*trans* conformation, every third, fifth, seventh, etc., CH_2 group has the same position as the first in the crystal lattice. As a result of the recurring methylene groups, an elementary cell is produced, which gives rise to short periodicity in the X-ray diagram. Subcell periodicity can be detected by X-ray scattering as strong reflections at relatively high angles. The arrangement of the molecular segments in the elementary cell can be deduced from the position of these reflections and their intensities. The chain structure of the macromolecule means that, in a crystal lattice, the atom intervals in the direction of the chain are different from those perpendicular to it. This anisotropy prevents the appearance of cubic lattices. The remaining six kinds of lattice (hexagonal, tetragonal, rhombohedral, orthorhombic, monoclinic, and triclinic) are observed, on the other hand, in long-chain macromolecules (Table 5-5). The direction of the molecular chain is termed the *c* direction. The value of $c = 0.2534$ nm in poly(ethylene) is exactly what results from the carbon–carbon bond length of 0.154 nm, and the C—C—C bond angle of 112° for the distance between every second CH_2 group in the

Table 5-5. Lattice Constants and Crystal Forms of Some Crystalline Polymers at 25° C (1 Å = 0.1 nm)

Polymer	Number of base units in the unit cell	Lattice constants in Å				Crystal system
		a	*b*	*c*	Helix	
Poly(ethylene)	2	7.36	4.92	2.534	—	Orthorhombic
st-Poly(vinyl chloride)	4	10.40	5.30	5.10	—	Orthorhombic
Poly(isobutene)	16	6.94	11.96	18.63	8_5	Orthorhombic
it-Poly(propylene) (α form)	12	6.65	20.96	6.50	3_1	Monoclinic
it-Poly(propylene) (β form)	?	6.47	10.71	?	3_1	Pseudohexagonal
it-Poly(propylene) (γ form)	3	6.38	6.38	6.33	3_1	Triclinic
st-Poly(propylene)	8	14.50	5.81	7.3	4_1	Orthorhombic
it-Poly(styrene)	18	22.08	22.08	6.63	3_1	Triclinic
it-Poly(vinyl cyclohexane)	16	21.9	21.9	6.50	4_1	Tetragonal
it-Poly(*o*-methyl styrene)	16	19.01	19.01	8.10	4_1	Tetragonal
it-Poly(butene-1) (mod. 1)	18	17.69	17.69	6.50	3_1	Triclinic
it-Poly(butene-1) (mod. 2)	44	14.85	14.85	20.60	11_3	Tetragonal
it-Poly(butene-1) (mod. 3)	?	12.49	8.96	?	?	Orthorhombic

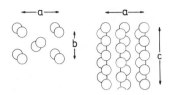

Figure 5-8. The arrangement of the CH_2 groups (shown as ○) in the crystal lattice of poly(ethylene). Because of chain folding, the chains proceed in an antiparallel direction. (After C. W. Bunn.)

chain when poly(ethylene) crystallizes into an all-*trans* conformation (Figure 5-8). st-Poly(vinyl chloride) likewise crystallizes into an all-*trans* conformation, but only every second CHCl group is in the same position as the first, so that the lattice constant is doubled, becoming $c = 0.51$ nm. In poly-(isobutene), on the other hand, the c value is not an integer multiple of 0.253 nm, so that the absence of an all-*trans* chain conformation can be concluded from this value alone. Indeed, poly(isobutene) adopts an 8_5 helical conformation in the crystal. The lattice constants are naturally very dependent on constitution and configuration, as can be seen from the four isomeric poly(butadienes) (Figure 5-9).

With increased temperatures, the measurements in the c direction, e.g., in poly(ethylene), remain constant, since the bond lengths and the valence angles of the chain remain essentially constant. However, since the forces between the molecules are affected by temperature, the a and b values must alter. In poly(ethylene), for example, for a temperature increase from $-196°$ C to $+138°$ C, the value of b increases by $\sim 7\%$.

In the packing of long-chain macromolecules already discussed, the lateral order plays a subordinate role. Effects of this kind are noticeable, however, if strong hydrogen bonds can be formed between the individual chains, as, e.g., in polyamides and proteins. A few representatives of this class of substance crystallize in the form of pleated-sheet structures, as is shown in Figure 5-10 for nylon 6 and nylon 6,6.

In the crystalline state, the molecular chains generally lie parallel to each other. They can, however, differ in terms of chirality, conformation, and orientation.

Two chains of the same chirality and conformation are isomorphous with respect to each other. Thus, two isotactic poly(propylene) helices, each with the same conformational sequence . . . $TG^+TG^+TG^+$ . . . are isomorphous. Two such chains are also isoclinal if the bond vectors in each chain always have the same positive or negative orientation. With anticlinal chains, however, the bond vectors of one chain have the reverse orientation to those of the other chain (Figure 5-11). Thus, although the chains of the polyamides 6 and 6,6 are parallel to each other, the former is anticlinal and the latter is isoclinal (Figure 5-10).

On the other hand, two chains with identical conformation but opposite

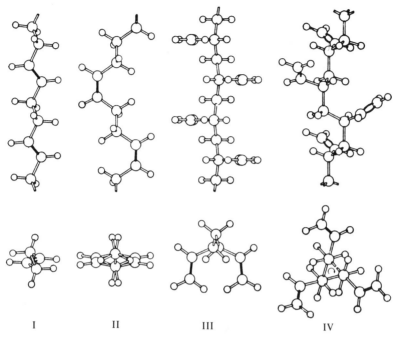

| | I | II | III | IV |
	1,4-*cis*	1,4-*trans*	1,2-syndiotactic	1,2-isotactic
a/nm	0.460	0.454	1.098	1.73
b/nm	0.950	—	0.660	
c/nm	0.860	0.49	0.514	0.65
$\rho/\text{g cm}^{-3}$	1.01	1.01	0.963	0.96
Crystal type	Monoclinic	Hexagonal	Orthorhombic	Orthorhombic

Figure 5-9. Conformations, lattice constants (a, b, and c), density ρ, and crystal forms of the four isomeric poly(butadienes) (after G. Natta and P. Corradini).

chirality are enantiomorphous to each other (Figure 5-11). An example of this is provided by isotactic chains with the conformations . . . $TG^+TG^+TG^+$ . . . and . . . $G^-TG^-TG^-T$ . . . Enantiomorphous chains may also be isoclinal or anticlinal.

5.3.3. Polymorphism

Polymorphism is defined as the occurrence of different crystal modifications of the same molecule or polymer possessing the same monomeric unit. The various modifications are characterized by different lattice constants or

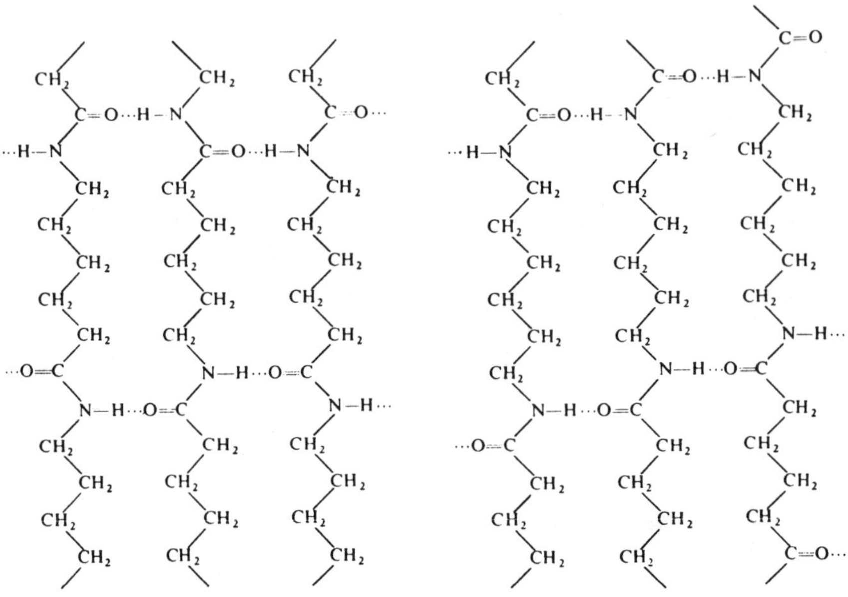

Polyamide 6

Polyamide 6,6

Figure 5-10. Pleated sheet structures of polyamide 6 [poly(caprolactam)] and polyamide 6,6 [poly(hexamethylene adipamide)].

lattice angles, and, consequently, different unit cells. The different unit cells result in microscopically perceptible differences in crystal shape, solubility, melting point, etc.

Polymorphism can result from conformational differences in a chain molecule or different packings of molecules with the same conformation. Such differences can be induced by slight alterations in the crystallization conditions, for example, by varying the crystallization temperature.

Polymorphism is observed relatively frequently in long-chain macro-molecules, for which approximately isoenergetic structures exist. The stable

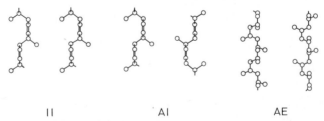

II   AI   AE

Figure 5-11. The relative arrangements of chains in a crystal lattice. II, isoclinal isomorphous; AI, anticlinal isomorphous; AE, anticlinal enantiomorphous.

Table 5-6. Thermodynamic Parameters for the Three
Modifications of it-Poly(butene-1)

Modification	T_M in °C	ΔH_M in J/mol	ΔS_M in J K^{-1} mol^{-1}
1	138	6700	16.3
2	130	4200	10.4
3	106.5 (?)	6300	16.5

crystal form of poly(ethylene), for example, possesses an orthorhombic lattice, but on elongation, triclinic and monoclinic modifications are observed. Three modifications are known in it-poly(propylene): α (monoclinic), β (pseudohexagonal), and γ (triclinic). Since the molecules are in a 3_1 helix conformation in all the modifications, differences in the packing of the chain must be responsible for this polymorphism. The three modifications appear at varying crystallization temperatures. In it-poly(butene-1), however, the various modifications correspond to different kinds of helix, so that variations in conformation must be important (see also Table 5-5). The enthalpy and entropy differences for these modifications are generally small (Table 5-6).

5.3.4. Isomorphism

The phenomenon by which various monomer units can replace each other in the lattice is termed *isomorphism*. Isomorphism is possible in copolymers if the corresponding unipolymers show analogous crystal modifications, similar lattice constants, and the same helix type. For example, according to Table 5-5, the γ form of it-poly(propylene) and modification 1 of it-poly(butene-1) possess triclinic crystal form, similar lattice constants for the c dimension, and the same helix type. The copolymers of propylene and butene-1 therefore show isomorphism. Isomorphism occurs particularly readily in helix-forming macromolecules, since the helix conformations lead to "channels" in the crystal lattice, which can easily accommodate different substituents.

5.3.5 Lattice Defects

Crystal lattices can possess a number of different defects. Some of these are characteristic of all nonmetallic solids; others are specific to crystalline macromolecular substances. Phonons, electrons, holes, excitons, site defects, interstitial defects, and displacements are commonly occurring lattice defects.

Typical crystalline macromolecular substance lattice defects result from end groups, kinks, jogs, Reneker defects, and chain displacements. Distortion of the whole crystal lattice can be conceived in terms of the paracrystal. The defects can be classified in terms of point, line, and network defects.

General Point Defects. Lattice atoms can oscillate thermally about their ideal positions. This oscillation can be conceived in terms of the oscillation of an elastic body with the energy $h\nu$. Such elastic bodies are called *phonons*. *Electrons* and *holes* are especially important with nonmetallic semiconducting solids. A semiconductor is considered to be perfect when it has an empty semiconducting band. An isolated electron in a perfect solid will, of course, produce a defect. Holes are quantum states in a normally filled semiconducting band. They behave in an electric field like a positive charge.

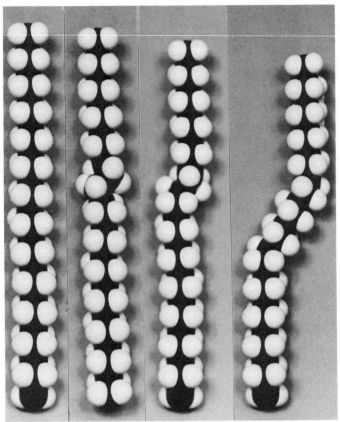

Figure 5-12. Some lattice defects in poly(ethylene). From left to right: all-*trans* conformation (defect-free), Reneker defect, kink, and jog.

Electrons and holes can be produced by thermal motion or the absorption of light. *Excitons* are electron/hole pairs. Excitons are produced when an electron takes up energy, but not a sufficient amount to escape the "hole" produced. Consequently, the electronic charge of an exciton is zero. The exciton can transport energy but cannot conduct an electric current. Empty lattice sites are called *site defects* or vacancies. Atoms residing in sites between lattice ponts are called *interstitial defects*.

Special Macromolecular Point Defects. End groups have a different chemical structure than that of the main-chain monomeric unit. They consequently produce a defect in the crystal lattice (see also Figure 5-21).

Kinks, jogs, and *Reneker defects* are conformational defects (see Figure 5-12). With kinks and jogs, a part of the chain is displaced perpendicular to the long axis by "false" conformations. This kind of defect is called a kink when the displacement is smaller than the interchain distance (example: . . . TTTTG$^+$TG$^-$TTTT . . .). If, on the other hand, the displacement is larger than the interchain distance (e.g., . . . TTTTG$^+$TTTTG$^-$TTTT . . .), then the defect is called a jog. Kinks and jogs shorten planar chains and twist helices.

Reneker defects result from conformational defects as well as from changes in bond angles (see Figure 5-12). As with kinks and jogs, the chain is also shortened here. Reneker defects can pass along the polymer chain without causing any change in the relative position of the chain in the crystal aggregate. On the other hand, spatious chain movements would result if kinks and jogs were to move through the lattice system.

Network defects occur when the lattice atom positions are randomly displaced from their ideal lattice positions. Network defects can be conceived in terms of the paracrystal model (Figure 5-13).

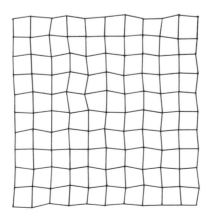

Figure 5-13. Paracrystal (schematic).

5.4. Morphology of Crystalline Polymers

5.4.1. Fringed Micelles

In the early days of macromolecular physics, crystalline reflections were observed alongside amorphous halos in the X-ray photograph of gelatine (a degradation product of the protein collagen), and this was interpreted according to the two-phase model as the coexistence of perfect crystalline regions and totally amorphous regions. Crystallite lengths of 10–80 nm were calculated from the line broadening of the reflections in these wide-angle X-ray exposures, and later from the positions of the reflections in small-angle X-ray scattering. The sizes of the crystallites were thus smaller than the maximum molecule length, which could be found from the molar mass. In poly(oxymethylenes), it was also noticed that, with increasing molar mass, the short periods resulting from the elementary cells remained, while the long periods vanished. This effect was ascribed to the absence of higher order, but higher order can only originate from lattices of high regularity. Since the then-known macromolecular substances could not be seen to crystallize either by direct observation or under the microscope, the alternative possibility— random lattice defects in crystals—seemed less probable. Thus, all these findings lead to the fringed-micelle model (Figure 5-14).

In this model, it is assumed that an individual molecular chain runs through several crystalline regions dispersed in an amorphous matrix. The model could explain the X-ray results and many other observations. Such observations and their explanations are: the smaller macroscopic density

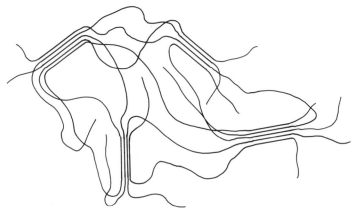

Figure 5-14. Fringed micelle model: One chain may run through several crystalline and amorphous regions.

compared to the unit cell density, as a consequence of the amorphous regions; the appearance of arcs in the X-ray diagrams of drawn polymers as a result of crystallite orientation; the finite melting range as a result of crystallites of varying sizes; the optical birefringence of drawn polymers caused by the orientation of molecular chains in amorphous regions; and the heterogeneity in relation to chemical and physical reactions as a result of the greater accessibility of the amorphous phase compared to the crystalline.

5.4.2 Polymer Single Crystals

In 1957 it was found, however, that solutions of ~0.1% of poly(ethylene) produced, on cooling, small, rhombohedral platelets, which were visible under the electron microscope (Figure 5-15). For a given solvent, the height of these platelets was always the same for a given crystallization temperature. Electron diffraction showed sharp point reflections that suggested single crystals. The interpretation of the electron-diffraction diagram led to a model in which the direction of the molecular chain is vertical to the surface of the platelets. Since the height of these rhombohedral platelets is smaller than the chain length, it follows that the chain must fold back on itself (Figure 5-16). Polymer single

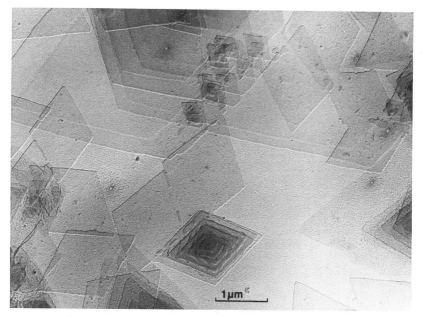

Figure 5-15. Single crystals of poly(ethylene) obtained from dilute solution. A spiral dislocation can be observed in lower center crystal (after A. J. Pennings and A. M. Kiel).

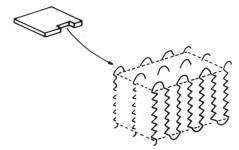

Figure 5-16. Schematic representation of chain folding in a poly(ethylene) single crystal (after W. D. Niegisch and P. R. Swan).

crystals have been observed in many crystallizable macromolecular substances, for example, in poly(oxymethylene), poly(acrylonitrile), nylon 6 (see Figure 5-1), poly(acrylic acid), cellulose derivatives, and amylose.

Studies on tearing also indicate chain folding in single crystals. According to Figure 5-16, the molecular lamellae must change direction at the diagonals in a chain folding. Lamellar cracking (cleavage) should therefore stop at the diagonals (Figure 5-17). Cracking causes the molecules to be drawn out perpendicular to the chain axis in the form of fibrils (Figure 5-18).

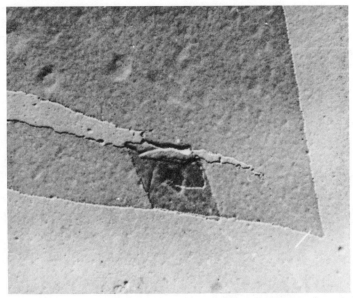

Figure 5-17. The termination of a dislocation in the direction of the plane of the chain at the diagonal of a polymer single crystal (after P. H. Lindenmeyer, V. F. Holland, and F. R. Anderson).

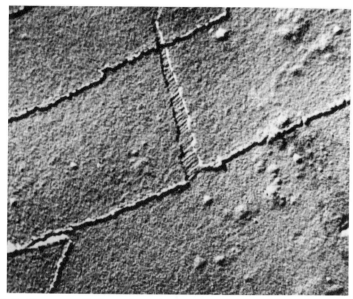

Figure 5-18. Tear perpendicular to the plane of the chain in a poly(ethylene) single crystal. The molecules are drawn out as fibrils at the point of tear (after P. H. Lindenmeyer, V. F. Holland, and F. R. Anderson).

Crystallites of this type are also called chain-folded crystals since the macromolecules in polymer single crystals crystallize with chain folding. Chain-folded crystals are not only produced from dilute solutions (Figure 5-15), they also occur as lamellar structures on crystallizing from the melt (Figure 5-19). Lamellar height is strongly increased (i.e., the relative proportion of surface layer material is decreased) when crystallization from the melt occurs under pressure (Figure 5-20). Thus, such "extended chain crystals" correspond to the equilibrium state. Chain folding must therefore result from kinetic factors (see also Chapter 10).

Despite extensive research, the exact nature of chain-fold crystal structure is not known. The distances between lamellar centers of gravity can, for example, be obtained from X-ray small-angle scattering measurements, and the thickness of the ordered lamellar layer can be determined by broad-angle X-ray scattering. According to these measurements, the difference between centers of gravity separations and lamellar thicknesses is only 0.1–1.0 nm. It is concluded from this that the lamellar surface layers have a quite regular structure. On the other hand, the single crystals are about 75%–85% X-ray crystalline. About 100% crystallinity is obtained for the residues if the surface layers are removed by oxidation with fuming nitric acid. This leads to the conclusion that the surface layers are moderately disordered ("amor-

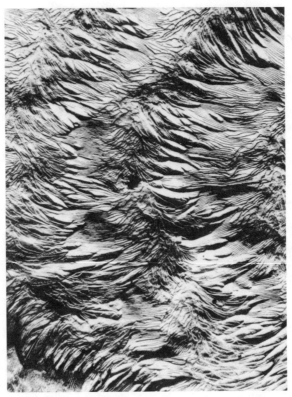

Figure 5-19. Lamellae of melt-crystallized poly(ethylene) (after P. H. Lindenmeyer, V. F. Holland, and F. R. Anderson).

phous"). In addition, neutron diffraction measurements on melt-crystallized poly(ethylene) containing small amounts of poly(deuteroethylene) show that the radius of gyration increases with the square root of the molar mass, and this also suggests that the chain folding in the surface layers is not regular. It is also questionable whether reentry of the chain to the fold surface occurs in the neighboring position (as shown in Figure 5-21) or in a more distant position, as supposed by what is known as the "switchboard" model. Other reports interpret the "amorphous" surface layer as consisting of physically adsorbed polymer layers.

The fold length or lamellar height can be increased by raising the crystallization temperature, as well as by increasing the pressure. The fold length for many polymers increases linearly with the reciprocal difference between the melting point and the crystallization temperature (see Figure

Figure 5-20. Extended chain crystals of poly(ethylene). Crystallization at 225° C and 4800 bar; $\overline{M}_w$ = 78.300 g/mol, $\overline{M}_n$ = 14,800 g/mol, 99% X-ray crystalline (after B. Wunderlich and B. Prime).

10-9). This supercooling effect also suggests a kinetic mechanism for chain folding.

The fold length is independent of the supercooling in the case of polyamides, however. In this case, the fold length is determined by the number of hydrogens involved in hydrogen bonding. Nylon 3, nylon 6,6, and nylon 6,12 have 16 hydrogen bonds (i.e., four repeating units) per fold length. Nylon 10,10 and nylon 12,12 possess only 12 hydrogen bonds (e.g., only three repeating units) per fold length.

Since the fold length increases sharply with increasing crystallization

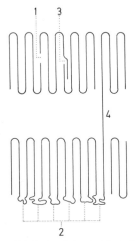

Figure 5-21. Some possible lattice defects with chain folding. 1, Chain ends; 2, disordered surface layer; 3, dislocations; 4, interlamellar linkages.

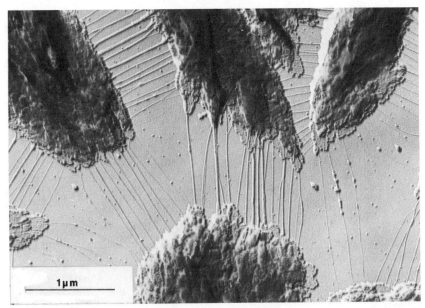

Figure 5-22. Interlamellar links between poly(ethylene) lamellae. Crystallization of mixtures of poly(ethylene) and paraffin wax (after H. D. Keith, F. J. Padden, and R. G. Vadimsky).

temperature, a polymer single crystal subsequently tempered at higher temperatures correspondingly increases in thickness. The material required to increase the fold length is taken from the interior of the crystal, thus producing holes there.

When crystallization is carried out from concentrated solution or from the melt, there is a high probability that parts of the same polymer molecule will crystallize into more than one lamella. Such interlamellar links, crystal bridges or tie molecules, were first shown to occur by jointly crystallizing poly(ethylene) with paraffin mixtures and subsequently dissolving the paraffins away (Figure 5-22). The number of interlamellar links increases with increasing molecular weight because the probability that parts of the same polymer molecule crystallize in more than one lamella increases with increasing molecular weight. Consequently, material crystallized from the melt always has a relatively high amorphous content. If the amorphous material and crystal bridges are removed, what are known as microcrystalline polymers are obtained.

5.4.3. Spherulites

In crystallization from the melt, polycrystalline regions sometimes occur which are called spherulites because of their spherical form and optical properties. Microtome sections show that their internal structure is radially symmetric. Circular structures of similar internal construction occur in the crystallization of thin films (Figure 5-23). They are therefore likewise termed *spherulites*, since they can be considered as cross sections of bulk-crystallized spherulites.

Spherulites with diameters between 5 μm and a few millimeters can be studied with an ordinary polarizing microscope, and those with diameters below 5 μm with an electron microscope or by small-angle light scattering. In

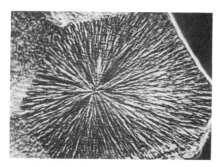

Figure 5-23. An it-poly(propylene) spherulite as seen under the phase-contrast microscope (left) and the polarization microscope (right) (after R. J. Samuels).

polarized light, spherulites show the typical Maltese cross that is caused by
birefringence effects (Figure 5-23). These effects occur because the speed of
the light varies within the different spherulitic regions. The Maltese cross
appears because the spherulites behave as crystals with radial optical
symmetry, and in this case there are four extinction positions (see Section 7.4).

The differences in the speed of the light result from differences in the
refractive index. If the highest refractive index is in the radial direction, one
talks of positive spherulites. Negative spherulites show the highest refractive
index in the tangential direction. Thus, information about the microstructure
of the spherulites can be gained from their optical properties.

To illustrate: In drawn poly(ethylene) fibers, the speed of light is less in
the direction of the fibers than in the direction perpendicular to this. Here,
light parallel to the fiber direction shows a higher refractive index. In drawn
poly(ethylene) fibers, the molecular axes are largely parallel to the fiber axis.
Since poly(ethylene) forms negative spherulites, the molecular axes must be at
right angles to the spherulite radius (Figure 5-24).

In poly(vinylidene chloride), the refractive index is lower along, rather
than perpendicular to, the molecular chain direction. Since the spherulites are
positive, then here, too, the molecular axes must be arranged tangentially to
the spherulite radius. This relationship occurs particularly in polymers with
strongly polar groups, e.g., in the polyesters and polyamides. In some cases,
the same material can form both positive and negative spherulites, possibly
even simultaneously. The negative spherulites of nylon 6,6, for example, have
a higher melting temperature than do the positive spherulites.

Spherulites show an imperfect crystalline structure, since the melting
point of the spherulite usually lies considerably below the thermodynamic
melting point (see Chapter 10). Even then, a further increase in X-ray
crystallinity can also be observed when the spherulites have filled the volume.
Localized orientation of the crystalline region leads to the characteristic

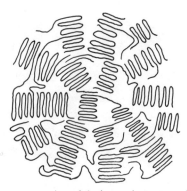

Figure 5-24. Schematic representation of the internal structure of a polyamide spherulite.

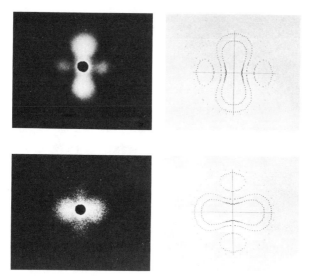

Figure 5-25. Narrow-angle light scattering of positive and negative spherulites. Photographs from incident and scattered light. Left, experimental; right, theoretical. Top, negative spherulites; bottom, positive spherulites (after R. J. Samuels).

optical properties of spherulites. If spherulites are cross-linked by radiation, the identity of the individual spherulite is retained even after they have been heated above the melting point. The birefringence of oriented sections of spherulites is lower, however, than in highly oriented fibers.

The orientation of the molecular axes in the spherulites can be followed particularly well through light scattering at very small angles. The distribution of vertically polarized scattered light can be calculated. In this case, for example, positive and negative spherulites show different scattering diagrams (Figure 5-25).

Spherulites make films and foils opaque when their diameters are greater than half in wavelength of the light and when, in addition, inhomogeneities exist in relation to the density or to the refractive index. Spherulitic poly(ethylene), for example, is opaque, but spherulitic poly(4-methyl-pentene-1) is glass clear (at room temperature), even when the latter has the same number of spherulites with the same dimensions as poly(ethylene).

5.4.4. Dendrites and Epitaxial Growth

Spherulites are produced because the overall crystallization rate is the same for all directions in space. However, the growth rates of crystals in the spherulites may be directionally dependent. Conversely, if the overall

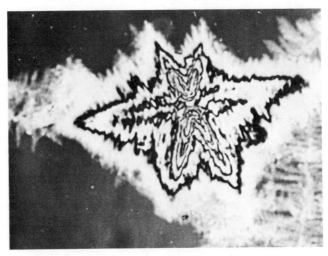

Figure 5-26. Poly(ethylene) dendrites crystallized from a dilute xylene solution at ~70° C (after B. Wunderlich).

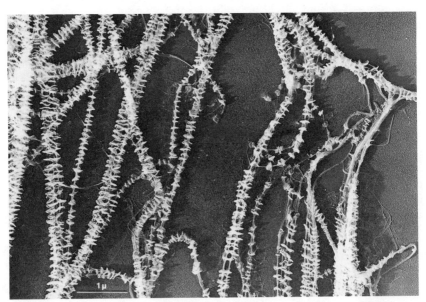

Figure 5-27. Shish-kebab structures of linear poly(ethylene) ($\overline{M}_w$ = 153 000, $\overline{M}_n$ = 12 000 g/mol). Crystallization from 5% xylene solutions with stirring at 102° C (after A. J. Pennings and A. M. Kiel).

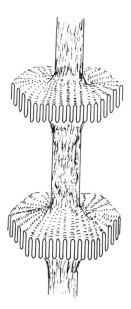

Figure 5-28. Schematic representation of arrangements of chains in shish-kebab structures (after A. J. Pennings, J. M. M. A. van der Mark, and A. M. Kiel).

crystallization rate is also direction dependent, then what are called dendrites are produced. Dendrites are structures that appear snowflake-like under the light or electron microscope (Figure 5-26). The amorphous material residing in the interior of the dendrites can be etched away by oxidation with nitric acid. The remaining crystalline component also has a lamellar structure of regular thickness.

Directionally dependent crystallization rates also lead to the "shish-kebab" (from the Armenian dish) structures (Figure 5-27). When the crystallization from dilute solution occurs with strong stirring, shish-kebab structures are produced. The macromolecules are oriented along the direction of flow and settle out parallel to each other. X-ray scattering, electron diffraction, and birefringence measurments show that the chains lie parallel to the fiber axis. The fibrils produced order themselves into nucleation bundles, but the shear gradient is strongly reduced between these nucleation bundles. The remaining macromolecules in the solution between the fibrils crystallize out onto these in the form of folded-chain lamellae. The lamellae here lie vertical with respect to the fibrils (Figure 5-28).

Shish-kebab formation is a special case of epitaxial growth. The oriented growth of one crystalline substance on another is defined as epitaxy.

5.5. *Mesomorphous Structures*

Crystals possess three-dimensional long-range order, and amorphous polymers have no order. Therefore, cases with one- or two-dimensional order must theoretically exist. They are called mesomorphous structures or meso phases, since they lie between the completely ordered and the completely disordered state.

Meso phases can be recognized and evaluated by X-ray measurements. Three-dimensionally crystalline polymers have a meridian at the equator, and also exhibit sharp diffraction reflexes in all other directions (see Figure 5-6). Other polymers only give sharp diffraction reflexes at the meridian. Examples of this behavior are poly(trimethylene terephthalamide) and poly(oxydiethylene-4,4'-dibenzoate). Such behavior is caused by longitudinal order in the chain direction, and lateral disorder in both other dimensions: the crystals have one-dimensional order and two-dimensional disorder, which is characteristic of nematic meso phases (Figure 5-29). Smectic and cholesterinic meso phases are other one-dimensional oriented ordered systems.

The molecules arrange themselves in parallel layers in smectic systems. In these cases, the molecular axis is perpendicular to the layer plane. Within the layers, the molecule may be arranged randomly or ordered with respect to other molecules. The molecules are also arranged parallel to each other but not in layers in nematic systems. The cholesteric state is midway between nematic and smectic: the molecules are arranged in layers, but with the molecular axis being parallel to the plane of the layer.

Smectic and nematic meso phases can be easily interconverted in the case of low-molar-mass compounds, and this gives melts of these compounds the characteristics of liquids. On the other hand, these melts are optically anisotropic because of their one-dimensional order, and so, have characteristic colors. Consequently, they are also known as "liquid crystals." Solutions of rod-shaped macromolecules exhibit similar ordered behavior; they are called tactoidal solutions. One-dimensional order can be induced in polymer melts by lowering the temperature below the melt or glass transition temperature, whereby the one-dimensional order is frozen in. The characteristic X-ray diagrams discussed above are then obtained.

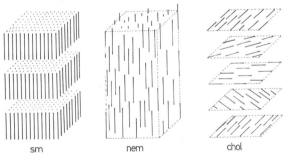

sm   nem   chol

Figure 5-29. Schematic representation of the positioning of the molecules in smectic (sm), nematic (nm), and cholesteric (chol) mesophases.

Some oriented polymers only exhibit sharp X-ray diffractions at the equator. In such cases, the macromolecules must be packed in two-dimensional lateral order. However, the monomeric units in each polymer chain are randomly arranged, because of either a random displacement of polymer chains with respect to their neighbors, or a random distribution of monomeric units within each chain or because of irregular chain structure with respect to tacticity. Poly(acrylonitrile) is an example of the last case, and poly(ethylene-*p*-carboxyphenoxyundecanoate) is an example of the first case.

Finally, it is possible that side chains in polymers with very long side chains occur in mesomorphous structures, whereas the main chain is amorphous below the glass transition temperature, and behaves like a normal liquid above it. This type of behavior is observed when the movements of the main and side chains are decoupled by flexible "spacer" groups joining them together. An example of this is a poly(methacrylate) with the monomeric unit:

$$+CH_2-C(CH_3)+ \cdot$$
$$|$$
$$COO(CH_2)_6O-(p-C_6H_4)-COO-(p-C_6H_4)-OCH_3$$

5.6. Amorphous State

Be definition, no long-range order of monomeric units is present in the completely amorphous state. Thus, amorphous materials can not be X-ray crystalline. Such a definition, or course, provides no information on the inter-chain-segmental order or on the relative positioning of the molecules themselves. Various ideal crystalline states can, of course, be defined, and the deviations from ideality can be estimated by suitable measurements. But there is no experimental method available which can measure deviations from the ideal random state, and the problem of how to define ideal randomness for molecular chains of finite length and thickness remains. Another problem is defining the closest possible packing for chainlike molecules. The closest packing for hard spheres and other simple bodies may be calulated from geometrical considerations. The problem is much more difficult with chain-like molecules because of their finite thickness and persistence, and it can be conceived that a certain number of vacancies must always be present. These vacancies give rise to a "free volume," which can be defined in various ways.

5.6.1. Free Volume

The density in the amorphous state is different from that of the hypothetical liquid. The specific volume of polymer–monomer mixtures at first decreases linearly with increasing polymer concentration (Figure 5-30).

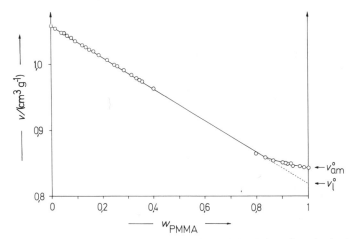

Figure 5-30. Specific volume of methyl methacrylate–poly(methyl methacrylate) mixtures as function of mass fraction of polymer at 25° C. $v_{am}^{o} = 0.842 \text{ cm}^3/\text{g}$, $v_{l}^{o} = 0.820 \text{ cm}^3/\text{g}$ (after D. Panke and W. Wunderlich).

At a certain polymer concentration, however, the viscosity of the mixture becomes so great that the polymer segments can no longer move freely. Because of this freezing-in process the specific volume of the amorphous polymer v_{am}^{o} is larger than the specific volume of the liquid polymer v_{l}^{o} would be at the same temperature. Conversely, the density of the liquid polymer is higher than the density of the solid: The solid polymer has vacant sites or what is called free volume. These vacant sites should be seen as having diameters of the order of those of atoms. The free-volume fraction f_{WLF} is calculated as

$$f_{WLF} = \frac{v_{am}^{o} - v_{l}^{o}}{v_{am}^{o}} \tag{5-9}$$

The same free volume also occurs in the Williams-Landel-Ferry dynamic glass-transition-temperature equation (see Section 10.5.2). The free-volume fraction is independent of the polymer type. It has a value of about 2.5% (Table 5-7).

In addition to the WLF free volume, a series of other free volumes can be defined and discussed. The vacant volume can be obtained from the specific volume of the amorphous polymer v_{am}^{o} measured at the temperature T and the specific volume v_{vdW}^{o} calculated from the van der Waals radii. The vacant-volume fraction f_{vac} is

$$f_{vac} = \frac{v_{am}^{o} - v_{vdW}^{o}}{v_{am}^{o}} \tag{5-10}$$

As is true for simple liquids, the vacant-volume fraction here is considerable (Table 5-7). With macromolecules, the vacant volume is not completely available for thermal motion since not all vacant sites are accessible to monomeric units on conformational grounds. The volume available for thermal expansion f_{exp} can be calculated from the specific volumes of the amorphous and crystalline polymer at 0 K:

$$f_{exp} = \frac{(v_{am}^{o})_0 - (v_{cr}^{o})_0}{(v_{am}^{o})_0} \tag{5-11}$$

Finally, a fluctuation volume f_{fluc} can be determined from sound-velocity measurements, and this describes the motion of the center of gravity of a molecule as a result of thermal motion (see Table 5-7).

5.6.2. Morphology of Homopolymers

The physical structure of solid amorphous homopolymers is a matter of considerable dispute. All proposed structures can be considered to derive from two limiting types: the coil model and the bundle model.

The coil model is based on the consideration that a randomly chosen chain grouping of a polymer in the amorphous state is always surrounded by like chain groupings. Consequently, the forces interacting between these groupings must be of equal magnitude. In addition, the chain groupings cannot distinguish between inter- and intramolecular interactions. So, the conditions prevailing should be similar to those for isolated coiled molecules in the unperturbed state (see Section 4.5.1).

In fact, neutron scattering studies show that coiled molecules have the same unperturbed dimensions as they would have in the appropriate dilute solution (see Figure 4-20). The experimental conclusion is independent of the method of preparation of the solid solution. In one case, such a solid solution was prepared by dissolving a protonated polymer in deuterated monomer and

Table 5-7. *Various Free-Volume Fractions of Amorphous Polymers Calculated from Crystalline Density at $0°C^{a}$*

| Polymer | f_{vac} | Free volume fraction | | |
		f_{exp}	f_{WLF}	f_{fluc}
Poly(styrene)	0.375	0.127	0.025	0.0035
Poly(vinyl acetate)	0.348	0.14	0.028	0.0023
Poly(methyl methacrylate)	0.335	0.13	0.025	0.0015
Poly(butyl methacrylate)	0.335	0.13	0.026	0.0010
Poly(isobutylene)	0.320	0.125	0.026	0.0017

^{a}Data from A. Bondi.

then polymerizing. A kind of spaghetti structure was intuitively expected for such a system. In another preparation, a protonated polymer and a deuterated polymer of otherwise identical constitution was mixed in dilute solution and the solution concentrated. In this case, a structure like the packing of balls of wool was intuitively expected. In both cases, however, the same dimensions were obtained. The evidence does not, of course, exclude a certain degree of short-range order.

The various bundle models actually presuppose various kinds and degrees of order in amorphous polymers. The models are based on various experimental conclusions and different theoretical considerations.

The spaghetti model, for example, predicts the density of the amorphous polymer to be about 65% of that of the crystalline polymer. A value of 85%–95% is experimentally found (see Table 5-3). Thus, it is concluded that a certain degree of order must prevail in the amorphous state, for example, as short parallel arranged lengths of chain or segments.

Such calculations, however, are based on crude geometrical models, and, so, cannot be considered to be rigorous. In fact, up to 88% of the positions of a primitive cubic lattice can be occupied, according to computer simulations, without the occurrence of ideal or perturbed chain bundles. So, space-filling problems do not hinder the dense packing of coiled molecules.

A parallel arrangement of chain segments is, of course, quite probable, and is already seen with alkanes as a result of chain persistence. The question is now: how far does such order extend? Is there long-range as well as short-range order? Data on molten poly(ethylenes) are quite conclusive; in this case, the X-ray data show greater intermolecular order than in the cases of poly(carbonates) and poly(ethylene terephthalates). Even in the poly(ethylene) case, the order does not extend more than about 2 nm in the chain direction or about 3 nm perpendicular to this direction. In addition, broad line nuclear magnetic resonance measurements show that the peaks are about 50 times smaller than expected for what is known as the meander model. This model presumes bundles with kinks and / or jogs as thermodynamically stable units. Thus, the meander model has no physical significance.

Spherically shaped structures can occasionally be electron microscopically observed with polymers. The diameters of these "nodules" vary between 2–4 nm for poly(styrene) and about 8 nm for poly(ethylene terephthalate). It is still a matter of controversy as to whether these structures are real or are experimental artifacts caused by insufficient focusing, surface effects produced by fracture or in the sample preparation, etc.

5.6.3. Morphology of Block Polymers

Two constitutionally different polymers are generally incompatible; their mixtures demix over experimentally accessible concentration ranges (see

Section 6). The blocks of block polymers will also tend to demix. The demixing cannot, however, be complete, since the blocks are, of course, jointed to each other. So, like blocks can, at most, aggregate. The shapes of these aggregates is determined by the principle of closest possible packing.

Isolated chains of a noncrystallizable polymer tend to adopt the random coil form. The same applies to both blocks of an amorphous biblock polymer. If the space requirements of both blocks are of the same magnitude, then all A blocks will be arranged in one layer, and all B blocks in another. Because of compatibility, the A layer faces another A layer and the B layer correspondingly faces another B layer. Thus, the block polymer forms lamellae of alternating A and B layers (Figure 5-31).

If, however, the space requirement of one block is very much larger than the other, then the smaller block cannot be packed in a lamellar layer— otherwise, the principle of closest possible packing or densest packing would be infringed. It is much more favorable if the smaller blocks adopt a spherical structure and the larger blocks surround these spheres as a matrix (Figure 5-32).

So, lamellar structures are formed when the space requirements of the blocks are approximately equal. Theoretically, such structures should be formed for 30%–70% contents of the smaller space requiring blocks: contents of 35%–65% have been experimentally found for styrene and butadiene block polymers (Table 5-8). With greatly differing block lengths, the excess component, on the other hand, should form a matrix for the other

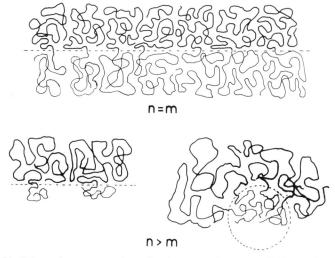

$n = m$

$n > m$

Figure 5-31. Schematic representation of volume requirements of blocks for two-block copolymers $A_n B_m$ with different block ratios n/m. The block ratios n/m refer to dimensions, not to number of monomeric units.

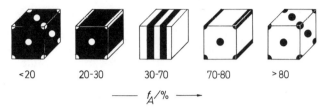

<20   20-30   30-70   70-80   >80

$f_A/\%$ ——→

Figure 5-32. Schematic representation of the morphology of bi and triblock polymers consisting of A blocks (white phase) and B blocks (black phase) (after G. Molau).

component, which adopts spherical shape. Between the ranges favorable for the formation of either lamellae or spheres-in-matrices, there is a stability region for the formation of cylinders, which can be considered as transition forms between lamellae and spheres. Forms other than lamellae, spheres, and cylinders are not to be expected since they transgress the principle of closest packing. Triblock polymers, $A_{m/2} B_n A_{m/2}$, can be treated like biblock polymers, $A_m B_n$; that is, half of the A blocks are first considered.

The three theoretically predicted domain types and their dependence on block length ratio are actually experimentally found (Figure 5-33). The experimentally found composition ratio for the occurrence of cylinders of 60/40 does not, of course, completely agree with the theoretically required values of 70/30 to 80/20. This effect, however, can be traced to the fact that the films were cast from solution. The coil dimensions change with the quality, or goodness, of the solvent, and these dimensions are more or less frozen in on removal of the solvent. Thus, the morphology of the block polymer depends not only on the block composition, but also on the preparation conditions (Table 5-9).

Similar solvent effects influence the morphology of block polymers of a crystallizable and a noncrystallizable component also. In principle, the poly(ethylene oxide) blocks of the biblock polymer, poly(styrene-*b*-ethylene

Table 5-8. Domain structure of the Component Not in Excess in Amorphous Poly(styrene-b-butadiene) Block Polymers. A is Roughly Proportional to $(M_A/M_B)^{1/2}$, Where M_A and M_B Are the Molar Masses of the A and B Blocks

Structure	Proportion of component not in excess (in %)		Molar mass dependence of the dimensions
	Theory	Experiment	
Spheres	0–20	0–20	Radius $= 1.33\ A M^{1/2}$
Cylinder	20–30	20–35	Radius $= 1.0\ A M^{1/2}$
Lamellae	30–70	35–65	Thickness $= 1.4\ A M^{1/2}$

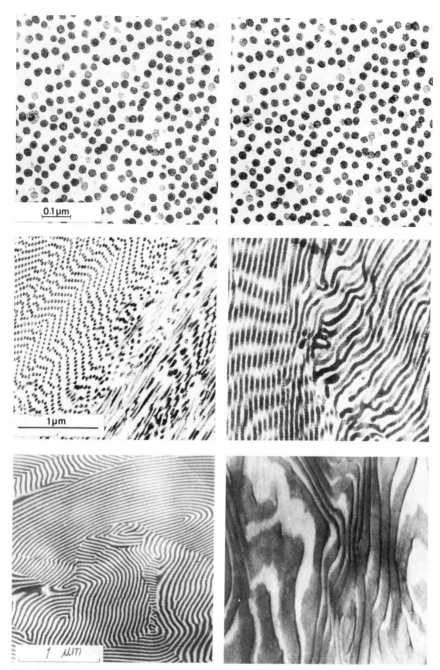

Figure 5-33. Electron micrographs of films of two-block and three-block copolymers of styrene and butadiene cut vertically (left) or parallel (right) to the film surface (M. Matsuo). Top row: SBS polymer with $S/B = 80/20$ mol/mol. Spherical domains of poly(butadiene) segments embedded in a matrix of poly(styrene) segments; center row: SB polymer with $S/B = 60/40$ mol/mol. Rods (cylinders) of poly(butadiene) segments in a matrix of poly(styrene) segments; bottom row: SBS polymer with $S/B = 40/60$ mol/mol. Lamellae of poly(butadiene) segments alternate with lamellae of poly(styrene) segments.

Table 5-9. Dependence of the Morphology of Amorphous Poly(styrene-b-butadiene-b-styrene) Triblock Polymers on the Solvent Used in Their Preparation (after M. Matsuo)

Styrene/Butadiene mol/mol	Solvent		
	Toluene	cyclohexane	butanone
80/20	Spheres	Network	Spheres
60/40	Cylinders	Cylinders	Spheres
40/60	Lamellae	Lamellae	Short cylinders

oxide) are crystallizable. Butyl phthalate is a good solvent for the poly-(styrene) blocks and a relatively poor one for the poly(ethylene oxide) blocks. On phase separation, lamallae are formed from both components, whereby those of poly(ethylene oxide) crystallize. Nitromethane, in contrast, is a good solvent for poly(ethylene oxide) and a poor one for poly(styrene). In this case, the poly(styrene) blocks form spherical domains in the poly(ethylene oxide) matrix (Figure 5-34).

5.7. Orientation

5.7.1. Definition

When fibers of films are drawn, molecules and/or crystal regions can arrange themselves in the direction of the elongation, and thus orient themselves. Since the degree of orientation is difficult to measure and the distribution function of the orientation is so far practically impossible to measure, the draw ratio is often taken as a measurement of the orientation. However, the draw ratio is not a good indication of the degree of orientation,

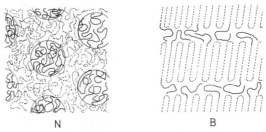

N   B

Figure 5-34. Morphology of biblock poly(styrene-*b*-ethylene oxide) copolymers cast from nitromethane (N) or butyl phthalate (B). —, Poly(styrene) blocks, ---, poly(ethylene oxide) blocks (after C. Sadron).

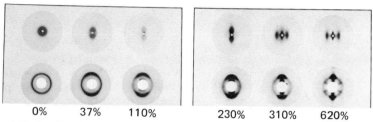

0%   37%   110%   230%   310%   620%

Figure 5-35. Small-angle (above) and wide-angle (below) X-ray interference by drawn poly-(ethylene) (after Hendus).

since, in the extreme case, only viscous flow can result from drawing. The draw conditions therefore have a great influence on the degree of orientation reached. In addition, the degree of orientation reached at a given elongation naturally depends a great deal on the previous history of the material.

Methods used to characterize orientation are wide-angle X-ray scattering, ir spectroscopy, small-angle light scattering, birefringence measurements, polarized fluorescence, and sound-velocity measurements. They refer to effects resulting partly from a combination of both types (combined effect).

5.7.2. X-Ray Diffraction

As the draw ratio increases, arcs first develop from the circular reflections at right angles to the draw direction and then point-shaped reflections in wide-angle X-ray pictures (Figure 5-35). The reciprocal length of an arc is a measure, therefore, of the extent of orientation of the crystallites, or, more precisely, the specific lattice planes. Arcs at various positions in the X-ray diagram correspond to the different lattice planes. Thus, an orientation factor f exists for each of the three spatial coordinates, and this is related to the angle of orientation β via

$$f = 0.5(3 \, \overline{\cos^2\beta} - 1) \tag{5-12}$$

where β is defined as the angle between the elongation direction and main optical axis of the unit. f will be equal to 1 for complete orientation in the chain direction ($\beta = 0$), equal to -0.5 for complete orientation perpendicular to the chain direction ($\beta = 90°$), and equal to 0 for random orientation. When the optical axes of the crystallites are at right angles to one another, then $f_a + f_b + f_c = 0$. Uniaxial elongated polymers are characterized by a single f value. The method is particularly suitable for low-to-medium degrees of orientation, since in some cases the crystallites will have become deformed at very high elongations.

5.7.3. Optical Birefringence

Every transparent material has three refractive indices n_x, n_y, and n_z along the three main axes. A material with $n_x = n_y = n_z$ is called isotropic. At least two of these refractive indices differ in anisotropic materials. The difference between each pair of these refractive indices is called the birefringence Δn.

Refractive indices vary according to the polarizability. For example, an alkane has greater polarizability along the chain than perpendicular to it because the electron mobility is greater along the chain.

Amorphous, nonoriented polymers are not optically birefringent, because their optically anisotropic monomeric units are randomly ordered with respect to one another. A birefringence first occurs when the chains are oriented or placed under strain. Generally, the following relation holds:

$$\Delta n = \Sigma \phi_i \, \Delta n_i + \Delta n_f + \Delta n_{st} \qquad (5\text{-}13)$$

Thus, each individual phase i contributes to the birefringence according to its volume fraction ϕ_i and birefringence Δn_i. These different phases can, for example, be the amorphous and crystalline phases of partially crystalline polymers, aggregates in block copolymers, fillers, or plasticized regions.

A form, or shape, birefringence Δn_f is produced when an electric field is distorted at the interface between two phases. In this case each phase must have dimensions of the order of the wavelength of light.

Amorphous polymers also become birefringent when placed under strain. The strain birefringence Δn_{st} depends on the strain applied and the anisotropy of the monomeric units. Strain birefringence is especially easy to see, even with unpolarized light, in the case of poly(styrene) with its strongly anisotropic phenyl groups. Strain birefringence is especially important in construction work with plastics since samples fail very readily at the points of highest strain.

Polarized light is generally required, however, for birefringence studies. The strongest interference colors is observed when the birefringent sample is at an angle of 45° to the oscillation direction of the polarizers. The order of the interference colors depends on the thickness of the sample and the difference in the refractive indices parallel and perpendicular to the strain direction. Refractive indices are determined by immersing the samples in inert liquids of known refractive index.

5.7.4. Infrared Dichroism

Light is absorbed when the direction of oscillation of its electrical vector has a component along the direction of oscillation of the absorbing group.

The intensity of the absorption band of an oriented polymer thus depends on the direction of the electrical vector of the incident ray relative to the direction of orientation. Consequently, the absorption will be different according to the direction of oscillation of the incident polarized light. The degree of orientation is measured using the dichroic ratio R:

$$R = \frac{A_{\parallel}}{A_{\perp}} = \frac{\ln(I_0/I_{\parallel})}{\ln(I_0/I_{\perp})} \tag{5-14}$$

where I_0 is the intensity of the incident light and $I_{\parallel}$ and $I_{\perp}$ are the intensities of the transmitted light parallel and at right angles to the direction of elongation, respectively. The orientation factor f is derived from the dichroic ratio R and the corresponding value R^{∞} for complete orientation, analogously to the Lorentz–Lorenz formula, as

$$f = \frac{(R-1)(R^{\infty}+2)}{(R^{\infty}-1)(R+2)} \tag{5-15}$$

R^{∞} can be calculated if it is known that the dipole oscillation of a specific group in a uniaxially drawn polymer occurs at right angles to the chain axis, as is the case, for example, with hydrogen bonds between amide groups in polyamides. This method yields information about both "amorphous" and "crystalline" bands. First, however, every change in the ir bands produced by elongation must be shown not to have come from conformational changes in the molecule during drawing.

5.7.5. Polarized Fluorescence

Most organic polymers do not fluoresce. Consequently, about $10^{-4}\%$ (weight) of a fluorescing organic dyestuff is mixed in with the polymer. In the analysis of the results, it is assumed that the dyestuff added does not alter the morphology of the polymer and that the axes of the dyestuff molecules correspond to those of the polymer molecules. Also, the chromophore groups are not considered to rotate during the lifetime of the excited state, which is probably true because of the high viscosity. Since the dyestuff molecule generally does not have access to the crystal lattice, the method is only suitable for amorphous regions.

For measurement, plane-polarized light is allowed to fall on the fluorescing groups. The fluorescent light is also polarized.

When the polarizer and analyzer are parallel to the strain direction, the observed intensity varies with the fourth power of the cosine of the angle β between the strain direction and the molecular axis:

$$I = \text{const} \times \langle \cos^4 \beta \rangle \tag{5-16}$$

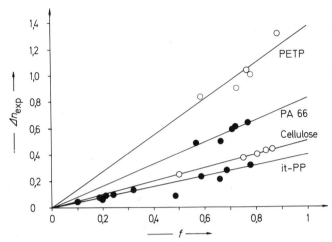

Figure 5-36. Optical birefringence Δn_{exp} as a function of orientation factor f from sound-velocity measurements of different polymers (after H. M. Morgan).

For fluorescent light observed with polarizers perpendicular to the strain direction and for uniaxial strain,

$$I_{\perp} = 0.5 \times \text{const} \times (\langle \cos^2 \beta \rangle - \langle \cos^4 \beta \rangle) \qquad (5\text{-}17)$$

5.7.6. Sound Propagation

The velocity of sound depends on the distances between the main-chain atoms and the intermolecular distances between the chains. In order to determine the orientation angle β from measurements of the sound velocity $\hat{c}$ in the fiber direction, it is necessary to know the sound velocities $\hat{c}_{\perp}$ and $\hat{c}_{\parallel}$ perpendicular and parallel to a polymer sample with completely oriented chains:

$$\frac{1}{\hat{c}^4} = \frac{1 - \langle \cos^2 \beta \rangle}{\hat{c}_{\perp}^2} + \frac{\langle \cos^2 \beta \rangle}{\hat{c}_{\parallel}^2} \qquad (5\text{-}18)$$

In a completely unoriented sample, $f = 0$ according to Equation (5-12) and thus $\cos^2 \beta = 1/3$. With these values, Equation (5-18) gives

$$\hat{c}_{\perp}^2 = \frac{2\hat{c}_u^2 \hat{c}_{\parallel}^2}{3\hat{c}_{\parallel}^2 - \hat{c}_u^2} \qquad (5\text{-}19)$$

Typical $\hat{c}_{\parallel}$ and $\hat{c}_{\perp}$ values are about 1.5 and 7–10 km/s, respectively.

The following procedure is adopted to determine the orientation angle. The velocity $\hat{c}_{\parallel}$ is either estimated or theoretically calculated. The velocity c_u is measured. Then $\hat{c}_\perp$ is calculated from Equation (5-19). From all experience to date, the inequality $3\hat{c}_\parallel^2 \gg \hat{c}_u^2$ applies, so that the calculated values of $\hat{c}_\perp$ is fairly insensitive to the assumed value of $\hat{c}_\parallel$. Equation (5-19) can thus be given as

$$\hat{c}_\perp^2 = \tfrac{2}{3}\hat{c}_u^2 \qquad (5\text{-}20)$$

and Equation (5-18) as

$$\frac{1}{\hat{c}^2} = \frac{1 - \langle \cos^2 \beta \rangle}{\hat{c}_\perp^2} \qquad (5\text{-}21)$$

Combining Equations (5-20) and (5-21), we obtain

$$\langle \cos^2 \beta \rangle = 1 - \frac{2\hat{c}_u^2}{3\hat{c}^2} \qquad (5\text{-}22)$$

Consequently, an expression for the orientation factor f can be obtained from Equations (5-22) and (5-12). f can be determined from the sound velocity in the fiber and in the unoriented sample:

$$f = 1 - \frac{\hat{c}_u^2}{\hat{c}^2} \qquad (5\text{-}23)$$

The method allows the orientation factor to be determined for films and fibers during the actual straining or stretching process. A linear relationship between the orientation factor determined by sound-velocity measurement and that obtained by birefringence has been found experimentally (Figure 5-36).

Literature

5.1. General Reviews

B. Wunderlich, *Macromolecular Physics*, Academic Press, New York, 3 Vols., 1973–1980.

J. E. Spruiell and E. S. Clark, X-Ray Diffraction, in R. Fara (ed.), *Polymers* (= Vol. 16B of L. Marton and C. Marton (eds.-in-chief), *Methods of Experimental Physics*), Academic Press, New York, 1980.

J. F. Rabek, *Experimental Methods in Polymer Chemistry*, Wiley, New York, 1980.

5.2. Determination of Crystallinity and Crystal Structures

S. Krimm, Infrared spectra of high polymers, *Fortschr. Hochpolym. Forsch.—Adv. Polym. Sci* **2**, 51 (1960).

E. W. Fischer, Electron diffration, in B. Ke, (ed.), *Newer Methods of Polymer Characterization*, Interscience, New York, 1964, p. 279.

W. O. Statton, Small angle X-ray studies of polymers, in B. Ke (ed.), *Newer Methods of Polymer Characterization*, Interscience, New York, 1964, p. 231.

B. K. Vainsthein, *Diffraction of X-Rays by Chain Molecules*, Elsevier, Amsterdam, 1966.

H. Brumberger (ed.), *Small Angle X-Ray Scattering*, Gordon and Breach, New York, 1967.

A. Elliott, *Infrared Spectra and Structure of Organic Long-Chain Polymers*, Arnold, London, 1969.

L. E. Alexander, *X-Ray Diffraction Methods in Polymer Science*, Wiley, New York, 1969.

S. Kavesh and J. M. Smith, Meaning and measurement of crystallinity in polymers: A review, *Polym. Eng. Sci.* **9**, 331 (1969).

M. Kakudo and N. Kasai, *X-Ray Diffraction by Polymers*, Kodansha, Tokyo, and Elsevier, Amsterdam, 1972.

G. H. W. Milburn, *X-Ray Crystallography, An Introduction to the Theory and Practice of Single-Crystal Structure Analysis*. Butterworths, London, 1972.

5.3. Crystal Structures

C. W. Bunn, *Chemical Crystallography*, Clarendon Press, Oxford, 1946.

F. Danusso, Macromolecular polymorphism and stereoregular synthetic polymers, *Polymer* (London) **8**, 281 (1967).

G. Allegra and I. W. Bassi, Isomorphism in synthetic macromolecular systems, *Adv. Polym. Sci.* **6**, 549 (1969).

R. Hosemann, The paracrystalline state of synthetic polymers, *Crit. Revs. Macromol. Sci.* **1**, 351 (1972).

A. I. Kitaigorodsky, *Molecular Crystals and Molecules*, Academic Press, New York, 1973.

H. Tadokoro, *Structure of Crystalline Polymers*, Wiley, New York, 1979.

5.4. Morphology of Crystalline Polymers

P. H. Geil, *Polymer Single Crystals*, Wiley, New York, 1963.

D. A. Blackadder, Ten years of polymer single crystals, *J. Macromol. Sci.* (*Rev.*) *C1*, 297 (1967).

J. Willems, Oriented overgrowth (epitaxy) of macromolecular organic compounds, *Experientia* **23**, 409 (1967).

L. Mandelkern, Thermodynamics and physical properties of polymer crystals formed from dilute solution, *Progr. Polym. Sci.* **2**, 163 (1970).

R. A. Fava, Polyethylene crystals, *J. Polym. Sci.* **D5**, 1 (1971).

R. H. Marchessault, B. Fisa, and H. D. Chanzy, Nascent morphology of polyolefins, *Crit. Rev. Macromol. Sci.* **1**, 315 (1972).

A. Keller, Morphology of lamellar polymer crystals, in C. E. H. Bawn (ed.), *Macromol. Sci.* MTP *Internatl. Rev. Sci., Phys. Chem. Ser. One*, Vol. 8, Butterworths, Baltimore, Md, 1972.

R. J. Samuels, *Structured Polymers Properties*, Wiley, New York, 1974.

D. C. Bassett, Chain-extended polyethylene in context: a review, *Polym.* [*London*] **17**, 460 (1976).

D. G. H. Ballard, G. W. Longman, T. L. Crowley, A. Cunningham, and J. Schelten, Neutron scattering . . . of semicrystalline Polymers, *Polymer* **20**, 399 (1979).

J.-I. Wang, and I. R. Harrison, X-Ray Diffraction, Crystallite Size and Lamellar Thickness by X-Ray Methods, *Methods Exp. Phys.* **16B**, 128 (1980).

V. J. McBrierty, and D. C. Douglass, Nuclear magnetic resonance of solid polymers, *Phys. Rep.* **63**, 61 (1980).

5.5. Mesophases

V. P. Shibayev and N. A. Plate, Liquid crystalline polymers, *Vysokomol. Soyed.* **A19**, 923–972 (1977); *Polym. Sci. USSR* **19**, 1065–1122 (1977).

S. P. Papkov, The liquid crystalline state of linear polymers. Review, *Polym. Sci. USSR* **19**, 1–19 (1978).

P. G. De Gennes, *The Physics of Liquid Crystals*, Clarendon Press, Oxford, 1974.

W. G. Miller, Stiff chain polymer lyotropic liquid crystals, *Ann. Rev. Phys. Chem.* **29**, 519 (1978).

A. Blumstein (ed.), *Liquid Crystalline Order in Polymers*, Academic Press, New York, 1978.

E. T. Samulski and D. B. DuPré, Polymeric liquid crystals, *Adv. Liq. Cryst.* **4**, 121 (1979).

J. L. White and J. F. Fellers, Macromolecular liquid crystals and their applications to high-modulus and tensile-strength fibers, *J. Appl. Polym. Sci.: Appl. Polym. Symp.* **33**, 137 (1978).

5.6. Amorphous State

R. N. Haward, Occupied volume of liquids and polymers, *J. Macromol. Sci. Revs.* **C4**, 191 (1970).

T. G. F. Schoon, Microstructure in solid polymers, *Brit. Polym. J*, **2**, 86 (1970).

G. S. Y. Yeh, Morphology of amorphous polymers, *Crit. Revs. Macromol. Sci.* **1**, 197 (1972).

R. E. Robertson, Molecular organization of amorphous polymers, *Ann. Rev. Mater. Sci.* **5**, 73 (1975).

R. F. Boyer, Structure of amorphous solids. Structure of the amorphous state in polymers, *Ann. N. Y. Acad. Sci.* **279**, 223–233 (1976).

J. A. Manson and L. H. Sperling, *Polymer Blends and Composites*, Plenum Press, New York, 1976.

A. Noshay and J. E. McGrath, *Block Copolymers: Overview and Critical Survey*, Academic Press, New York, 1976.

G. Allen and S. E. B. Petrie (eds.), *Physical Structure of the Amorphous State*, Marcel Dekker, New York, 1977.

B. R. M. Gallot, Preparation and study of block copolymers with ordered structures, *Adv. Polym. Sci.* **29**, 85 (1978).

5.7. Orientation

G. L. Wilkes, The measurement of molecular orientation in polymeric solids, *Adv. Polym. Sci.* **8**, 91 (1971).

C. R. Desper, Technique for measuring orientation in polymers, *Crit. revs. Macromol. Sci.* **1**, 501 (1973).

I. M. Ward (ed.), *Structure and Properties of Oriented Polymers*, Halsted Press, New York, 1975.

B. Jasse and J. L. Koenig, Orientational measurements in polymers using vibrational spectroscopy, *J. Macromol. Sci. (Revs. Macromol. Chem.)* **C17**, 61 (1979).

Part II
Solution Properties

Chapter 6
Solution Thermodynamics

6.1. Basic Principles

According to the second law of thermodynamics, the Gibbs energy ("free energy") G is related to the enthalpy H, the entropy S, and the thermodynamic temperature T by

$$G = H - TS = U + pV - TS \qquad (6\text{-}1)$$

where U is the internal energy, p is the pressure, and V is the volume. The Helmholtz energy A is given by

$$A = U - TS = G - pV \qquad (6\text{-}2)$$

$\Delta G \approx \Delta A$ frequently holds for isobaric processes in condensed systems since the change in volume is often (but not always) negligibly small.

The change of Gibbs energy of a system caused by the addition of 1 mol of component i to an infinite system is called the partial molar Gibbs energy $\tilde{G}_i^m$, or the chemical potential μ_i:

$$\left(\frac{\partial G}{\partial n_i} \right)_{T,p,n_{j \neq i}} \equiv \tilde{G}_i^m \equiv \mu_i \qquad (6\text{-}3)$$

Differentiation of the chemical potential of component i gives (see textbooks on chemical thermodynamics)

$$\partial \tilde{G}_i^m = \frac{\partial \tilde{G}_i^m}{\partial p} \, dp + \frac{\partial \tilde{G}_i^m}{\partial T} \, dT + \frac{\partial \tilde{G}_i^m}{\partial n_i} \, dn_i$$

$$= \tilde{V}_i^m \, dp - \tilde{S}_i^m \, dT + RT \, d(\ln a_i) \qquad (6\text{-}4)$$

$\tilde{V}_i^m$ is the partial molar volume of component i of relative activity a_i. The complete differential form of G^m is

$$dG^m = \sum_i \tilde{G}_i^m \, dn_i + \sum_j n_i \, d\tilde{G}_i^m \qquad (6\text{-}5)$$

The left-hand side of this equation must be identical to the first term of the right-hand side, according to equation (6-3). The so-called Gibbs–Duhem relationship thus states

$$\sum_i n_i \, d\tilde{G}_i^m = \sum_i n_i \, d\mu_i = 0 \qquad (6\text{-}6)$$

With $dp = 0$ and $dT = 0$ for an isothermal–isobaric process, one obtains from Equation (6-4) after integrating and converting to chemical potentials

$$\mu_i = \mu_i^\circ + RT \ln a_i = \mu_i^\circ + RT \ln x_i \gamma_i \qquad (6\text{-}7)$$

The integration constant μ_i° is the chemical potential of the pure substance. The relative activity is often further separated into the mole fraction x_i and the activity coefficient γ_i. The contribution from the mole fraction is often called the ideal component or function, and that coming from the activity coefficient is the excess function:

$$\Delta \mu_i = \mu_i - \mu_i^\circ = RT \ln x_i + RT \ln \gamma_i = \Delta \mu_i^{id} + \Delta \mu_i^{exc} \qquad (6\text{-}8)$$

Solutions or mixtures can be classified into four types according to the magnitude and sign of the excess functions: ideal, athermal, regular, and irregular (or real) solutions. In ideal solutions, the total contribution to the Gibbs energy of mixing comes solely from the ideal entropy of mixing (see Section 6.3.1). The enthalpy of mixing is also zero with athermal solutions, but the entropy of mixing is different from the ideal entropy of mixing. There is no excess entropy of mixing with regular solutions, but in this case the enthalpy of mixing is not zero. An enthalpy of mixing as well as an excess entropy of mixing are observed with irregular solutions.

The pseudoideal or theta solution is an important special case of irregular solutions in macromolecular science. The enthalpy of mixing and the excess entropy of mixing exactly compensate each other at a certain temperature with the dilute theta solution. Theta solutions at this theta temperature thus behave like ideal solutions. In contrast to ideal solutions, however, the enthalpy of mixing is not zero and the entropy of mixing differs considerably from the ideal entropy of mixing. Thus, an ideal solution exhibits ideal behavior at all temperatures, the pseudoideal solution only behaves ideally at the theta temperature. Consequently, the theta temperature corresponds to the Boyle temperature of real gases.

6.2. Solubility Parameter

6.2.1. Basic Principles

Thermodynamic analysis allows solutions to be classified *after* thermodynamic parameters have been determined. It cannot, however, predict the solubility or miscibility of two substances without the aid of additional assumptions. A prediction of this type is possible with the concept of the solubility parameter, which is based on the following considerations.

The transfer from the liquid to the gaseous state requires overcoming an interaction energy of $z\epsilon_j/2$ per molecule and consequently $N_L z\epsilon_j/2$ per mole. This is, however, exactly equal to the negative internal molar energy of vaporization $(\Delta E_{vap})_j$. The ϵ_j is the energy per bond. One molecule has z neighbors. The corresponding quantity related to the molar volume V^m is called the cohesive energy density:

$$e_j = \frac{\Delta E_{vap\,j}}{V_j^m} = -0.5 \frac{N_L \epsilon_j z}{V_j^m} \tag{6-9}$$

The solubility parameter is defined as the square root of the cohesive energy density:

$$\delta_j \equiv e_j^{0.5} \tag{6-10}$$

Interaction energies ϵ are related to each other in the following manner. Mixing solvent 1 and polymer 2 produces two solvent–polymer 1–2 bonds for every broken solvent–solvent 1–1 and polymer–polymer 2–2 bond. The change in interaction energy during the mixing process is consequently

$$\Delta\epsilon = \epsilon_{12} - 0.5(\epsilon_{11} + \epsilon_{22}) \tag{6-11}$$

$$-2\Delta\epsilon = (\epsilon_{11}^{0.5})^2 - 2\epsilon_{12} + (\epsilon_{22}^{0.5})^2 \tag{6-12}$$

From quantum mechanics it is known that the interaction energy of two different spherical molecules due to dispersion forces is equal to the geometric mean of the mutual interaction energies of the molecules themselves, i.e.,

$$\epsilon_{12} = -(\epsilon_{11}\epsilon_{22})^{0.5} \tag{6-13}$$

The minus sign occurs in Equation (6-13) because the interaction energy ϵ_{12} represents a geometric mean of two normally negative interaction energies ϵ_{11} and ϵ_{22}.

Inserting Equation (6-13) into Equation (6-12), we obtain

$$\Delta\epsilon = -0.5(|\epsilon_{11}|^{0.5} - |\epsilon_{22}|^{0.5})^2 \qquad (6\text{-}14)$$

Assuming equal molar volumes of solvent and polymer monomeric units, combination of Equations (6-10) and (6-14) gives

$$\frac{0.5z N_L \Delta\epsilon}{V^m} = -0.5(\delta_1 - \delta_2)^2 \qquad (6\text{-}15)$$

The difference in solubility parameters thus yields a measure of the interaction between solvent and solute with respect to the mutual interactions between like components. Now, if $\epsilon_{11} \gg \epsilon_{12}$ and/or $\epsilon_{22} \gg \epsilon_{12}$ then there will be practically no interaction between solute and solvent. The difference $|\delta_1 - \delta_2|$ will then be very large. With equal interactions between 1–1, 2–2, and 1–2, on the other hand, $\delta_1 - \delta_2 = 0$ and good solubility is obtained. There must therefore be a maximum permitted difference $|\delta_1 - \delta_2|$ at which it is still just possible to have mixing. The experimentally obtained maximum differences vary, according to the polarity of the solvent, between ± 0.8 and ± 3.4 (Table 6-1).

Solubility parameters are traditionally given without units, although, strictly speaking, the figures given in literature have the units $(cal/cm^3)^{1/2}$. Note that $1\ (cal/cm^3)^{1/2} = 2.05(J/cm^3)^{1/2}$.

The concept of solubility parameters is an attempt to quantify the old rule-of-thumb, "like dissolves like." It must of necessity fail when the interaction forces differ greatly in nature. Recently the solubility parameter has been separated into three component parameters in order to refine the quite crude process of trying to predict the possible mixability of a polymer and a solvent. The three components describe the interaction between dispersion, dipole, and hydrogen-bonding forces:

$$\delta^2 = \delta_d^2 + \delta_p^2 + \delta_h^2 \qquad (6\text{-}16)$$

Table 6-1. δ-Regions for Polymers

Polymer	Solubility parameter δ of solvents that dissolve the polymer	
	Apolar solvents	Polar solvents[a]
Poly(styrene)	9.3 ± 1.3	9.0 ± 0.9
Poly(vinyl chloride-co-vinyl acetate)	10.2 ± 0.9	10.6 ± 2.8
Poly(vinyl acetate)	10.8 ± 1.9	11.6 ± 3.1
Poly(methyl methacrylate)	10.8 ± 1.2	10.9 ± 2.4
Cellulose trinitrate	11.9 ± 0.8	11.2 ± 3.4

[a] Alcohols, esters, ethers, ketones.

As expected, the contribution due to dispersion forces δ_d varies only very slightly from system to system. Consequently, solubility diagrams are constructed so that δ_h values are plotted against δ_p values. Each solvent will have its own δ_d value. Points are then drawn on the plot representing the different solvents for the polymer in question (for example, drawn in color). Finally a contour is drawn with the aid of the δ_d values for all solvents that actually dissolve the polymer. Normally, the solubility increases with increasing δ_d for like values of δ_p and δ_h. Thus, if a substance has a δ_d value which is within the contour for this numerical value, then it is a solvent for the polymer.

6.2.2. Experimental Determination

Solubility parameters δ_1 can be determined directly via Equation (6-9). One has simply to subtract the work against the external pressure from the experimental negative enthalpy of vaporization. All δ_1 values compiled in Table 6-2 were determined this way.

Macromolecules cannot be vaporized without degradation, because of the high cohesive energy density per molecule. Their solubility parameters are therefore often not determined directly, but are assumed to be equal to those of low-molecular-weight compounds. Alternatively, they can be estimated from the swelling of the cross-linked parent polymer. Cross-linked polymers swell more, the greater the interaction between polymer and solvent (see Section 6.6.7). A plot of the degrees of swelling against the solubility parameter δ_1 of the solvents thus gives a maximum at a certain δ_1, which corresponds to the δ_2 of the polymer (Figure 6-1).

Intrinsic viscosities $[\eta]$ can be measured for soluble polymers in various solvents. $[\eta]$ increases with increasing polymer–solvent interaction (Chapter 9.9.6). If the $[\eta]$ values are plotted against the solubility parameter of the solvents used, the maximum corresponds to the solubility parameter δ_2 of the polymer.

The viscosity method for soluble polymers and the swelling method for cross-linked network polymers yield quite unambiguous values for polymer solubility parameters, so long as one is confined to a series of structurally similar solvents. For example, the data in Figure 6-1 apply to aliphatic hydrocarbons as well as to long-chain esters and ketones. Cycloaliphatic hydrocarbons and short-chain esters such as ethyl acetate deviate significantly from the curves shown.

The solubility parameters of polymers may also be estimated from those of low-molar-mass analogs. The solubility parameters of a homologous series are plotted against the ratio V_e^m / V_c^m, and extrapolated to vanishingly small

Table 6-2. The Solubility Parameters [$in \ (cal/cm^3)^{1/2}$]

Solvent	δ_1	δ_d	δ_p	δ_h
Heptane	7.4	7.4	0	0
Cyclohexane	8.18	8.18	0	0
Benzene	9.05	8.99	0.5	1.0
Carbon tetrachloride	8.65	8.65	0	0
Chloroform	9.33	8.75	1.65	2.8
Dichloromethane	9.73	8.72	3.1	3.0
1,2-Dichloroethane	9.42	8.85	2.6	2.0
Acetone	9.75	7.58	5.1	3.4
Butanone	9.30	7.77	4.45	2.5
Cyclohexanone	10.00	8.65	4.35	2.5
Ethyl acetate	9.08	7.44	2.6	4.5
Propyl acetate	8.74	7.61	2.2	3.7
Amyl acetate	8.49	7.66	2.1	3.3
Acetonitrile	11.95	7.50	8.8	3.0
Pyridine	10.60	9.25	4.3	2.9
Diethyl ether	7.61	7.05	1.4	2.5
Tetrahydrofuran	9.49	8.22	2.7	3.9
p-Dioxane	9.65	8.93	0.65	3.6
1-Pentanol	10.59	7.81	2.2	6.8
1-Propanol	11.85	7.75	3.25	8.35
Ethanol	12.90	7.73	4.3	9.4
Methanol	14.60	7.42	6.1	11.0
Cyclohexanol	10.69	7.75	3.8	6.3
m-Cresol	11.52	9.14	2.35	6.6
Nitrobenzene	11.25	9.17	6.2	2.0
Dimethyl acetamide	10.24	8.2	5.6	5.0
Hexamethyl phosphamide	11.35	9.0	4.2	5.5
Dimethyl formamide	12.14	8.5	6.7	5.5
Dimethyl sulfoxide	13.04	9.0	8.0	5.0
Water	23.43	6.0	15.3	16.7

end group molar volume, V_e^m, or infinitely high main-chain group molar volume, V_c^m, (Figure 6-2).

6.2.3. Applications

The solubility of a polymer can be estimated from its δ_2 value in many cases.

Apolar substances have low solubility parameters, whereas those of polar substances are high, since the heat of vaporization is higher for the latter. Apolar, noncrystalline polymers will therefore dissolve well in solvents with low δ_1 values. Predictions about solubility on the basis of the solubility parameter are still quite permissible for polar, noncrystalline polymers in

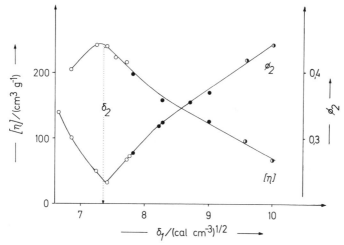

Figure 6-1. Influence of goodness of solvent, as measured by the solubility parameter, δ_1, of the solvent, on the intrinsic viscosity. $[\eta]$, of dissolved natural rubber and on the volume fraction, ϕ_2, of the cross-linked natural rubber polymer in aliphatic hydrocarbons, ($\circ$), long-chain esters, ($\bullet$), and long-chain ketones, ($\circleddash$). After data from G. M. Bristow and W. F. Watson. The solubility parameter is given in the traditional physical units.

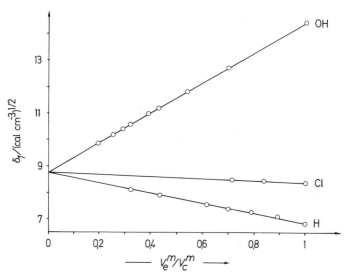

Figure 6-2. Solubility parameters, δ_1, of alkanes, $H(CH_2)_n H$, 1-chloroalkanes, and 1-hydroxy-alkanes as a function of the ratio of the end group molar volume (V_e^m) to the main-chain group molar volume (V_c^m) (after B. A. Wolf).

polar solvents (see Table 6-3). It is more difficult in the case of crystalline polymers or apolar polymers in polar solvents, and vice versa, since Equation (6-13), which was derived for pure dispersion forces, no longer applies in these cases.

Dilute solutions of poly(styrene) ($\delta_2 = 9.3$), for example, are readily obtained with butanone ($\delta_1 = 9.3$) and dimethylformamide ($\delta_1 = 12.1$), but not with acetone ($\delta_1 = 9.8$). That is, in liquid acetone, the acetone molecules form dimers through dipole–dipole interactions. In these dimers, the keto groups are shielded by the methyl groups, and so are no longer able to solvate the phenyl groups of the poly(styrene). The addition of cyclohexane ($\delta_1 = 8.2$) decreases the tendency of the acetone molecules toward association, and thus frees keto groups for solvation. For the same reason, it is also possible to have 40% solutions of poly(styrene) in acetone. Butanone, on the other hand, is "internally diluted" by the additional CH_2 group, and is therefore a solvent over the whole concentration range.

Similar reasoning applies to mixtures of solvents. The combination of a nonsolvent with a lower, and a nonsolvent with a higher, solubility parameter

Table 6-3. The Solubilities and Solubility Parameters of Polymers

Solvent		Solubility of the polymer[a]			
Name	δ_1	Poly(isobutene) $\delta_2 = 7.9$	Poly(methyl methacrylate) $\delta_2 = 9.1$	Poly(vinyl acetate) $\delta_2 = 9.4$	Poly(hexamethylene adipamide) $\delta_2 = 13.6$
Decafluorobutane	5.2	−	−	−	−
Neopentane	6.25	+	−	−	−
Hexane	7.3	+	−	−	−
Diethyl ether	7.4	−	−	−	−
Cyclohexane	8.2	+	−	−	−
Carbon tetrachloride	8.62	+	+	−	−
Benzene	9.2	+	+	+	−
Chloroform	9.3	+	+	+	−
Butanone	9.3	−	+	+	−
Acetone	9.8	−	+	+	−
Carbon disulfide	10.0	−	−	−	−
Dioxane	10.0	−	+	+	−
Dimethyl formamide	12.1	−	+	+	(+)
m-Cresol	13.3	−	+	+	+
Formic acid	13.5	−	+	−	+
Methanol	14.5	−	−	−	−
Water	23.4	−	−	−	−

[a] + means soluble, − means insoluble, (+) means soluble at high temperatures.

Table 6-4. Solubility of Polymers in Mixtures of Nonsolvents

Polymer		Solutions possible with mixtures of			
Type	δ_2	Nonsolvent I	δ_1	Nonsolvent II	δ_{11}
at-Poly(styrene)	9.3	Acetone	9.8	Cyclohexane	8.2
at-Poly(vinyl chloride)	9.53	Acetone	9.8	Carbon disulfide	10.0
at-Poly(acrylonitrile)	12.8	Nitromethane	12.6	Water	23.4
Poly(chloroprene)					
(radically polymerized)	8.2	Diethyl ether	7.4	Ethyl acetate	9.1
Nitrocellulose	10.6	Ethanol	12.7	Diethyl ether	7.4

than that of the polymer often gives a good solvent for the polymer (Table 6-4). Conversely, a mixture of two solvents can be a nonsolvent. Poly-(acrylonitrile) ($\delta_2 = 12.8$), for example, dissolves in both dimethylformamide ($\delta_1 = 12.1$) and malodinitrile ($\delta_1 = 15.1$), but not in a mixture of the two.

In order to dissolve crystalline polymers, it is necessary to consider the Gibbs energy of fusion. This additional energy expenditure is not taken into account in the concept of the solubility parameter. Crystalline polymers therefore often dissolve only above their melting temperatures and in solvents with roughly the same solubility parameter. Unbranched, highly crystalline poly(ethylene) ($\delta_2 = 8.0$) only dissolves in decane ($\delta_1 = 7.8$) at temperatures close to the melting point of $\sim 135°$ C.

The crystallinity of polymers is also responsible for the curious effect where a polymer at constant temperature first dissolves in a solvent and later, at the same temperature, precipitates out again. In these cases, the original polymer is of low crystallinity and therefore dissolves well. On dissolution, the chains become mobile. A crystalline polymer–solvent equilibrium is rapidly achieved with precipitation of polymer of higher crystallinity than the original material.

6.3. Statistical Thermodynamics

6.3.1. Entropy of Mixing

In ideal solutions it is assumed for the pair interactions that no energy is released on replacing a unit of group 1 by a unit of group 2, that is, $\Delta \epsilon$ in Equation (6-11) equals zero. Consequently, the enthalpy of mixing of an ideal solution is also equal to zero.

Since all energies are, by definition, equal in magnitude with ideal solutions, all environment-dependent entropy contributions can contribute nothing to the overall change in entropy. Consequently, the translational entropy, the internal rotation entropy, and the vibrational entropy are not changed by mixing. But the molecules of the components of the solution can be ordered relative to one another in many different ways. Thus, the many different arrangement combinations lead to an entropy contribution on mixing ΔS_{comb} (often called the "configurational entropy"). This entropy contribution can be calculated from the Ω possible arrangements of molecules (or monomeric units 1 and 2) with respect to each other as long as the molar volumes are the same in each case (see textbooks of statistical thermodynamics):

$$\Delta S^{id} \approx \Delta S_{comb} = k \ln \Omega = k \ln \frac{(N_1 + N_2)!}{N_1! \, N_2!} \qquad (6\text{-}17)$$

With the aid of the Stirling approximation

$$\ln N! \approx N \ln N - N \qquad (6\text{-}18)$$

Equation (6-17) converts to

$$\Delta S_{comb} = -k \, (N_1 \ln x_1 + N_2 \ln x_2) = \Delta S \qquad (6\text{-}19)$$

or, on converting to mole fractions

$$\Delta S_{comb}^{m} = -R(x_1 \ln x_1 + x_2 \ln x_2) \qquad (6\text{-}20)$$

With low-molar-mass substances, 1 and 2, the mole fractions x_1 and x_2 are those of the molecules, since each molecule occupies a "cell" in the lattice that the solution is considered to be (Fig. 6-3). In contrast, the mole fractions are with respect to monomeric units when macromolecules are considered,

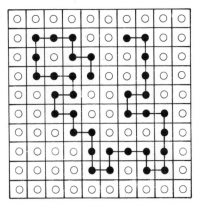 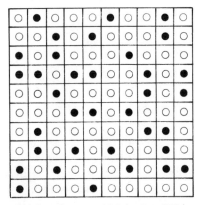

Figure 6-3. Arrangement of low-molar-mass (left) and high-molar-mass (right) solute (●) in solvent (○) for a two-dimensional lattice.

since in this case, it is the monomeric unit which occupies a lattice position. The monomeric unit mole fraction is then given by

$$x_2 = \frac{N_2 X_2}{N_1 X_1 + N_2 X_2} = \frac{N_2 X_2}{N_1 + N_2 X_2} \tag{6-21}$$

since, by definition, the degree of polymerization of the solvent, X_1, is unity. When the spatial requirements of solvent molecule and monomeric unit are equal, the mole fractions can be directly replaced by volume fractions. If the spatial requirements are not equal, the volume fraction is to be used instead of the mole fraction, which, however, will not be further pursued here.

6.3.2. Enthalpy of Mixing

In calculating the enthalpy of mixing, it is assumed that the distribution of molecules or monomeric units is not influenced by the enthalpy of mixing. This assumption allows entropies and enthalpies of mixing to be calculated independently of each other. The enthalpy of mixing ΔH is given by the difference between the enthalpies of solution H_{12} and the enthalpies H_{11} and H_{22} of the pure components

$$\Delta H = H_{12} - (H_{11} + H_{22}) \tag{6-22}$$

The enthalpies H_{12}, H_{11}, and H_{22} are calculated as follows: An interaction energy ϵ_{ij} exists between every two monomeric units. Each unit thus contributes $0.5\,\epsilon_{ij}$. Furthermore, every unit is surrounded by z neighbors. Generally, a molecule consists of X monomeric units. Consequently, for the $X_1 N_1$ monomeric units of all of the N_1 solvent molecules, H_{11} is given by the following, with the definition $\phi_1 = N_1 X_1 / N_g$ for the volume fraction:

$$H_{11} = N_1 X_1 z (0.5\epsilon_{11}) = z(0.5\epsilon_{11}) N_g \phi_1 \tag{6-23}$$

where N_g is the total number of lattice sites. The enthalpy of the pure polymer is obtained analogously:

$$H_{22} = N_2 X_2 z (0.5\epsilon_{22}) = z(0.5\epsilon_{22}) N_g \phi_2 \tag{6-24}$$

In calculating the enthalpy of the solution it is necessary to take into account the interaction energies between each segment unit and its z neighbors. There are $X_1 N_1$ solvent segment units, so a total of $z X_1 N_1$ interactions have to be considered. A solvent molecule can be surrounded by other solvent units with the interaction energy $0.5\epsilon_{11}$ per unit and/or by solute units with the interaction energy $0.5\epsilon_{12}$ per unit. The relative contribution of these two possible interactions is given by the volume fraction of the two kinds of unit in the solution. A corresponding contribution also comes from the solute units. The enthalpy of the solution is therefore

$$H_{12} = X_1 N_1 z (0.5\epsilon_{11}\phi_1 + 0.5\epsilon_{12}\phi_2) + X_2 N_2 z (0.5\epsilon_{22}\phi_2 + 0.5\epsilon_{12}\phi_1) \qquad (6\text{-}25)$$

With the mole fraction, as defined by Equation (6-21), and under the assumption that volume and mole fractions are equal, one gets from insertion of Equations (6-23)–(6-25) into Equation (6-22)

$$\phi_i \equiv x_i^u \equiv \frac{N_i X_i}{N_i X_i + N_j X_j} = \frac{N_i X_i}{N_g} \qquad (6\text{-}26)$$

and

$$\Delta H = z N_g \phi_1 \phi_2 (\epsilon_{12} - 0.5\epsilon_{11} - 0.5\epsilon_{22}) = z N_1 X_1 \phi_2 \Delta \epsilon \qquad (6\text{-}27)$$

Next, an interaction parameter χ is defined as follows: $\Delta\epsilon$ is the average energy gain per contact. Each solvent unit, however, is surrounded by z neighbors, and every solvent molecule possesses X_1 segment units. Thus, we have, with respect to the thermal energy kT,

$$\chi \equiv \frac{z X_1 \Delta \epsilon}{kT} \qquad (6\text{-}28)$$

Thus what is known as the Flory–Huggins parameter χ is, by definition, a measure of the interaction energy $\Delta\epsilon$. However, $\Delta\epsilon$ is in reality a measure of the Gibbs energy and not of the enthalpy. Consequently, χ also contains an entropy contribution, which is often found to depend on the concentration. A linear dependence of the interaction parameter on the volume fraction ϕ_2 of the solute can be assumed as a first approximation (see also Fig. 6-4):

$$\chi \equiv \chi_0 + \sigma \phi_2 \qquad (6\text{-}29)$$

With relationships (6-28) and (6-29), Equation (6-27) becomes

$$\Delta H = k T N_1 \phi_2 (\chi_0 + \sigma \phi_2) \qquad (6\text{-}30)$$

or with respect to amount of substance, $n_1 = N_1/N_L$, or mole fractions, $x_1 = n_1/(n_1 + n_2)$, the molar enthalpy of mixing is given by

$$\Delta H^m = \frac{\Delta H}{n_1 + n_2} = RT x_1 \phi_2 (\chi_0 + \sigma \phi_2) \qquad (6\text{-}31)$$

6.3.3. Gibbs Energy of Mixing for Nonelectrolytes

Using the relationships $\phi_2 = N_2 X_2/N_g$, $N_g \equiv n_g N_1$, $N_g = N_1 + N_2$, $k = R/N_L$, $\Delta G^m = \Delta G/n_g$, and $x_1 = \phi_1$, combination of Equation (6-1)

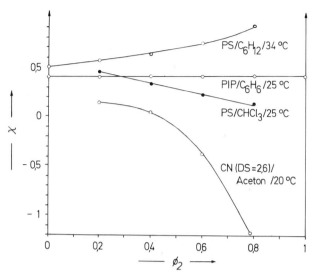

Figure 6-4. Dependence of the Flory–Huggins interaction parameter, χ, on volume fraction, ϕ_2, of the polymer for poly(styrene) in cyclohexane and chloroform and for *cis*-1,4-poly(isoprene) in benzene as well as for cellulose nitrate in acetone.

with (6-20) and (6-30) leads to

$$\frac{\Delta G^m}{RT} = X_1^{-1}(\phi_1\phi_2\chi_0 + \phi_1\phi_2^2\sigma + \phi_1 \ln \phi_1 + X_1 X_2^{-1}\phi_2 \ln \phi_2) \qquad (6\text{-}32)$$

$\Delta G^m / RT$ is plotted according to Equation (6-32) as a function of the volume fraction ϕ_2 of monomeric units of solute (Figure 6-5). We may have $X_1 = X_2 = 1$ for mixtures of some low-molecular-weight compounds. The molar Gibbs energy is always negative for these interaction parameters (e.g., $\chi_0 = 0.5$) and has a minimum at $\phi_2 = 0.5$. Such mixtures can never demix.

The function becomes asymmetric when the degree of polymerization X_2 goes from 1 to 100 for the same interaction parameter $\chi_0 = 0.5$. This behavior is caused by the last term in Equation (6-32), that is, by the entropy term. This behavior of high-molecular-weight polymer solutions deviates from the behavior of low-molecular-weight solution systems and is essentially caused by the difference in molecular size between the low-molecular-weight solvent and the high-molecular-weight solute.

If the value of the interaction parameter χ_0 increases from 0.5 to 1.2 at the same degree of polymerization $X_2 = 100$, then the molar Gibbs energy of mixing will even be positive between $\phi_2 = 2 \times 10^{-9}$ and $\phi_2 = 0.3$. Since there are two regions of concentration where the molar Gibbs energy of mixing is negative, however, a phase separation of the dilute starting solution into two

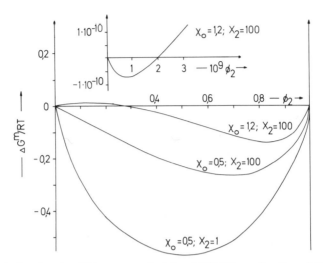

Figure 6-5. Reduced molar Gibbs energy of mixing $\Delta G^m/RT$ as a function of the volume fraction ϕ_2, of solute monomeric unit with different interaction parameters χ_0 and degrees of polymerization X_2 of the solute in low-molecular-weight solvents ($X_1 = 1$). Calculations according to Equation (6-32) with $\sigma = 0$.

solutions occurs. One of these solutions is very dilute, the other very concentrated in solute (see also Section 6.6).

The simple Flory–Huggins theory discussed above is based on a series of questionable assumptions: lattice sites of equal size for solvent segments and polymer monomeric units, uniform distribution of the monomeric units in the lattice, random distribution of the molecules, and the use of volume fractions instead of surface-area fractions in deriving the enthalpy of mixing. Proposed improvements, however, have led to more complicated equations or to worse agreement between theory and experiment. Obviously, various simplifications in the Flory–Huggins theory are self-compensating in character.

6.3.4. Gibbs Energy of Mixing for Polyelectrolytes

The molar Gibbs energy of mixing of polyelectrolytes ΔG_{el}^m is composed of the free energy of mixing of the uncharged polymer ΔG^m [see Equation (6-32)], the contribution ΔG_{Coul}^m for the Coulombic interaction between the polyion and the gegenion, and the contribution ΔG_{mm}^m for ionic interactions within the macromolecule itself:

$$\Delta G_{el}^m = \Delta G^m + \Delta G_{Coul}^m + \Delta G_{mm}^m \qquad (6-33)$$

The magnitude of ΔG_{mm} is determined by the distribution of the ions within the macromolecule. This distribution is not yet experimentally determinable. A specific model is therefore assumed for the distribution of ions in the macromolecule. For example, the model of a rod is very suitable for true rod-forming macromolecules (viruses, nucleic acids) or for long-chain macromolecules at high degrees of ionization. In the case of the latter, the many like charges repeal one another along the whole length of the chain, resulting in rigidity and rodlike behavior. For coiled molecules at low degrees of ionization or rigid spheres (e.g., globulin), on the other hand, spherical models are more suitable.

6.3.5. Chemical Potential of Concentrated Solutions

According to Equation (6-3), the chemical potential of the solvent is defined as the derivative of the Gibbs energy of mixing with respect to amount of solvent. Consequently, with the condition $\phi_1 = 1 - \phi_2$, differentiation of Equation (6-32) gives

$$\Delta\mu_1 = RT[(\chi_0 - \sigma + 2\sigma\phi_2)\phi_2^2 + \ln(1 - \phi_2) + (1 - X_1 X_2^{-1})\phi_2] \quad (6\text{-}34)$$

and analogously, for the chemical potential of the solute,

$$\Delta\mu_2 = RT[(\chi_0\phi_1 + 2\sigma\phi_2\phi_1 - 1) X_2 X_1^{-1}\phi_1 + \phi_1 + \ln\phi_2] \quad (6\text{-}35)$$

In a polymer–solvent system the chemical potential of the solvent decreases slowly at first and then more rapidly to negative values with increasing volume fraction of polymer if χ_0 is zero (Figure 6-3). The initial shape of the curve is increasingly flatter with increasing value of the interaction parameter. In the chosen example, there is a practically horizontal portion of the curve between $\phi_2 > 0.05$ and $\phi_2 < 0.14$ after an initial slow decrease and before the strong decrease of $\Delta\mu_1/RT$ to negative values for $\chi_0 = 0.605$. The value of $\Delta\mu_1/RT$ passes through a weak minimum (not recognizable in Figure 6-6) before passing through a strong maximum for still higher values of χ_0. Thus, $\chi_0 = 0.605$ is a critical interaction parameter value for the chosen example. The critical concentration is defined as that volume fraction of the solute at which maximum, minimum, and inflection point coincide.

Chemical potentials for every volume fraction can be taken from a plot of $\Delta G^m = f(\phi_2)$ (see Figures 6-5 and 6-11). Equations (6-5) and (6-6) give, of course,

$$d(\Delta G^m) = \Delta\mu_1 dn_1 + \Delta\mu_2 dn_2 \quad (6\text{-}36)$$

With $\phi_1 = N_i X_i/N_g$ and $n_i = N_i/N_L$, integration leads to

$$\Delta G^m = n_1 \Delta\mu_1 + n_2 \Delta\mu_2 = \frac{\phi_1 N_g X_1^{-1} \Delta\mu_1 + \phi_2 N_g X_2^{-1} \Delta\mu_2}{N_L} \quad (6\text{-}37)$$

$$N_L \Delta G^m = N_g X_1^{-1} \Delta\mu_1 + N_g\phi_2(X_2^{-1} \Delta\mu_2 - X_1^{-1} \Delta\mu_1) \quad (6\text{-}38)$$

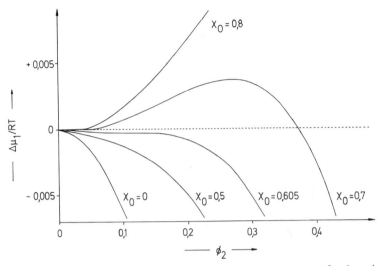

Figure 6-6. Reduced chemical potential $\Delta \mu_1 / RT$ of the solvent (degree of polymerization $X_1 = 1$) as a function of the volume fraction ϕ_2 of monomeric units of a monodisperse solute with degree of polymerization $X_2 = 100$ and with different interaction parameters χ_0. Calculations according to Equation (6-32) with $\sigma = 0$.

The equation for the tangent to the curve of $\Delta G^m = f(\phi_2)$ at a point $\phi_2^\S$ is

$$Y = A + B\phi_2^\S \tag{6-39}$$

The slope B is given by differentiation of Equation (6-38)

$$B = \left(\frac{\partial \Delta G^m}{\partial \phi_2} \right)_{Ng}^\S = N_g (X_2^{-1} \Delta \mu_2^\S - X_1^{-1} \Delta \mu_1^\S) \tag{6-40}$$

A is obtained from the argument that at the point $\phi_2^\S$ the value of ΔG^m is also given by Equation (6-38), i.e., by

$$A = X_1^{-1} N_g \Delta \mu_1^\S \tag{6-41}$$

Consequently, for Y

$$Y = X_1^{-1} N_g \Delta \mu_1^\S + N_g (X_2^{-1} \Delta \mu_2^\S - X_1^{-1} \Delta \mu_1^\S) \phi_2^\S \tag{6-42}$$

Thus, for the limiting cases of $\phi_2 \to 0$ and $\phi_2 \to 1$

$$\lim_{\phi_2 \to 0} Y = X_1^{-1} N_g \Delta \mu_1^\S \quad \text{and} \quad \lim_{\phi_2 \to 1} Y = X_2^{-1} N_g \Delta \mu_2^\S \tag{643}$$

Thus, if a tangent is drawn to the $\Delta G = f(\phi_2)$ curve at a volume fraction of $\phi_2 = \phi_2^\S$, then the extrapolation of this tangent to the ΔG^m axis for values of $\phi_2 = 0$ and $\phi_2 = 1$ gives quantities from which the chemical potentials of solvent and solute, respectively, are obtained.

6.3.6. *Chemical Potential of Dilute Solutions*

Since the volume fraction of the solute ϕ_2 is very small in dilute macromolecular solutions, the expression $\ln(1 - \phi_2)$ can be expanded in a series

$$\ln(1 -\phi_2) = -\phi_2 - (\phi_2^2/2) - (\phi_2^3/3) - \cdots \qquad (6\text{-}44)$$

and terminated after the second term. When this is inserted into Equation (6-34), we obtain the following for high-molar-mass solutes in low-molar-mass solvents in a small volume element:

$$\frac{\Delta \mu_1^{\text{exc}}}{RT} = (\chi_0 - \sigma - 0.5) \phi_2^2 \qquad (6\text{-}45)$$

Equation (6-45) contains the *excess* chemical potential instead of the chemical potential itself because the ideal term $X_1 X_2^{-1}$ has been omitted. This omission was necessary because the considerations have to be restricted to a small volume of solution where the segment distribution is uniform enough so that the lattice theory can be applied. Although in this equation the quantity $(\chi_0 - \sigma)$ is derived from the enthalpy of mixing and the factor 0.5 from the entropy of mixing, it is convenient to replace this combination of terms by another with a new enthalpy parameter κ and a new entropy parameter ψ:

$$\frac{\Delta \mu_1^{\text{exc}}}{RT} = (\kappa - \psi) \phi_2^2 \qquad (6\text{-}46)$$

The chemical potential gives the partial molar Gibbs energy of dilution. The partial molar dilution enthalpy and the partial molar dilution entropy are given by

$$\Delta \tilde{H}_1^m = RT\kappa\phi_2^2, \qquad \Delta \tilde{S}_1^m = R\psi\phi_2^2 \qquad (6\text{-}47)$$

In the theta state (see Section 6.1), $\Theta = \Delta \tilde{H}_1^m / \Delta \tilde{S}_1^m$, and, according to Equation (6-47) with $T = \Theta$, we also have

$$\Theta = \kappa T / \psi \qquad (6\text{-}48)$$

The combination of Equations (6-34), (6-46), and (6-48) thus gives an expression for the temperature dependence of the Flory–Huggins interaction parameter:

$$(\chi_0 - \sigma) = (0.5 - \psi) + \psi \left(\frac{\Theta}{T} \right) \qquad (6\text{-}49)$$

The expression $(\chi_0 - \sigma)$ takes on the value of 0.5 for $X_2/X_1 \gg 0$ in the theta state. It is smaller than 0.5 in good solvents, since then $T > \Theta$.

6.4. Virial Coefficients

6.4.1. Definitions

The chemical potential of solutions of nonelectrolytes can always be written in terms of a series of positive integral powers of the concentration

$$\Delta\mu_1 = -RT\tilde{V}_1^m(A_1c_1 + A_2c_2^2 + A_3c_2^3 + \cdots) \qquad (6\text{-}50)$$

The proportionality coefficients of this series are called the first, second, third, . . . , virial coefficients.*

On comparing with the expression for the osmotic pressure Π, we obtain the following (see textbooks of chemical thermodynamics):

$$\Delta\mu_1 = -\Pi\,\tilde{V}_1^m = -\frac{RTc_2\tilde{V}_1^m}{M_2} \qquad (6\text{-}51)$$

showing that the first virial coefficient is equal to the reciprocal molar mass of the solute. The second virial coefficient is a measure of the excluded volume (see Section 6.4.2).

It is important to pay attention to definitions of the virial coefficients when comparing their values reported in the literature. Equation (6-50), when used with osmotic pressures, corresponds to the expression

$$\frac{\Pi}{c_2} = RTM_2^{-1} + RTA_2c_2 + RTA_3c_2^2 + \cdots \qquad (6\text{-}52)$$

Instead of this, the following definition is often used:

$$\frac{\Pi}{c_2} = RTM_2^{-1} + A_2c_2 + A_3c_2^2 + \cdots \qquad (6\text{-}53)$$

Equation (6-52) generally will be used here. An apparent molar mass M_{app} can be defined as a molar mass calculated from experimental data at finite concentrations from an equation applicable to infinite dilution only:

$$M_{app}^{-1} \equiv \frac{\Pi}{RTc_2} = M_2^{-1} + A_2c_2 + A_3c_2^2 + \cdots \qquad (6\text{-}54)$$

---

*The name virial coefficient comes from the virial theorem, which was much used toward the end of the 19th century. This theorem states

$$\text{average of } (mv^2/2) = -\text{average of } 0.5(Xx + Yy + Zz)$$

Here m is the mass of the particles; v is their velocity; x, y, and z are their coordinates; and X, Y, and Z are the components of the forces which act upon them. The expression on the right-hand side was called "virial" because forces were considered [vis(Latin) = force]. The virial could be expanded into a series whose coefficients were consequently the virial coefficients.

Consequently, virial coefficients can be determined from the concentration dependence of the reciprocal apparent molar mass. But, since the various methods for measuring the molar mass yield various averages of it (see Chapters 8 and 9), the virial coefficients obtained will be average values which vary according to the method used to determine them. Virial coefficients obtained from osmotic-pressure measurements (and all other measurements based on colligative methods) will give the average

$$A_2^{\Pi} = \sum_i \sum_j w_i w_j A_{ij} \tag{6-55}$$

while light scattering measurements, for example, give the average

$$A_2^{LS} = \frac{\sum_i \sum_j w_i M_i w_j M_j A_{ij}}{(\sum_i w_i M_i)^2} \tag{6-56}$$

6.4.2. Excluded Volume

The second virial coefficient depends on the excluded volume u. The macromolecules arrange themselves with little mutual interference since the total excluded volume $N_2 u$ is much smaller than the total volume V. The total number of possible ways of arranging these N_2 macromolecules is calculated from the partition function Ω,

$$\Omega = \text{const} \times \prod_{i=0}^{N_2-1} (V - iu) \tag{6-57}$$

Polymer–solvent interactions can be considered in terms of the concept of the effective excluded volume. Consequently, ΔH is equal to zero and the addition of solvent occurs "athermally":

$$\Delta G = -T\Delta S = -kT \ln \Omega = -kT \ln \left[\text{const} \times \prod_{i=0}^{N_2-1} (V - iu) \right] \tag{6-58}$$

The volume available to the second molecule is $V - u$, and to the third molecule is $V - 2u$, etc. Solving for the logarithmic expression, we obtain a sum instead of a product:

$$\Delta G = -kT \left[N_2 \ln V + \sum_{i=0}^{N_2-1} \ln \left(1 - \frac{iu}{V} \right) \right] + \text{const}' \tag{6-59}$$

In dilute solutions, $iu/V \ll 1$. The logarithm can be expanded into a series: $\ln(1 - y) = -y - \ldots$, and the following is obtained:

$$\Delta G = -kT \left(N_2 \ln V + \sum_{i=0}^{N_2-1} \frac{iu}{V} \right) + \text{const}' \qquad (6\text{-}60)$$

Since u/V is constant, the sum term yields, for $N_2 \to \infty$,

$$\sum_{i=0}^{N_2-1} i = N_2^2/2 \qquad (6\text{-}61)$$

The osmotic pressure is given by Equations (6-51) and (6-3):

$$\Pi = -(\tilde{V}_1^m)^{-1} \left(\frac{\partial G}{\partial n_1} \right)_{N_2,p,T} = -\frac{N_L}{\tilde{V}_1^m} \left(\frac{\partial G}{\partial N_1} \right)_{N_2,p,T}$$

$$= -\frac{N_L}{\tilde{V}_1^m} \left(\frac{\partial G}{\partial V} \right)_{N_2,p,T} \left(\frac{\partial V}{\partial N_1} \right)_{N_2,p,T}$$

$$= -\left(\frac{\partial G}{\partial V} \right)_{N_2,p,T} \qquad (6\text{-}62)$$

On considering Equation (6-62), inserting Equation (6-61) into (6-60), and differentiating with respect to V with $N_2/V = c_2 N_L/M_2$, we obtain the following:

$$\frac{\Pi}{c_2} = \frac{RT}{M_2} + \frac{RTN_Lu}{2M_2^2} c_2 + \ldots \qquad (6\text{-}63)$$

Equating the coefficients with those in Equation (6-52), we obtain for the second virial coefficient

$$A_2 = \frac{N_L u}{2M_2^2} \qquad (6\text{-}64)$$

Expressions for the excluded volume of rigid particles have already been derived in Section 4.4.4. The second virial coefficient for unsolvated spheres is given by Equations (6-64) and (4-41) as

$$A_2 = \frac{4v_2}{M_2} \qquad (6\text{-}65)$$

where v_2 is the specific volume. Thus, the second virial coefficient for spheres is reciprocally proportional to the molar mass and is zero for infinitely large molar mass.

The second virial coefficient for unsolvated rods is obtained from Equations (6-64) and (4-43) with $M_2 = (\pi R_2^2) L N_L/v_2$:

$$A_2 = \frac{v_2^2}{2\pi R_2^3 N_L} \qquad (6\text{-}66)$$

where R is the radius of the rod. Thus, the second virial coefficient of rod-shaped molecules is independent of length and molar mass.

The dependence of the second virial coefficient of coil-shaped molecules is difficult to calculate since the excluded volume is a complicated function of the molar mass (see Section 4.4.5). The usual method is to replace the excluded volume of the molecule u in Equation (6-64) by the excluded volume of the chain segment u_{seg}; the molar mass M_2 of the molecule is also replaced by the formula molar mass M_u of the chain segment. The molar mass dependence of the excluded volume is expressed in terms of a function h(z) whose coefficients have been evaluated theoretically:

$$A_2 = \frac{N_L u_{seg}}{2 M_u^2}(1 - 2.865z + 14.278z^2 - \cdots) \qquad (6\text{-}67)$$

When $u_{seg} = 0$ and $z = 0$, $A_2 = 0$ also, that is, A_2 is zero under theta conditions (see also Section 4.5.2.2). A_2 decreases with increasing molar mass since z increases less rapidly with increasing molar mass than the square of the molar mass. The dependence of A_2 on the molar mass can be given by a power expression:

$$A_2 = K_A M_2^{a_A} \qquad (6\text{-}68)$$

where K_A and a_A are empirical constants determined for each polymer–solvent–temperature system.

6.5. Association

6.5.1. Basic Principles

Macromolecules can associate with each other to larger but still soluble entities in solution when certain conditions are met. This process will be called multimerization. Reversible multimerization is further defined as association, and irreversible multimerization is defined as aggregation. In the literature, however, these concepts are often not differentiated.

Association can be investigated by group-specific or molecule-specific methods. Group-specific methods are concerned with the behavior of specific groups, i.e., hydrogen-bonding groups, etc. However, these methods are often not sufficiently sensitive. One associating group per molecule is, of course, sufficient to cause association of a macromolecule. This corresponds to a group concentration of only 0.1% for a degree of polymerization of 1000. But the accuracy of group-specific methods is often not better than ±1%.

Molecule-specific methods are concerned with the molecular or particle mass. The particle mass will double for a total dimerization, that is, there is a 100% change. Consequently, molecule-specific methods are much more sensitive than group-specific methods. However, molecule-specific methods only show changes when intermolecular association occurs, and they yield no information on the molecular causes of association.

In molecule-specific methods, the apparent molar mass is measured as a function of the concentration (see Section 6.4.1). All the expressions so far given, however, are only suitable for the case where an increase in the weight concentration leads to an equally large increase in the amount (double the weight concentration gives double the amount). This assumption does not hold for multimerization, where an increase in the weight concentration leads to a relatively smaller concentration of kinetically independent particles. The concentration dependence of the apparent molar mass is consequently given by two different terms. The association term describes the change in concentration of independent particles relative to the change in the weight concentration. This association term also occurs in the absence of any polymer–solvent interaction when a polymer associates, i.e., it occurs also under theta conditions. The virial coefficients, on the other hand, take all other interactions into account. The following applies instead of Equation (6-54):

$$M_{\text{app}}^{-1} = (M_{\text{app}})_{\theta}^{-1} + \left(\frac{1}{c^2} \sum_i \sum_j (A_2)_{ij} c_i c_j \right) c + \cdots \tag{6-69}$$

$(M_{\text{app}})_{\theta}$, is, of course, concentration dependent. The exact form of the concentration dependence depends on the stoichiometry of the association and on the effective associating unit.

Two simple cases can be distinguished for the stoichiometry. *Open association* is the term given to a consecutive process:

$$M_I + M_I \rightleftharpoons M_{II}$$

$$M_{II} + M_I \rightleftharpoons M_{III}$$

$$M_{III} + M_I \rightleftharpoons M_{IV}, \quad \text{etc.} \tag{6-70}$$

Thus, all possible "multimers" are in equilibrium with the "unimers."

Closed association is concerned with an "all or nothing" process with only two types of particles being involved:

$$N M_I \rightleftharpoons M_N \tag{6-71}$$

The effective associating unit can be the molecule or a segment of the molecule. The number of associogenic groups is independent of the molecular size in *molecule-related* association. An example of this is the association of end groups. Linear molecules have only two end groups per molecule. Consequently each molecule has two associogenic groups. In this case, the equilibrium association constant must obviously be related to the content.

Segements of several monomeric units are responsible for association in *segment-related* association. Examples are, e.g., syndiotactic sequences of sufficient length in an "atactic" polymer. The number of these associogenic segments will increase with increasing molar mass. In this case, the equilibrium association constant is related to the weight concentration.

Polymers generally have a molar mass distribution. When they associate, a particle-size distribution is produced, which will be different from the molar mass distribution and will vary with the nature of the effective unit. The relationships between the molar mass distributions and the particle mass distributions can be derived by statistical methods. Only the results are given here.

The number-average molar mass of the N-mer is exactly N times as large as the number average of the unimer in *molecule-related* association

$$(\overline{M}_N)_n = N(\overline{M}_1)_n \qquad (6\text{-}72)$$

The mass-average molar mass of the N-mer is given, however, for any distribution function by

$$(\overline{M}_N)_w = (\overline{M}_1)_w + (N-1)(\overline{M}_1)_n \neq N(\overline{M}_1)_w \qquad (6\text{-}73)$$

For *segment-related* association

$$(\overline{M}_N)_n = (\overline{M}_1)_n + (N-1)(\overline{M}_1)_w \qquad (6\text{-}74)$$

which only holds for the Schulz–Flory distribution, and the following, which applies to all distributions:

$$(\overline{M}_N)_w = N(\overline{M}_1)_w \qquad (6\text{-}75)$$

Consequently, the polydispersity $(\overline{M}_N)_w/(\overline{M}_N)_n$ for segment-related association as well as for molecule-related association is always smaller than the polymolecularity $(M_1)_w/(\overline{M}_1)_n$. A linear relationship between these two quantities exists for molecule-related, but not segment-related, association:

$$\frac{(\overline{M}_N)_w}{(\overline{M}_N)_n} - 1 = N^{-1}\left(\frac{(\overline{M}_1)_w}{(\overline{M}_1)_n} - 1\right) \qquad (6\text{-}76)$$

The distribution becomes narrower because the variation in molecular size now takes place within the particle.

6.5.2. Open Association

A series of particles M_I, M_{II}, M_{III}, $\cdots$ is produced by open association. The total amount is therefore

$$[M] = [M_I] + [M_{II}] + [M_{III}] + \cdots \tag{6-77}$$

The equilibrium constant for molecule-related open association is defined as

$$({}^n K_{N-1})_0 = [M_N]/[M_{N-1}][M_1] \tag{6-78}$$

If the association occurs, for example, via the end groups, then it can be assumed that the equilibrium constant defined above is independent of the degree of association N of the multimers produced:

$$ {}^n K_0 = ({}^n K_1)_0 = ({}^n K_{II})_0 = ({}^n K_{III})_0 = \cdots \tag{6-79}$$

Inserting Equations (6-78) and (6-79) into (6-77), we obtain

$$[M] = [M_1][1 + {}^n K_0[M_1] + ({}^n K_0[M_1])^2 + \cdots] \tag{6-80}$$

Since, according to Equation (6-78), ${}^n K_0[M_1] = [M_{II}]/[M_1]$ still holds, and since the content of dimers must be smaller than that of monomers, ${}^n K_0[M_1]$ is always smaller than 1. Consequently, Equation (6-80) can, according to the rules valid for such series, be given as

$$[M] = [M_1](1 - {}^n K_0[M_1])^{-1} \tag{6-81}$$

The total molar concentration is given as

$$[M] = \frac{c}{(\overline{M}_n)_{app,\theta}} \tag{6-82}$$

Combining Equations (6-81) and (6-82), we obtain

$$[M_1]^{-1} = {}^n K_0 + (M_n)_{app,\theta} c^{-1} \tag{6-83}$$

and the weight concentration is given by

$$c = c_I + c_{II} + c_{III} + \cdots \tag{6-84}$$

With

$$[M_i] = c_i(\overline{M}_i)_n^{-1} \tag{6-85}$$

and Equations (6-78), (6-79), and (6-72), we have

$$c = [M_1](\overline{M}_1)_n\{1 + 2({}^n K_0[M_1]) + 3[{}^n(K_0[M_1])^2 + \cdots]\} \tag{6-86}$$

or, for ${}^n K_0[M_1] < 1$

$$c = \frac{[M_1](\overline{M}_1)_n}{(1 - {}^n K_0[M_1])^2} \tag{6-87}$$

Combination of Equations (6-87) and (6-83) gives

$$(M_n)_{app,\Theta} = (\overline{M}_1)_n + {}^nK_0(M_1)_n \frac{c}{(M_n)_{app,\Theta}} \qquad (6\text{-}88)$$

Analogous calculations give for the apparent mass-average molar mass in the theta state

$$(M_w)_{app,\Theta} = (\overline{M}_1)_w + 2[{}^nK_0(\overline{M}_1)_n] \frac{c}{(M_n)_{app,\Theta}} \qquad (6\text{-}89)$$

It is seen from Equation (6-88) that if the apparent number-average molar mass in the theta state is plotted against $c/(\overline{M}_n)_{app,\Theta}$, the true number-average molar mass is obtained from the ordinate intercept and the equilibrium constant for the association can be calculated from the slope. Equation (6-89) shows, however, that neither the mass average molar mass nor the association equilibrium constant can be determined from data on the apparent mass-average molar mass alone when dealing with molecule-related association; the corresponding apparent number-average molar mass must also be known.

Figure 6-7 shows the concentration dependence of the normalized reciprocal apparent number-average molar mass of some poly(oxyethylenes), H$\ce{-}$(OCH$_2$CH$_2$)$_n$OH, in benzene. The high-molar mass material, H 6000, describes the course expected from Equation (6-54) (i.e., no association). Association is obvious with the lower-molar–mass material. But it cannot be assumed

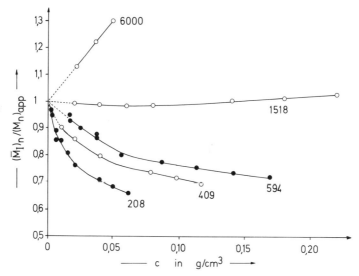

Figure 6-7. Concentration dependence of normalized inverse apparent number-average degrees of polymerization of α-hydro-ω-hydroxy-poly(oxyethylenes) in benzene at 25°C. Numbers indicate the number-average molar-masses of the unimers (H.-G. Elias and H. Lys).

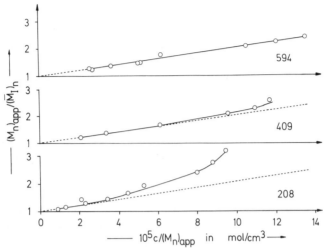

Figure 6-8. Plots of the data of Figure 6-4 under assumption of molecule-based open associations with particle-independent equilibrium constants of multimerization. Numbers indicate the number-average molar mass.

from such plots that the association decreases with increasing molar mass. In fact, a plot according to Equation (6-88) shows that the initial slope of the function $(\overline{M}_n)_{app,\theta} / (\overline{M}_I)_n = f(c / (\overline{M}_n)_{app,\theta})$ has the same value independent of the molar mass of the unimer. The equilibrium association constants must consequently also be equal in value. Increasing departure from linearity with decreasing molar mass in Figure 6-8 indicates negative second virial coefficients.

6.5.3. Closed Association

Equilibrium constants for molecule-related closed association are defined by

$$^{n}K_c \equiv \frac{[\mathbf{M}_N]}{[\mathbf{M}_I]^N} \tag{6-90}$$

There is no closed expression for the concentration dependence of the apparent molar masses. Generally, the molar masses of the unimers, the equilibrium constants $^{n}K_c$, and the degree of association N are obtained by iteration.

The calculated dependence of the concentrations c_N of the multimer and the unimers c_1 on the total concentration c shows a more or less accentuated kink (Figure 6-9). This kink is generally known as the critical micelle concentration (cmc). As can be seen for the concentration dependence of the

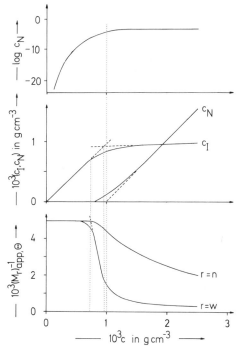

Figure 6-9. Calculated concentration dependences of unimer and multimer concentrations as well as of the inverse apparent number- and mass-average molar masses for closed, molecule-based associations in the theta state (H.-G. Elias and J. Gerber). Calculations for $(\overline{M}_1)_n = (\overline{M}_1)_w = 200\,g/mol$, $N = 21$, and $^nK_c = 10^{45}\,(dm^3/mol)^{N-1}$. Critical micelle concentrations are determined by extrapolation of dotted lines.

apparent molar masses, such "critical micelle concentrations" are also observed there. The position of this critical micelle concentration depends on the measurement method used. The cmc is not a well-defined physical quantity, and it is certainly not the concentration at which associates first appear (see also Figure 6-9).

Solutions of detergents are especially prone to closed association. Closed association has also been observed for poly(γ-benzyl-L-glutamate) in various organic solvents (Figure 6-10). It has been shown in these cases that the Gibbs energy of association depends also on the reciprocal number-average molar masses of the unimers. Thus, the association must occur here via the end groups. The apparent contradiction between this result and the occurrence of closed association is explained on the basis of the formation of ring-shaped associates. Since the molecules occur in the helix form and consequently are relatively rigid, a relationship must exist between the degree of association and the equilibrium association constant. This relationship has actually been observed.

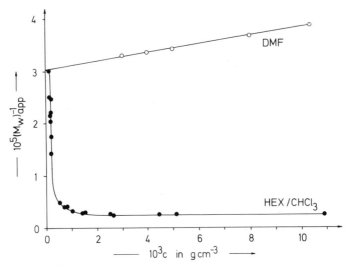

Figure 6-10. Concentration dependence of the inverse apparent mass-average molar mass of a poly(γ-benzyl-L-glutamate) in dimethylformamide at 70°C and in hexane–chloroform (v/v = 47.7/52.3) at 25°C (H.-G. Elias and J. Gerber).

6.5.4. Bonding Forces

The association of poly(oxyethylene) (POE) and poly (γ-benzyl-L-glutamate) (PBLG) discussed above occurs through the same bonding as is met with corresponding low-molar-mass compounds. According to IR and PMR measurements, however, multimerization occurs exclusively via hydrogen bonding: via hydroxyl groups in the case of POE and via amino groups in the case of PBLG.

But with polymers, other types of bonding can occur. In certain solvents, and at higher concentrations, poly(γ-benzyl-L-glutamates) do not form end-group associated ringlike entities but form laterally arranged bundlelike structures. Mesomorphous structures such as formed by certain stiff low-molar-mass molecules are produced (see also Section 5.5).

On the other hand, stereocomplexes and solvatophobic bonds are specific to macromolecular multimerization. Certain polymers with mutually complementary stereostructures form what are called stereocomplexes whose stoichiometry is dependent on the stereosequence length. Examples of such stereocomplexing are provided by the pairs poly(γ-benzyl-D-glutamate) with poly(γ-benzyl-L-glutamate) and isotactic with syndiotactic poly(methyl methacrylates) (PMMA). The specific enthalpy of stereocomplex formation with it- and st-PMMA is, for example, a linear function of the syndiotactic diad mass fraction, with a maximum at $w_{st} = 0.58$ (Figure 6-11). If, however,

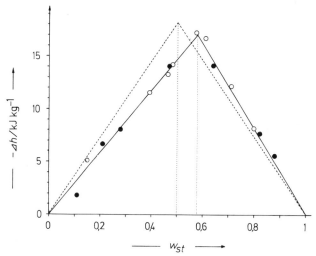

Figure 6-11. Negative specific enthalpy for the stereocomplex formation from st- and it-poly(methyl methacrylates) in *o*-xylol (O) or dimethyl formamide (●) at 25°C as a function of the mass fraction of the syndiotactic polymer. The dashed line gives the relationship for a plot against the syndiotactic heptad content (after data from W. Siemens and G. Rehage).

only the syndiotactic heptads are considered, then the plot is symmetrical and the maximum occurs at exactly $w_{hept} = 0.5$. Thus, syndiotactic complexes can only form with these samples if there are at least seven units in the stereosequence.

Association through solvatophobic bonds only occurs in ordered solvents. Such solvents may be benzene, with money-roll-ordered benzene molecules, or water, with its hydrogen-bonded network structure. If a solvatophobic solute is introduced to such a solvent, then a more ordered solvent molecular structure forms around the solute, as, for example, "iceberg structures" in the case of water. When the solute associates, some of these icebergs melt and, so, association by solvatophobic bonding is characterized by an entropy increase. In other kinds of association, on the other hand, there is a decrease in entropy, since the arranging of unimers to multimers is accompanied by a loss in translational entropy. Solvatophobic bonding is encountered, for example, in the association of syndiotactic poly(propylene) in benzene (see Section 4.3.2) and in the association of detergents and certain proteins in water. The name, "hydrophobic bonding" has become established for solvatophobic bonding in aqueous solution. This name is, however, badly chosen, since the phenomenon depends not on the hydrophobicity of the solute, but on the degree of order in the solvent. In addition, solvatophobic bonding is not the normal bonding occurring between *functional groups*. For these reasons, hydrophobic bonding is also known as entropy bonding.

6.6. Phase Separation

6.6.1. Basic Principles

If a phase separation occurs in a system, the chemical potential of each component in each phase must be the same at equilibrium. Thus, for a binary system of components 1 and 2,

$$\mu_1' = \mu_1'' \quad \text{and} \quad \mu_2' = \mu_2'' \tag{6-91}$$

and consequently

$$\Delta\mu_1' = \mu_1' - \mu_1^0 = \mu_1'' - \mu_1^0 = \Delta\mu_1''$$
$$\Delta\mu_2' = \mu_2' - \mu_2^0 = \mu_2'' - \mu_2^0 = \Delta\mu_2'' \tag{6-92}$$

The values of $\Delta\mu_1$ and $\Delta\mu_2$, however, are given as ordinate intercepts of the tangents to the $\Delta G^m = f(\phi_2)$ curve (see Section 6.3.5). The required equivalence of the chemical potentials in both phases can only be fulfilled if two points of the curve possess a common tangent (see Figure 6-9). But there is only one common tangent to a curve with two minima. The points of contact A and B of this tangent to the $\Delta G^m = f(\phi_2)$ curve determine the compositions ϕ' and ϕ'' of the two phases.

Only systems with $\phi_2 < \phi_2'$ and $\phi_2 > \phi_2''$ are stable. Every other system with the composition $\phi_2' < \phi_2 < \phi_2''$ will separate into two phases. The border between the stable and instable regions is called a binodal. Binodals are obtained by equating the chemical potentials of each polymer in both phases. The calculations are complex, however, since polymers contain macro-molecules of very different degrees of polymerization and Equations (6-91) or (6-92) must be evaluated for every single degree of polymerization.

The instable region is further subdivided into a metastable and an instable region, whereby the borders are called spinodals. Since the condition $(\partial^2 \Delta G^m / \partial \phi_2^2) > 0$ applies to the metastable region, this region is stable to phases of vanishingly small difference in compositions. However, the metastable region is instable to the system of composition $\phi_2' < \phi_2 < \phi_2'''$ and $\phi_2''' < \phi_2 < \phi_2''$ (Figure 6-12).

The spinodals are characterized by the inflection points of the function $\Delta G^m = f(\phi_2)$, that is, by

$$\partial^2 \Delta G^m / \partial \phi_2^2 = \partial \Delta\mu_1 / \partial \phi_2 = 0$$

Using this condition, Equation (6-34) gives for the spinodals, with $X_1 = 1$ and $\sigma = 0$

$$\frac{\partial \Delta\mu_1}{\partial \phi_2} = RT[2\chi_0\phi_2 - (1 - \phi_2)^{-1} + (1 - X_2^{-1})] = 0 \tag{6-93}$$

The volume fraction at which the maximum, minimum, and point of

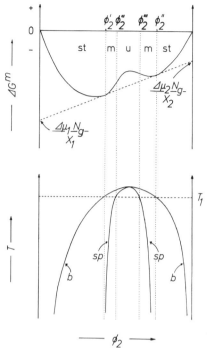

Figure 6-12. Schematic representation of the molar Gibbs energy (above) and the demixing temperature (below) as a function of the volume fraction of the solute for a partially miscible system. st, Stable region; m, metastable region; u, instable region; b, binodals, sp, spinodals; T_1, temperature for which the upper diagram is applicable.

inflection of the curve $\Delta\mu_1 = f(\phi_2)$ coincide is defined as the critical point. Thus, differentiation of Equation (6-93) gives

$$\frac{\partial^2 \Delta\mu_1}{\partial\phi_2^2} = RT[2\chi_0 - (1 - \phi_2)^{-2}] = 0 \qquad (6\text{-}94)$$

Equations (6-93) and (6-94) are each solved for χ_0. Remembering that $(1 + X_2^{0.5})(1 - X_2^{0.5}) = 1 - X_2$ and taking the negative root of $[X_2/(1 - X_2)^2]^{0.5}$, we find that the critical point is given by

$$(\phi_2)_{\text{crit}} = (1 + X_2^{0.5})^{-1} \qquad (6\text{-}95)$$

The higher the degree of polymerization of the solute, therefore, the lower will be the values of the critical volume fraction.

The critical value for the Flory–Huggins interaction parameter is obtained from the combination of Equations (6-94) and (6-95):

$$(\chi_0)_{\text{crit}} = \frac{(1 + X_2^{0.5})^2}{2X_2} \approx 0.5 + X_2^{-0.5} \qquad (6\text{-}96)$$

The critical interaction parameter has a value of 0.5 for infinitely high degrees of polymerization.

6.6.2. Upper and Lower Critical Solution Temperatures

To a good approximation, the temperature dependence of the Flory–Huggins interaction parameter can be given by

$$\chi_0 = \alpha + (\beta/T) \tag{6-97}$$

where α and β are system-dependent constants. The term β is generally positive (endothermal mixing). Consequently, χ_0 decreases with increasing temperature. Above a certain temperature ("upper critical temperature," UCST), complete solution occurs (Figures 6-13 and 6-14).

There are also systems that form one phase below a certain temperature

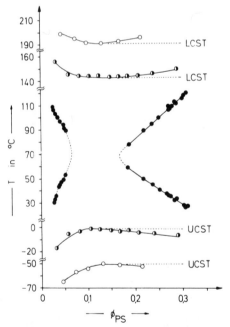
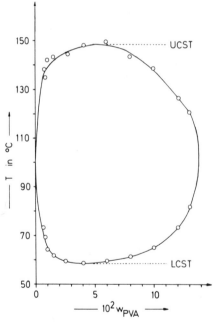

Figure 6-13. Precipitation temperatures of poly(styrene) in acetone as a function of the volume fraction of the solute. Measurement for molar masses of 4800 (○), 10 300 (●) and 19 800 (◑). LCST is the lower critical solution temperature, UCST is the upper critical solution temperature (K. G. Siow, G. Delmas and D. Patterson).

Figure 6-14. Precipitation temperatures of poly[(vinyl alcohol)₉₃-co-(vinyl acetate)₇] in water as a function of the mass fraction of the solute. $\overline{M}_n = 140\ 000$ g/mol (G. Rehage).

("lower critical solution temperature," LCST). They demix above the LCST (Figs. 6-13 and 6-14). The LCST corresponds to an entropically, the UCST to an enthalpically induced demixing. In principle, each polymer/solvent system should have both a LCST and a UCST, but one or the other critical demixing temperature may not be measurable for practical reasons.

The terms UCST and LCST have nothing to do with the absolute value of the demixing temperature. For some systems, the UCST, in fact, is at a higher temperature than the LCST (Figure 6-13), but for others, it is lower (Figure 6-14). Thus, depending on the positions of these two demixing temperatures, one observes an "hour-glass" plot (Figure 6-13) or a closed miscibility loop (Figure 6-14). The same polymer can exhibit one or the other type of behavior according to the nature of the solvent used to dissolve it. For example, poly(oxyethylene) has a closed miscibility loop for solutions in water, but shows hour glass behavior when dissolved in *t*-butyl acetate.

The case of UCST > LCST is observed with water-soluble polymers. Examples of these are poly(vinyl alcohol) (see Figure 6-14), poly(vinyl methyl ether), methyl cellulose, and poly(L-proline). The heating of aqueous solutions of these polymers causes a decreasing solvation of the polymer and thus a demixing. In some cases, closed miscibility loops can be observed.

The case of UCST < LCST occurs quite generally with solutions of macromolecules at temperatures above the boiling point of the solvent and at pressures of several bars. In these systems, a contraction occurs on mixing the dense polymer with the highly expanded solvent. This leads to negative entropies of mixing and, consequently, to lower critical solution temperatures. Since this effect is characteristic of all solutions of macromolecules, the "goodness" of the solvent must pass through a maximum between the upper and the lower critical solution temperature, that is, χ_0 must pass through a minimum with temperature.

6.6.3. Quasibinary Systems

All the points so far considered are concerned with genuine binary systems. A system is "binary" if both the solute and the solvent are monodisperse. But macromolecular substances generally possess a molar mass distribution: consequently they only form quasibinary systems with pure solvents.

The phase-separation behavior of quasibinary systems is different from that of binary systems. This phenomenon can be most simply described in terms of the cloud-point curves of ternary systems consisting of a solvent and two monodisperse solutes. The cloud-point curve corresponds to the special case of phase separation where the volume of one of the phases tends toward zero.

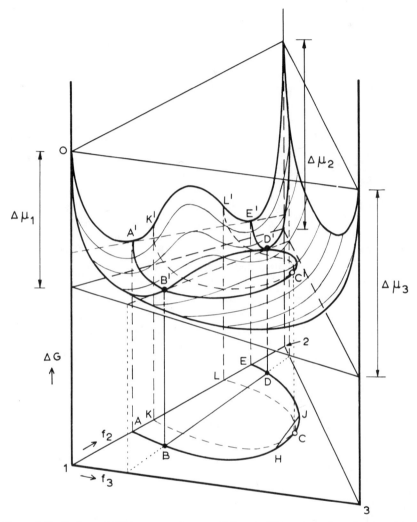

Figure 6-15. Surface of Gibbs energy for a partially miscible system of a solvent and two molecular homogeneous polymers. The binary system 1–2 exhibits partial miscibility, the systems 2–3 and 1–3 are completely miscible (R. Koningsveld).

Such a ternary system has a Gibbs energy surface instead of the Gibbs energy curve associated with genuine binary systems (Figure 6-15). Also, a tangent plane occurs instead of a tangent. If a tangent plane is moved over the Gibbs energy surface of a ternary system of limited miscibility, two series of contact points (e.g., *B'* and *D'*) are produced. The line *A'B'C'D'E'* and its projection *A BCDE* to the base surface are each called binodals. The binodals describe the boundary between stable and metastable mixtures and give the compositions of the coexisting phases. These compositions are related via the joining lines AE, BD, HJ, etc. The compositions of the coexisting phases are identical at the critical point C' (or C).

The Gibbs energy surface is altered by a change in temperature, and consequently, the positions of the binodals also change (see Figure 6-16). The critical points move along the joining line $C-C_5-C'$. The maximum of this joining line occurs at the critical point C for the pure polymer P_2. Consequently, the critical point can only be identical with the maximum in the cloud-point curve in the case of a genuine binary system. In contrast, the critical points of polymer mixtures lie lower than the cloud-point-curve maximum, that is, at larger polymer volume fractions. The point of intersection of the coexistence curve with the cloud-point curve gives the critical point (see Figure 6-17). Quantitative calculations give the critical volume fraction of a quasibinary system as [compare Equation (6-95) for binary systems]

$$(\phi_2)_{\text{crit}} = (1 + \overline{X}_w \overline{X}_z^{-0.5})^{-1} \tag{6-98}$$

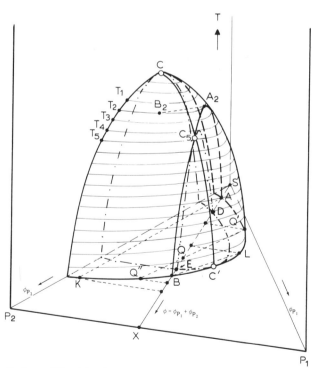

Figure 6-16. Surface of binodals of a ternary liquid system with a two-phase region, CC_5C', Connecting line for critical points; AA_2C_5B, quasibinary line (cloud curve) (R. Koningsveld).

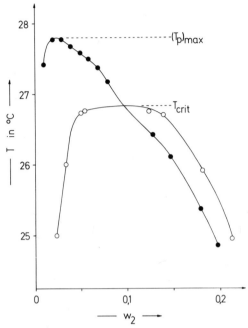

Figure 6-17. Dependence of the cloud point temperature T of a polydisperse polystyrene ($\overline{M}_z : \overline{M}_w : \overline{M}_n = 2.4 : 1.65 : 1$ and $\overline{M}_n = 210{,}000$ g/mol) on the mass fraction w_2 of the polymer (●–●): The curve (○–○) gives the curve of coexistence for a 6% initial solution at different temperatures T, i.e., the mass fractions w_2 at which the polymer is found in the two coexistent phases (after G. Rehage and D. Moller).

The critical interaction parameter is given as

$$(\chi_0)_{crit} = 0.5(1 + \overline{X}_z^{0.5}\,\overline{X}_w^{-1})(1 + \overline{X}_z^{-0.5}) \qquad (6\text{-}99)$$

The difference between the volume fraction of the solute at the maximum of the cloud-point curve and the critical volume fraction can be used as a measure of the polymolecularity. The same holds for the difference between the maximum cloud-point temperature and the critical demixing temperature.

6.6.4. Fractionation

Fractionation of polymers according to molar mass represents the most significant analytical application of phase-separation phenomena. The polymers of highest molar mass separate out first on lowering the temperature of a quasibinary endothermic dilute solution system. Of course, this "precipitation" represents the formation of a highly concentrated "gel phase" and a dilute "sol phase." Successive decreases in temperature lead to further fractions, and the amounts and molar masses of these are determined. The fractionation process should be so carried out that the fractions obtained reproduce the original molar mass distribution as closely as possible. Computer calculations on the basis of the Flory–Huggins theory show that this is best achieved by separating initially into, e.g., five fractions and then separating each fraction into three subfractions (or vice versa). The fractions produced do not necessarily possess a much narrower molecular-weight distribution than the parent sample. The fractions may even have a much wider molar mass distribution than the original material.

Since precipitation temperatures may lie in experimentally unfavorable regions, precipitation fractionation is often carried out by addition of precipitant to the polymer solution at constant temperature. It is advantageous to use a 1% solution of the polymer in a poor solvent as initial solution and to add a weak nonsolvent as precipitant. To achieve a good fractionation, it is best to heat a precipitated system to redissolution and then cool the well-stirred solution to the original temperature. Further fractions are obtained by adding more precipitant.

6.6.5. Determination of Theta States

The critical temperature is identical with the maximum of the cloud-point curve for binary systems (see Section 6.6.3). The dependence of the critical temperature on the degree of polymerization is given by combining Equations (6-49) and (6-96) (with $\sigma = 0$):

$$\frac{1}{T_{crit}} = \frac{1}{\Theta} + \frac{1}{\Theta\psi}\left(\frac{1}{X_2^{0.5}} + \frac{1}{2X_2}\right)$$
(6-100)

Consequently, the critical temperature of mixing is identical with the theta temperature at infinitely high degrees of polymerization. The theta temperature is therefore the critical temperature of mixing of a polymer of infinitely high degree of polymerization.

The dependence of the critical temperature of mixing on the degree of polymerization required by Equation (6-100) is found for quasibinary systems as well as for binary systems (Figure 6-15). The slope of line describing this dependence is determined by the entropy term ψ. If ψ is very small, the critical mixing temperature lies far from the theta temperature. For example, poly(chloroprene) has an entropy term of $\psi = 0.05$ and a theta temperature of 298.2 K. The critical temperature of mixing is $-73°$C for a molecular weight of 700 000 g/mol. Thus, under certain conditions, a quite wide temperature range must be studied to determine the critical temperature of mixing.

The method for determining the theta temperature based on Equation (6-100) requires several samples of known degree of polymerization and is

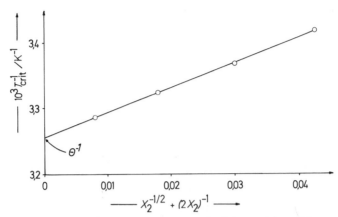

Figure 6-18. Determination of the theta temperature of poly(styrene) in cyclohexane from the critical temperature dependence on the degree of polymerization (after A. R. Schultz).

very time consuming. The theta temperature for a single sample of unknown molecular weight can be determined by another process. The method was first evolved for the determination of theta mixtures and was later used to determine the theta temperatures of binary and quasibinary systems.

The process uses very dilute solutions, normally in the concentration range of ϕ_2 between 10^{-5} and 10^{-2}. In what is called the cloud-point titration method, dilute polymer solutions are titrated with nonsolvent at constant temperature to the first cloud point. The volume fraction ϕ_3 of nonsolvent to give the first cloud point is plotted against the logarithm of the volume fraction ϕ_2 of the polymer at the cloud point. The extrapolated straight line obtained for a homologous series at a point on the $\phi_2 = 1$ axis was found experimentally and theoretically to be equivalent to $(\phi_3)_\Theta$ (Figure 6-19). $(\phi_3)_\Theta$ corresponds to the solvent–precipitant theta mixture for the polymer at this temperature.

A first cloud point determined by lowering the temperature can also be obtained in an analogous manner to that produced by addition of precipitant. The dependence on concentration of the first cloud-point temperature is given by

$$T_p^{-1} = \Theta^{-1} + \text{const} \times \log \phi_2 \tag{6-101}$$

Thus, the reciprocal of the theta temperature is obtained at $\phi_2 = 1$.

If the measurements are carried out with the same solvent–precipitant–temperature system on copolymers of different composition, then the values of $(\phi_3)_\Theta$ for the contents w_A of the various monomeric units of the copolymers lie on a straight line (Figure 6-20). It has been shown experimentally that the method yields data concerning the average composition of the copolymer,

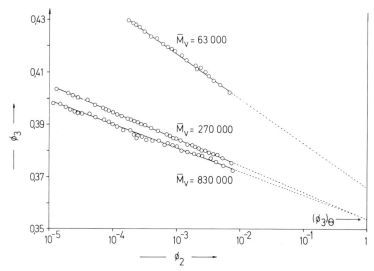

Figure 6-19. Dependence of the volume fraction ϕ_3 of the precipitant on the logarithm of the volume fraction ϕ_2 of the solute at the point of incipient turbidity for polystyrenes of different viscosity-average molar masses in a benzene-isopropanol system at 25°C (after A. Stasko and H.-G. Elias).

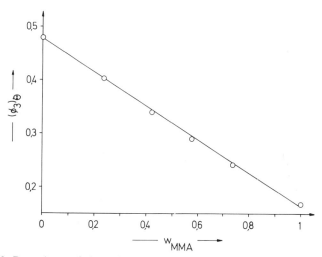

Figure 6-20. Dependence of the $(\phi_3)_\Theta$ values on the average mass fraction w_{MMA} of the methylmethacrylate monomeric units in poly(styrene-co-methyl methacrylates). Solvent: methyl-isopropyl ketone; precipitant: *n*-hexane; temperature: 25°C (after H.-G. Elias and U. Gruber).

provided that copolymer compositions vary only slightly from molecule to molecule. The values of $(\phi_3)_\Theta$ obtained are thus independent of whether the copolymers are alternating, random, branched, block, or graft copolymers. The method can therefore be used to elucidate or confirm the composition of copolymers. It is of particular interest in the study of graft copolymers, for example, since addition of a small amount of homopolymer with a lower $(\phi_3)_\Theta$ yields only this lower value and not that corresponding to the average composition of the mixture. If, on the other hand, another solvent–precipitant–temperature system is chosen in which this homopolymer now has a higher $(\phi_3)_\Theta$ value than the copolymer, then the copolymer composition can be determined independent of the amount of homopolymer present.

The molar mass distribution can, at least in principle, be determined by the cloud-point titration or turbidimetric titration method. In turbidimetric titration, a precipitant is added continually, with stirring, to a very dilute ($\sim 0.01\%$) solution, and the increase in turbidity is observed as a function of the quantity of precipitant added. The turbidity curve which is obtained is a qualitative measure of the molar mass distribution. These curves are difficult to evaluate quantitatively, however, since the turbidity continually changes because of the coagulation of the droplets during titration. The turbidity is therefore not entirely due to the molar mass and concentration of the polymer.

6.6.6. Phase Separation with Solutions of Rods

Coil-like macromolecular solutions generally demix into two phases: a low concentration sol phase and a higher concentration gel phase. Solutions of rods can form mesophases (see Section 5.5), and, so, at least three phases occur on demixing. The three phases are: a dilute isotropic phase, an anisotropic liquid-crystalline mesophase, and a heterogeneous phase consisting of a dispersion of the mesophase in the isotropic phase. If more than one mesophase can be formed, a fourth (anisotropic) phase consisting of the two mesophases is observed (Figure 6-21).

The critical conditions for phase separation of solutions of rods have been theoretically derived. To a first approximation, the critical volume fraction is given by

$$(\phi_2)_{\text{crit}} = \frac{8}{\Lambda}\left(1 - \frac{2}{\Lambda}\right)$$

where Λ is the rod axis ratio or aspect ratio. A critical volume fraction of about 0.053 is obtained for the sample shown in Figure 6-21.

Such phase separations have been observed for a series of rigid macromolecules, as, for example, solutions of poly(γ-benzyl-L-glutamates) in

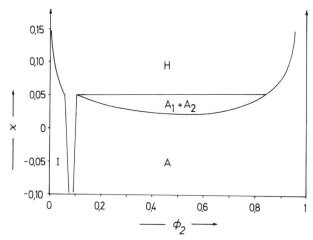

Figure 6-21. Interaction parameter, χ, as a function of the volume fraction, ϕ_2, of rigid rods with an axis ratio of 150 in a solvent. The phases: isotropic solutions (I), anisotropic mesophases (A), heterogeneous phases (H), and mixed phases of two mesophases ($A_1 + A_2$) are formed. The interaction parameter can be replaced by experimentally directly measurable quantities such as the reciprocal temperature [see Equation (6-94)] or the salt concentration in solvent/salt mixtures.

helicogenic solvents, solutions of poly(p-benzamide) in solutions of N,N-dimethyl acetamide containing LiCl, etc.

6.6.7. Incompatibility

In a mixture of two different polymers, polymer 1 takes on the role of solvent for polymer 2. The degrees of polymerization X_1 and X_2 are of the same order or magnitude in a system polymer 1–polymer 2. Consequently, for $\sigma = 0$, Equation (6-34) becomes

$$\frac{\Delta \mu_1}{RT} = \chi_0 \phi_2^2 + \ln(1 - \phi_2) \tag{6-103}$$

Thus, the total entropy contribution comes from the relatively small logarithmic term. So, on mixing two polymers, the entropy gain is small, since the great chain lengths only allow few contact points and this reduces the configurational entropy drastically. Thus, the compatibility is predominantly determined by the enthalpy of mixing.

Numerical comparison of both terms in Equation (6-103) shows that the chemical potential will already be positive for very small concentrations. Consequently, mixtures of two polymers are generally thermodynamically incompatible. Here, "incompatible" does not mean that the two polymers are not mixable over the whole of the concentration range. It only means that the systems show incompatibilty in concentration ranges that are important in practice.

The theoretical prediction has been experimentally verified. Of a total of 281 pairs tested, 239 pairs were definitely incompatible. Even very similar polymer pairs such as poly(styrene)/poly(p-methyl styrene) are incompatible. On the other hand, nitrocellulose is compatible with quite a number of polymers.

In general the same principles apply to mixtures of two polymers in a common solvent. Theoretical calculations of the spinodals show that the incompatibility depends on the interaction parameter χ_{23} (polymer–solvent 3) at high polymer concentrations. Conversely, the difference between the interaction parameters χ_{12} and χ_{13} becomes important at low concentrations. If these interaction parameters differ greatly, a very strong solvent influence on the incompatibility will be observed in dilute solutions. The poly(styrene)/poly(vinyl methyl ether) system is, for example, compatible in toluene, benzene, or perchloroethylene, but not in chloroform or methylene chloride. Conversely, at high polymer concentrations, incompatibility in one solvent is normally accompanied by incompatibility in all other solvents.

Incompatible polymer mixtures can be recognized, in many cases, by purely optical means, such as the opaque appearance in the solid state. On the other hand, transparent samples are not a guarantee of compatibility of two polymers, since an opacity only occurs for sufficiently large refractive index differences between regions of sufficient size (see also Section 14.3). Consequently, incompatibility in transparent samples may require identification by electron microscopy. If the regions of pure polymer in the polymer mixture are of sufficient size, two different glass-transition temperatures can often be measured. In the case of incompatibile polymer mixtures, such different glass-transition temperatures remain invariant with the composition of the mixture.

Incompatibility is not always undesirable in practice. It is even utilized in the case of block copolymers (see Section 5) and high-impact-strength polymers (see Section 35.3).

6.6.8. Swelling

The equilibrium swelling of a chemically cross-linked macromolecule depends on the thermodynamic goodness of the solvent (Figure 6-22). Swelling occurs to an equilibrium value, since the solvent tries to fully dissolve the gel. However, elastic restraining forces resulting from the chemical cross-

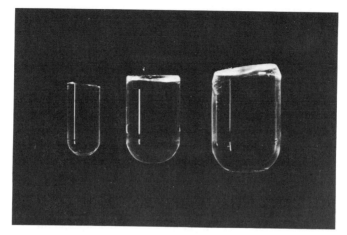

Figure 6-22. Influence of the solvent power on the swelling of weakly cross-linked samples of polystyrene (cross-linked with divinylbenzene). From left to right: unswollen sample, swelling in the poor solvent cyclohexane (χ_0 high), swelling in the good solvent benzene (χ_0 low).

linking are effective. The following holds for equilibrium swelling:

$$\Delta G = \Delta G_{mix} + \Delta G_{el} = 0 \qquad (6\text{-}104)$$

where ΔG_{mix} is the Gibbs energy of mixing and ΔG_{el} is the Gibbs energy of elasticity. The chemical potential of the solvent in the gel is

$$\Delta \mu_1^{gel} = N_L \left(\frac{\partial \Delta G_{mix}}{\partial N_1} \right)_{p,T,N_2} + N_L \left(\frac{\partial \Delta G_{el}}{\partial N_1} \right)_{p,T,N_2} = 0 \qquad (6\text{-}105)$$

with

$$N_L \left(\frac{\partial \Delta G_{el}}{\partial \alpha} \right)_{N_2,p,T} \left(\frac{\partial \alpha}{\partial N_1} \right)_{N_2,p,T} = N_L \left(\frac{\partial \Delta G_{el}}{\partial N_1} \right)_{N_2,p,T} \qquad (6\text{-}106)$$

The three differentials in Equations (6-105) and (6-106) can be evaluated as follows:

1. The chemical potential of the solvent

$$\Delta \mu_1 = \left(\frac{\partial \Delta G}{\partial n_1} \right)_{p,T,n_2} = N_L \left(\frac{\partial \Delta G}{\partial N_1} \right)_{p,T,N_2} \qquad (6\text{-}107)$$

is given by Equation (6-34). For $\sigma = 0$ and with $R/N_L = k$, the following is obtained for a polymer of infinitely high degree of polymerization (i.e., cross-linked, $X_2 \to \infty$) and solvent of low molecular weight ($X_1 = 1$):

$$\left(\frac{\partial \Delta G_{mix}}{\partial N_1} \right)_{p,T,N_2} = kT[\chi_0 \phi_2^2 + \ln(1 - \phi_2) + \phi_2] \qquad (6\text{-}108)$$

2. Following the derivation in Section 11.3.3, the Gibbs energy of elasticity depends on the effective chain content ν_e in the cross-linked system before cross-linking as well as on the expansion factor $\alpha = \alpha_x = \alpha_y = \alpha_z$. From Equation (11-34), the following is obtained with $\Delta G_{el} = -T\Delta S_{el}$:

$$\Delta G_{el} = 0.5 kT\nu_e(3\alpha^2 - 3 - \ln\alpha^3) \tag{6-109}$$

or, after differentiation

$$\left(\frac{\partial\Delta G_{el}}{\partial\alpha}\right)_{p,T,N_2} = 0.5 kT\nu_e(6\alpha - 3\alpha^{-1}) \tag{6-110}$$

3. A cross-linked polymer of volume V_0 in the unswollen state swells to the volume V. For an isotropic example $\alpha^3 = V/V_0 = \phi_2^{-1}$. Also $\phi_2 = V_0/(V_1 + V_0) = V_0(N_1 V_1^m N_L^{-1} + V_0)$ is valid when the volumes are additive. Inserting all these relationships into the expression for α and differentiating, we obtain

$$\left(\frac{\partial\alpha}{\partial N_1}\right)_{p,T,N_2} = \frac{V_1^m}{3\alpha^2 V_0 N_L} \tag{6-111}$$

Combining Equations (6-105), (6-106), (6-108), (6-110), and (6-111) and transforming, we obtain

$$\chi_0\phi_2^2 + \ln(1 - \phi_2) + \phi_2 = -(\nu_e V_1^m V_0^{-1} N_L^{-1})(\phi_2^{1/3} - \tfrac{1}{3}\phi_2) \tag{6-112}$$

If the Flory–Huggins interaction parameter is known from other methods, the effective number of cross-linked network chains can be calculated from the observed volume fraction of polymer in the gel.

Equation (6-112) describes the behavior of weakly swollen, weakly cross-linked polymers quite well. The contributions of the polymer–solvent interaction and the dilution by solvent are negligibly small in comparison to the elasticity term. Highly cross-linked polymers swell to the same (small) extent in different good solvents.

6.6.9. Crystalline Polymers

All the derivations considered so far are only valid for demixing into liquid phases. A phase separation of a crystalline polymer, however, may result in the formation of a solid, crystalline phase and a liquid phase (as long as the phase separation occurs below the melting point of the crystalline polymer). According to Equation (6-4), the change in chemical potential of the solute in the solid phase is

$$d\mu_i^{sol} = \bar{V}_i^m \, dp - \bar{S}_i^m \, dT + RT \, d(\ln a_i) \tag{6-113}$$

By definition, the activity of a pure crystalline substance is equal to one:

$$d\mu_i^{cr} = V_i^m \, dp - S_i^m \, dT \tag{6-114}$$

The chemical potentials in both phases are equal at solution equilibrium,

$$d(\mu_i^{sol} - \mu_i^{cr}) = (\tilde{V}_i^m - V_i^m)dp - (\tilde{S}_i^m - S_i^m)dT + RT \, d(\ln a_i) = 0 \tag{6-115}$$

$dp = 0$ and $dT = 0$, and, consequently, $d(\ln a_i) = 0$, i.e., $a_i = \text{const}$ for an isothermal–isobaric process. Crystalline substances cannot, therefore, exceed a certain saturation level at solution equilibrium. Consequently, in contrast to amorphous material, crystalline substances have only limited solubility. The saturation concentration can be raised by additives (salting in) or lowered by additives (salting out), since the saturation limit depends on the activity and the activity can be changed by additives.

The melting temperature T_{cr} of a monodisperse polymer–solvent system depends strongly on the concentration of the polymer. The chemical potential of the polymer in the crystalline phase is identical with the molar Gibbs energy of fusion ΔG_M^m:

$$\mu_2^{cr} - \mu_2^0 = \Delta G_M^m = \Delta H_M^m - T_{cr} \, \Delta S_M^m = \Delta H_M^m \left(1 - \frac{T_{cr} \Delta S_M^m}{\Delta H_M^m} \right) \tag{6-116}$$

The equation $T_M^\infty = \Delta H_M^m / \Delta S_M^m$ holds for the melting temperature of the undiluted polymer, so Equation (6-116) becomes

$$\mu_2^{cr} - \mu_2^0 = \Delta H_M^m \left(1 - \frac{T_{cr}}{T_M^\infty} \right) \tag{6-117}$$

Assuming the enthalpy and the entropy of fusion to be temperature independent, and since at equilibrium

$$\mu_2^{cr} - \mu_2^0 = \mu_2^{sol} - \mu_2^0 \tag{6-118}$$

then inserting Equations (6-118) and (6-35) into (6-117), we find

$$\frac{1}{T_{cr}} = \frac{1}{T_M^\infty} + \frac{R}{\Delta H_M^m} [X_2 X_1^{-1} (\chi_0 + 2\sigma\phi_2)(1 - \phi_2)^2$$
$$+ (1 - X_2 X_1^{-1})(1 - \phi_2) + \ln \phi_2] \tag{6-119}$$

According to Equation (6-119), the melting point should decrease with increasing solvent concentration when phase separation is into liquid and crystalline phases. This type of behavior is observed, for example, with solutions of poly(ethylene) in xylene (Figure 6-23).

The Flory–Huggins interaction parameter can, however, exceed its critical value above a certain temperature for solutions in poor solvents. In this case, separation into two liquid phases also occurs with crystalline

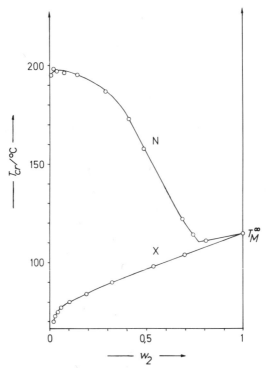

Figure 6-23. Crystallization temperature T_{cr} of a polyethylene in xylene (X) or nitrobenzene (N) as a function of the mass fraction w_2 of the polymer (after R. B. Richards).

polymers, as is shown, for example, by poly(ethylene) in nitrobenzene (Figure 6-23).

Literature

Section 6.1. Basic Principles

P. J. Flory, *Principles of Polymer Chemistry*, Cornell University Press, Ithaca, New York, 1953.

H. Tompa, *Polymer Solutions*, Butterworth, London, 1956.

H. Morawetz, *Macromolecules in Solution*, Wiley–Interscience, New York, 1965.

P. J. Flory, *Statistical Mechanics of Chain Molecules*, second ed., Wiley–Interscience, New York, 1975.

V. N. Tsvetkov, V. Ye. Eskin, and S. Ya. Frenkel, *Structure of Macromolecules in Solution*, Butterworth, London, 1970.

G. C. Berry and E. F. Casassa, Thermodynamic and hydrodynamic behavior of dilute polymer solutions, *Macromol. Rev.* **4**, 1 (1970).

H. Yamakawa, *Modern Theory of Polymer Solutions*, Harper and Row, New York, 1971.

S. von Taparicza and J. Prausnitz, Thermodynamik von Polymerlösungen: Eine Einführung, *Chem.-Ing.-Tech.* **47**, 552 (1975).

D. W. van Krevelen, *Properties of Polymers*, second ed., Elsevier, Amsterdam, 1976.

H. Eisenberg, *Biological Macromolecules and Polyelectrolytes in Solution*, Clarendon Press, Oxford, 1976.

Section 6.2. Solubility Parameter

J. L. Gardon, Cohesive-energy density, in *Encyclopedia of Polymer Science and Technology*, Vol. 3, H. F. Mark, N. G. Gaylord, and N. M. Bikales (eds.), Wiley–Interscience, New York, 1966, p. 833.

D. Patterson, Free volume and polymer solubility, a qualitative review, *Macromolecules* **2**, 672 (1969).

A. F. M. Barton, Solubility parameters, *Chem. Rev.* **75**, 731 (1975).

A. Yayarsi and M. Yaseen, Heat of vaporization. Its determination and application in evaluation of the solubility parameter—A review, *Prog. Org. Coatings* **7**, 167 (1979).

Section 6.3. Statistical Mechanics

D. Patterson, Thermodynamics of non-dilute polymer solutions, *Rubber Chem. Technol.* **40**, 1 (1967).

H. Sotobayashi and J. Springer, Oligomere in verdünnten Lösungen, *Adv. Polym. Sci.* **6**, 473 (1969).

K.-J. Liu and J. E. Anderson, Proton magnetic resonance studies of molecular interactions in polymer solutions, *J. Macromol. Sci.* **C5**, 1 (1970).

A. Veis (ed.), *Biological Polyelectrolytes, Biol. Macromol. Ser.*, Vol. **3**, Marcel Dekker, New York, 1970.

J. J. Hermans (ed.), *Polymer Solution Properties, Part I, Statistics and Thermodynamics*, Dowden, Hutchinson and Ross, Stroudsbury, Pennsylvania, 1978.

D. J. R. Laurence, Interactions of polymers with small ions and molecules, in *Physical Methods in Macromolecular Chemistry* (B. Carroll, ed.), Dekker, New York, Vol. 2, 1972.

B. E. Conway, Solvation of synthetic and natural polyelectrolytes, *J. Macromol. Sci. C (Rev. Macromol. Chem.)* **6**, 113 (1972).

A. Katchalsky, Polyelectrolytes (IUPAC-International Symposium on Macromolecules, Leiden 1970), *Pure Appl. Chem.* **26**(3–4), 327 (1971).

F. Oosawa, *Polyelectrolytes*, Marcel Dekker, New York, 1971.

N. Ise, The mean activity coefficient of polyelectrolytes in aqueous solutions and its related properties, *Adv. Polym. Sci.* **7**, 536 (1971).

W. V. Smith, Fractionation of polymers, *Rubber Chem. Technol.* **45**, 667 (1972).

E. Sélégny, M. Mandel, and U. P. Strauss (eds.), *Polyelectrolytes* (*Charged and Reactive Polymers*, Vol. 1), Reidel, Dordrecht, Holland, 1974.

L. Rebenfeld, P. J. Makarewicz, H. D. Weigmann, and G. L. Wilkes, Interaction between solvents and polymers in the solid state, *J. Macromol. Sci.-Revs. Macromol. Chem.* **C15**, 279 (1976).

R. A. Orwoll, The Polymer–Solvent Interaction Parameter χ, *Rubber Chem. Technol.* **50**, 451 (1977).

Section 6.5. Association

H.-G. Elias, Association and aggregation as studied via light scattering, in *Light Scattering from Polymer Solutions* (M. B. Huglin, ed.), Academic Press, London, 1972.

H.-G. Elias, Association of synthetic polymers, in *Order in Polymer Solutions* (Midland Macromol. Monographs, Vol. 2), K. Solc (ed.), Gordon and Breach, New York, 1975.

T. Wagenknecht and V. A. Bloomfield, Equilibrium mechanisms of length regulation in linear protein aggregates, *Biopolymers* **14**, 2297 (1975).

F. Oozowa and S. Asakura, *Thermodynamics of the Polymerisation of Protein*, Academic Press, London, 1975.

Z. Tuzar and P. Kratochvil, Block and graft copolymer micelles in solution, *Adv. Colloid Interf. Sci.* **6**, 201 (1976).

Section 6.6. Phase Separation

M. J. R. Cantow (ed.), *Polymer Fractionation*, Academic Press, New York, 1967.

R. Konigsveld, Preparative and analytical aspects of polymer fractionation, *Adv. Polym. Sci.* **7**, 1 (1970).

W. V. Smith, Fractionation of polymers, *Rubber Chem. Technol.* **45**, 667 (1972).

B. A. Wolf, Zur Thermodynamik der enthalpisch und entropisch bedingten Entmischung von Polymerlösungen, *Adv. Polym. Sci.* **10**, 109 (1972).

S. Krause, Polymer compatibility, *J. Macromol. Sci. C* (*Rev. Macromol. Chem.*) **7**, 251 (1972).

S. Krause, *Symposium on Microencapsulation* (Chicago 1973), Plenum Press, New York, 1974.

H.-G. Elias, Cloud point and turbidity titrations, in *Fractionation of Synthetic Polymers*, L. H. Tung (ed.), Marcel Dekker, New York, 1977.

R. Koningsveld, Phase equilibria in polymer systems, *Brit. Polym. J.* **7**, 435 (1975).

S. Krause, Polymer–Polymer compatibility, in Polymer Blends, D.R. Paul and S. Newman (eds.), Academic Press, New York, 1978.

T. Kondo, Microcapsules: Their preparation and properties, in *Surface and Colloid Science*, Vol. 10, E. Matijevic (ed.), Plenum Press, New York (1978).

O. Olabisi, L. M. Robeson, and M. T. Shaw, *Polymer–Polymer Miscibility*, Academic Press, New York, 1978.

Chapter 7

Transport Phenomena

Matter, energy, charge, momentum, etc., can be transported with observable effects. The transport of matter occurs by diffusion in the earth's gravitational field, by sedimentation in a centrifugal field, and, for example, by electrophoresis in an electrical field. The viscosity of gases is due to the transfer of momentum. Energy is, for example, transported by heat conduction.

The basic principles for the transport of macromolecules in dilute solution will be discussed in this section; these principles will be utilized for molar mass determinations in Section 9. In addition, the viscosities of melts and concentrated solutions will also be treated, but the determination of molar masses by relative viscosities will not be treated here, but in Section 9. It is also more appropriate to treat heat conduction with the thermal properties of polymers in Section 10.

7.1. Effective Quantities

When matter is transported in solution, the effective mass and the effective volume are not identical to the mass and volume of the "dry" macromolecule. A transported protein molecule such as myoglobin (see Figure 4-14) drags associated solvent along with it, which contributes to the frictional resistance.

The effective mass m_h of a moving molecule consists of the mass m_2 of the "dry" macromolecule of molecular weight M_2 and the mass $m_1^\square$ of the solvent carried along with the macromolecule. If the mass of the solvent carried is expressed as a multiple $\Gamma_h = m_1^\square / m_2$ of the polymer mass, the following is then obtained for the effective mass of the molecule:

$$m_h = m_2 + m_1^\square = m_2(1 + \Gamma_h) = \frac{M_2(1 + \Gamma_h)}{N_L} \qquad (7\text{-}1)$$

The effective volume V_h is composed, analogously, of the volume of the dry macromolecule and that of the solvent moving with the macromolecule, where the volumes are replaceable by the specific volumes v_2 and v_1:

$$V_h = V_2 + V_1^\square = v_2 m_2 + v_1^\square m_1 = \frac{M_2(v_2 + \Gamma_h v_1^\square)}{N_L} \qquad (7\text{-}2)$$

The specific volume v_1 of the solvent in the macromolecule is different from the specific volume v_1 of the pure solvent, since part of the solvent which is in the macromolecule will undergo a specific interaction (solvation) with the macromolecule. The other part will be dragged along purely mechanically. The total volume V of the solution of a total mass m_1 of solvent and mass m_2 of dry macromolecule is thus given as

$$V = m_2 v_2 + (m_1 - m_1^\square)v_1 + m_1^\square v_1^\square \qquad (7\text{-}3)$$

and, with $\Gamma_h = m_1^\square / m_2$, as

$$V = m_2 v_2 + m_1 v_1 + \Gamma_h m_2(v_1^\square - v_1) \qquad (7\text{-}4)$$

In very dilute solution, Γ_h is constant, and does not depend on concentration. The partial specific volume of the solute is obtained in this case by differentiation of Equation (7-4).

$$\tilde{v}_2 = \left(\frac{\partial V}{\partial m_2} \right)_{p, T, m_1} = v_2 + \Gamma_h(v_1^\square - v_1) \qquad (7\text{-}5)$$

Combining Equations (7-2) and (7-5), we obtain

$$V_h = \frac{M_2}{N_L}(v_2 + \Gamma_h v_1) \qquad (7\text{-}6)$$

The hydrodynamic volume therefore depends very much on the factor Γ_h. This factor measures the amount of solvent transported with the macromolecule by solvation and/or purely mechanically.

7.2. Diffusion in Dilute Solution

7.2.1. Basic Principles

Diffusion processes can be subdivided into various classes, e.g., translational, rotational, thermal, etc. Translational diffusion consists of the isothermal equilibration of matter between two phases of differing concentration. The rotation of molecules and particles around their own axis is termed rotational diffusion. Thermal diffusion is the equilibration of matter under the influence of a temperature gradient.

The diffusion of a solute is described by its flux J_d, which is defined as the change in mass with time on passing through an interface of area A:

$$J_d \equiv dm/(A\,dt) \tag{7-7}$$

The flux occurs from a higher to a lower concentration, and so, according to Fick's first law of diffusion, it is proportional to the concentration gradient (see textbooks on physical chemistry):

$$J_d = -D(\partial c/\partial r) \tag{7-8}$$

The proportionality constant D is called the diffusion coefficient.

With solution of polymers, it is mostly the change with time of the concentration, and not the change in mass with time or, consequently, the flux, which is most often measured. Since the law of conservation of mass must hold, the following is directly obtained:

$$\partial c/\partial t = -(\partial J_d/\partial r) \tag{7-9}$$

Combining this equation with (7-8) gives Fick's second law of diffusion:

$$\frac{\partial c}{\partial t} = \frac{\partial[D(\partial c/\partial r)]}{\partial r} \tag{7-10}$$

If the diffusion coefficient D does not depend on concentration c, and thus also not on the path r, then Equation (7-10) transforms to

$$\frac{\partial c}{\partial t} = D\frac{\partial^2 c}{\partial r^2} \tag{7-11}$$

7.2.2. Experimental Methods

A relatively simple solution to Fick's second law of diffusion is obtained when the experiment starts with two solutions of concentration c' and c'' separated by an infinitely thin interface ($dr = 0$). In addition, the experiment should only be observed over such a time period that the initial concentrations remain unchanged at the top and bottom of the diffusional chamber. Integration of Equation (7-11) gives

$$c(r,t) = \left(\frac{c'-c''}{2}\right)\left\{1 - \frac{2}{\pi^{0.5}}\int_{r_0}^{r}\exp\left[-\frac{(r-r_0)^2}{4Dt}\right]dr\right\} \tag{7-12}$$

The sharp boundary necessary for these limiting conditions is realized experimentally by several types of diffusion cells. In slide cells, a slide separates the lower chamber filled with the denser solution from the upper chamber with the less dense solution (mostly solvent). The slide is withdrawn at the beginning of the diffusion experiment and the diffusion followed by, e.g., the broadening of the concentration gradient curve. Another diffusion cell works on the lower-layer principle. Here, the chamber is first half-filled with the less dense solution. The dense solution is then added in such a way that it forms a layer at the base of the chamber. This is done by adding the solution through a tube that reaches to the base of the chamber. Addition continues until the separating layer between the two solutions reaches the center of the observation window. In this position, slits are fixed at right angles to the cell window, through which the separating layer can be drawn by suction and thus sharpened.

The progress of the diffusion can be followed, e.g., by light absorption measurements (visible light, uv, ir) or by interference measurements. Both methods register a concentration-dependent quantity as a function of the position at a constant time (or vice versa), thus fulfilling the conditions of Equation (7-12). The concentration gradients dc/dr, however, can be found as a function of the position by means of a suitable optical method, the Schlieren optics method. To monitor experiments by Schlieren optics, Equation (7-12) has to be differentiated:

$$\frac{\partial c}{\partial r} = -\left[0.5 \frac{c' - c''}{(\pi Dt)^{0.5}} \right] \exp \left[-\frac{(r - r_0)^2}{4Dt} \right] \qquad (7\text{-}13)$$

The diffusion coefficients found using Equation (7-12) or (7-13) usually depend also on the concentration of the initial solution. All the equations quoted so far, on the other hand, relate to infinitely dilute solutions. The dependence on the concentration of the diffusion coefficients in the case of dilute solutions can usually be given as

$$D_c = D(1 + k_D c) \qquad (7\text{-}14)$$

The constant k_D contains a hydrodynamic and a thermodynamic term. It generally decreases with increasing molecular weight in a polymer homologous series. The diffusion coefficient D_c measured at a finite concentration c relates to the average between the two initial concentrations, i.e., to an average concentration of $c_0/2$, where c_0 is the initial concentration of one solution and the other solution is pure solvent.

The diffusion coefficients D of polymolecular materials are obtained as average values which differ according to the evaluation method. Particularly often used are the mass average (also known as the moment average):

$$\bar{D}_w = \sum_i w_i D_i \bigg/ \sum_i w_i \equiv D_w^{(1)} \qquad (7\text{-}15)$$

and the area average (or D_1 average):

$$\bar{D}_A = \left(\sum_i w_i D_i^{-0.5} / \sum_i w_i \right)^{-2} \equiv D_w^{(-0.5)} \tag{7-16}$$

When the molar mass distribution is not too wide, the mass average and area average for random coils and flexible rods are, within the experimental error, practially identical. A whole series of other averages, besides these two averages, may also be defined (see also Section 8).

7.2.3. *Molecular Quantities*

According to the Einstein–Sutherland equation, the diffusion coefficient D is related to the molecular frictional coefficient f_D:

$$D = kT/f_D = RT/f_D N_L \tag{7-17}$$

On the other hand, the frictional coefficient of an unsolvated sphere of homogeneous density, f_{sphere}, in a solvent of viscosity η_1 is, according to Stokes:

$$f_{sphere} = 6\pi\eta_1 r_{sphere} \tag{7-18}$$

When the sphere is solvated, the radius, r_{sphere}, is replaced by the hydrodynamically effective radius, r_h. In addition, deviations from spherical shape can be described by an asymmetry factor $f_A = f_D/f_{sphere}$. Thus, the coefficient of friction of a solvated particle of any given shape is given by

$$f_D = f_A(6\pi\eta_1 r_h) \tag{7-19}$$

The hydrodynamic radius can be expressed in terms of the hydrodynamic volume of Equation (7-6) and the frictional coefficient can be given as in Equation (7-17). Consequently, the diffusion coefficient can be obtained as

$$D = \left(\frac{RT}{6\pi\eta_1 N_L f_A} \right) \left[\frac{3M_2}{4\pi N_L} \left(\tilde{v}_2 + \Gamma_h v_1 \right) \right]^{-1/3} \tag{7-20}$$

The diffusion coefficient D thus depends on the known or measurable quantities R, T, N_L, η_1, v_2, and v_1 and on three unknowns: the molar mass M_2, the asymmetry factor f_A, and the parameter Γ_h. It cannot be interpreted molecularly without further assumptions. For unsolvated solid spheres, $f_A = 1$ and $\Gamma_h = 0$. Consequently, in a homologous series of such spherical molecules, the diffusion coefficient D decreases with $M^{-1/3}$. In a homologous series of molecules of different shapes, the molar mass dependence of f_A and Γ_h also has to be taken into account. In general, it is found empirically for such a homologous series that

$$D = K_D M_2^{a_D} \tag{7-21}$$

Table 7-1. Diffusion Coefficient D of Macromolecules in Dilute Solution

Macromolecule[a]	Molar mass in g/mol	Solvent	Temperature, in °C	10^7 D in cm^2/s
Ribonuclease	13 683	Water	20	11.9
Hemoglobin	68 000	Water	20	6.9
Collagen	345 000	Water	20	0.69
Myosin	493 000	Water	20	1.16
Deoxyribonucleic acid	6 000 000	Water	20	0.13
PMMA	34 100	Acetone	20	17.4
PMMA	280 000	Acetone	20	4.65
PMMA	580 000	Acetone	20	1.15
PMMA	935 000	Acetone	20	0.85
PMMA	200 000	Butyl chloride	35.6	7.18

[a] PMMA = poly(methyl methacrylate).

K_D and a_D are shape- and solvation-dependent constants. a_D can be expressed through the exponents of the relationships between molecular weight and other hydrodynamic quantities [see Equation (8-61)].

The diffusion coefficients D of macromolecules in dilute solutions have values of $\sim 10^{-7}$ cm^2/s (Table 7-1). The diffusion coefficients of the proteins ribonuclease, hemoglobin, collagen, and myosin, as well as deoxyribonuclease, were measured in dilute aqueous salt solutions. They were transformed into "pure water" diffusion coefficients by assuming that any differences are caused by differing viscosities only, and not by changes in the asymmetry or solvation coefficients. The diffusion coefficients of macromolecules in the melt are much lower, being $\sim 10^{-13}-10^{-12}$ cm^2/s.

In low-molar-mass liquids, the diffusion coefficients of dissolved substances initially decrease strongly with increasing solvent viscosity (Figure 7-1). They then remain practically constant for solvents consisting of polymers of not too high molar mass. This allows the conclusion that the diffusion of the solute is only dependent on the segment mobility of the solvent molecules and not on the mobility of the molecular centers of gravity.

7.3. Rotational Diffusion and Streaming Birefringence

The longitudinal axes of nonspherical particles have a random distribution of angles with respect to each other in dilute solutions. This angular distribution of the longitudinal axes is disturbed by externally applied shear gradients or other fields. The longitudinal axes orient themselves parallel or perpendicular to the direction of the field, according to the type of field and its

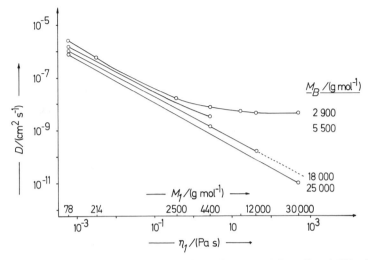

Figure 7-1. Dependence of the diffusion coefficients of *cis*-1,4-poly(butadienes), BR, of molar masses M_{BR} = 2900, 5500, 18,000, and 25,000 g/mol on the viscosity η_1 of the solvent at 40° C. Benzene, pentachloropropane, and *trans*-1,4-poly(pentenamers) of molar masses between 2500 and 30 000 g/mol were used as solvents (after M. Hoffman).

interaction with the particles. Orientation can be forced, for example, by an electric field (Kerr effect) or a magnetic field (Cotton–Mouton effect). When the field is no longer applied, the particles again move into their randomly oriented equilibrium positions by means of rotational diffusion. An analogous equation to that for the normal diffusion coefficient [Equation (7-17)] holds for the rotational diffusion coefficient D_r:

$$D_r = \frac{RT}{f_r N_L} \tag{7-22}$$

where f_r is the rotational frictional coefficient.

For the rod-shaped particles of tobacco mosaic virus of length 280 nm at room temperature, for example, rotational diffusion coefficients of 550 s^{-1} were found. The rods thus require 0.0018 s in order to return to their equilibrium (randomly oriented) positions.

The particles can also be oriented purely mechanically by a shear gradient, as well as by electric or magnetic fields. This type of flow with a linear shear can be produced when the liquid to be studied is placed in a narrow space between two concentric cylinders (Figure 7-2). One of the two cylinders turns (rotor), while the other remains stationary (stator). Owing to the partial orientation of the anisotropic molecules, the refractive indices n differ according to whether the long axis of the particles is parallel or at right

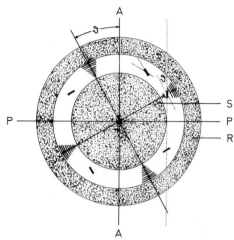

Figure 7-2. Schematic representation of the flow birefringence of ellipsoidal particles between the stator S and the rotor R. *A–A* or *P–P*: polarization planes of the analyzer or polarizer. ϑ is the extinction angle.

angles to the direction of flow. The difference $n = n_\perp - n_\parallel$ of these two refractive indices is termed *birefringence.*

A dark cross against a light background is observed if the rotating solution is studied under crossed Nicol prisms. The effect comes about as follows: If horizontal polarized light enters an isotropic solution, complete extinction takes place. Under the same conditions, solutions with partially oriented anisotropic particles cause extinction only at those points where the optical axis of the anisotropic particles is parallel to the polarization plane of the polarizer or the analyzer (Figure 7-2). Extinction can therefore occur in four positions. As can be seen from Figure 7-2, all indicated particles form an angle ϑ to a tangent to the direction of flow. However, only at four regions in the circle, which form a cross, are they also parallel to the polarization planes *A–A* or *P–P*, as is shown for the particle in the upper right-hand quadrant. It is found experimentally that the cross lies below an angle at 45° to the two polarization planes at low shear gradients, and below an angle of 0° at very high shear gradients. The smaller of the angles between the polarization planes and the black cross is termed the extinction angle ϑ. Consequently, ϑ varies from 45° at low to 0° at high shear gradients. The extinction angle is therefore an indication of the orientation of the particles in the field of flow. The strength of the birefringence is an indication of the degree of orientation.

The orientation of the molecules counteracts the rotational diffusion. The smaller the molecules, the faster will be the rotational diffusion. For a given ratio of the molecular axes, therefore, there is a lower limit for which a particle length can be measured by birefringence. It is at ~20 nm. Shorter molecules than this require such high shear gradients that the flow becomes turbulent and the assumptions necessary for measurement and analysis no longer hold. Rigid, long molecules, on the other hand, require only low gradients. Thus, for example, for tobacco mosaic virus a strong effect is observed even at shear gradients of ~ 5 s^{-1}. With the flexible coils of poly(styrene), on the other hand, only a low streaming birefringence can be seen even at $10^4 s^{-1}$. The method is therefore mainly applied in the case of rigid macromolecules, and then provides a means of measuring the particle lengths.

7.4. Electrophoresis

The movement of an electrically charged particle of mass m and charge Q under the influence of a uniform electric field of strength E is called electrophoresis. These particles may be biological cells, colloids, macromolecules, or low-molar-mass substances. Even electrically neutral particles can be made electrophoretically mobile through formation of suitable complexes. An example is the formation of borate complexes by polysaccharides:

$$H_2BO_3^+ + \begin{array}{c} HO \\ \diagdown \\ HO \diagup \end{array} R\sim\sim\sim \;\rightleftharpoons\; \left[\begin{array}{c} HO \\ \diagdown \\ HO \diagup \end{array} B\begin{array}{c} O \\ \cdots \diagdown \\ \cdots \diagup \\ O \end{array} R\sim\sim \right]^- + H_2O \quad (7\text{-}23)$$

The particles move in a solvent, typically aqueous salt solutions, in free electrophoresis or Tiselius electrophoresis. In carrier electrophoresis, the particles move in a swollen carrier [for example, paper, starch gels, crosslinked poly(acrylamide)].

The particle movement is induced by a force QE. The movement is resisted by the frictional force $f(dl/dt)$. Here, f is the frictional coefficient and dl/dt is the particle velocity. The resultant of these two forces is given by Newton's second law of motion as $m(d^2l/dt^2)$. Thus

$$m\frac{d^2l}{dt^2} = QE - \frac{dl}{dt} \quad (7\text{-}24)$$

or, integrated once,

$$\frac{dl}{dt} = \frac{QE}{f}\left[1 - \exp\left(-\frac{f}{m}t \right) \right] \quad (7\text{-}25)$$

The quotient f/m is about 10^{12}–10^{14} s^{-1} for molecular particles. Consequently, equation (7-25) reduces, for times greater than 10^{-11} s, to

$$\frac{dl}{dt} = \frac{QE}{f} \quad (7\text{-}26)$$

The electrophoretic mobility μ is defined as the velocity of movement induced by the effect of an electric field of 1 V/cm. On inserting into the Einstein–Sutherland Equation (7-17), one obtains, with Equation (7-26),

$$\mu = \frac{dl/dt}{E} = \frac{N_L QD}{RT} \tag{7-27}$$

In the laboratory, electrophoresis is used for analysis and to separate charged particles on the basis of their different electrophoretic mobilities. In the analysis of a mixture of proteins, the apparent protein content A, for example, depends on the ionic strength Γ as well as on the total protein concentration. Consequently, the apparent content of A is plotted against c/Γ and extrapolated to $c/\Gamma \to 0$.

In industry, electrophoresis is used in the electro-dip-coating process or electrophoretic coating, especially in the automobile industry for painting car bodies. Here the metal component is connected as the anode in an electric circuit. On applying an electric field, the negatively charged particles (mostly latex particles) move to the anode and are discharged as a film there. An electroosmosis effect subsequently occurs and water molecules are expelled from the film. The solids content of the polymer film is increased up to 95%. A final electrophoresis can be applied to remove the last traces of water and dissolved ions. In contrast to other automated coating procedures, electro-dip-coating allows even relatively inaccessible corners and edges to be uniformly coated. Additionally, the solvent base of the paints used is water, so the costly installations required for the recovery of organic solvents from the vapor phase are not needed. Consequently, electro-dip-coating is being increasingly used for coating automobile bodies.

7.5. Viscosity

7.5.1. Concepts

Consider a ribbon of infinite length moving with the velocity v (cm/s) through a liquid contained between two infinitely long, parallel plates which are at a distance y (cm) apart. In the immediate vicinity of the plates the liquid is at rest, whereas in the immediate vicinity of the ribbon it will be moving at the velocity v. A velocity gradient $\dot{\gamma} = dv/dy$ with the units s^{-1} thus exists between plate and ribbon. $\dot{\gamma}$ is also called the shear rate.

Because of the problem of sealing at both ends, this so-called ribbon viscometer can only be realized for materials of extremely high viscosity. The properties of a ribbon viscometer are shown to a good approximation by a rotation viscometer of the Couette type (see Section 9.5.2). In Couette viscometers, a rotor revolves around a stator (or vice versa). The viscous

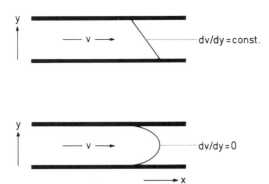

Figure 7-3. Definition of the shear gradient dv/dy in a current with velocity v between two parallel plates of a stator and a rotor (above), or in a capillary (below). With the plates, the shear gradient dv/dv = constant between the plates. With the capillary, dv/dy = 0 at the center of the capillary.

liquid lies between the rotor and the stator. If the space between them is sufficiently narrow, the shear rate is constant across the whole distance between rotor and stator (Figure 7-3).

With the concentric lamellar flow of liquids in a capillary, however, dv/dv is not constant, but is a function of the distance y from the capillary wall. The shear gradient is the greatest near the capillary wall, while at the center of the capillary $\dot\gamma = dv/dy = 0$ (Figure 7-3).

The velocity gradient, measured perpendicular to the direction of flow, is the difference in velocity of two lamellae flowing past each other. Forces are active in the direction of flow at the contact surface between the two layers. They are called shear, thrust, or tangential forces. The ratio of the shear forces K to the contact surface A is called the shear stress σ_{21}.

With liquids of low viscosity there is, according to Newton's law, a direct proportionality between the shear gradient $\dot\gamma$ and the shear stress σ_{21} for many liquids of low viscosity:

$$\sigma_{21} = \eta\dot\gamma = \eta \frac{dv}{dy} \qquad (7\text{-}28)$$

The proportionality factor η is called the viscosity and its inverse value $1/\eta$ is called fluidity.

Liquids that follow Newton's law are called Newtonian liquids. In non-Newtonian liquids, the quantity η, which can be calculated from the quotient, $\sigma_{21}/\dot\gamma$, also changes with the velocity gradient, or with the shear stress. Newtonian behavior is usually observed for the limiting case $\dot\gamma \to 0$ or $\sigma_{21} \to 0$. Melts and macromolecular solutions often exhibit non-Newtonian behavior. Non-Newtonian liquids are classified as dilatant, Bingham body, pseudoplastic, thixotropic, or rheopectic liquids.

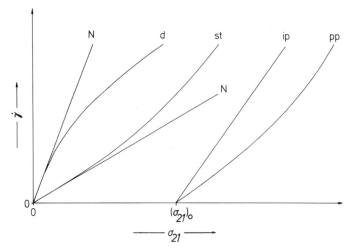

Figure 7-4. A plot of shear rate $\dot{\gamma}$ as a function of shear stress σ_{21} for Newtonian (N), dilatant (d), and pseudoplastic (st) liquids, and ideal plastic (ip) and pseudoplastic (pp) variants of Bingham bodies. $(\sigma_{21})_0$ is the yield value.

In dilatant liquids, $\dot{\gamma}$ increases less than proportionally with σ_{21}, while in pseudoplastic liquids $\dot{\gamma}$ increases more than proportionally (Figure 7-4). Expressed another way: In pseudoplastic liquids, the apparent viscosity $\eta_{app} = \sigma_{21}/\dot{\gamma}$ falls as the shear stress increases (shear thinning), whereas in dilatant liquids it rises (shear thickening). The fluidities, on the other hand, increase with rising σ_{21} in pseudoplastic liquids and fall in dilatant liquids. At $\gamma \rightarrow 0$, both dilatant and pseudoplastic liquids show a Newtonian behavior. Shear thickening is rare in melts and macromolecular solutions, but occurs often in dispersions.

Shear thinning is shown by flow-oriented, asymmetric, rigid particles and/or shear-deformed flexible coils. In the first case, the pseudoplastic viscosity should vary with the square of the concentration because of the interactions between two particles, and in the second case only with the concentration itself.

Plastic bodies are also called Bingham bodies. They exhibit a stress limit to flow (Figure 7-4). The limiting value is defined as the minimum value of σ_{21} above which $\dot{\gamma}$ begins to vary with σ_{21}, i.e., above $(\sigma_{21})_0$. It is also called yield value. Ideal plastic bodies show Newtonian behavior above the flow limit. Pseudoplastic bodies, on the other hand, show pseudoplastic behavior above $(\sigma_{21})_0$. The plasticity or flow limit is interpreted as the breaking up of molecular associations. Plasticity is particularly desirable in paints.

The flow properties of the various classes of non-Newtonian liquids can be summarized in a generalized flow curve (Figure 7-5). This flow curve is

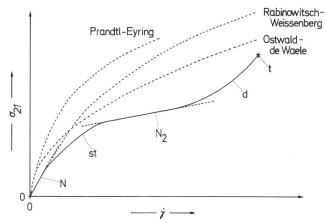

Figure 7-5. Generalized flow curve with the first Newtonian region (N), pseudoplastic region (st), second Newtonian region (N_2), dilatant region (d), and onset of turbulence or melt break (t).

remarkably similar to the stress–strain diagrams in tensile strength experiments (see Section 11). Following the Newtonian region (the proportionality region) is the region of pseudoplasticity, which in turn is followed by the dilatant region, which is finally followed by a region of turbulence. A more or less extended second "Newtonian region" lies between the pseudoplastic and dilatant regions.

There have been no lack of efforts to describe general flow behavior by an empirical flow law. Such equations are generally applicable only over a limited shear stress range (see Figure 7-5). These equations include

$$\dot{\gamma} = a\sigma_{21}^m \qquad \text{(Ostwald–de Waele)} \qquad (7\text{-}28)$$

$$\dot{\gamma} = b \, \sinh{(\sigma_{21}/d)} \qquad \text{(Prandtl–Eyring)} \qquad (7\text{-}29)$$

$$\dot{\gamma} = f\sigma_{21}^3 + g\sigma_{21}^3 \qquad \text{(Rabinowitsch–Weissenberg)} \qquad (7\text{-}30)$$

where a, b, d, f, g, and m are empirical constants.

The shear stress and the shear rate can often vary over several orders of magnitude. Consequently, they are plotted mostly in their logarithmic form, and not directly against each other (Figure 7-6). The resulting flow curves have a slope of unity at very low shear stresses. In this Newtonian region, the flow curve intercepts the ordinate at the value of $1/\eta_0$ at $\log \sigma_{21} = 0$, that is at $\sigma_{21} = 1$. Since such a logarithmic plot corresponds to a linearization of the Ostwald–de Waele equation, the slope of such a flow curve plot gives what is known as the flow exponent m. According to Figures 7-5 and 7-6, the flow exponent can, of course, only be regarded as constant over a limited shear stress range. The flow exponent very often has a value of between 2 and 3.

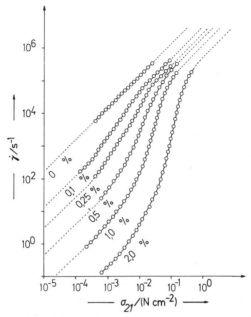

Figure 7-6. Flow curves of solutions of different concentration of a cellulose nitrate ($M = 294\ 000$ g/mol) in butyl acetate at 20° C (after K. Edelmann).

The second Newtonian range occurring at very high shear stresses is often difficult to measure; many rheologists even doubt its existence. Since a general flow law has not been discovered to date, the flow behavior cannot, of course, be characterized by one or two parameters. The difficulty is compounded in that the apparent viscosity of some liquids is time dependent.

When the shear stress changes in Newtonian, dilatant, or pseudoplastic liquids, as well as in Bingham bodies or fluids above the flow limit, the corresponding shear gradient or the corresponding viscosity is reached almost instantaneously. In some liquids, however, a noticeable induction time is necessary, i.e., the viscosity also depends on time. If, at a constant shear stress or constant shear gradient, the viscosity falls as the time increases, then the liquid is termed *thixotropic*. Liquids are termed *rheopectic* or *antithixotropic*, on the other hand, when the apparent viscosity increases with time. Thixotropy is interpreted as a time-dependent collapse of ordered structures. A clear molecular picture for rheopexy is not available.

If the rate of flow of low-molar-mass liquids is increased very sharply, additional velocity components occur because of the surface roughness of the walls. As the flow rate increases, these disturbances of laminar flow finally become so large that they are no longer dampened by the viscosity of the liquid. The individual liquid lamellae no longer flow in a parallel manner, and the flow becomes turbulent. The onset of turbulence is described by the Reynolds number, which gives the ratio of inertial forces to viscous forces.

The same effect is observed in melts of macromolecular substances. Since, in this case, the liquids are elastic, however, additional elastic oscillations of small liquid particles occur. The unevenness of the oscillations causes an elastic turbulence. This occurs at much lower flow velocities than the normal turbulence, i.e., at low Reynolds numbers. Elastic turbulence can also be recognized by the fact that the flow rate increases much more sharply with increasing pressure in the elastic turbulence region than in the laminar flow region; in normal liquids the increase in the flow rate is less in the turbulence region than in the laminar flow region. Elastic turbulence becomes apparent in the processing of plastics at what is called the melt break.

7.5.2. Methods

When making measurements on highly viscous solutions and melts, care must be taken to ensure that the systems are in thermal equilibrium. In many cases it is sufficient to anneal the systems for about a week at the measuring temperature. In a few studies, however, the equilibrium was only reached after half a year.

Viscosities of shear gradients and stresses can be measured with a range of instruments. The most important of these are rotation and capillary viscometers; both are available commercially.

Rotation Viscometers. In rotation viscometers, a rotor moves against a stator (see Figure 9-23). The Epprecht viscometer is particularly suitable for measurements on highly viscous solutions. Here, the angle of rotation of a torsion wire on which the stator is suspended is a measure of the torque produced by the rotating rotor on the liquid. Since all the other quantities (cylinder radius, width of the slit, gap between rotor and stator, number of rotations in unit time) are kept constant, the viscosity is easily calculated.

The Brookfield viscometer is of simpler construction than the Epprecht viscometer. In this case, a rotating metal plate or cylinder is immersed in the liquid and the frictional braking forces that act on it are measured.

In cone and plate viscometers, a cone revolves on a metal plate. The angle between the plate and the cone is kept as low as possible (smaller than 4°) so that the shear gradient remains uniform.

To calculate viscosity, shear stress, and shear gradient, it is usually necessary to make a series of corrections for the finite length of the cylinder, the variation of the shear gradient with distance, etc. These corrections vary from instrument to instrument.

Capillary Viscometers. Those used for measurements on concentrated solutions and melts usually consist solely of one capillary tube in a pressure chamber. Because of the high driving pressure, the capillaries are often of metal.

When a liquid flows through a capillary with radius R and length L under a pressure p, a force $\pi r^2 p$ is exerted on the liquid column at a radius r. It is counteracted by a frictional force $2\pi r L \sigma_{21}$. At equilibrium, therefore,

$$\pi R^2 p = 2\pi r L \sigma_{21} = 0 \tag{7-31}$$

or, solved for the shear stress σ_{21},

$$\sigma_{21} = \frac{pr}{2L} \tag{7-32}$$

The σ_{21} calculated using Equation (7-32) is equal for both Newtonian and non-Newtonian solutions because p, r, and L depend on the measuring apparatus only and not on the properties of the liquid being investigated. σ_{21} is proportional to the radius and has a maximum value $\sigma_{21}(max) = pR/2L$ for capillary radius R.

If Equation (7-32) is inserted into Newton's equation, Equation (7-28), with $y = R$, the result is $dv = (pR/2\eta L)dR$. If η is Newtonian, then integration with the boundary condition $v_R = 0$ for the velocity v at distances $y \leq$ from the capillary wall yields

$$v = \frac{(R^2 - y^2)p}{4\eta L} \tag{7-33}$$

The flow in a capillary is envisaged as the telescoping movement of concentric hollow cylinders at different velocities. A volume q flows per unit time through such a hollow cylinder with radii y and $y + dy$:

$$q = 2\pi y v \, dy \tag{7-34}$$

The total flow volume is given by integration over the flow volumes of all the hollow cylinders,

$$Q = \int_{y=0}^{y=R} 2\pi y v \, dy \tag{7-35}$$

and, with Equation (7-33),

$$Q = \int_{y=0}^{y=R} 2\pi y \frac{R^2 - y^2}{4\eta L} p \, dy$$

$$= \frac{\pi p}{2\eta L} \int_{y=0}^{y=R} (R^2 - y^2) y \, dy$$

$$= \frac{\pi p}{2\eta L} \left(\frac{R^2 y^2}{2} - \frac{y^4}{4} \right) \Big|_0^R = \frac{\pi p R^4}{8\eta L} \tag{7-36}$$

Equation (7-36) is the Hagen–Poiseuille law. By inserting Equation (7-32) and (7-36) into (7-28) and solving for dv/dR, we obtain the following for the maximum shear gradient dv/dR at the capillary surface:

$$\frac{dv}{dR} = \frac{4Q}{\pi R^3} = \dot{\gamma} \tag{7-37}$$

The "viscosity" calculated from Equation (7-37) for non-Newtonian fluids is an apparent viscosity only. The calculation of the correct viscosity requires a number of corrections. In non-Newtonian liquids, $\dot{\gamma}$ is a more complicated function of σ_{21} than is given by Newton's law, i.e., $dv/dR \neq \dot{\gamma}$. Equation (7-35), therefore, can only be partially integrated:

$$Q = 2\pi \left| \frac{y^2 v}{2} \right|_2^R - \int_0^R y^2 \frac{dv}{dy}\, dv \tag{7-38}$$

In both limiting cases, the first term will be zero. Considering that $\dot{\gamma} = \sigma_{21}/\sigma_{app} = f(\sigma_{21})$ and after introducing the shear stress $(\sigma_{21})_R$ at the capillary surface, the second term gives

$$Q = \pi \int_0^R y^2 f(\sigma_{21})\, dv = \frac{\pi R^3}{(\sigma_{21})_R^3} \int_0^R \sigma_{21}^2 f(\sigma_{21})\, d\sigma_{21} \tag{7-39}$$

According to Equation (7-37), $dv/dR \neq 4Q/\pi R^3 = \dot{\gamma}$ is valid for non-Newtonian liquids. Equation (7-39) therefore becomes

$$\dot{\gamma} = \frac{4}{(\sigma_{21})_R^3} \int_0^R \sigma_{21}^2 f(\sigma_{21})\, d\sigma_{21} \tag{7-40}$$

Thus $d\dot{\gamma}/d\sigma$ becomes

$$\frac{d\dot{\gamma}}{d\sigma_{21}} = \frac{4}{(\sigma_{21})_R^3} \sigma_{21}^2 f(\sigma_{21}) = \frac{4}{(\sigma_{21})_R^3} \sigma_{21}^2 \dot{\gamma} \tag{7-41}$$

from which one obtains

$$\frac{1}{4} \sigma_{21} \frac{d\dot{\gamma}}{d\sigma_{21}} = \frac{\sigma_{21}^3}{(\sigma_{21})_R^3} \dot{\gamma} \tag{7-42}$$

In analogy to the equation dv/dR for Newtonian liquids, $dv/dR = A\dot{\gamma}$ is assumed for non-Newtonian liquids. Since a factor $\sigma_{21}^3/(\sigma_{21})_R^3 \neq 1$ occurs in Equation (7-42), A is separated according to $A = a + \sigma_{21}^3/(\sigma_{21})_R^3$. On inserting into Equation (7-42), we find

$$\frac{dv}{dR} = a\dot{\gamma} + \frac{\sigma_{21}^3}{(\sigma_{21})_R^3} \dot{\gamma} = a\dot{\gamma} + \frac{1}{4} \sigma_{21} \frac{d\dot{\gamma}}{d\sigma_{21}} \tag{7-43}$$

Equation (7-43) must also apply to Newtonian liquids. Since it is true here that $\dot\gamma = \sigma_{21}/\eta$ and $d\dot\gamma/d\sigma_{21} = 1/\eta$, this also gives $dv/dR = a(\sigma_{21}/\eta) + (1/4)(\sigma_{21}/\eta)$ and, with $dv/dR = \sigma_{21}/\eta$, it follows that $a = 3/4$. Equation (7-43) thus becomes the Weissenberg equation

$$\frac{dv}{dR} = \frac{3}{4}\dot\gamma + \frac{1}{4}\sigma_{21}\frac{d\dot\gamma}{d\sigma_{21}} \tag{7-44}$$

or, with $dv/dR = \sigma_{21}/\eta$ and $\dot\gamma = \sigma_{21}/\eta_{app}$

$$\frac{1}{\eta} = \frac{3}{4}\frac{1}{\eta_{app}} + \frac{1}{4}\frac{d\dot\gamma}{d\sigma_{21}} \tag{7-45}$$

where $d\dot\gamma/d\sigma_{21}$ is taken from a graph of $\sigma_{21} = f(\gamma)$.

On examining data from the literature, it should be noted that it is frequently not the shear gradient $\dot\gamma$ but, according to Kroepelin, the average shear gradient G across the whole diameter of the capillary which is given [see Equation (7-37)]:

$$G = \frac{8Q}{3\pi R^3} = \frac{2}{3}\dot\gamma \tag{7-46}$$

Technical Viscometers. These usually do not enable shear stresses and shear gradients to be calculated, since the conditions of measurement usually cannot be varied. Hence, the values calculated from these instruments are only relative values. Their advantage is simple construction and rapid measurement.

Calibrated vessels with a hole at the base are called Ford cups. The liquid runs out under its own pressure. The time taken for a standard quantity of liquid to flow out is an indication of the viscosity. Since the height of the liquid and, subsequently, the pressure change during the measurement, the shear stress also varies with time. Ford cups are particularly used in the paint industry. Apparatus for determining the melt index (also called the grade value or grade number) works on a similar principle. Here, the amount of melt which flows out in a given time under given conditions is measured. The melt index is thus proportional to the fluidity and not the viscosity.

With Höppler viscometers, the time required for a rolling ball to run along an inclined tube is measured. In Cochius tubes, the time taken for an air bubble to rise is a measure of the viscosity. Here, the true viscosities, shear stresses, and shear gradients are also difficult to determine.

7.5.3. Viscosities of Melts and Highly Concentrated Solutions

If the logarithmus of the viscosity of concentrated polymer solutions is plotted against the logarithmus of the concentration, a curve is obtained

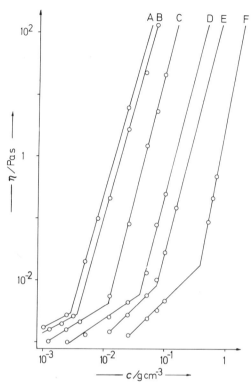

Figure 7-7. Viscosity η of solutions of poly(isobutylene) of different molar masses in toluene at 25°C as a function of the concentration c. $\overline{M}_w = 7\,270\,000$ (A); $3\,550\,000$ (B); $1\,250\,000$ (C); $328\,000$ (D); $139\,000$ (E); $40\,600$ g/mol (F) (After J. Schurz and H. Hochberger.).

which approximates to two straight lines of differing slope (Figure 7-7). The straight lines intersect at a critical concentration, c_{crit}, which is dependent on the solvent quality. This critical concentration is interpreted as the concentration at which polymer chain entanglements and/or close packing of molecular segments first occurs. The slope of the line is about 2–4 below the critical concentration and 5–6 above it.

The concept of entanglements or close packing leads one to expect a dependence of the critical concentration on the molar mass of the polymer. Actually, a proportionality between c_{crit} and M^α is experimentally observed. Theoretical considerations suggest a value for the exponent α of unity for entanglements and 0.5 for close packing of coil-like macromolecules.

Theoretically, the viscosity of rod-shaped macromolecules for a shear gradient of zero should be proportional to both the volume fraction ϕ and the molar mass M of the polymer:

$$\eta \propto \phi^{5/3} M^8 \qquad (7\text{-}47)$$

A critical molar mass is analogously found for melts. Below the critical molar mass, the melt viscosity at zero shear gradient is directly proportional to the molar mass:

$$\eta = K \langle M \rangle_w \tag{7-48}$$

In contrast, the viscosity above the critical molar mass is given by (see Fig. 7-8):

$$\eta = K' \langle M \rangle_w^{3.4} \tag{7-49}$$

Deviations from this behavior are often observed for finite shear gradients (Figure 7-8). Correlations for constant shear stress, however, are presumably more realistic than for constant shear gradient.

Of course, entanglements can only occur when the molecular chains are sufficiently long, that is, when a sufficiently large number N_{crit} of chain links is present. But the critical chain link number is not a general constant; it is still dependent on the macromolecular constitution (Table 7-2). The tendency to entangle increases with decreasing stiffness of the macromolecule. A measure of the stiffness is the ratio $\langle R_\theta^2 \rangle / \langle M \rangle_w$, where R_θ is the radius of gyration in the

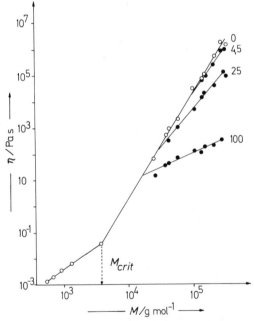

Figure 7-8. Melt viscosity η of unbranched poly(ethylenes) as a function of the molar mass M at 190°C and shear stress $\sigma_{21} = 0$ ($\bigcirc$), or $\sigma_{21} = 4.5, 25,$ and 100 N/cm^2 (all $\bullet$)(after H. P. Schreiber, E. B. Bagley, and D. C. West); M_c occurs at log $M = 3.6$.

Table 7-2. Critical Chain-Link Number, N_{crit}, of Various Polymers

Polymer	N_{crit}
Poly(ethylene)	286
1,4-*cis*-Poly(isoprene)	296
at-Poly(vinyl acetate)	570
Poly(isobutylene)	609
at-Poly(styrene)	730
Poly(dimethyl siloxane)	784

[a]Data from T. G. Fox.

unperturbed state. The entanglement tendency is also related to the volume fraction ϕ_2 and specific volume v_2 of the polymer in order to compare viscosities of melts with those of concentrated solutions. A new parameter, z_w is thus defined:

$$z_w = N_{crit}\phi_2\langle R_\theta^2\rangle/(\langle M\rangle_w v_2) \qquad (7\text{-}50)$$

If a plot of $\log\eta = f(\log z_w)$ is made, intercept points are obtained which are independent of the macromolecular constitution (Figure 7-9).

The mass averages of the molar mass and the chain-link number are used in these relationships since experiments have shown that the viscosities of polymers of various molar mass distributions at zero shear stress depend on the mass average molar mass. At finite shear stresses, polymers of the same molar mass average have higher viscosities when the distribution is narrower, since the lower-molar-mass component appears to function as a lubricant. But it has not been completely established which is the correct molar mass average to use.

The temperature dependence of the viscosity of polymer melts typically does not follow an Arrhenius relationship unless temperatures are far above T_G ($T > T_G + 100$ K). If a viscosity ratio η_R (not to be confused with the ratio η/η_1 in dilute solutions; cf. Chapter 9.9) is defined by means of the viscosities η and densities ρ at the measurement temperature T and reference temperature T_1,

$$\eta_R = \frac{\eta\rho T}{\eta_1\rho_1 T_1} \qquad (7\text{-}51)$$

then the overall temperature dependence can be expressed by the semi-empirical Williams–Landel–Ferry equation (WLF equation)

$$\log\eta_R = \frac{-B(T - T_1)}{C + (T - T_1)} \qquad (7\text{-}52)$$

(see also the derivation in Section 10.5.2). Here, B and C are constants specific

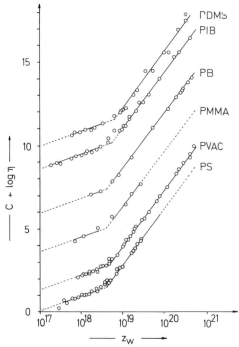

Figure 7.9. Dependence of the melt viscosity η of polymers on the parameter z_w (see text) at $\sigma_{21} \to 0$. For easier comparison, the η values of the different types of polymer have all been multiplied by a constant factor of C. PDMS, Poly(dimethyl siloxane); PIB, poly(isobutylene); PB, poly(butadiene); PMMA, poly(methyl methacrylate); PVAC, poly(vinyl acetate); PS, poly(styrene) (after T. G. Fox).

to the material. The lowest possible reference temperature is the glass transition temperature T_G.

7.6. Permeation through Solids

7.6.1. Basic Principles

If a gas at two different pressures is separated by a membrane or a film, then the gas will permeate through the membrane until the pressures on each side are equal. If the pressure difference, Δp, is kept constant, then, according to Henry's law, which holds for the simplest case of permanent gases, a concentration difference for the gas on each side of the membrane is formed according to

$$\Delta w = S \, \Delta p \qquad (7\text{-}53)$$

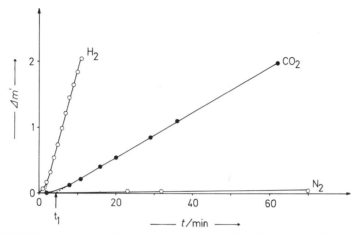

Figure 7-10. Time dependence of gas permeation through a film of a styrene copolymer at 25° C. With the chosen experimental conditions, the change $\Delta m'$ is proportional to the change in mass Δm (after P. Goeldi and H.-G. Elias).

The proportionality constant S is called the solubility coefficient. If the concentration difference is given in terms of the mass fractions, then S has the physical units of reciprocal pressure.

The gas requires a certain time t_1 after application of pressure before permeating through an unloaded membrane (Figure 7-10). Thus, Equation (7-7) converts to

$$\Delta m = J_d A (t - t_1) \tag{7-54}$$

The lag time t_1 is related to the diffusion coefficient of the gas in the membrane, as is shown by a lengthy theoretical calculation:

$$D = L_m^2/(6t_1) \tag{7-55}$$

where L_m is the thickness of the membrane. Thus, the diffusion coefficient of the gas in the membrane can be calculated from the lag time.

The gas permeability reaches a stationary state after a time of $t \approx 3t_1$. If Fick's first law of diffusion applies, then the diffusion coefficient will be concentration dependent. dc in Equation (7-8) can be replaced by the concentration difference Δw and dr can be replaced by the membrane thickness L_m. Equation (7-8) converts to

$$-J_d = D\Delta w / L_m \tag{7-56}$$

Combination with Henry's law leads to

$$-J_d = DS(\Delta p / L_m) = P(\Delta p / L_m) \tag{7-57}$$

Then the following is obtained from Equations (7-54), (7-55), and (7-57):

$$\Delta m = \frac{PA \, \Delta p}{L_m} \left(t - \frac{L_m^2}{6D} \right) \qquad (7\text{-}58)$$

The permeability coefficient P can be obtained from the permeation with time of a gas through a membrane of known surface area A and constant thickness L_m when there is a constant pressure difference Δp. The solubility coefficient S can then be calculated from the permeability and diffusion coefficients.

The temperature dependence of the diffusion coefficients is governed by the activation energy of diffusion, E_D^+:

$$D = D_0 \exp \left(-E_D^+ / RT \right) \qquad (7\text{-}59)$$

Here, there is a differentiation between two possible modes of diffusion: activated diffusion and diffusion through pores. With activated diffusion, the gas diffuses through the membrane by means of a series of activated diffusion steps. The diffusion coefficient increases with increasing temperature. The diffusion decreases slightly with increasing temperature, on the other hand, when the diffusion is through pores. This occurs because the viscosity of the gas increases with temperature. For example, nitrogen diffuses by activated diffusion through poly(ethylene) but diffuses through pores in the case of pergamin (Table 7-3).

The opposing effects of the two diffusion modes is utilised in the lamination of two different kinds of film. For example, oxygen under a pressure difference of about 1 bar diffuses through 0.025-mm-thick aluminum films with pores of 1 μm at a rate of about 5×10^{-5} cm^3/s. After laminating with poly(ethylene) film of about 0.025-mm thickness, the rate of diffusion decreases to 5×10^{-13} cm^3/s.

The temperature dependence of the solubility coefficient is governed by the enthalpy of solution, ΔH:

$$S = S_0 \exp(-\Delta H / RT) \qquad (7\text{-}60)$$

Table 7-3. Permeability Coefficients for Nitrogen in
Poly(ethylene) and Parchment at Various Temperatures

T in °C	10^7 P in cm^2 s^{-1} bar^{-1}	
	Poly(ethylene)	Pergamin
0	0.25	11.2
30	2.1	9.4
50	7.4	9.3
70	22.0	8.4

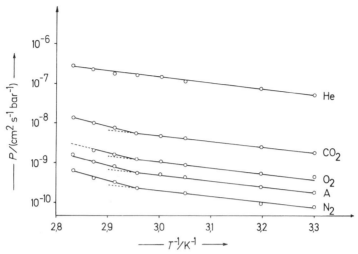

Figure 7-11. Temperature dependence of the permeability coefficient of various gases through Barex (see also text) (after H. Yasuda and T. Hirotsu).

The solubility coefficient generally decreases with increasing temperature, whereas the diffusion coefficient generally increases. Consequently, the permeability coefficient, which is the product of both, may either increase or decrease with decreasing temperature. As expected, a change in the permeability behavior or gases is often observed at the glass transition temperature (Figure 7-11).

A compensation effect is often observed for the relationship between log D_0 and E_D^+ in the case of elastomers, that is, polymers above the glass transition temperature (Figure 7-12). Values for glassy polymers (not shown) lie significantly lower than the line shown, whereas the D_0 values for the diffusion of hydrogen in elastomers are higher.

Special problems occur in the permeation of liquids through films when the liquids swell the film. Equation (7-58) is replaced by

$$m = KAt^n \tag{7-61}$$

which governs the time dependence of the mass permeating through in the steady state. A is the permeation area and K is a constant. The exponent has the value 0.5 in what is known as case I and the value 1 in what is known as case II. Case I corresponds to behavior according to Fick's law of diffusion; the mobility of the permeating molecule is, in this case, much less than the relaxation rates of the polymer segments. The situation is exactly reversed in case II: the mobility of the solvent is much larger than the relaxation rates of the polymer segments. The permeability behavior is characterized by a sharp interfacial boundary between a swollen zone advancing with constant velocity

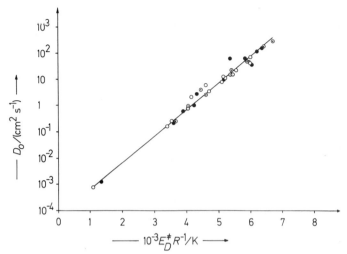

Figure 7-12. Compensation effect between the action constant D_0 and the activation energy $E_D^{\ddagger}$ of diffusion of nitrogen, ●, oxygen, ○, and carbon dioxide, ⊙, through various polymers above their glass transition temperatures.

and an interior glassy core. What is known as anomalous diffusion, with $0.5 < n < 1$, occurs between these two limiting cases.

7.6.2. Constitutional Influences

The permeability coefficients of gases in polymers can vary between exceptionally wide limits (Table 7-4). They are generally lower for polymers below the glass transition temperature than they are above this temperature; but a sharp boundary does not exist. In individual cases, considerable differences occur: compared with poly(vinyl alcohol), the permeability of oxygen in cellulose is 900 times larger, in poly(vinyl chloride) it is 5×10^4 times larger, in poly(ethylene) it is 7×10^5 times larger, and in poly(dimethyl siloxane) it is even 10^8 times larger. These permeability differences are industrially very important. For example, the packaging of carbonated liquids requires low carbon dioxide and oxygen permeabilities. In contrast, fresh fruit packaging, vegetable packaging, and fish packaging need high oxygen permeabilities, as do membranes for artificial lungs.

The permeation of gases through membranes or films is determined by both the diffusion coefficient D and the solubility coefficient S [see Equation (7-58)]. The diffusion coefficient decreases in general with increasing molecular weight of the gas for a given membrane (Table 7-5). Since the solubility coefficient S depends on the interaction between the gas and the membrane

Table 7-4. Permeability Coefficients of Gases and Water Vapor in Various Polymers at 30° C. The Gas Passage, Δm, Is Measured in cm^3. Barex Is a Graft Copolymer of Acrylonitrile/Methyl Acrylate on Nitrile Rubber. Lopac Is a Copolymer of Methacrylonitrile and Styrene.

Polymer	$10^7\ P$ in $cm^2\ s^{-1}\ bar^{-1}$				
	H_2	He	O_2	CO_2	H_2O
Poly(dimethyl siloxane)		233	605	3240	40000
Poly(oxy-2,6-dimethyl phenylene)	113	78	16	76	4060
Poly(tetrafluoroethylene)	10		4.9	13	33
Poly(styrene)	23	19	2.6	10.5	1200
Poly(propylene); ρ=0.907 g/cm^3	41	38	2.2	9.2	65
Polycarbonate	12		1.4	8.0	1400
Butyl rubber	7.3		1.3	5.2	120
Cellulose acetate	3.5	16	0.8	2.4	6800
Poly(ethylene); ρ=0.964 g/cm^3		1.14	0.40	1.8	12
Poly(vinyl chloride)	1.7	2.05	0.045	0.16	275
Polyamide 6		0.53	0.038	0.16	275
Poly(ethylene terephthalate)		1.32	0.035	0.17	175
Barex			0.0054	0.018	660
Poly(vinylidene chloride)		0.31	0.0053	0.029	1.0
Lopac			0.0035	0.011	340
Poly(methacrylonitrile)			0.0012	0.0032	410
Poly(acrylonitrile)		0.55	0.0003	0.0018	300

material, however, there is no general relationship between the molecular weight of the gas and the permeability coefficient P.

The diffusion coefficient will be lower, the greater the distance the gas must travel in traversing the membrane. Thus bulky monomeric units, fillers, and crystalline regions lower diffusion coefficients (detour factor). The more flexible the chains of the membrane material, the less activation energy will be needed for the diffusion, and the greater also will be the diffusion coefficient.

Table 7-5. Permeability Coefficient P, Diffusion Coefficient D, and Solubility Coefficient S of Various Gases in Vulcanized cis-1,4-Poly(isoprene) at 25° C

Gas	M in g/mol	$10^7\ P$ in $(cm^2/s)/bar$	$10^7\ D$ in cm^2/s	S in $(cm^3/cm^3)/bar$
H_2	2	3.4	85	0.040
N_2	28	0.51	15	0.035
O_2	32	1.5	21	0.070
CO_2	44	10.0	11	0.90

[a]Data from R. M. Barrer.

The permeation of liquids, particularly water, plays an important part in the weather resistance of plastics. It is also a problem in the drying of polymers. If polymer solutions are evaporated, it is frequently found that a considerable part of the solvent cannot be removed from the polymer even above the solvent boiling point. Such inclusion can amount, for example, to 20% for CCl_4 in poly(styrene), and up to 10% for dimethylformamide in poly(acrylonitrile). Allowances must be made for this inclusion in the interpretation of analytical data. Inclusion occurs because the permeation of liquids (solvents, monomers, etc.) through polymers is very low below their glass transition temperature. It is best avoided by freeze-drying the polymer solution. In this case, $\sim 1\%-10\%$ solutions of the polymer in solvents with sufficiently high melting points and which evaporate readily (sublime) as a solid (e.g., benzene, dioxane, water, formic acid) are completely frozen, and then, below the melting point, the solvent is sublimed out under vacuum. This method removes the solvent almost completely. Since the remaining polymer has a large surface area, however, and is therefore very prone to absorbing moisture, it is advisable to sinter the polymer carefully after the freeze-drying. Inclusion can also be avoided by adding certain nonsolvents to the solvent before evaporating; these nonsolvents should form azeotropes with the solvent. Another possibility is to dissolve the polymer in a poor solvent and precipitate it with a strong precipitant.

Literature

Section 7.2. Diffusion in Dilute Solution

J. Klein, The self-diffusion of polymers, *Contemp. Phys.* **20**, 611 (1979).

Section 7.3. Rotational Diffusion and Streaming Birefringence

V. N. Tsvetkov, Floe birefringence, *in*: *Newer Methods of Polymer Characterization* (B. Ke. ed.), Wiley-Interscience, New York, 1964.
H. Janeschitz-Kriegl. Flow birefringence of elastico-viscous polymer systems. *Adv. Poly. Sci.* **6**, 170 (1969).

Section 7.4. Electrophoresis

R. L. Yeates, *Electropainting*, Draper, Teddington, 1966.
K. Weigel, *Electrophorese-Lacke*, Wiss. Verlagsges., Stuttgart, 1967.
J. R. Cann and W. B. Goad, *Interacting Macromolecules, The Theory and Practice of Their Electrophoresis, Ultracentrifugation and Chromatography*, Academic Press, New York, 1970.

W. Machu, Die Elektrotauchlackierung, Verlag Chemie, Weinheim, 1973.

Ö. Gaál, G. A. Medgyesi, and L. Vereczkey, *Electrophoresis in the Separation of Biological Macromolecules*, Wiley, New York, 1980.

Section 7.5. Viscosity

W. Philippoff, *Viskosität der Kolloide*, D. Steinkopff, Dresden, 1942.

M. Reiner, *Deformation and Flow*, K. H. Lewis & Co., London, 1949.

F. R. Eirich (ed.), *Rheology, Theory and Applications*, 5 vols., Academic Press, New York, 1956–1969.

J. R. Van Wazer, J. W. Lyons, K. Y. Kim, and R. E. Colwell, *Viscosity and Flow Measurement*, Interscience, New York, 1963.

A. Peterlin, Non-Newtonian viscosity and the macromolecule, *Adv. Macromol. Chem.* **1**, 225 (1968).

S. Middleman, *The Flow of High Polymers*, Wiley–Interscience, New York, 1968.

G. C. Berry and T. G. Fox, The viscosity of polymers and their concentrated solutions, *Adv. Polym. Sci.* **5**, 261 (1968).

G. W. Scott-Blair, *Elementary Rheology*, Academic Press, London, 1969.

V. Semjonov, Schmelzviskositäten hochpolymerer Stoffe, *Adv. Polymer. Sci.* **5**, 387 (1968),

J. D. Ferry, *Viscoelastic Properties of Polymers*, 10th ed., Wiley, Chichester, 1980.

J. A. Brydson, *Flow Properties of Polymer Melts*, Iliffe, London, 1970.

O. Plajer, Praktische Rheologie für Kunststoffschmelzen, Zechner and Hüthig, Heidelberg, 1970.

J. Schurz, Viskositätsmessungen an Hochpolymeren, Kohlhammer, Stuttgart, 1972.

W. W. Graessley, The entanglement concept in polymer rheology, *Adv. Polym. Sci.* **16**, (1974).

K. Walters, *Rheometry*, Halsted, New York, 1975.

R. Darby, *Viscoelastic Fluids*, Marcel Dekker, New York, 1976.

J. W. Hill and J. A. Cuculo, Elongated Flow Behavior of Polymeric Fluids, *J. Macromol. Sci. Revs.* **C14**, 107 (1976).

E. Dschagarowa and G. Mennig, Verminderung des Reibungswiderstandes von Flüssigkeiten im turbulenten Bereich mittels hochpolymerer Zusätze, *Fortschr. Ber. VDI Z.*, Series 7, No. 41, VDI-Verlag, Düsseldorf, 1976.

L. E. Nielsen, *Polymer Rheology*, Marcel Dekker, New York, 1977.

R. B. Bird, O. Hassager, R. C. Armstrong, and C. F. Curtiss, *Dynamics of Polymer Liquids*, 2 Vols., Wiley, New York, 1977.

E. Boudreaux, Jr., and J. A. Cuculo, Polymer flow instability: A review and analysis, *J. Macromol. Sci. Rev. Macromol. Chem.* **C16**, 39 (1977/78).

R. S. Lenk, *Polymer Rheology*, Appl. Sci. Publ., Barking, Essex, 1978.

C. J. S. Petrie, *Elongational Flows (Research Notes in Mathematics*, Vol. 29), Pitman, London 1979.

Section 7.6. Permeation through Solids

J. Crank and G. S. Park (eds.), *Diffusion in Polymers*, Academic Press, London, 1968.

H. J. Bixler and O. J. Sweeting, Barrier properties of polymer films, in *The Science and Technology of Polymer Films*, Vol. II (O. J. Sweeting, ed.), Wiley–Interscience, New York, 1971.

C. E. Rogers and D. Machin, The concentration dependence of diffusion coefficients in polymer-penetrant systems, *Crit. Rev. Macromol. Sci.* **1**, 245 (1972).

V. Stannett, H. B. Hopfenberg, and J. H. Petropoulos, Diffusion in polymers in *Macromolecular Science* (Vol. 8 of Physical Chemistry Series 1) (C. E. H. Bawn, ed.), MTP International Review of Science, 1972.

H. B. Hopfenberg (ed.), *Permeability of Plastic Films and Coatings*, Plenum Press, New York, 1974.

V. T. Stannett, W. J. Koros, D. R. Paul, H. K. Lonsdale, and R. W. Baker, Recent advances in membrane science and technology, *Adv. Polym. Sci.* **32**, 69 (1979).

Chapter 8

Molar Masses and Molar Mass Distributions

8.1. Introduction

In synthesizing polymers *in vivo* and *in vitro*, molecular homogenous ("monodisperse") polymers (i.e., those in which every macromolecule has the same molar mass or "molecular weight") occur only under quite specific conditions. The overwhelming majority of polymer syntheses proceed more or less randomly, and the resulting macromolecular substances have more or less broad molar mass distributions. The kind of molar mass distribution obtained depends on the nature of the polymerization, which may be either thermodynamically or kinetically controlled. Each kind of distribution is characterized by a definite relationship between the mole fraction x and the degree of polymerization X. Consequently, it is possible in many cases to deduce the kind of polymerization involved from the type of distribution function obtained.

Instead of determining the complete degree of polymerization distribution, it is often considered to be sufficient to measure various moments of the distribution, that is, various degree of polymerization averages. The relationships between the various moments or averages are also characteristic of the distribution type.

The distribution function, the moments, and the averages can be given in terms of the molar masses rather than in terms of the degrees of polymerization. Theoretically, a description in terms of the degrees of polymerization is more desirable, but molar masses are obtained experimentally.

8.2. Statistical Weights

Various statistical weights can be assigned to individual species i of a degree of polymerization or molar mass distribution. The statistical weight g can, for example, be a number or a mass.

From a mechanistic point of view, all events are given in terms of the reacting amount of substance (in, e.g., moles) n_i or the number of molecules N_i, where

$$n_i = N_i/N_L \qquad (8\text{-}1)$$

The mass m_i of all molecules of species i, not their number, is measured by fractionation. The mass m_i of all i molecules is given by the number N_i of i molecules and molar mass $(m_{mol})_i$ of an individual molecule

$$m_i = N_i(m_{mol})_i \qquad (8\text{-}2)$$

With the definition of molar mass, we have

$$M_i = (m_{mol})_i N_L \qquad (8\text{-}3)$$

Consequently, with Equations (8-2) and (8-3)

$$M_i = \frac{m_i}{N_i} N_L = \frac{m_i}{n_i} \qquad (8\text{-}4)$$

Mass and amount of substance are consequently related via the molar mass. Equation (8-4) is valid only for molecularly homogenous fractions. For polymolecular fractions, the number average molar mass, $\langle M_n \rangle_i$, must be used instead of the molar mass, M_i (see Section 1.1):

$$m_i = n_i \langle M_n \rangle_i \qquad (8\text{-}5)$$

Higher and lower statistical weights can be defined in analogy to Equation (8-4). For example,

$$(n-1) \text{ stat. weight:} \quad (n-1)_i = n_i M_i^{-1} = m_i M_i^{-2} = (m-2)_i \quad (8\text{-}6)$$

$$\text{number stat. weight:} \quad n_i = n_i M_i^0 = m_i M_i^{-1} = (m-1)_i \quad (8\text{-}7)$$

$$\text{mass stat. weight:} \quad m_i = n_i M_i = m_i M_i^0 = m_i \quad (8\text{-}8)$$

$$z \text{ stat. weight:} \quad z_i = n_i M_i^2 = m_i M_i = (m+1)_i \quad (8\text{-}9)$$

$$(z+1) \text{ stat. weight:} \quad (z+1)_i = n_i M_i^3 = m_i M_i^2 = (m+2)_i \quad (8\text{-}10)$$

The expressions preceding the first equals sign are normally used; they represent symbols and not mathematical operations. Of course, the symbols $(m+1)_i$ or $(n+2)_i$ can be used instead of the symbol z_i, but this alternative symbolism has not become established. When dealing with polymolecular species, that average corresponding to the statistical weight by which it is multiplied should be used for the molar mass, M_i [see also Equation (8-5)]:

$$(z + 1) = z_i \langle M_z \rangle_i = m_i \langle M_w \rangle_i \langle M_z \rangle_i = n_i \langle M_n \rangle_i \langle M_w \rangle_i \langle M_z \rangle_i \quad (8\text{-}11)$$

This procedure applies generally to each statistical weight and to each property considered.

Mole fraction, x_i, mass fraction, w_i, z fraction, Z_i, etc., can also be used instead of amount of substance, mass, etc. Consequently, from their definitions and the equations above, we have

$$x_i = \frac{n_i}{\sum_i n_i} \quad (8\text{-}12)$$

$$w_i = \frac{m_i}{\sum_i m_i} = x_i \frac{\langle \overline{M}_n \rangle_i}{\overline{M}_n} \quad (8\text{-}13)$$

$$Z_i = \frac{z_i}{\sum_i z_i} = w_i \frac{\langle \overline{M}_w \rangle_i}{\overline{M}_w} = x_i \frac{\langle M_n \rangle_i \langle M_w \rangle_i}{\overline{M}_n \overline{M}_w} \quad (8\text{-}14)$$

Here, $\langle \overline{M}_n \rangle_i$, $\langle \overline{M}_w \rangle_i$, $\langle \overline{M}_z \rangle_i$, etc., are number, weight, and z averages of the individual i fractions, whereas $\overline{M}_n$, $\overline{M}_w$, $\overline{M}_z$, etc., are the corresponding averages of the complete sample. Equations (8-12)–(8-14) are obtained from the respective summations, i.e., from

$$\sum_i m_i = \sum_i [n_i (\overline{M}_n)_i] = n \overline{M}_n = m \quad (8\text{-}15)$$

As well as statistical weights relative to molecules and particles, there are also others relative to their dimensions. Thus, there are not only number, mass, and z averages, but also length, surface, and volume averages. Statistical weights based on dimensions can be converted into those of particle number or mass if the shapes of the particles are known. Thus, for spheres of differing sizes having various radii R_i and volumes V_i but constant density ρ, the masses of the spheres are given by

$$m_i = N_i V_i \rho = N_i (4\pi R_i^3/3)\rho = N_i (\rho/3) R_i A_i \quad \text{(spheres)} \quad (8\text{-}16)$$

And, correspondingly, for rods of lengths L_i and radii R, which are constant and insignificant with respect to lengths:

$$m_i = N_i V_i \rho = N_i (\pi R^2 L_i)\rho \approx N_i (\rho/2) R A_i \quad \text{(rods)} \quad (8\text{-}17)$$

As well as disks of radii R_i and constant heights H, which are insignificant with respect to the radii:

$$m_i = N_i V_i \rho = N_i (\pi R_i^2 H)\rho \approx N_i (\rho/2) H A_i \quad \text{(disks)} \quad (8\text{-}18)$$

8.3. Molar Mass Distributions

8.3.1. Representation of the Distribution Functions

Distribution functions can be classified as discontinuous or continuous. Discontinuous distribution functions are subdivided into frequency distributions and cumulative distributions. Continuous distribution functions are further classified as differential and integral distribution functions.

In a *discontinuous* distribution function, the *frequency* distribution gives the distribution of the statistical weights of the components i as a function of the property E. For example, E may be the degree of polymerization. Typical property distributions are

$$x_i(E_i), \qquad w_i(E_i), \qquad Z_i(E_i), \qquad \text{etc.}$$

The *cumulative* distributions give the summations over all statistical weights. They give the probability of finding the property E for values less than E_i. Typical cumulative distributions are then

$$\sum_{k=1}^{k=i} n_k(E_k), \qquad \sum_{k=1}^{k=i} w_k(E_k), \qquad \sum_{k=1}^{k=i} Z_k(E_k), \qquad \text{etc.}$$

Frequency and cumulative distributions are stepwise distributions (see Figure 8-1). They can be transformed into continuous distributions by the process described in Section 2.4.3.

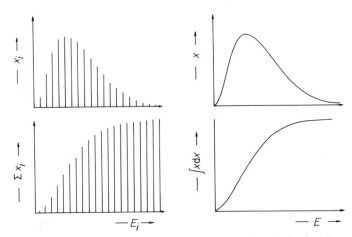

Figure 8-1. Frequency distribution (top left) and a cumulative distribution (bottom left), together with a differential distribution (top right) and an integral distribution (bottom right) of the mole fraction x of a property E (diagrammatic). Read $\int x\,dE$ instead of $\int x\,dx$.

Discontinuous distribution functions can, of course, be transformed into continuous distribution functions when the difference between two neighboring properties is very small compared with the whole range of property values. Frequency distributions convert to the corresponding differential distributions and cumulative distributions convert to integral distributions,

$$X(E), \qquad W(E), \qquad Z(E), \qquad \text{etc.}$$

and cumulative distributions convert to integral distributions,

$$\int_0^E x(E')dE', \qquad \int_0^E w(E')dE', \qquad \int_0^E Z(E')dE', \qquad \text{etc.}$$

In the German scientific tradition, only molar distributions are described as "frequency distributions." This terminology can lead to confusion since weight and z distributions can also be given in terms of frequency distributions.

8.3.2. Types of Distribution Functions

Distribution functions are mostly named after the person who discovered them. Only the mathematical consequences of the distribution functions will be discussed in this chapter. The application to mechanisms or molar mass averages will be discussed in Chapters 16–23.

8.3.2.1. Gaussian Distribution

The Gaussian distribution is the best known distribution. It represents the error law about the arithmetic mean. Because of its frequent appearance, the Gaussian distribution is also called the normal distribution in mathematics. In contrast, a certain form of the Schulz–Flory distribution is often called the normal distribution in macromolecular science.

The mole fraction differential distribution of the property E is given in the form of a Gaussian function as

$$x(E) = \frac{1}{\sigma_n(2\pi)^{1/2}} \exp\left[\frac{(E - \bar{E}_m)^2}{2\sigma_n^2}\right] \qquad (8\text{-}19)$$

where $\bar{E}_m$ is the median value, i.e., that value for which

$$\int_{-\infty}^{+E_m} dx = 0.5$$

Since the Gaussian distribution is symmetric about the median [see Figure (8-2)] the median is equivalent to the number average, $\bar{E}_n = \langle E_n \rangle$, of the

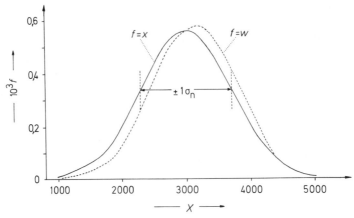

Figure 8-2. The mole fraction $x(—)$ or mass fraction $w(--)$ as a function of the degree of polymerization X for a Gaussian molar differential distribution of the degree of polymerization. Calculations based on $\overline{X}_w = 3170$ and $\overline{X}_n = 3000$.

property. The following considerations will be confined to the degree of polymerization X as property.

σ_n in Equation (8-19) acts as a curve-fitting parameter. Since it describes the width of the distribution, and, thus, the deviation from the mean value, it is also called the standard deviation. The deviation of an X_i value from the mean value, $\overline{X}_n = \langle X_n \rangle$, is given as the mean "error," s_n, of the individual value:

$$s_n = \left[\frac{n_i(X_i - \overline{X}_n)^2}{\sum_i n_i} \right]^{0.5} \tag{8-20}$$

If Equation (8-20) is solved and summed, the following is obtained with $\sum_i s_n^2 = \sigma_n^2$:

$$\sigma_n^2 \sum_i n_i = \sum_i n_i X_i^2 - 2\overline{X}_n \sum_i n_i X_i + \overline{X}_n^2 \sum_i n_i \tag{8-21}$$

Dividing by $\sum_i n_i$, inserting the expressions for the mass- and number-average degrees of properties [see Equations (8-44) and (8-45)], and solving, we obtain

$$\sigma_n = (\overline{X}_w \overline{X}_n - \overline{X}_n^2)^{0.5} \tag{8-22}$$

Thus, the standard deviation of the molar distribution of the degree of polymerization can be calculated from the number and mass averages. Further, the standard deviation is an absolute measure of the width of a Gaussian distribution (and only of a Gaussian distribution), since the molar fraction of 0.6826 lies within the limits $\overline{X}_n \pm 1\sigma_n$, and the molar fraction 0.9544 lies within $\overline{X}_n \pm 2\sigma_n$. Thus, for $\overline{X}_w = 3170$ and $\overline{X}_n = 3000$ in the case of the example shown in Figure 8-2, $\sigma_n = 714$, according to Equation (8-22).

Consequently, 68.26% of all molecules lie within the range of $\bar{X}_n = 3000 \pm 714$.

The differential molar Gaussian distribution loses its symmetry when the distribution is plotted in terms of mass fractions instead of molar fractions (Figure 8-2). The curve symmetry occurs again if a mass fraction Gaussian distribution is plotted in terms of the mass fraction. Of course, this differential mass fraction Gaussian distribution must be given in terms of the mass-average degree of polymerization and the standard deviation σ_w:

$$w(X) = \frac{1}{\sigma_w(2\pi)^{1/2}} \exp\left[-\frac{(X - \bar{X}_w)^2}{2\sigma_w^2} \right]$$ (8-23)

and the standard deviation is given as

$$\sigma_w = (\bar{X}_z \bar{X}_w - \bar{X}_w)^{0.5}$$ (8-24)

Consequently, when stating a type of distribution function, the property on which it is based, i.e., amount of substance, mass, etc., should also be given.

8.3.2.2. Logarithmic Normal Distribution

The differential logarithmic normal distribution has the same mathematical form as the Gaussian distribution with the small difference that the logarithm of the property occurs in place of the property itself:

$$x(X) = \frac{1}{(2\pi)^{0.5} X\sigma_n^*} \exp\left[-\frac{(\ln X - \ln \bar{X}_M)^2}{2(\sigma_n)^2} \right]$$ (8-25)

Here, the curve is symmetric about $\ln \bar{X}_M$. The median of the curve $\bar{X}_M$ is not identical with the number average $\bar{X}_n$ (see below). The function corresponds to the error distribution about the geometric mean. The *ratio* of the degrees of polymerization is therefore important with the logarithmic normal distribution, in contrast to the Gaussian distribution, where the *difference* is important.

Differential logarithmic normal distributions can be generalized, for example, for the mass distribution of the degree of polymerization:

$$w(X) = \frac{1}{(2\pi)^{0.5}\sigma_w^*} \frac{X^A}{B\bar{X}_M^{A+1}} \exp\left[-\frac{(\ln X - \ln \bar{X}_M)^2}{2(\sigma_w^*)^2} \right]$$ (8-26)

with $B = \exp[0.5(\sigma_w^*)^2(A + 1)^2]$.

Two special cases are used in macromolecular science:

Lansing–Kraemer distribution: $\qquad A = 0 \qquad B = \exp[0.5(\sigma_w^*)^2]$
Wesslau distribution: $\qquad\qquad\quad A = -1 \qquad B = 1$

A molar logarithmic normal distribution as described by Equation (8-25)

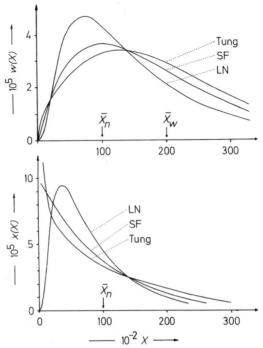

Figure 8-3. Dependence of the mole fraction x and mass fraction w on the degree of polymerization X for molar distributions according to a Schulz–Flor (SF), Tung, and logarithmic normal (LN) distribution. Schematic representation for a sample with $\langle X_w \rangle = 20\,000$ and $\langle X_n \rangle = 10,000$.

is shown in Figure 8-3. The logarithmic normal distribution is thus a skewed distribution when the degree of polymerization is chosen as the abscissa. The curve maximum does not occur at the same point as the number-average degree of polymerization.

If the molar logarithmic normal distribution is plotted in terms of the mass fraction, and not the molar fraction, the shape of the curve remains essentially unchanged (Figure 8-3). Equally, the curve maximum is not identical with either the number- or weight-average degree of polymerization.

Logarithmic normal distributions give straight-line plots when the integral distribution is plotted on paper where the ordinate is graduated in summation probability units and the abscissa is logarithmic (Figure 8-4).

The relationships between the median value and the various averages can be derived from Equation (8-26):

$$\overline{X}_n = \overline{X}_M \exp\left[(2A + 1)(\sigma_w^*)^2/2\right] \qquad (8\text{-}27)$$

$$\overline{X}_w = \overline{X}_M \exp\left[(2A + 3)(\sigma_w^*)^2/2\right] \qquad (8\text{-}28)$$

$$\overline{X}_z = \overline{X}_M \exp\left[(2A + 5)(\sigma_w^*)^2/2\right] \qquad (8\text{-}29)$$

and analogously for the viscosity-average degree of polymerization (with the exponent a_η of the viscosity–molar mass relationship; see Chapter 9):

$$\overline{X}_\eta = \overline{X}_M \exp\left\{[2(A + a_\eta) + 1](\sigma_w^*)^2/2\right\} \qquad (8\text{-}30)$$

Equations (8-27)–(8-29) consequently lead to

$$\exp(\sigma_w^*)^2 = \frac{\overline{X}_w}{\overline{X}_n} = \frac{\overline{X}_z}{\overline{X}_w} \qquad (8\text{-}31)$$

Thus, the ratio of any two simple averages, in sequential order, of the degree of polymerization is constant in logarithmic normal distributions, and this constancy is independent of which two averages are chosen.

8.3.2.3. Poisson Distribution

A Poisson distribution occurs when a constant number of polymer chains begin to grow simultaneously and when the addition of monomeric units is random and occurs independently of the previous addition of other monomeric units. Consequently, Poisson distributions may occur with what are called "living polymers" (see Sections 15 and 18).

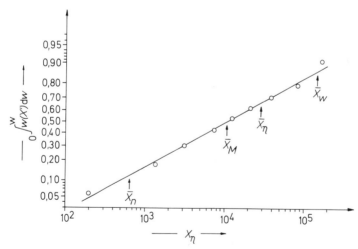

Figure 8-4. Integral mass distribution for a Wesslau distribution plotted on logarithmic–cumulative frequency graph paper. The mass fraction w_i and viscosity average, $\langle X_\eta \rangle_i$, (instead of, more correctly, $\langle X_w \rangle_i$), degree of polymerization of the fractions ($\bigcirc$) were measured. The number average and mass average degrees of polymerization were calculated via Equations (8-27) and (8-28) from the degree of polymerization at the median, $\langle X_M \rangle$, the viscosity average degree of polymerization, $\langle X_\eta \rangle$, of the whole sample and the known viscosity–molar mass relationship exponent α_η.

For the differential molar degree of polymerization distribution, the Poisson distribution gives

$$x = \frac{\nu^{X-1} \exp(-\nu)}{\Gamma(X)} \tag{8-32}$$

where $\nu = \overline{X}_n - 1$ and $\Gamma(X)$ is the gamma function. Further, the number-average degree of polymerization is related to the weight-average degree by (see Chapter 18)

$$\frac{\overline{X}_w}{\overline{X}_n} = 1 + \left(\frac{1}{\overline{X}_n}\right) - \left(\frac{1}{\overline{X}_n}\right)^2 \tag{8-33}$$

Consequently, the ratio $\overline{X}_w / \overline{X}_n$ in the Poisson distribution depends on the number-average degree of polymerization and on no other parameter. The ratio $\overline{X}_w / \overline{X}_n$ tends to a value of unity with increasing degree of polymerization. The Poisson distribution is consequently a very narrow distribution.

8.3.2.4. Schulz–Flory Distribution

The process whereby growing chains whose number is constant in time randomly add monomer till the growth center of individual chains is destroyed by termination provides the basis for the Schulz–Flory distribution. Thus, the originally existing individual chain growth centers do not necessarily remain active. This is in contrast to the case leading to a Poisson distribution. Also, individual chain growth centers need not all commence growth at the same time in the Schulz–Flory distribution. It is only required that the growth center concentration remain constant. Polycondensation and most free radical polymerizations are processes of this type. Such processes were considered to be the normal processes in the early days of polymer science, and thus the distributions so produced were called "normal distributions." This terminology is different from mathematical terminology with respect to a "normal distribution" (see Section 8.3.2.1). The "most probable distribution" often referred to in the literature is concerned with a Schulz–Flory distribution where $\overline{X}_w / \overline{X}_n = 2$.

What is known as the degree of coupling k must be known before the Schulz–Flory distribution can be calculated. The degree of coupling is defined as the number of independently growing chains required to form one dead chain. The degree of coupling is two when two growing free radical chains form one dead chain by recombination:

$$P_i + P_{x-i} \rightarrow P_x \tag{8-34}$$

The differential molar distribution is given as (Chapter 20)

$$x = \frac{\beta^{k+1} X^{k+1} \overline{X}_n \exp(-\beta X)}{\Gamma(k+1)} \tag{8-35}$$

and, from this, the differential mass fraction distribution is given by

$$w = \frac{\beta^{k+1}X^k \exp(-\beta X)}{\Gamma(k+1)}; \qquad \beta = \frac{k}{\overline{X}_n} \qquad (8\text{-}36)$$

An "exponentially" decreasing curve is obtained for $k = 1$ when the mole fraction is plotted against the degree of polymerization (see Figure 8-3). For this reason, and not because an exponential function appears in Equation (8-35), the Schulz–Flory distribution is called an exponential distribution.

The simple one-moment degree of polymerization averages are related to each other by

$$\frac{\overline{X}_n}{k} = \frac{\overline{X}_w}{k+1} = \frac{\overline{X}_z}{k+2} \qquad (8\text{-}37)$$

Consequently, the distributions become increasingly narrower for increasing degrees of coupling. The Schulz–Flory distribution can be distinguished from the logarithmic normal distributions via Equations (8-37) and (8-31). For the Schulz–Flory distribution, the following, of course, holds:

$$(\langle M_n \rangle + \langle M_z \rangle)/\langle M_w \rangle = 2 \qquad (8\text{-}38)$$

whereas, this relationship only holds for logarithmic distributions when $\langle X_z \rangle : \langle X_w \rangle : \langle X_n \rangle = 3:2:1$ because of the requirements of Equation (8-31).

8.3.2.5. Kubin Distribution

The Kubin distribution is an empirical generalized exponential distribution with the empirical constants, γ, β, and λ:

$$w = \gamma\beta^{(\lambda+2)/\gamma}\{\Gamma[(\lambda+2)/\gamma]\}^{-1}X^{\lambda+1}\exp(-\beta X^\gamma) \qquad (8\text{-}39)$$

It includes the following special cases:

$\gamma = 1$;   normal exponential distribution

$\gamma = 1$ $\lambda = k - 1$ Schulz–Flory distribution

$\gamma = B$ $\lambda = B - 2$ Tung distribution

where B is also an empirical constant.

8.4. Moments

The term *moment* comes from mechanics. The first moment of a force $\nu^{(1)}$ is defined as the product of the force (e.g., g) and the distance (e.g., E) along the axis from the point of application of the force. The second moment $\nu^{(2)}$ is the product of the force and the square of the distance. If several forces are applied at various distances, then the moments are determined by summation of the vector products. Also, the first and second moments, as well as any desired moment, may be defined with respect to any desired reference point E_0:

$$\nu_g^{(q)}(E) \equiv \frac{\Sigma_i g_i (E_i - E_0)^q}{\Sigma_i g_i} \tag{8-40}$$

$\nu_g^{(q)}(E)$ is consequently the qth moment of E values with respect to E_0. Moments can, of course, be used not only for the relationships between force and distance, but also for the relationships between any desired quantities.

The statistical weight g can, for example, be a number or a mass (see also Section 8.2).

The order q can be positive or negative, as desired. It may be an integer or a fraction and can take real or imaginary values. Consequently, a moment generally is expressed in physical units different from the property. The property may be the degree of polymerization, the molar mass, the sedimentation coefficient, molecular dimensions, or any other desired property.

In principle, E_0 may take on any chosen value. However, since, for example, negative degrees of polymerization do not exist, it is frequently convenient to use moments with respect to a reference value of zero and to give such moments a special symbol:

$$\mu_g^{(q)}(E) = \frac{\Sigma_i g_i E_i^q}{\Sigma_i g_i} \tag{8-41}$$

The introduction of moments of degrees of polymerization or molar mass considerably simplifies equations describing complicated averages of these quantities.

8.5. Averages

8.5.1. General Relationships

In contrast to moments, averages always possess the same physical units as the properties on which they are based. Averages are consequently first-order moments or such combinations of moments of different order that the resulting physical units are the same as those of the property.

Most of the averages so far considered are composed of one or two moments. They can be described by the generalized formula

$$\langle X_{g(p,q)} \rangle = \left[\frac{\mu_g^{p+q-1}(X)}{\mu_g^{q-1}(X)} \right]^{1/p} = \left(\frac{\Sigma_i G_i X_i^{p+q-1}}{\Sigma_i G_i X_i^{q-1}} \right)^{1/p} \tag{8-42}$$

Equation (8-42) contains four important special cases:

1. When $p = q = 1$, Equation (8-42) reduces to that of a simle one-moment average.
2. When $q = 1$ and $p \neq q$, a one-moment exponent average is obtained.
3. When $q \neq 1$ and $p \neq q$, a two-moment exponent average results.
4. When $p = 1$ and $p \neq q$, an average of two-moment order results.

8.5.2. Simple One-Moment Averages

Simple one-moment averages are defined by

$$\langle X_g \rangle = \frac{\sum_i g_i X_i}{\sum_i g_i} = \sum_i G_i X_i \tag{8-43}$$

According to the nature of the statistical weight they are called the number-average ($g = n$), the mass or "weight" average ($g = m$), the z average ($g = z$), etc. For historic reasons, and to avoid confusion with the index m for "mol," the mass-average degree of polymerization is usually given the symbol $\bar{X}_w$ and not the symbol $\bar{X}_m$. Thus, the number-average molar mass is given by

$$\langle X_n \rangle = \frac{\sum_i n_i \langle X_i \rangle_n}{\sum_i n_i} = \sum_i x_i \langle X_i \rangle_n = \frac{\sum_i c_i}{\sum [c_i / \langle X_i \rangle_n]} = \frac{c}{\sum [c_i / \langle X_i \rangle_n]} \tag{8-44}$$

and the mass-average molar mass is given by

$$\langle X_w \rangle = \frac{\sum_i m_i \langle X_i \rangle_w}{\sum_i m_i} = \frac{\sum_i n_i \langle X_i \rangle_n \langle X_i \rangle_w}{\sum_i n_i \langle X_i \rangle_n} = \frac{\sum_i c_i \langle X_i \rangle_w}{\sum_i c_i} = \sum_i w_i \langle X_i \rangle_w$$

$$= \sum_i x_i \langle X_i \rangle_n \langle X_i \rangle \tag{8-45}$$

The z-average molar mass is given by

$$\langle X_z \rangle = \frac{\sum_i z_i \langle X_i \rangle_z}{\sum_i z_i} = \frac{\sum_i m_i \langle X_i \rangle_w \langle X_i \rangle_z}{\sum_i m_i \langle X_i \rangle_w} = \frac{\sum_i n_i \langle X_i \rangle_n \langle X_i \rangle_w \langle X_i \rangle_z}{\sum_i n_i \langle X_i \rangle_n \langle X_i \rangle_w}$$

$$= \sum_i z_i \langle X_i \rangle_z = \sum_i w_i \langle X_i \rangle_w \langle X_i \rangle_z = \sum_i x_i \langle X_i \rangle_n \langle X_i \rangle_w \langle X_i \rangle_z \tag{8-46}$$

According to these equations, the following always holds:

$$\langle X_{z+1} \rangle \geq \langle X_z \rangle \geq \langle X_w \rangle \geq \langle X_n \rangle \geq \langle X_{n-1} \rangle \tag{8-47}$$

as can be readily seen by transforming Equation (8-22)

$$\langle X_w \rangle / \langle X_n \rangle = 1 + \sigma_n^2 / \langle X_n \rangle^2 \tag{8-48}$$

Table 8-1. *Mass fractions w_i and Degrees of*
Polymerization, X_i, of the Three Components,
A, B, and C of a Hypothetical Mixture

i	w_i	$\langle X_n \rangle_i$	$\langle X_w \rangle_i$	$\langle X_z \rangle_i$
A	0.2	100	150	200
B	0.5	300	500	800
C	0.3	500	700	900

This expression can never be less than unity. Analogous inequalities may also be written for $\langle X_z \rangle / \langle X_w \rangle$, $\langle X_{z+1} \rangle / \langle X_z \rangle$ etc.

A numerical example may clarify these ratios. Assume there are three fractions A, B, and C, which are mixed together according to their mass fractions w_i. Each fraction also has a degree of polymerization distribution which is characterized by individual number, mass, and z averages (Table 8-1). The number, mass, and z averages of the mixture are, according to Equations (8-44)–(8-46),

$$\langle X_n \rangle = \frac{0.2 + 0.5 + 0.3}{0.2/100 + 0.5/300 + 0.3/500} \approx 234$$

$$\langle X_w \rangle = 0.2 \times 150 + 0.5 \times 500 + 0.3 \times 700 = 490$$

$$\langle X_z \rangle = \frac{0.2 \times 150 \times 200 + 0.5 \times 500 \times 800 + 0.3 \times 700 \times 900}{0.2 \times 150 + 0.5 \times 500 + 0.3 \times 700} \approx 806$$

8.5.3. One-Moment Exponent Averages

The general formula for a one-moment average is

$$\langle X_g \rangle = \left(\sum_i G_i X_i^q \right)^{1/q} \tag{8-49}$$

The best known of these one-moment exponent averages is what is known as the viscosity-average molar mass:

$$\langle M_\eta \rangle = \left(\sum_i w_i M_i^{a_\eta} \right)^{1/a_\eta} \tag{8-50}$$

where a_η is the exponent in the intrinsic viscosity–molar mass relationship ($[\eta] = K_\eta M^{a_\eta}$). Strictly speaking, the viscosity average is a mass–viscosity average, since the statistical weight on which it is based is a mass. Analogous averages with various statistical weights exist for sedimentation, diffusion, etc.

8.5.4. Multimoment Averages

According to Equation (8-42), averages from two moments are also possible. The order of the moments [that is, $(p + q - 1)$ in the numerator and $(q - 1)$ in the denominator] must be combined with the exponent $1/p$ in such a manner that the total expression has the same physical units as the property. Since the physical units on both sides of the equation must be the same, the so-called exponent rule can be directly obtained from Equation (8-42),

$$1 = (p + q - 1)\frac{1}{p} - (q - 1)\frac{1}{p} \tag{8-51}$$

or, in general form: The product sum of the exponents must always equal unity. The rule is based on dimensional analysis and is consequently independent of whatever shape the macromolecules may adopt.

The exponent rule is especially significant in combination with the rule that states that the relationships between two variables can always, at least over a limited range of values, be written as an exponential relationship. It has been found empirically that over wide ranges of the molar mass the following relationships between the molecular weight and the sedimentation coefficient s, the diffusion coefficient D, or the intrinsic viscosity $[\eta]$ are valid:

$$s = K_s M^{a_s} \tag{8-52}$$

$$D = K_D M^{a_D} \tag{8-53}$$

$$[\eta] = K_\eta M^{a_\eta} \tag{8-54}$$

The molar mass can be obtained from any pair of the three quantities (for derivation, see Chapter 9)

$$\langle M_{sD} \rangle = A_{sD} K_{sD} s D^{-1} \tag{8-55}$$

$$\langle M_{s\eta} \rangle = A_{s\eta} K_{s\eta} s^{3/2} [\eta]^{1/2} \tag{8-56}$$

$$\langle M_{D\eta} \rangle = A_{D\eta} K_{D\eta} D^{-3} [\eta]^{-1} \tag{8-57}$$

The quantities K_{sD}, K_s, and K_D are accessible through independent measurements and are independent of the molar mass. They are consequently called physical constants. On the other hand, A_{sD}, $A_{s\eta}$, and $A_{D\eta}$ are model constants, since they are based on certain assumptions. If, for example, the frictional coefficients from sedimentation and diffusion are of equal magnitude (see Chapter 9), then $A_{sD} = 1$. The model constants can, of course, influence the numerical value of the molar mass, but they have no effect on the composition of the average from the various individual molecular species contributions. Consequently, model constants can always be assumed to have a value of unity until evidence to the contrary is obtained.

If is is assumed that the properties s, D, and $[\eta]$ can each be obtained as simple mass averages, then Equations (8-52)–(8-57) give

$$\langle M_{s_w D_w} \rangle = A_{sD} \left(\sum_i w_i M_i^{a_s} \right) \left(\sum_i w_i M_i^{a_D} \right)^{-1} ; \qquad a_s - a_D = 1 \qquad (8\text{-}58)$$

$$\langle M_{s_w \eta_w} \rangle = A_{s\eta} \left(\sum_i w_i M_i^{a_s} \right)^{3/2} \left(\sum_i w_i M_i^{a_\eta} \right)^{1/2} ; \qquad \tfrac{3}{2}a_s + \tfrac{1}{2}a_\eta = 1 \qquad (8\text{-}59)$$

$$\langle M_{D_w \eta_w} \rangle = A_{D\eta} \left(\sum_i w_i M_i^{a_D} \right)^{-3} \left(\sum_i w_i M_i^{a_\eta} \right)^{-1} ; \qquad -3a_D - a_\eta = 1 \qquad (8\text{-}60)$$

According to dimensional analysis, the product of the physical constants must be unity.

From Equations (8-58)–(8-60), we obtain the following for the exponent rule:

$$a_\eta = 2 - 3a_s = -(1 + 3a_D) \qquad (8\text{-}61)$$

The averages appearing in Equations (8-58)–(8-60) can be transformed into other averages or moments with the help of these relationships. It can be seen from Table 8-2 that the averages of such two-moment averages are determined by the composition of the property averages as well as by the nature of the composition of the property averages as well as by the nature of the molecular-weight dependence of the property. For example, if $a_\eta = 2$, combining the mass average of the sedimentation coefficient with the mass average of the diffusion coefficient gives the number-average molar mass. In some cases, the average of a two-moment average consists of a combination of two simple averages; in other cases it is simpler to describe a two-moment average via moments.

Consequently, for the same width of the distribution, the numerical values of these two-moment molar mass constants depend on the value of the

Table 8-2. *Moments and Averages of Molar Masses for Some Combinations of s, D, and [η]*

Combination	Exponent a_η	Moment or average
s_n and D_n	As desired	$\mu_n^{(1)} \equiv \langle M_n \rangle$
s_w and D_w	2	$\mu_n^{(1)} \equiv \langle M_n \rangle$
s_w and D_w	0.5	$\mu_n^{(1)} \mu_n^{(0.5)} / \mu_n^{(0.5)} = \sum_i x_i M_i^{1.5} \Big/ \sum_i x_i M_i^{0.5}$
s_w and D_z	As desired	$\mu_w^{(1)} \equiv \langle M_w \rangle$
s_w and $[\eta]_w$	2	$(\mu_w^{(2)})^{0.5} = (\langle M_w \rangle \langle M_z \rangle)^{0.5}$
s_w and $[\eta]_w$	0.5	$(\mu_w^{(0.5)})^2 = \left(\sum_i w_i M_i^{0.5} \right)^2 \equiv \langle M_\eta \rangle_\Theta$

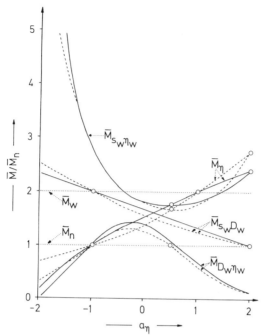

Figure 8-5. Calculated $\langle M \rangle / \langle M_n \rangle$ ratios as a function of the exponents a_η for a Schulz–Flory (—) or a generalized logarithmic normal (--) molar mass distribution for, in each case, $\langle M_w \rangle / \langle M_n \rangle = 2$. $\langle M \rangle$ may be $\langle M_{S_w \eta_w} \rangle$, $\langle M_\eta \rangle$, $\langle M_{S_w D_w} \rangle$, or $\langle M_{D_w \eta_w} \rangle$ (after H.-G. Elias, R. Bareiss and J. G. Watterson).

constant a_η (Figure 8-5). For a given homologous series, a_η is in turn a function of the particle shape and its interaction with the solvent. For example, rigid rods have a value of $a_\eta = 2$; for spheres, $a_\eta = 0$. Values of between 0.5 and 0.9 are normally obtained for random coils (see Section 9.8).

Thus these two moment averages possess the apparent paradox that for polydisperse materials, an absolute method of determining the molar mass gives different values according to the nature of the solvent (that is, according to a_η). A method of determining the molar mass is considered to be absolute when all parameters can be directly measured and no assumptions need be made about the chemical and physical structure. This applies, for example, to Equation (8-55), where the quantities s, D, and $K_{sD} = RT/(1 - v_2\rho_1)$ can be directly measured and A_{sD} can be set equal to 1.

Two-moment averages can vary considerably with the exponent a_η even though the molar mass distribution remains the same (Figure 8-5). In some cases, different averages have identical values irrespective of the width of the distribution. For example, with $a_\eta = 1$, $\langle M_\eta \rangle = \langle M_w \rangle$ and with $a_\eta = -1$,

$\langle M_{s_w D_w} \rangle$ is always equal to $\langle M_w \rangle$. When $a_\eta = 0.5$, $\langle M_{s_w \eta w} \rangle$ always equals $\langle M_\eta \rangle$.

8.5.5. Molar Mass Ratios

The width of a molecular weight distribution can, because of Equation (8-47), always be described by the ratio of two molecular weight averages. The molar mass ratio ("polydispersity index") Q is given as

$$Q = \langle M_w \rangle / \langle M_n \rangle = U + 1 \qquad (8\text{-}62)$$

U is also known as the "molecular inhomogeneity."

With monomolecular substances, $Q = 1$ and $U = 0$. Molar mass ratios and molecular inhomogeneities can, of course, be defined as the quotients of other averages besides the number and mass averages.

The width of the molar mass distribution increases with increasing Q and U. However, Q and U are not very sensitive to the distribution width for narrow molar mass distributions, as can be seen from Equation (8-48). Samples with different degrees of polymerization (or molar masses) but equal molar mass ratios have different standard deviations (Figure 8-6). Conversely, polymers with equal degree of polymerization standard deviations, but different degrees of polymerization, also have different molar mass ratios. In addition, it must be kept in mind that the standard deviation is a relative and not an absolute measure of the distribution width. To be an absolute measure of the distribution width, the fraction of material encompassed by the standard deviation must be independent of the width of the distribution. This

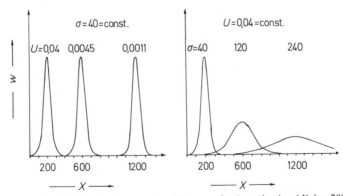

Figure 8-6. Gaussian distribution for samples of degrees of polymerization $\langle X_n \rangle = 200$, 600, or 1200 of constant standard deviation σ (and consequently, variable molecular inhomogeneity U) or constant molecular inhomogeneity (and variable standard deviation).

is only true for a Gaussian distribution; it does not hold for other distribution types.

8.5.6. Copolymers

With copolymers, each molecule may differ in terms of its relative composition as well as in terms of the absolute number of A and B monomeric units. The relative composition of the copolymer is given in terms of the mass fraction, w_A, of monomeric units of type A:

$$w_A = \frac{\sum_{N_A^0} N_A^S N_A^0 M_A^0}{\sum_{N_i} N_i (N_A^0 M_A^0 + N_B^0 M_B^0)} \tag{8-63}$$

whereby N_A^s is the number of homosequences of A monomeric units per molecule, N_A^0 is the number of A monomeric units per homosequence, N_i is the total number of copolymer molecules, and M_A^0 is the molar mass of the monomeric unit A.

The number average of the molar masses of the homosequences A or B and the number average of the copolymer molar mass are

$$\langle M_n^A \rangle = \frac{\sum_{N_A^0} N_A^S N_A^0 M_A^0}{\sum_{N_A^0} N_A^S}; \qquad \langle M_n^B \rangle = \frac{\sum_{N_B^0} N_B^S N_B^0 M_B^0}{\sum_{N_B^0} N_B^S} \tag{8-64}$$

$$\langle M_n \rangle = \frac{\sum_i N_i (N_A^0 M_A^0 + N_B^0 M_B^0)}{\sum_i N_i} \tag{8-65}$$

Thus, the number average molar mass consists of the sum of the molar masses of the homosequences, as, for example, for a two-block polymer:

$$\langle M_n \rangle = \langle M_n^A \rangle + \langle M_n^B \rangle \tag{8-66}$$

The number average molar mass of the copolymer can analogously be obtained from the mass fraction of A monomeric units and the number average molar mass of A homosequences.

The mass ("weight") average molar mass is correspondingly

$$\langle M_w \rangle = \frac{\sum_i N_i (N_A^s M_A^0 + N_B^s M_B^0)^2}{\sum_i N_i (N_A^s M_A^0 + N_B^s M_B^0)} \tag{8-67}$$

If the mass average of the A and B sequence molar masses is defined as

$$\langle M_w^A \rangle = \frac{\sum_{N_A^0} N_A^S (N_A^0 M_A^0)^2}{\sum_{N_B^0} N_A^S (N_A^0 M_A^0)}; \qquad \langle M_w^B \rangle = \frac{\sum_{N_B^0} N_B^S (N_B^0 M_B^0)^2}{\sum_{N_B^0} N_B^S (N_B^0 M_B^0)} \tag{8-68}$$

then, it follows directly from Equation (8-67) that the mass average molar mass is not the simple sum of the mass averages of the segmental molar masses.

Literature

L. H. Peebles, *Molecular Weight Distributions in Polymers*, Wiley–Interscience, New York, 1971.

H.-G. Elias, R. Bareiss, and J. G. Watterson, Mittelwerte des Molekulargewichtes und anderer Eigenschaften, *Adv. Polym. Sci.—Fortschr. Hochpolym. Forsch.* **11**, 111 (1973).

H.-G. Elias, Polymolecularity and polydispersity in molecular weight determinations, *Pure Appl. Chem.* **43**(1–2), 115 (1975).

K. W. Min, On the application of fractional moments in determining average molecular weight, *J. Appl. Polym. Sci.* **22**, 589 (1978).

C. W. Pyun, Ratios of average molecular weights and molecular weight distributions in polymers, *J. Polym. Sci.—Polym. Phys. Ed.* **17**, 2111 (1979).

Chapter 9

Determination of Molar Mass and Molar Mass Distributions

9.1. Introduction

Molar mass determination methods can be classed as absolute, equivalent, or relative. Absolute methods allow the molar mass to be directly calculated from the measured quantities without the need for assumptions concerning the physical and/or chemical structure of the polymers. In contrast, equivalent methods require a knowledge of the chemical structure of the macromolecules. Relative methods depend on the chemical and physical structure of the solute as well as on the solute–solvent interaction; these methods require calibration against another molecular mass determination method.

Absolute methods include all scattering processes (light, small-angle x-ray, neutron scattering), equilibrium sedimentation and colligative (membrane osmometry, ebullioscopy, cryoscopy, and vapor phase osmometry) methods. Each group analysis is an equivalent method, since the chemical nature and number of end groups is necessary to the calculation of the molar mass. The most important relative methods are viscometry and gel permeation chromatography.

In the predominant number of cases, a molar mass, and not a relative molecular mass, is determined. In this book, the physical units of the molar mass will always be given as g/mol macromolecule, since the numerical values then are identical to those for relative molecular masses. The unit of kg/mol

Table 9-1. Approximate Working Ranges of the More Important Methods of Determining Molar Mass

Molar mass average	Method	Type[a]	Molar mass range, g/mol
$\langle M_n \rangle$	Ebulliometry, cryoscopy, vapor-phase osmometry, isothermal distillation	A	$<10^4$
$\langle M_n \rangle$	End-group analysis	E	10^2–3×10^4
$\langle M_n \rangle$	Membrane osmometry	A	5×10^3–10^6
$\langle M_n \rangle$	Electron microscopy	A	$>5 \times 10^5$
$\langle M_w \rangle$	Equilibrium sedimentation	A	10^2–10^6
$\langle M_w \rangle$	Light scattering	A	$>10^2$
$\langle M_w \rangle$	Equilibrium sedimentation in a density gradient	A	$>5 \times 10^4$
$\langle M_w \rangle$	Small-angle X-ray scattering	A	$>10^2$
$\langle M_{sD} \rangle$	Sedimentation combined with diffusion	A	$>10^3$
$\langle M_\eta \rangle$	Dilute solution viscometry	R	$>10^2$
$\langle M_{\text{GPC}} \rangle$	Gel Permeation chromatography	R	$>10^3$

[a]A, Absolute method; E, equivalent method; R, relative method.

recommended by IUPAC is thus smaller by a factor of 1000 than the relative molecular mass unit. In addition, a unit of 1 dalton = 1 g/mol is used in the biochemical literature. (See Table 9-1.)

The choice of method depends primarily on the information required, and secondarily on the field of study, the amount of substance available, the time required, and, when necessary, on the effort required to purify the samples. Determinations are generally made at various concentrations. Then the apparent molar mass is calculated with the aid of an "ideal" theoretical relationship—that is, a relationship that only applies strictly at infinite dilution. This apparent molar mass must then be extrapolated to zero concentration to obtain the true molar mass. Apparent and true molar masses may differ considerably. Coil-shaped macromolecules of number-average molar mass of 10^6 g/mol can, for example, have an apparent number-average molar mass in good solvents of 555 000 g/mol at a concentration of 0.01 g/ml and 110 000 g/mol only, on the other hand, when the concentration is 0.1 g/ml.

9.2 Membrane Osmometry

9.2.1. Semipermeable Membranes

Equations applicable to membrane osmometry, as also in the case of ebulliometry, cryoscopy, and vapor phase osmometry, can be rigorously derived from the second law of thermodynamics in the form

$$dG = V\,dp - S\,dT \tag{9-1}$$

In membrane osmometry, the pressure difference between a solution and the pure solvent is measured for the case where the solvent is separated from the solution by a semipermeable membrane, i.e., a membrane permeable only to solvent molecules. Since the experiment is carried out isothermally, Equation (9-1) becomes, with $dT = 0$

$$\Delta G = V \Delta p = V\Pi \tag{9-2}$$

when, for small pressure differences, differentials are replaced by differences. The manometrically determined pressure difference Δp is called the osmotic pressure Π. Differentiation of Equation (8-2) with respect to amount of solvent n_1 yields

$$\frac{\partial \Delta G}{\partial n_1} = \Pi \frac{\partial V}{\partial n_1} \tag{9-3}$$

From the well-known laws governing differences in chemical potential of the solvent $\Delta \mu_1$ and the partial molar volume $\tilde{V}_1^m$, one obtains

$$-\Delta \mu_1 = \mu_{1(p)} - \mu_1 = \Pi \tilde{V}_1^m \tag{9-4}$$

The chemical potential difference can be replaced by the solvent activity a_1 and, in very dilute solutions, the mole fraction x_1 of the solvent or that of the solute x_2:

$$\Pi \tilde{V}_1^m = -RT \ln a_1 \cong -RT \ln x_1 = -RT \ln (1 - x_2) \approx RTx_2 \tag{9-5}$$

In dilute solution, where $n_2 \ll n_1$ and $V_2 \ll V_1$, and consequently $x_2 = V_1^m c_2 / M_2$, since $x_2 = n_2 / (n_2 + n_1)$ and $n_2 = m_2 / M_2$, then $c_2 = m_2 / V_2 + V_1$ and $V_1^m = V_1 / n_1$, and we can derive from Equation (9-5) the van't Hoff equation as a limiting law for infinite dilution where $V_1^m \approx \tilde{V}_1^m$:

$$\lim_{c_2 \to 0} \frac{\Pi}{c_2} = \frac{RT}{M_2} \tag{9-6}$$

For solutions of nonassociating nonelectrolytes at finite concentrations, Π / c_2 is given as an ascending series of positive powers of solute concentration [see also Equation (6-54)]:

$$\frac{\Pi}{RTc_2} \equiv (M_2)_{app}^{-1} = A_1 + A_2 c_2 + A_3 c_2^2 + \cdots \tag{9-7}$$

where A_1, A_2, etc., are the first, second, etc. virial coefficients. Comparison of the coefficients of Equations (9-6) and (9-7) gives A_1 as $(M_2)^{-1}$. The coefficient A_2 is obtained by measuring the osmotic pressure at different concentrations. The factor RTA_1 is given as the ordinate intercept at $c_2 \to 0$ in a plot of reduced osmotic pressure Π / c_2 against c_2 (Figure 9-1). With associating solutes, complicated expressions occur on the right-hand side of Equation (9-7) (see Section 6.5). Since the measured osmotic pressure Π is inversely

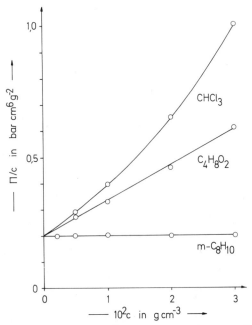

Figure 9-1. The concentration dependence of the reduced osmotic pressure Π/c of a poly(methyl methacrylate) in chloroform, dioxane, and *m*-xylene at 20°C (according to G. V. Schulz and H. Doll).

proportional to the molar mass, the method becomes more and more inaccurate with increasing molar masses. The upper limit for reasonable accuracy is at molar mass of 1–2 million.

With a polydisperse solute, the molar mass M_2 that occurs in Equation (9-6) is the number-average molar mass of the solute. In a multicomponent system, the observed osmotic pressure Π is given as the sum of all the osmotic pressures Π_i:

$$\Pi = \sum_i \Pi_i = RT \sum_i \frac{c_i}{M_i} \qquad (9\text{-}8)$$

The sum $\Sigma_i(c_i/M_i)$ in Equation (9-8) is also contained in the definition of the number-average molar mass $\langle M_n \rangle = \Sigma_i c_i / \Sigma_i[c_i/(M_i)_n]$ given in equation (8-44). If this expression is inserted into Equation (9-8), it is seen that osmotic pressure measurements give the number-average molar mass:

$$\Pi = \frac{RT\Sigma_i c_i}{\langle M_n \rangle} = \frac{RTc}{\langle M_n \rangle} \qquad (9\text{-}9)$$

In the case of polyelectrolytes, the membrane is not permeable to the polyions because of their size, or to the gegenions because of the need to preserve electroneutrality. As there are many gegenions per polyion, the following approximation of Equation (9-6) holds for low contents where $[M_E] = c_2/M_2$:

$$\Pi = RTN_z[M_E] \qquad (9\text{-}10)$$

N_z is the effective degree of ionization, that is, it is the fraction of gegenions that contribute to the osmotic pressure. Consequently, N_z is smaller than the total number of gegenions. It is practically constant at high degrees of ionization.

9.2.2. Experimental Methods

In the simplest case, the osmotic pressure Π is measured in a single-cell osmometer with a horizontally arranged membrane (Figure 9-2). Π is then identified as the manometrically measured difference in pressure Δp_{eq} at equilibrium.

The osmotic pressure is obtained from h_s and h_1, the heights reached, and the densities ρ_s and ρ_1 of the solution S and solvent 1, with the notation $\Delta h = h_s - h_1$ and $\Delta\rho = \rho_{s\cdot} - \rho_1$, together with the relationship $\Pi = \Delta p_{eq}$, which applies to the semipermeable membranes, as

$$\Pi = \Delta p_{eq} = h_s\rho_s - h_1\rho_1 = \Delta h\rho_1 - h_s\,\Delta\rho = \Delta h\rho_s + h_1\,\Delta\rho \qquad (9\text{-}11)$$

To calculate the osmotic pressure Π it is necessary to know the absolute height h_2 (or h_1) and the density difference $\Delta\rho$, as well as the difference in

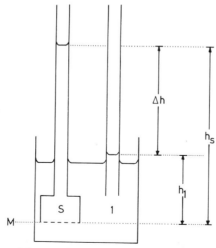

Figure 9-2. Calculation of the osmotic pressure Π from the heights of the solution h_s and of the solvent h_1 above a horizontally arranged membrane.

heights reached Δh and the solvent density. When using osmometers with vertically arranged membranes, the center of the membrane, to a good approximation, can be taken as the reference point in measuring the absolute height h_2.

At the beginning of an osmotic experiment, the difference in heights Δh observed after filling both chambers of the osmometer does not correspond to the osmotic pressure at equilibrium. The equilibrium pressure is only observed after solvent molecules permeate the membrane. If Δh is greater than the equilibrium osmotic pressure, the solvent molecules permeate from the solution chamber into the solvent chamber, and in the reverse direction if Δh is smaller than the equilibrium osmotic pressure. The time taken to reach equilibrium increases with the amount of solvent that must be displaced, i.e., increases with the diameter of the capillaries. Since, experimentally, problems such as dirt in the capillaries, etc., limit the size of capillary that one can go down to, and since the membranes must be tight (semipermeability), the establishment of osmotic equilibrium can take days or weeks. Other problems such as poor solvent drainage in the capillaries, adsorption of solute on the membrane, partial permeation of solute through the membrane, etc., can interfere with the attainment of a true osmotic equilibrium. The absence or presence and allowance for these complications must be individually established.

The time required to reach equilibrium is much reduced through the use of novel technology in commercially available automatic membrane osmometers. If, for example, the capillary height in the solution chamber increases because solvent permeates from the solvent chamber, this is immediately compensated by the application, via a servomechanism, of a pressure on the solution chamber, such that the capillary heights above solvent and solution remain the same. Since this method involves the transport of only very small amounts of liquid, equilibrium is reached after only 10–30 min.

The osmotic pressure can, alternatively, be calculated from the rate of attaining equilibrium. The rate of approach to equilibrium is proportional to the displacement from equilibrium:

$$\frac{d(p - \Pi)}{dt} = -k(p - \Pi) \tag{9-12}$$

or, integrated,

$$\ln \frac{p_1 - \Pi}{p_2 - \Pi} = \frac{t_2 - t_1}{t_{0.5}} \ln 2 = \alpha \ln 2 \tag{9-13}$$

The terms p_1 and p_2 are the osmotic pressures at the times t_1 and t_2. Here $t_{0.5}$ is the half-time for solvent passage through the membrane and is determined in a preliminary experiment. The antilogarithmic form of Equation (9-13) is solved for Π:

$$\frac{p_1 - \pi}{p_2 - \Pi} = 2^\alpha \qquad (9\text{-}14)$$

$$\Pi = \frac{2^\alpha p_2 - p_1}{2^\alpha - 1} \qquad (9\text{-}15)$$

Films of regenerated cellulose, for example, Cellophane 600, Gel cellophane, Ultracella, grades fine and ultrafine, are used as membranes for organic solvent systems. For aqueous solutions, membranes of cellulose acetate (for example ultrafine filter) or nitrocellulose (collodium) are suitable. Corrosive solvents, such as formic acid, etc., require the use of glass membranes.

9.2.3. Nonsemipermeable or Leaky Membranes

According to the necessary theoretical assumptions, the membrane used should be strictly semipermeable, i.e., it should be permeable to solvent molecules and completely impermeable to solute. In the case of native proteins, for example, this requirement is easily fulfilled. Native proteins are predominantly monodisperse and have a compact structure. As long as the pore diameter of the membrane is less than the protein molecule diameter, the membrane is strictly semipermeable. Since protein molecules mostly have a molecular diameter of more than 5 nm, it is not too difficult to find a suitable membrane, e.g., in this case a cellulose acetate-based membrane for aqueous solutions. Coil-forming macromolecules, on the other hand, do have a large coil diameter, but only a very small chain diameter. Thus they can very easily pass through membranes with the relatively small pore size of 1–2 nm. This permeation occurs all the more readily the lower the molar mass. In polymolecular substances, therefore, some of the substance can permeate. At osmotic equilibrium (with the so-called static techniques) all permeating species will distribute themselves according to their activities in a Donnan equilibrium on both sides of the membrane. The observed osmotic pressure does not correspond to the theoretical osmotic pressure of the original substance at any concentration. For the limiting value of the reduced osmotic pressure Π/c_2 for $c_2 \to 0$, one obtains the molar mass of the nonpermeable part.

Partial or complete permeation of the solute can frequently be recognized when measurements are made "from below" (capillary height difference Δp_0 at $t = 0$ smaller than Π) since the observable pressure Δp goes through a maximum before reaching the equilibrium value (Figure 9-3). The effect is the

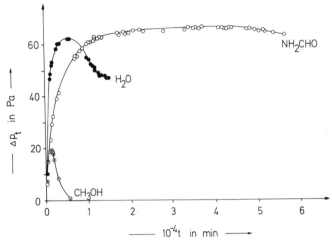

Figure 9-3. The time dependence of the hydrostatic heads Δp_t of solutions of poly(ethylene glycol) ($c = 2 \times 10^{-4}$ g/cm³; $\langle M_n \rangle = 4000$, $\langle M_w \rangle = 4300$ g/mol molecule) in formamide, methanol, or water on cellophane 600 membranes (cellulose hydrate) at 25°C. The theoretically expected osmotic pressure in an ideal solution at this concentration is $\Pi_{id} = 127$ Pa (according to H.-G. Elias).

result of the simultaneous permeation of the solvent molecules into the solution chamber and of the permeable solute material into the solvent chamber. Since at short observation times virtually no solute can permeate, it is often assumed that, even for permeating solutes, the true osmotic pressure is observed with automatic osmometers because of the short measuring times. This assumption is wrong.

With nonsemipermeable membranes, both the solute and the solvent can pass through the membrane. Let J_v be the volume flow and J_D be the flow due to diffusion (permeation). Both these flows can result from either a hydrostatic pressure difference Δp or an osmotic pressure Π:

$$J_v = L_p \, \Delta p + L_{pD} \Pi \tag{9-16}$$

$$J_D = L_{Dp} \, \Delta p + L_D \Pi \tag{9-17}$$

The pressure difference at zero volume flow ($J_v = 0$) is, of course, measured by dynamic osmometry. Thus, with Equation (9-16), we have

$$(\Delta p)_{J_v} = - (L_{pD}/L_p) = s\Pi \tag{9-18}$$

The negative ratio of what are known as the two phenomenological coefficients, L_{pD} and L_p, is called the selectivity, reflection, or Staverman coefficient.

With semipermeable membranes, $-L_{pD} = L_p$, and the Staverman coefficient s is unity. On the other hand, $-L_{pD} < L_p$ for nonsemipermeable membranes, and the Staverman coefficient is smaller than 1. In the limit of completely permeable membranes, the Staverman coefficient has the value zero. Thus, the osmotic pressure Π is not measured by dynamic osmometry at zero volume flow; a smaller value of $s\Pi$ is always measured instead.

The Staverman coefficient s has not, to date, been theoretically calculated. Experimentally, it has been found with cellophane 600 membranes and poly(oxyethylenes) of different molar masses that the Staverman coefficient s decreases more or less linearly from a value of 1 for molar masses greater than 6000 g/mol to $s = 0$ for $M = 62$ g/mol.

Thus, in order to determine the molar mass of a partially permeating solute, the solution is first dialyzed with the same membrane as is to be used for osmometry. The nondialyzing part is then studied by membrane osmometry, and the dialyzing part is then studied by, for example, vapor phase osmometry. The molar mass of the original sample is then calculated from the mass fractions and molar masses of both the dialyzing and the nondialyzing parts.

9.3. Ebulliometry and Cryoscopy

The boiling temperatures of a solution and of the pure solvent are different because of the difference in activities. At equilibrium, Equation (9-1) changes from the form $d\Delta G = \Delta V\, dp - \Delta S\, dt$ into (since $d\Delta G = 0$)

$$\Delta V\, dp = \Delta S\, dT \qquad (9\text{-}19)$$

For a reversible isothermal isobaric process, on the other hand, the second law of thermodynamics applies in the form $\Delta S = (\Delta H)_{T,p}/T$. If this equation is inserted into Equation (9-19), one obtains after rearrangement

$$\Delta H = T\,\Delta V \frac{dp}{dT} \qquad (9\text{-}20)$$

At the boiling temperature, the volume of vapor is much greater than the volume of liquid: $\Delta V = V_{vap} - V_{liq} \approx V_{vap}$. On introducing this expression and the ideal gas law $pV_{vap}^m = RT_{bp}$, where T_{bp} is the boiling point, into Equation (9-20), one obtains, with subscript bp for thermodynamic parameters at the boiling temperature, and converting to content units:

$$\Delta H_{bp}^m = T_{bp}\frac{dp}{dt}\frac{RT_{bp}}{p} \qquad (9\text{-}21)$$

or with Raoult's law $x_2 = \Delta p/p_1$ and with $x_2 = n_2/(n_1 + n_2) \approx n_2/n_1 = m_2 M_1/m_1 M_2 = m_2 M_1/M_2\rho_1 V_1 \cong c_2 M_1/M_2\rho_1$, after rearranging,

$$\frac{\Delta T_{bp}}{c_2} = \left(\frac{RT_{bp}^2 M_1}{\rho_1 \Delta H_{bp}^m} \right) \frac{1}{M_2} = E \frac{1}{M_2} \qquad \text{(for } c_2 \rightarrow 0) \qquad (9\text{-}22)$$

or, written analogously to Equation (9-6),

$$\frac{\Delta T_{bp}}{c_2} \left(\frac{\rho_1 \Delta H_{bp}^m}{T_{bp} M_1} \right) = \frac{RT_{bp}}{M_2} \qquad \text{(for } c_2 \rightarrow 0) \qquad (9\text{-}23)$$

In order to obtain the largest possible boiling temperature elevation ΔT_{bp} for a solute of given molar mass M_2, the ebullioscopic constant E of the solvent must be large. The solvent should have a high boiling temperature T_{bp}, a large molar mass M_1, and a low heat of vaporization ΔH_{bp}.

According to the derivation, Equation (9-23) only applies to solutions at infinite dilution. For finite concentrations, one can, in analogy to the procedure adopted for membrane osmometry measurements, develop a series with virial coefficients. In polymeric solutes, the number-average molar mass is measured in ebulliometry. (The proof is analogous to that given for osmotic-pressure measurements.)

An analogous expression can be derived for the lowering of the freezing temperature ΔT_M in cryoscopic measurements of infinitely dilute solutions:

$$\frac{\Delta T_M}{c_2} = \frac{RT_g^2 M_1}{\rho_1 \Delta H_M^m} \frac{1}{M_2} \qquad \text{(for } c_2 \rightarrow 0) \qquad (9\text{-}24)$$

where T_g is the freezing temperature of the solvent and ΔH_M^m is the molar heat of fusion.

9.4. Vapor Phase Osmometry

Vapor pressure osmometric (thermoelectric, vaporometric) measurements depend on the following principle: A drop of a solution with a nonvolatile solute resides on a temperature sensor, i.e., a thermistor. The surrounding region is saturated with solvent vapor. Initially, the drop and vapor are at the same temperature. Since the vapor pressure of the solution is lower than that of the pure solvent, solvent vapor condenses on the solution drop. Because heat of condensation is released, the temperature of the drop rises until the difference in temperature ΔT_{th} between the drop and the solvent vapor again eliminates the difference in vapor pressure, so that the chemical potential of the solvent in both phases is equal. An analogous equation to that which applies in ebulliometry is applicable in this case to the relationship between the temperature difference ΔT_{th} and the number-average molar mass $\langle M_n \rangle = M_2$ of the solute,

$$\frac{\Delta T_{th}}{c_2} = \frac{RT^2}{L_1 \rho_s} \frac{1}{\langle M_n \rangle} \qquad \text{(for } c_2 \rightarrow 0) \tag{9-25}$$

where L_1 is the latent heat of vaporization of the solvent per mass and ρ_s is the density of the solvent.

Thus, like ebulliometry or cryoscopy, the method would have a strong thermodynamic basis if heat transfer other than that due to vapor condensation could be prevented. Vapor and drop are, however, in contact with one another, and the temperature thus tends to equilibrate in time by convection, radiation, and conduction. This again causes renewed condensation of solvent vapor, which proceeds until a final steady state with a temperature difference ΔT is reached. Equation (9-25) becomes, with $\Delta T = k_E \Delta T_{th}$:

$$\frac{\Delta T}{c_2} = k_E \frac{RT^2}{L_1 \rho_s} \frac{1}{\overline{M}_n} = K_E \frac{1}{\langle M_n \rangle} \qquad \text{(for } c_2 \rightarrow 0) \tag{9-26}$$

Since k_E cannot be derived theoretically, K_E is usually determined by calibration with substances of known molar mass. As with all the other molar-mass-determination methods, only apparent molar masses M_{app} are obtained for finite concentrations (see Sections 6.4 and 6.5) because of the effects of the virial coefficients and/or association, M_{app} must therefore be extrapolated to the concentration $c_2 \rightarrow 0$. In vapor phase osmometry, low amounts of nonvolatile impurities interfere with the result, but volatile impurities do not, since they pass into the vapor phase.

9.5. Light Scattering

9.5.1. Basic Principles

The light scattering method of determining molar mass measures the light scattered at an angle to the incident light. The light scattered from large particles can be seen by the naked eye; it is called the Tyndall effect. Incident light of intensity L_0 on passing through a scattering medium, will have its intensity diminished by the amount I_s of scattered light according to Beer's law:

$$I_0 - I_s = I = I_0 \exp(-\tau r) \tag{9-27}$$

where r is the path length through the medium and τ is the extinction coefficient of the scattered light (radiation). The total intensity $(I_0 = I + I_s)$ remains constant. This is a case of conservation of extinction and not of intensity loss as is the case with absorption by colored solutions. In pure liquids and dilute macromolecular solutions, the scattered light intensity I_s is only 1/10 000–1/30 000 of the incident intensity I_0. The intensity of light scattered I_s thus cannot be measured with sufficient accuracy simply by measuring the difference. Therefore I_s is directly measured by a photo-multiplier–photocell system, and accuracy is achieved.

In light scattering, it is necessary to differentiate between "large" and "small" particles or molecules. The dimensions of small particles or molecules are much smaller than the wavelength λ_0 of the incident light, i.e., smaller than about $0.05–0.07\ \lambda_0$.

9.5.2. Small Particles

Visible light has an electric vector E perpendicular to the propagation direction, and this vector varies sinusoidally with time. For the field strengths of the vertical (E_v) and horizontal (E_h) components of plane polarized light at one particular point in space, one can write (see texts on theoretical physics)

$$E = E_0 \cos \omega t$$

$$E_v = E_{0v} \cos \omega t, \qquad E_{0v} = E_0 \cos \phi \qquad (9\text{-}28)$$

$$E_h = E_{0h} \cos \omega t, \qquad E_{0h} = E_0 \sin \phi$$

where E_0 is the amplitude of the electric vector, ω is the angular frequency, t is the time, and ϕ is the angle between the vector E and the vertical.

The electric field interacts with every particle in the path of the light beam, producing a dipole moment p at the interaction site, since the electrons in the electron shells of atoms composing these particles are displaced in the opposite direction to the electric field. The electric field strength and the induced dipole moment are proportional to each other; the proportionality constant is called the polarizability α:

$$p = \alpha E \qquad (9\text{-}29)$$

Assuming that the particles are small (no intramolecular interference, see Section 9.5.5), that they are independent of each other (ideal gas or infinitely dilute solution), and that there is no loss of light intensity due to absorption, Equations (9-28) and (9-29) may be combined:

$$p = \alpha E_0 \cos \omega t \qquad (9\text{-}30)$$

Equation (9-30) states that the induced dipole follows the oscillating electric field with the same frequency. An oscillating dipole also emits electromagnetic radiation, i.e., scattered light. According to Equation (9-30), the scattered light has the same wavelength as the incident light. The energy of light is measured by its intensity, that is, the energy absorbed per second on a surface area. According to Poynting's theorem, this energy is proportional to the time-averaged mean square of the amplitude, i.e., it is proportional to $\bar{E}^2 \equiv \langle E^2 \rangle$. From Equation (9-28), accordingly, for the vertical and horizontal components of plane polarized light, we have

$$I_{0,v} = \text{const} \times \langle E_v^2 \rangle = \text{const} \times E_{0v}^2 \langle \cos^2 \omega t \rangle$$

$$I_{0,h} = \text{const}$$

(9-31)

The intensity $i_{s,v}$ of the vertically polarized light scattered by a molecule is analogously given by the amplitude $E_{s,v}$ of the scattered light:

$$i_{s,v} = const \times E_{s,v}^2 \qquad (9\text{-}32)$$

and also analogously for the horizontal component of the polarized scattered light. The field E_s of the vertical or horizontal components of the polarized scattered light can be obtained from the following considerations. The first derivative of the dipole moment with respect to time, dp/dt, corresponds to an electric current that would produce a magnetic field of constant strength. The second derivative d^2p/dt^2 corresponds to an oscillating field, such as that produced by an oscillating dipole. We have

$$E_s = const' \times \frac{d^2p}{dt^2} \qquad (9\text{-}33)$$

An expression for d^2p/dt^2 is obtained by double differentiation of Equation (9-30):

$$\frac{d^2p}{dt^2} = \alpha E_0 \omega^2 \cos \omega t \qquad (9\text{-}34)$$

The proportionality constant, $const'$, in (9-33) is composed of two factors, $1/r$ and $\sin \vartheta_x$. Here x indicates the horizontally (h) or the vertically (v) polarized component of scattered light.

The factor $1/r$ follows from the law of conservation of energy. The scattered light is distributed in all directions about the oscillating dipole. The total energy flux, i.e., the amount of energy scattered per time, must be constant. The energy flux per area is equal to the intensity; consequently the intensity varies with $1/r^2$. Since the intensity is proportional to the square of the field strength, or amplitude, the amplitude itself must be proportional to $1/r$.

The factor $\sin \vartheta_x$ is obtained by the following reasoning: Although the scattering envelope of the particle is spherical, the amplitude is direction dependent (Figure 9-4). Consider vertically polarized light of intensity $(I_v)_0$ falling in the x direction on the particle. A dipole is induced which oscillates in the z direction. The greatest amplitude of the scattered light is observed

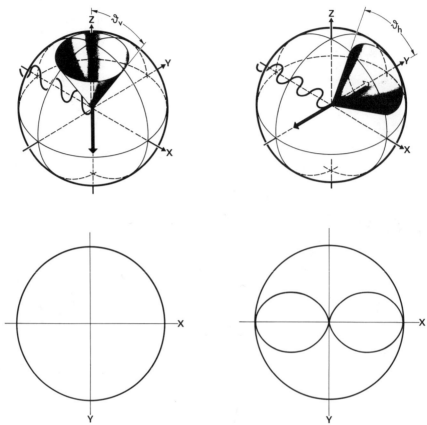

Figure 9-4. The scattering profile of small particles with vertically (v) and horizontally (h) polarized incident light. Upper row: the oscillating dipole position and definition of the angles ϑ_v and ϑ_h. Lower row: polar plot of the light-scattering intensity (indicated by arrows) in the *xy* plane.

perpendicular to the dipole axis, whereas the amplitude is zero along the dipole axis. The amplitude is proportional to (sin ϑ_v), where ϑ_v is the angle between the dipole axis and the direction of observation. Consequently, in the *xy* plane, the intensity $i_{s,v}$ of vertically polarized light is independent of the observation angle.

Inserting these expressions and Equation (9-34) into Equation (9-33), we obtain the following for vertically polarized incident light:

$$E_{s,v} = (1/r)\,(\sin\,\vartheta_v)\,(\hat{c})^{-2}\,\alpha E_{0v}\omega^2\,\cos\,\omega t \qquad (9\text{-}35)$$

The right-hand side of this equation was also divided by the square of the speed of light $\hat{c}$ to maintain dimensionality. Combining Equations (9-31), (9-32), and (9-35), considering that the frequency $\omega/2\tau = \hat{c}\lambda_0$, and using an average oscillation period, we obtain

$$i_{s,v}/I_{0,v} = 16\pi^4\alpha^2(\sin^2\vartheta_v)r^{-2}\lambda_0^{-4} \tag{9-36}$$

With horizontally polarized incident light, the dipoles oscillate in the y direction. The largest amplitude is again seen perpendicular to the dipole axis. In the y direction, the intensity is zero (Figure 9-4). The amplitude is proportional to $\sin\vartheta_h$, where ϑ_h is the angle between the dipole axis and the direction of observation. Analogously to Equation (9-36), the scattered-light intensity of horizontally polarized light is

$$i_{s,h}/I_{0,h} = 16\pi^4\alpha^2(\sin^2\vartheta_h)r^{-2}\lambda_0^{-4} \tag{9-37}$$

or in terms of the total intensity, with $I_{0,v} = I_{0,h} = 0.5I_0$,

$$\frac{i_s}{I_0} = \frac{i_{s,v} + i_{s,h}}{I_0} = \frac{16\pi^4\alpha^2(\sin^2\vartheta_v + \sin^2\vartheta_h)}{2r^2\lambda_0^4} \tag{9-38}$$

and with $\sin^2\vartheta_v + \sin^2\vartheta_h = 1 + \cos^2\vartheta$, where ϑ is the angle between the observer and the incident light,

$$i_s r^2/I_0 = 16\pi^4\alpha^2\lambda_0^{-4}[(1 + \cos^2\vartheta)/2] \tag{9-39}$$

The factor $(1 + \cos^2\vartheta)/2$ describes the angular function for the reduced light scattering $i_s r^2/I_0$ for unpolarized light. Part of its contribution comes from the vertical component $(1/2)$ and part from the horizontal component $(\cos^2\vartheta)/2$. The reduced intensity of scattered light for horizontally polarized light will be zero at an observation angle of $90°$; consequently only vertically polarized or unpolarized light is used for light-scattering measurements.

The derivations given above assumed small, isotropic particles. With isotropic particles, vertically polarized incident light leads only to vertically polarized scattered light, and horizontally polarized incident light gives only horizontally polarized scattered light. With anisotropic particles (e.g., benzene molecules), however, depolarization of the scattered light occurs. In this case, therefore, vertically polarized incident light gives both vertically and horizontally polarized scattered light. One corrects for this effect by including a correction factor, the so-called Cabannes factor. With macromolecular solutions, the Cabannes factor is usually very close to 1.

All the parameters in Equation (9-39) except the polarizability α can be measured directly. α is the excess polarizability, that is, the difference between the polarizability of the solute and that of the displaced solvent in dilute solutions. The polarizability of gases is related to the dielectric constant ϵ via $\epsilon - 1 = 4\pi\alpha(N/V)$, where N is the number of molecules in the volume V.

Correspondingly, the difference in relative permittivity (dielectric constant) for solution and solvent must be considered in dilute solutions:

$$\epsilon - \epsilon_1 = 4\pi\alpha(N/V) = \Delta\epsilon \qquad (9\text{-}40)$$

From this and the Maxwell relation $\epsilon = n^2$, with the definition $N/V \equiv cN_L/M_2$, we obtain

$$\alpha = \frac{M_2(n^2 - n_1^2)}{4\pi c N_L} \qquad (9\text{-}41)$$

The refractive index n of dilute solutions can be expanded as a series in concentration c: $n = n_1 + (dn/dc)c + \cdots$, where n_1 is the refractive index of solvent. With $(dn/dc)^2 c^2 \ll 2n_1(dn/dc)c$, the square of the refractive index is given as

$$n^2 = n_1^2 + 2n_1\frac{dn}{dc}c \qquad (9\text{-}42)$$

On combining Equations (9-39)–(9-42) and with $c = (N/V)M_2/N_L$, we obtain

$$R_\vartheta = \frac{i_s r^2(N/V)}{I_0} = \frac{4n_1^2\pi^2(dn/dc)^2[(1 + \cos^2\vartheta)/2]cM_2}{N_L\lambda_0^4} \qquad (9\text{-}43)$$

The left-hand side of Equation (9-43) corresponds to the reduced scattered light intensity by all N molecules in the volume V, and is called the Rayleigh ratio R_ϑ. Defining the optical constant κ as

$$\kappa \equiv 4\pi^2 n_1^2\left(\frac{dn}{dc}\right)^2 N_L^{-1}\lambda_0^{-4}\frac{(1 + \cos^2\vartheta)}{2} \qquad (9\text{-}44)$$

allows Equation (9-43) to be written as

$$R_\vartheta = \kappa c M_2 \qquad \text{(for } c \to 0\text{)} \qquad (9\text{-}45)$$

According to the derivation, Equation (9-45) is valid for infinitely dilute solutions. It is the basis of molar mass determinations by the light-scattering method. The molar mass M_2 obtained here is the mass-average molar mass $\langle M_w \rangle$, as shown in the following derivation:

The Rayleigh ratio for a mixture of i polymer homologous macromolecules of different molar masses is, with Equation (9-45),

$$\bar{R}_\vartheta = \sum_i (R_\vartheta)_i = \sum_i \kappa c_i M_i = \kappa \sum_i c_i M_i \qquad (9\text{-}46)$$

since the refractive index increment dn/dc is independent of molar mass for molar masses above $\sim 20\,000$, and, according to Equation (9-44), κ is independent of both M and c. A comparison of the sum $\Sigma_i c_i M_i$ with the definitions of average molar mass [Equations (8-44) and (8-45)] shows that the following can be written at high molar masses ($M_E \ll M_w$) with $\Sigma c_i = c$ and $c_i/c = w_i$:

$$\bar{R}_\vartheta = \kappa c \langle M_w \rangle \tag{9-47}$$

9.5.3. Copolymers

Copolymers show in general both a molar mass and a sequence distribution. Since the individual molecules i do not have the same composition, they will also have different refractive index increments $Y_i = (dn/dc)_i$. The summation of Equation (9-43) proceeds in this case not as in Equation (9-46), but as

$$\bar{R}_\vartheta = \kappa' \sum_i Y_i^2 c_i M_i \tag{9-48}$$

κ' is defined analogously to κ [see Equation (9-44)]:

$$\kappa' = 4\pi^2 n_1^2 N_L^{-1} \lambda_0^{-4} [(1 + \cos^2 \vartheta)/2] \tag{9-49}$$

The summation must proceed both over molecules of the same average composition and different molar masses, and as molecules of the same molar mass but different average composition. On conventional analysis of data according to Equation (9-45) an apparent molar mass $(M_w)_{app}$ would, on account of Equation (9-48), be obtained instead of the true molar mass $\langle M_w \rangle$, even when $c \rightarrow 0$, i.e.,

$$\bar{R}_\vartheta = \kappa c (M_w)_{app} = \kappa' Y_{cp}^2 c (M_w)_{app} \tag{9-50}$$

Here, Y_{cp} is the refractive index increment for the whole polymer. On combining Equations (9-48) and (9-50), we obtain the following after introducing the mass contribution $w_i = c_i/c$ of the molecular species i:

$$(M_w)_{app} = Y_{cp}^{-2} \sum_i Y_i^2 w_i M_i \tag{9-51}$$

The refractive index increment Y_i of the molecular species i must now be related to the refractive index increments Y_A and Y_B of the unipolymers A and B. The refractive index n_{cp} of a copolymer of the monomeric units A and B depends on the refractive indices n_A and n_B of the unipolymers, as well as on the mass contributions w_A and w_B:

$$n_{cp} = n_A w_A + n_B w_B, \qquad w_A + w_B \equiv 1 \tag{9-52}$$

For a copolymer molecule of composition i, analogously,

$$n_i = n_A w_{A,i} + n_B w_{B,i} \tag{9-53}$$

The difference between the refractive indices and the refractive index of the solvent n_1 can be used instead of the refractive indices n_i:

$$n_{cp} - n_1 = (n_A - n_1)w_A + (n_B - n_1)w_B \tag{9-54}$$

or, after dividing both sides by the copolymer concentration c,

$$\frac{n_{cp} - n_1}{c} = \frac{n_A - n_1}{c} w_A + \frac{n_B - n_1}{c} w_B \tag{9-55}$$

The fractions represent the refractive index increments $Y = dn/dc$, assuming that the refractive indices of the solutions vary linearly with concentration:

$$Y_{cp} = Y_A w_A + Y_B w_B \tag{9-56}$$

Analogously, for the ith molecular species

$$Y_i = Y_A w_{A,i} + Y_B w_{B,i} \tag{9-57}$$

With $w_B = 1 - w_A$ and $w_{B,i} = 1 - w_{A,i}$, combining equations (9-56) and (9-57), we obtain

$$Y_i - Y_{cp} = (Y_A - Y_B)(w_{A,i} - w_A) = \Delta Y \Delta w_{A,i} \tag{9-58}$$

If Equation (9-58) is inserted into Equation (9-51), one obtains

$$(M_w)_{app} = \sum_i w_i M_i + 2 \frac{\Delta Y}{Y_{cp}} \sum_i w_i M_i (\Delta w_{A,i}) + \left(\frac{\Delta Y}{Y_{cp}}\right)^2 \sum_i w_i M_i (\Delta w_{A,i})^2 \tag{9-59}$$

In this equation, the first sum corresponds to the mass-average molar mass $\langle M_w \rangle = \Sigma_i w_i M_i$. The second and third sums contain the first and second moments of the z distribution $\nu_z^{(1)}$ and $\nu_z^{(2)}$ of $\Delta w_{A,i}$ (see Section 8.4), since, with $w_i = m_i / \Sigma m_i$, the following can be written after multiplying numerator and denominator by $\Sigma_i Z_i$, remembering that $\Delta w_{A,i} = E_i - \bar{E}$

$$\sum_i w_i M_i (\Delta w_{A,i}) = \frac{\Sigma_i m_i M_i (\Sigma w_{A,i})}{\Sigma_i Z_i} \frac{\Sigma_i Z_i}{\Sigma_i m_i} = \nu_z^{(1)} \langle M_w \rangle \tag{9-60}$$

$$\sum_i w_i M_i (\Delta w_{A,i})^2 = \nu_z^{(2)} \langle M_w \rangle$$

With this relationship, Equation (9-59) becomes

$$(M_w)_{app} = \langle M_w \rangle \left[1 + 2\nu_z^{(1)} \frac{Y_A - Y_B}{Y_{cp}} + \nu_z^{(2)} \left(\frac{Y_A - Y_B}{Y_{cp}} \right)^2 \right] \tag{9-61}$$

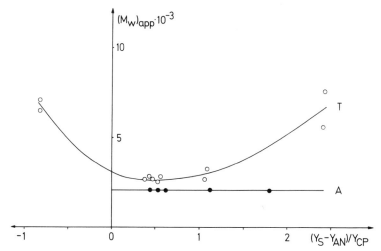

Figure 9-5. The dependence on solvent of the apparent mass average molar mass (at $c \rightarrow 0$) of one industrially (T) and one azeotropically (A) produced copolymer of styrene and acrylonitrile with refractive index increments of the polystyrene (Y_S), the polyacrylonitrile (Y_{AN}), and the copolymer (Y_{CP}) (according to H. Benoit). The $(\langle M_w \rangle)_{app}$ used here is extrapolated to $c \rightarrow 0$, and so is not the $(M_w)_{app}$ used here is extrapolated to $c \rightarrow 0$, and so is not the $(M_w)_{app}$ of Section 6.5 or the M_{app} of Section 9.5.4.

This equation states that light scattering measurements on chemically nonuniform copolymers or polymer mixtures do not give a mass-average, but an apparent mass-average molar mass $(M_w)_{app}$. The apparent mass-average molar mass also depends on the refractive index increments Y_A, Y_B, and Y_{cp} of the homopolymers A and B and the copolymer. Since, in a series of solvents, these refractive index increments differ from one another, one can carry out a series of light scattering measurements in solvents with widely different refractive indices. Then, when $(M_w)_{app} = f[(Y_A - Y_B)/Y_{cp}]$ is plotted, the true mass-average molar mass $\langle M_w \rangle$ is obtained as intercept for $(Y_A - Y_B)/Y_{cp} = 0$ (see Figure 9-5). The moments $\nu_z^{(1)}$ and $\nu_z^{(2)}$ can be calculated from the shape of the curve. In the case of constitutionally uniform copolymers, such as are obtained in an azeotropic copolymerization (see Chapter 22), the first and second moments $\nu_z^{(1)}$ and $\nu_z^{(2)}$ will be zero since $\Delta w_{A,i} = w_{A,i} - w_A = 0$. Copolymers of constitutional uniformity thus give molar masses that do not depend on the refractive index of the solvent when light scattering measurements are carried out in various solvents (see Figure 9-5).

9.5.4. Concentration Dependence

The derivations given in Section 9.5.2. refer to small, isotropic, randomly distributed molecules that move independently of one another, e.g., in a vacuum. The total intensity of scattered light is here given as the sum of the intensities scattered by the individual molecules. In liquids, the Brownian motions of the molecules are not independent of each other. Because of intermolecular interference, the measured total intensity of scattered light is less than the sum of the individual intensities.

In pure liquids, Brownian (i.e., thermal) motion of the molecules leads to fluctuations in time and place of the density of the liquid. In solutions, there is also a fluctuation in solute concentration. It can be assumed that fluctuations in solvent density and solute concentration are independent of each other. In this case, the intensity of scattered light i_s by the solute is given simply by subtracting the intensity of scattered light by the pure solvent i_{solv} from that by the solution i_{soln},

$$i_s = i_{soln} - i_{solv} \tag{9-62}$$

Equation (9-39) contains the light scattering intensity emitted by one molecule. The Rayleigh ratio R_ϑ for a system consisting of N scattering molecules in a volume V is defined as $R_\vartheta = i_s r^2 (N/V)/I_0$ [see also Equation (9-43)]. Correspondingly, Equation (9-39) can be written as

$$R_\vartheta = \frac{i_s r^2 (N/V)}{I_0} = \frac{16\pi^4 \alpha^2 (N/V)}{\lambda_0^4} \left(\frac{1 + \cos^2 \vartheta}{2} \right)$$

The total volume is then divided into q volume elements. Each volume element should possess dimensions smaller than the wavelength of light, but still be large enough to contain several scattering molecules. Every volume element should possess a polarizability of α^s, which varies by a certain amount $\Delta\alpha$ about the average polarizability $\bar{\alpha}$ of the whole system. For the square of the polarizability of a volume element, therefore,

$$(\alpha^s)^2 = (\bar{\alpha} + \Delta\alpha)^2 = (\bar{\alpha})^2 + 2\Delta\alpha(\bar{\alpha}) + (\Delta\alpha)^2 \tag{9-64}$$

The average polarizability α is the same for all volume elements and does not, therefore, contribute to the polarizability arising from random fluctuations. The average fluctuation $\Delta\alpha$ is zero; consequently, the contribution to the light scattering of the total system is only the mean square of the fluctuation, i.e., $(\Delta\alpha)^2$. The q volume elements make a contribution q times as large, so that Equation (9-63) becomes

$$R_\vartheta = 16\pi^4 \overline{(\Delta\alpha)^2} q \lambda^{-4} [(1 + \cos^2\vartheta)/2] \tag{9-65}$$

According to Equation (9-40), the polarizability is related to the optical relative permittivity. So, using the considered number of volume elements q instead of the concentration N/V, we find $(\Delta\epsilon)^2 = (4\pi q)^2 (\Delta\alpha)^2$, and Equation (9-65) becomes

$$R_\vartheta = \pi^2 \overline{(\Delta\epsilon)^2} q^{-1} \lambda_0^{-4} [(1 + \cos^2\vartheta)/2] \tag{9-66}$$

The mean square of the fluctuation in relative permittivities can be expressed in terms of the corresponding concentration fluctuations:

$$\overline{(\Delta\epsilon)^2} = \left(\frac{\partial\epsilon}{\partial c}\right)^2 \overline{(\Delta c)^2} \tag{9-67}$$

The mean square of the fluctuation in concentrations is given by the probability p that the individual squares occur:

$$\overline{(\Delta c)^2} \equiv \frac{\int_0^\infty p(\Delta c)^2 d(\Delta c)}{\int_0^\infty p\, d(\Delta c)} \tag{9-68}$$

The probabilities p are obtained from the concentration dependence of the fluctuation in the Gibbs energy ΔG. For fluctuations that are not too large, ΔG can be developed in a Taylor series that is terminated after the second term:

$$\Delta G = \left(\frac{\partial G}{\partial c}\right)_{p,T}(\Delta c) + \frac{1}{2!}\left(\frac{\partial^2 G}{\partial c^2}\right)_{p,T}(\Delta c)^2 + \cdots \tag{9-69}$$

The fluctuations occur at constant temperature and pressure about the equilibrium concentration. It follows, therefore, that $\partial G/\partial c = 0$. The probability p of finding a given value of Δc is thus given from equation (9-69) as

$$p = \exp\left(\frac{-\Delta G}{kT}\right) = \exp\left[\frac{-(\partial^2 G/\partial c^2)(\Delta c)^2}{2kT}\right] \tag{9-70}$$

One obtains the following after inserting equation (9-70) in (9-68) and replacing sums by integrals:

$$\overline{(\Delta c)^2} = \frac{\int_0^\infty \{\exp[-(\partial^2 G/\partial c^2)(\Delta c)^2/2kT]\}(\Delta c)^2 d(\Delta c)}{\int_0^\infty \exp[-(\partial^2 G/\partial c^2)(\Delta c)^2/2kT]d(\Delta c)}$$

$$\tag{9-71}$$

$$= \frac{\int_0^\infty x^2\exp(-ax^2)dx}{\int_0^\infty \exp(-ax^2)dx} = \frac{A}{B}$$

with $x = \Delta c$ and $a = (\partial^2 G/\partial c^2)/2kT$. The solution to both integrals is known:

$A = (1/4a)(\pi/a)^{0.5}$ and $B = (1/2)(\pi/a)^{0.5}$. So, equation (9-71) becomes

$$\overline{(\Delta c)^2} = \frac{kT}{(\partial^2 G/\partial c^2)_{p,T}} \tag{9-72}$$

Further, the following holds:

$$\left(\frac{\partial^2 G}{\partial c^2}\right)_{p,T} = \frac{-\partial\mu_1/\partial c}{V_1^m c q} \tag{9-73}$$

Combination of Equations (9-66), (9-67), (9-72), and (9-73) leads to

$$R_\vartheta = \frac{\pi^2 k T V_1^m c (\partial\epsilon/\partial c)^2}{\lambda_0^4 (-\partial\mu_1/\partial c)} \left(\frac{1 + \cos^2\vartheta}{2}\right) \tag{9-74}$$

With Maxwell's relationship $\epsilon = n^2$, it is possible to replace $\partial\epsilon/\partial c$ by its equivalent $\partial n^2/\partial c$. From Equation (9-42) this then gives $\partial\epsilon/\partial c = 2n_1(dn/dc)$.

The change in chemical potential with concentration is obtained from Equations (6-50) and (6-51) with $c = c_2$:

$$\frac{-\partial\mu_1}{\partial c} = RT\tilde{V}_1^m(A_1 + 2A_2 c + 3A_3 c^2 + \cdots) \tag{9-75}$$

The following is obtained from Equations (9-74), (9-75), and (6-50) with $\tilde{V}_1^m \approx V_1^m$ for the angle $\vartheta = 0$:

$$\frac{4\pi^2 n_1^2 (\partial n/\partial c)^2}{N_L \lambda_0^4} \frac{c}{R_0} = \frac{\kappa c}{R_0} = \frac{1}{M_2} + 2A_2 c + 3A_3 c^2 + \cdots \tag{9-76}$$

With solutions of nonassociating solutes, the apparent molar mass $M_{app} \equiv R_0/\kappa c$, according to Equation (9-76), decreases regularly with increase in concentration c. In the case of polymolecular solutes, M_2 is a mass average [see Equation (9-47)]. The virial coefficients A_2 and A_3 in Equation (9-76) are complex average values, and are only identical with the virial coefficients determined using number-average methods when the solute is unimolecular. In solutions of associating solutes, the right-hand side of Equation (9-76) becomes a complicated expression (see Section 6.5).

9.5.5. Large Particles

All the derivations presented so far related to molecules with small dimensions in comparison to the wavelength λ_0 of incident light. If the dimensions are greater than $\sim(0.1-0.05)\lambda_0$ then the molecule behaves as if it has many scattering centers. The ratio of phases emitted by these centers is fixed because the light is coherent. Interference can therefore occur between the light waves emitted by the different scattering centers. A schematic

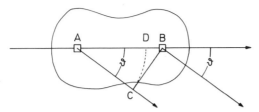

Figure 9-6. Diagrammatic representation of the out-of-phase displacement of the light scattered from two scattering enters A and B of a large particle.

representation of this effect is shown in Figure 9-6. The light waves scattered by the centers A and B with the same angle ϑ at any given instance lead to a path length difference Δ, which depends on the cosine of the scattering angle ϑ:

$$\Delta = \overrightarrow{DB} = \overrightarrow{AB} - \overrightarrow{AD} = \overrightarrow{AB}(1 - \cos \vartheta) \qquad (9\text{-}77)$$

The path length difference is thus zero for $\vartheta = 0$ and increases with increasing ϑ (Figure 9-7). The ratio z of the scattering intensity at two different observation angles ϑ and $180 - \vartheta$ is thus a measure of the interference that occurs. z is called the dissymmetry, and, experimentally, it is usually measured at the angles $45°$ and $135°$. In this case, $z = R_{45}/R_{135}$. The dissymmetry is a measure of the size of the particles, but it also depends on their shape and molar mass distribution, and so a quantitative evaluation of the effect of interference requires additional information or assumptions (Figure 9-8). According to Figure 9-8, however, the influence of molar mass distribution can usually be ignored for coil-shaped macromolecules of not too broad molar mass distribution.

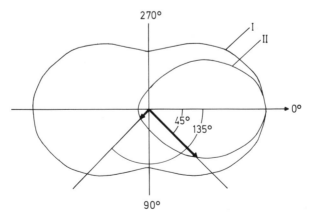

Figure 9-7. A plot of the scattering of unpolarized incident light; (I) small particles, (II) dilute solution of monodisperse spheres with diameter $\lambda_0/2$.

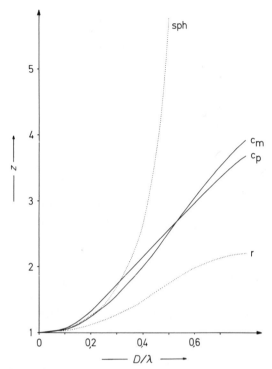

Figure 9-8. Dependence of the dissymmetry coefficient z of scattered light at angles of $45°$ and $135°$ on the ratio of D/λ for spheres (sph), unimolecular random coils (c_m), polymolecular ($\langle M_w \rangle / \langle M_n \rangle = 2$) random coils ($c_p$), and rods ($r$). Here λ_0 is the wavelength of light in the medium of refractive index n, and D corresponds to the diameter of the spheres, the length of the rods, and the chain end-to-end distance $\langle L^2 \rangle^{0.5}$ of coiled macromolecules.

The scattering function $P(\vartheta)$ is more useful. $P(\vartheta)$ is defined as the angular dependence of the scattering intensity of large particles relative to small particles. One can also write $P(\vartheta) = R_\vartheta / R_0$. With Equation (9-77), according to the definition, $P(\vartheta) = 1$ for $\vartheta = 0$. Thus, the equations derived in Sections 9.5.2–9.5.4 are also applicable to large particles when $\vartheta = 0$. The mass average molar mass $\langle M_w \rangle$ for large particles can also be obtained from Equation (9-77) when the light-scattering intensity at $\vartheta = 0$ is extrapolated to zero concentration.

Experimentally, the intensities of scattered light or the Rayleigh ratios R_ϑ are measured at different angles ϑ and then extrapolated to zero angle. The derivation of the correct mathematical expression for the variation of light-scattering intensity with observation angle ϑ for any desired particle is complicated and not given here. The result for unpolarized incident light is

$$P(\vartheta) = 1 - \frac{1}{3}\left(\frac{4\pi}{\lambda_0'}\right)^2 \langle R_G^2 \rangle \sin^2 \frac{\vartheta}{2} + \cdots \qquad (9\text{-}78)$$

Here $\lambda_0' = \lambda_0/n$ is the wavelength of light in the scattering medium. According to Equation (9-78), the mean square radius of gyration $\langle R_G^2 \rangle$ is obtained from $P(\vartheta)$ measurements at small observation angles. Increasingly lower observation angles must be used for increasingly larger particles. Of course, a value of $\langle R_G^2 \rangle$ alone contributes nothing to our knowledge of the shape of the particle. However, since the dissymmetry is affected by both the size and the shape of the particle (Figure 9-8), a comparison of $P(\vartheta)$ [or $\langle R_G^2 \rangle$, which can be calculated from $P(\vartheta)$] with z leads to elucidation of the shape of the particle. If the molar mass and the specific volume are known, the radius of gyration for rigid particles can be calculated (see Section 4.5), and deductions about the shape can be made from a comparison of calculated and observed results.

With the scattering function $P(\vartheta)$, the concentration dependence of the reduced intensities of scattered light is given as

$$\frac{\kappa c}{R_\vartheta} = \frac{1}{\langle M_w \rangle P(\vartheta)} + \frac{2A_2}{Q(\vartheta)} c + \cdots \qquad (9\text{-}79)$$

Here, $Q(\vartheta)$ is another scattering function applicable to finite concentrations c. The second virial coefficient A_2 can thus be found from the dependence on concentration c of the $\kappa c / R_\vartheta$ values as zero angle, and the light scattering function $P(\vartheta)$ (and from it the radius of gyration) can be obtained from the angular dependence of $\kappa c / R_\vartheta$ at zero concentration. Both extrapolations, at $c \to 0$, or $\vartheta \to 0$, yield the mass average molar mass.

According to Zimm, both extrapolations can be carried out in the same plot. In a Zimm plot, $\kappa c / R_\vartheta$ is plotted against $\sin^2(\vartheta/2) + kc$, where k is an arbitrarily chosen constant whose sole purpose is to give a good spread to the grid-shaped plot (see Figure 9-9). Quite often, Zimm plots do not possess the simple grid shape shown in Figure 9-9. In particular, linearity of $\kappa c / R_\vartheta = f(\sin^2 \vartheta/2)$ for $c = 0$ is only to be expected from random coils having a most probable Schulz–Flory distribution ($\langle M_w \rangle / \langle M_n \rangle = 2$).

With a polymolecular sample, $P(\vartheta)$ and, therefore, R_G^2 are mean values. According to Equation (9-79), the mean value of the light scattering function $\bar{P}(\vartheta)$ is $\bar{P}(\vartheta) = R_\vartheta[(\kappa c \langle M_w \rangle)^{-1}]$. If the corresponding expression for the i species is inserted and summed, then we obtain

$$\bar{P}(\vartheta) = \frac{R_\vartheta}{\kappa c \langle M_w \rangle} = \frac{\Sigma_i \kappa c_i M_i P_i(\vartheta)}{\Sigma_i \kappa c_i M_i} = \frac{\Sigma_i c_i M_i P(\vartheta)}{\Sigma_i c_i M_i} = \frac{\Sigma_i m_i M_i P_i(\vartheta)}{\Sigma_i m_i M_i} \qquad (9\text{-}80)$$

and then, with the definition $z_i \equiv m_i M_i$ (see Chapter 8.2), we find

$$\bar{P}(\vartheta) = \frac{\Sigma_i z_i P_i(\vartheta)}{\Sigma_i z_i} \equiv \bar{P}_z(\vartheta) \qquad (9\text{-}81)$$

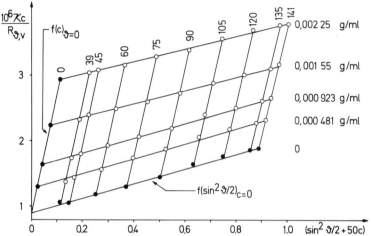

Figure 9-9. The Zimm plot of a poly(vinyl acetate) in butanone at 25°C (1 ml = 1 cm³).

Thus, the light scattering function $\bar{P}(\vartheta)$ and also, via Equation (9-81), the mean square radius of gyration are z averages. The molar masses calculable from the radii of gyration with the aid of a calibration function represent, according to the shape of the particle, different averages: i.e., $\langle M_z \rangle$ for coils in the theta state, $(\langle M_{z+1} \rangle \langle M_z \rangle)^{0.5}$ for rods, etc.

9.5.6. Experimental Procedure

Solutions used for light scattering measurements must be absolutely dust free. Dust particles are large, and thus contribute greatly to the observed light scattering. Dust can be removed through filtration through millipore filters and/or high-speed centrifugation. The presence of dust is usually recognized from the marked deviations observed in the Zimm plot at angles below $\sim 45°$–$60°$.

Commercially available light scattering photometers use mercury lamp light of a definite wavelength, which is selected by the use of color filters. Light rays are made parallel by passing them through lens systems (Figure 9-10). Recently, laser beams have been used. The beam is then incident on the solution-containing cell, which has parallel exit and entry "windows." The intensity is measured with a photomultiplier–photocell arrangement. As can be seen from Figure 9-10, the incident rays "sees" scattering volumes of different size at varying observation angles ϑ. Correction of the observed intensity for this effect is accomplished by normalization of the "seen" volume by multiplying by the sine of the observation angle. For precise measurements, other corrections, which depend on instrument design, must be made, e.g.,

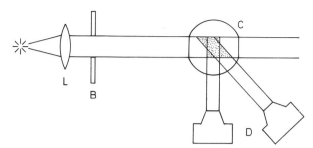

Figure 9-10. Diagram of a light-scattering photometer with light source; lens, L; collimator, B; measuring cell, C; and detector, D (photocell with secondary photomultiplier).

corrections for shape of cell and collimator, for stray light (scattered from the walls), for multiple scattering, etc. Corrections may also be necessary for depolarization and fluorescence. In order to obtain the scattering intensity of the solute, the scattering intensity of the solvent is subtracted from the corrected light scattering intensity of the solution. Here, it is assumed that density and concentration fluctuations occur independently.

Scattering intensities of sufficient magnitude are only observed when the absolute values of the refractive index increments $Y = dn/dc$ are more than $\sim 0.05 \text{ cm}^3/\text{g}$. To a rough approximation, refractive index increments increase linearly with the refractive index n_1 of the solvent, the slope being given by the partial specific volume $\bar{v}_2$ of the solute according to the Gladstone–Dale rule

$$\frac{dn}{dc} = \bar{v}_2 n_2 - \bar{v}_2 n_1 \qquad (9\text{-}82)$$

In general, refractive index increments of polymer solutions rarely exceed $0.2 \text{ cm}^3/\text{g}$. For solutions where $c = 0.01 \text{ g}/\text{cm}^3$, therefore, even in the most favorable circumstances, the difference in the refractive index for solution and solvent is only 0.002 unit. To determine molar masses to $\pm 2\%$, it is necessary to know the refractive index increments to within $\pm 1\%$, since they appear squared in Equation (9-76). The difference in refractive indices must therefore be known to better than $\pm 2 \times 10^{-5}$. Because of temperature fluctuations during individual measurements, the refractive indices of solution and solvent are not measured separately; instead the difference is measured directly in special differential refractometers.

9.6. Small-Angle X-Ray and Neutron Scattering

The theory of light scattering applies to all wavelengths. Consequently, it is also valid for small-angle X-ray scattering (SAXS) and neutron scattering (SANS). The form of Equation (9-76) remains the same in each case. Only the expression for the optical constant κ requires alteration. Whereas light scattering is concerned with the different polarizabilities of the molecules, X-ray scattering and neutron diffraction deal with different electron densities and differing collision cross sections of atoms, respectively. The term κ is given for the various methods as

$$\kappa_{LS} = \frac{4\pi^2 n_1^2 (dn/dc)^2}{N_L \lambda_0^4} \tag{9-83}$$

$$\kappa_{SAXS} = \frac{e^4 (\Delta N_e)^2}{m_e^2 \hat{c}^4 N_L} \tag{9-84}$$

$$\kappa_{SANS} = \frac{N_L N_p^2 (b_H - b_D)^2}{M_u^2} \tag{9-85}$$

Here, e and m_e are the charge and mass of the electron, ΔN_e is the difference between the number of electrons in 1 g of polymer and the number of electrons in the same volume of solvent, $\hat{c}$ is the velocity of light, N_p is the number of exchanged protons per monomeric unit of formula molecular weight M_u, and b_H and b_D are the coherent collision or scattering amplitudes of the hydrogen and the deuterium atoms, respectively. It is assumed in Equation (9-85) that the experiments are carried out with hydrogen-containing polymers in their deuterium-containing analogs.

With light scattering, the wavelength of the incident light is greater than the molecular dimensions. For X-ray scattering, the wavelength is smaller. According to Equation (9-78), the light-scattering intensity of a given particle at the angle ϑ is reduced by the factor $[\sin^2(\vartheta/2)]/(\lambda')^2$. Equation (9-78) is also valid for X-ray scattering. Consequently, the light scattering intensity reduction for an incident wavelength of $\lambda_0 = 436$ nm and a solvent with $n_1 = 1.45$ (i.e., $\lambda' = 436/1.45 = 300$ nm) at an angle of $90°$ would be the same as that observed with X-ray scattering ($\lambda_0 = 0.154$ nm) at $\vartheta = 0.03°$.

According to Guinier, the scattering function for small-angle X-ray scattering measurements can be approximated by

$$P(s) = \exp\left(-\frac{4\pi^2}{3\lambda_0^2} \langle R_e^2 \rangle \vartheta^2\right) \tag{9-86}$$

$\langle R_e^2 \rangle$ is the mean square radius of gyration of the distribution of electrons and not of the weight. Like the scattering function $P(\vartheta)$, $P(s)$ is also normalized to a value of unity at the angle $0°$. The scattering function is Gaussian in form. Experimentally, it is often only applicable at $\vartheta \to 0$, with large deviations occurring at greater angles. In contrast to light scattering, large associates do not influence the accuracy of small-angle X-ray scattering, since they only cause extremely small-angle scattering. Small-angle X-ray scattering can be used to determine the radius of gyration of molecules down to a molar mass of about 300 g/mol.

9.7. Ultracentrifugation

9.7.1. Phenomena and Methods

Dissolved particles of density ρ_2 travel through a solvent of density ρ_1 under the influence of a centrifugal field. They sediment in the direction of the centrifugal field when $\rho_2 > \rho_1$, and move to the center of rotation when $\rho_2 < \rho_1$. Under otherwise constant conditions, the rate of sedimentation (or flotation) depends on the mass and shape of the particles as well as on the solution viscosity. Therefore all of these quantities can, in theory, be determined from the rate of sedimentation.

Sedimentation works against diffusion caused by Brownian motion. With a sufficiently weak centrifugal field (relative to particle and density differences), a stage will be reached where the rate of sedimentation equals the rate of diffusion, and a state of equilibrium sedimentation occurs. For given experimental conditions, the sedimentation equilibrium depends on the mass of the dissolved molecule and is therefore a method for determining molecular weights.

Sedimentation rate and sedimentation equilibrium experiments are carried out on an instrument first developed by Th. Svedberg (see Figure 9-11). Such ultracentrifuges reach speeds of ~70 000 revolutions per minute, which corresponds to a gravitational field of ~350 000g (g is the gravitational field due to the earth). Rotational velocities v_r (in revolutions per minute) can be converted into angular velocities ω (in radians per second) with the relation $\omega = 2\pi v_r/60$. Solutions under study are placed in special cells having quartz or sapphire windows, and these fit into a rotor (Figure 9-11, 4) made of duraluminum or titanium. The rotor is driven by an electric motor via a gear system (6). The rotor speed is continuously compared with a reference speed of a synchronized motor with differential gear drive (15). In this way, the rotor speed is kept constant. For protection against accidental damage, the rotor is suspended in a steel chamber (3) by a thin steel axis. In order to keep frictional heating to a minimum, the steel chamber is evacuated to ~10^{-6} bar by means of a rotational (14) and an oil-diffusion (16) vacuum pump. With the aid of a cooling system (13) and a heating system (not shown), the temperature of the rotor is regulated at ~$\pm 0.1°$C. Changes occurring in the cells can be monitored by an optical system whereby light from a source (1) passes through lenses (2, 5, 9, 10) and is then registered directly on photographic

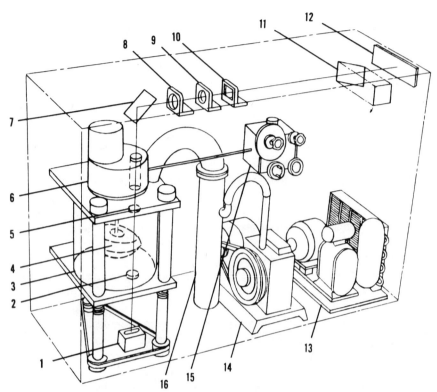

Figure 9-11. Diagram of an analytical ultracentrifuge. The numbers are identified in the text.
(Spinco ultracentrifuge made by Beckman Instruments.)

plates (12), observed directly after being reflected by a mirror (11) in the required direction. Three optical monitoring systems are in use: interference optics, Schlieren optics, and optical absorption. In interference optics the number or displacement of interference lines is measured. The number of interference lines is proportional to the difference in refractive indices and, thereby, the difference in concentrations. In Schlieren optics, the change in concentration c with distance r is optically differentiated with the aid of a special optical system, so that the concentration gradient dc/dr is observed as a function of distance r. The optical absorption of visible or uv light is measured and registered in absorption optics. In the newer absorption optics systems, the optical absorption is measured directly at each point by a photoelectric cell, thus obviating the need for the tedious intermediate use of a photographic plate.

Experiments are carried out in sector cells in order to avoid convection currents during sedimentation. If, for example, the walls of the cell are not arranged radially to the rotational center, moving particles collide with the

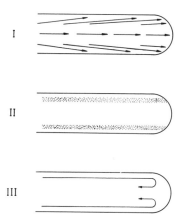

Figure 9-12. Schematic representation of the formation of convection currents in parallel-sided ultracentrifuge cells at the times I, II, and III.

wall (Figure 9-12, I), where they are reflected and form a layer of higher concentration close to the wall (II). Such a concentration distribution leads to a radial convection currents (III).

At time t_0, the cell is filled with a solution that is homogeneous with respect to concentration (Figure 9-13). All the molecules begin to move under the influence of gravity. After the time t, there is a layer of pure solvent at the meniscus (m) and molecules settle at the bottom (b). The boundary layer between solvent and sedimenting solution is not sharp, because of back-diffusion. Thus, a curve $c = f(r)$ is obtained instead of a concentration jump. The sectoring of the cell causes a dilution effect. With increasing duration of the experiment, the concentration in the constant-concentration zone of the cell becomes increasingly smaller. A gradient curve is obtained on differentiation (third row in Figure 9-13). The rate of displacement of the gradient curve is a measure of the sedimentation velocity.

9.7.2. Basic Equations

At every point of the cell a flow of $J = cv$ (i.e., the product of concentration c and molecular mobility v) occurs during sedimentation. The amount of solute that flows from a volume element A at distance r_A from the rotational center into a volume element B at distance r_B must be the same as the change with time of the concentration of the rest of the solute:

$$(rJ)_A - (rJ)_B = \frac{\partial}{\partial t} \int_{r_A}^{r_B} rc \, dr \qquad (9-87)$$

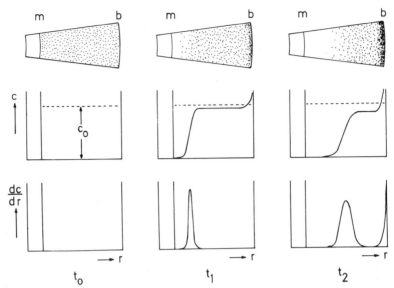

Figure 9-13. Schematic representation of the course of sedimentation in sectored cells at the times t_0, t_1, and t_2.

On dividing both sides by $\Delta r = r_B - r_A$ and using the limit for $\Delta r \to 0$, we obtain the following:

$$\left(\frac{\partial c}{\partial t}\right) = -\frac{1}{r}\left[\frac{\partial (rJ)}{\partial r}\right]_t \qquad (9\text{-}88)$$

Sedimentation causes a flow J_s in the direction of the centrifugal field. The flow J_d due to diffusion attempts to maintain concentrational homogeneity, and works against J_s. For the resulting flow, we have

$$J = J_s + J_d = cv_s + cv_d \qquad (9\text{-}89)$$

The flow due to diffusion J_d is given by Equation (7-8):

$$J_d = -D\,\frac{\partial c}{\partial r} \qquad (9\text{-}90)$$

The molecular sedimentation velocity v_s is proportional to the centrifugal field $\omega^2 r$; the proportionality coefficient is called the sedimentation coefficient s:

$$v_s = s\omega^2 r \qquad (9\text{-}91)$$

The term s is defined here as the sedimentation velocity in a unit field:

$$s \equiv \frac{1}{\omega^2 r}\frac{dr}{dt} \qquad (9\text{-}92)$$

A sedimentation coefficient of the value 1×10^{-13} s is called a Svedberg unit (1 S).

Combination of Equations (9-88)–(9-91) gives what is called the Lamm differential ultracentrifuge equation:

$$\left(\frac{\partial c}{\partial t}\right)_r = \frac{-\partial[s\omega^2 r^2 c - rD(\partial c/\partial r)]}{r\,\partial r} \tag{9-93}$$

9.7.3. Sedimentation Velocity

For sedimentation velocity experiments, angular velocities ω are chosen to be so high that the diffusion term $rD(\partial c/\partial r)$ in Equation (9-93) is much smaller than the sedimentation term $s\omega^2 r^2 c$. A forced migration of 1 mol of molecules with the velocity dr/dt produces a resistance F_s:

$$F_s = f_s N_L \frac{dr}{dt} \tag{9-94}$$

The proportionality constant f_s is called the frictional coefficient. Additionally, an effective centrifugal force F_r acts on the molecule of hydrodynamic volume V_h. The force F_r is the resultant of the centrifugal field force $m_h\omega^2 r$ and the buoyancy $V_h\rho_1\omega^2 r$ due to the solvent

$$F_r = m_h\omega^2 r - V_h\rho_1\omega^2 r \tag{9-95}$$

If $F_s = F_r$ and Equations (7-1) and (7-6) are inserted for m_h and V_h, respectively, then with $\rho_1 = 1/v_1$ and Equation (9-94), we have

$$M_2 = \frac{f_s s N_L}{1 - \bar{v}_2\rho_1} \tag{9-96}$$

Thus, s values alone are not a measure of the molar mass, since the molar mass also depends on the frictional coefficient f_s and the buoyancy term $(1 - \bar{v}_2\rho_1)$ (see Table 9-2). Frictional coefficients are determined by the shape and degree of solvation of the particles.

The frictional coefficient can be eliminated by the following procedure. If frictional coefficients are equal for both sedimentation and diffusion, as is observed experimentally, then the Svedberg equation is obtained from the combination of Equation (9-96) with Equation (7-17):

$$M_2 = \frac{sRT}{D(1 - \bar{v}_2\rho_1)} \tag{9-97}$$

Another means of eliminating the frictional coefficient f_s comes from viscosity measurements. According to Equation (7-19), the frictional coefficient f_D is related to an asymmetry factor f_A and the Stokes frictional

Table 9-2. Sedimentation Coefficient s and Frictional Coefficient f_s (as the ratio f_s/f_{sphere}) of Macromolecule with Molar Mass M_2

Substance	$10^{-4} M_2$ in g/mol	Solvent	Temperature in °C	$10^{13} s$ in s	f_s/f_{sphere}
Poly(styrene)	9	Butanone	20	12	1.38
Poly(styrene)	96	Butanone	20	22	3.75
Poly(styrene)	500	Butanone	20	45	5.24
Poly(vinyl alcohol)	6.5	Water	25	1.54	3.5
Cellulose	590	Cuoxam	20	17.5	13.1
Ribonuclease	1.27	Dilute salt solution	20	1.85	1.04
Myoglobin	1.67	Dilute salt solution	20	2.04	1.11
Tobacco mosaic virus	5900	Dilute salt solution	20	17.4	2.9

coefficient for spheres. On extending Equation (7-19) with $M_2^{0.5}$, we obtain the form

$$f_s = f_D = f_A 6\pi\eta_1 \left(\frac{\langle R_G^2 \rangle}{M_2} \right)^{0.5} M_2^{0.5} \tag{9-98}$$

which is similar to the expression (discussed in Section 9.9.6) for the intrinsic viscosity $[\eta]$:

$$[\eta] = \Phi \left(\frac{\langle R_G^2 \rangle}{M_2} \right)^{3/2} M_2^{0.5} \tag{9-99}$$

The Mandelkern–Flory–Scheraga equation is obtained by combining Equations (9-96), (9-98), and (9-99):

$$M_2 = \left[\frac{N_L \eta_1}{\Phi^{1/3}(6\pi f_A)^{-1}(1 - \tilde{v}_2 \rho_1)} \right]^{3/2} [\eta]^{1/2} s^{3/2} \tag{9-100}$$

The following symbols are very often used in the literature:

$$P = 6\pi f_A \tag{9-101}$$

$$\beta = \Phi^{1/3} P^{-1} \tag{9-102}$$

The factor f_A describes, first, the relationship between the radius of gyration and the radius most suitable in describing the molecule, and, second, all deviations from the Stokes frictional coefficient of an unsolvated sphere. The relationship $\langle R_G^2 \rangle = (3/5)r^2$ is valid for spheres. Thus, $f_A = (5/3)^{0.5}$ for unsolvated spheres. Consequently, P takes on a value of 24.34. The Einstein equation allows Φ to be calculated for spheres:

$$[\eta] = \frac{2.5}{\rho_2} = \frac{2.5 V_2}{m_2} = \frac{2.5(4\pi r^3/3)}{M_2/N_L} = \frac{2.5 \cdot 4\pi N_L (5/3)^{3/2}}{3} \left(\frac{R_G^3}{M_2}\right) = \Phi \, \frac{R_G^3}{M_2}$$

(9-103)

Thus, Φ is given as 13.57×10^{24} (mol macromolecule)$^{-1}$ and is based on the radius of gyration. $[\eta]$ is measured in cm^3/g. If Φ is based on the radius of gyration, then $\Phi = 6.30 \times 10^{22}$ is obtained with $[\eta]$ measured in 100 cm^3/g. Other numerical values for P and β are obtained with other molecular shapes (see Table 9-3).

Both Equations (9-97) and (9-100) yield mixed averages of the molar mass (see Section 8.5.4).

In these derivations, it was implicitly assumed that s and D are independent of the concentration. Equation (9-97), therefore, only applies at infinite dilution. However, sedimentation and diffusion coefficients are measured at finite concentrations, and so must be extrapolated to $c \to 0$. The extrapolation formula for sedimentation coefficients is obtained from the reasoning that, according to Equation (9-96), the s values, and so also the s_c values, are inversely proportional to the frictional coefficients f_s. Since f_s is proportional to the viscosity η, and this in turn is proportional to the concentration, $1/s_c$ must be directly proportional to the concentration. This dependence is usually formulated as

$$\frac{1}{s_c} = \frac{1}{s}(1 + k_s c)$$

(9-104)

Since the sedimentation velocity depends on the molar mass, for paucimolecular substances with two, three, etc. kinds of molecules but different molar masses, two, three, etc. gradient curves are observed in the

Table 9-3. *Calculated Constants* Φ, P, *and* β^a

Molecular shape	r_a/r_b	$10^{22}\Phi$ in 100/mol	P	$10^6\beta$
Spheres, unsolvated	1	13.57	24.34	2.11
Ellipsoids, oblate	1	—	—	2.12
	2	—	—	2.13
	3–300	—	—	$1.81(r_a/r_b)^{0.126}$
Ellipsoids, prolate	1–0.067	—	—	2.12–2.14
	0.05–0.0033	—	—	2.15
Coils, theta state	—	4.21	5.20	2.73
Semirigid chains	—	—	—	2.81

$^a\Phi$ is given relative to the radius of gyration, whereas β is relative to the chain end-to-end distance. r_a is the axial rotation axis radius and r_b is the equatorial radius of the ellipsoid. $[\eta]$ is in 100 cm^3/g.

sedimentation plot. Sedimentation measurements are therefore used very frequently in protein chemistry to test the homogeneity of materials. In polydisperse substances, on the other hand, the distribution of sedimentation coefficients can be determined from the time broadening of the gradient curves. For this, gradient curves are obtained at various times for a given initial concentration, and the sedimentation coefficients corresponding to masses of 5%, 10%, 20%, . . . , 80%, 90%, 95% materials are determined. Since the gradient curves can also be broadened by diffusion, however, the sedimentation curves thus obtained are extrapolated to infinite time. At infinite time, only the different sedimentation velocities play a role, and back-diffusion is no longer important. A function $w_i = f(s_i)$ results which is converted into the molar mass distribution $w_i = f(M_i)$ by means of the empirical relation $s = K_s M^{a_s}$. Such measurements are best made in Θ solvents; otherwise too many corrections are required for thermodynamic nonideality, etc.

9.7.4. Equilibrium Sedimentation

In equilibrium sedimentation, the concentration at every point in the ultracentrifuge cell no longer varies with time, i.e., $(\partial c/\partial t)_r = 0$. Equation (9-93) therefore becomes

$$\frac{s}{D} = \frac{\partial c/\partial r}{\omega^2 rc} \tag{9-105}$$

Combining Equations (9-105) and (9-97) gives the following for the molar mass at equilibrium sedimentation:

$$M_2 = \frac{RT}{\omega^2(1 - \tilde{v}_2\rho_1)} \frac{dc/dr}{rc} \tag{9-106}$$

The molar mass M_2 can be determined from Equation (9-106) if the concentrations c and the concentration gradients dc/dr at a distance r from the rotational center are known. With polymolecular substances, this molar mass M_2 is a mass-average $\langle M_w \rangle$. That is, after transforming Equation (9-106), the mean concentration gradient is given by

$$\overline{\left(\frac{dc}{dr}\right)} = \frac{\omega^2 r(1 - \tilde{v}_2\rho_1)}{RT} \sum_i c_i M_i \tag{9-107}$$

and with the definition of the mass average molar mass $\langle M_w \rangle \equiv \Sigma_i c_i M_i / \Sigma_i c_i$ and $\Sigma_i c_i = c$, we have

$$\langle M_w \rangle = \frac{RT}{\omega^2(1 - \tilde{v}_2\rho_1)} \frac{\overline{dc/dr}}{rc} \tag{9-108}$$

Equation (9-108) is, strictly speaking, only applicable at infinite dilution. At finite concentrations, it yields an apparent molar mass $(M_w)_{app}$, which is extrapolated in the usual manner by plotting $1/(M_w)_{app} = f(c)$ to $c \to 0$.

9.7.5. Sedimentation Equilibrium in a Density Gradient

Up to now, it has been assumed that the solvent in the sedimentation experiment consists of a single component. If, however, the solvent system consists of a mixture of two substances of widely different densities (e.g., CsCl in water or mixtures of benzene and CBr_4), the solvent components will sediment to different extents. At equilibrium, the solvent system possesses a density gradient. One density ρ_b, applies at the bottom of the cell, and another, ρ_m, at the meniscus. The density of the solute, ρ_2, should lie between these two densities $(\rho_m < \rho_2 < \rho_b)$. The macromolecules will then sediment from the meniscus toward the base of the cell and float from the base toward the meniscus (Figure 9-14). At equilibrium, the macromolecules will take up a position (designated by §) at which the density ρ_g exactly corresponds to the density of the macromolecule in solution ($\rho_g = \rho_2 \approx 1/v_2^\S$). This position is at a distance $r^\S$ from the center of rotation.

Quantitative evaluation begins with the equilibrium sedimentation equation (9-106) in the form for $r \approx r^\S$:

$$\frac{d \ln c}{dr} = M_2^\S \omega^2 r^\S \frac{1 - \tilde{v}_2 \rho}{RT} \tag{9-109}$$

where ρ is now the density of the variable-density system at any given point r.

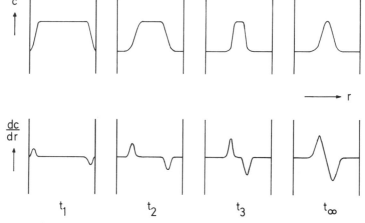

Figure 9-14. Diagrammatic representation of the establishment of a sedimentation equilibrium in a density gradient (c is the concentration of the macromolecular solute).

Reference is made, not to the distance from the rotational center, but to the distance from point r, so that dr is replaced by $d(r - r^\S)$. Equation (9-109) then becomes

$$\frac{d \ln c}{d(r - r^\S)} = M_2^\S \omega^2 r^\S \frac{1 - (\rho/\rho^\S)}{RT} \qquad (9\text{-}110)$$

To a first approximation, the density in the variable-density system varies in the neighborhood of r as

$$\rho = \rho^\S + \left(\frac{d\rho}{dr}\right)^\S (r - r^\S) \qquad (9\text{-}111)$$

Since the concentrations of the components forming the density gradient are much higher than the macromolecular concentration, the density gradient is practically unaffected by the presence of the macromolecules. Combination of (9-110) and (9-111) leads to

$$\frac{d \ln c}{d(r - r^\S)} = - \frac{M_2^\S \omega^2 r^\S (d\rho/dr)^\S (r - r^\S)}{RT\rho^\S} \qquad (9\text{-}112)$$

or, on integration,

$$\ln \frac{c}{c^\S} = - \frac{M_2^\S \omega^2 r^\S (d\rho/dr)^\S (r - r^\S)^2}{2RT\rho^\S} \qquad (9\text{-}113)$$

or, on transforming,

$$c = c^\S \exp \left[- \frac{M_2^\S \omega^2 r^\S (d\rho/dr)^\S (r - r^\S)^2}{2RT\rho^\S} \right] = \exp \left[- \frac{(r - r^\S)^2}{2\sigma^2} \right] \qquad (9\text{-}114)$$

where we have

$$\sigma^2 = \frac{RT\rho^\S}{M_2^\S \omega^2 r^\S (d\rho/dr)^\S} \qquad (9\text{-}115)$$

Equation (9-114) corresponds to a Gaussian distribution function (see also Section 8.3.2.1). The molar mass can be calculated from the position of the inflection point of the function $c = f(r - r^\S)$. For proteins in $CsCl/H_2O$, the lower molar mass limit giving a meaningful measurement is 10 000–50 000 g/mol molecule. The limit is essentially governed by the length of the ultracentrifuge cell (~ 1.2 cm) and the optimal values of $r - r^\S$ for this length.

The molar mass given in Equation (9-115), however, is not the molar mass of the unsolvated molecule, which appears in equations (9-106) and (9-107). The measurements are, of course, made in mixed solvents, where one component solvates the molecule more than the other. If only component 1 solvates, then the mass of the solvated macromolecule is composed of the mass of the "dry" macromolecules m_2 and the mass of the solvating solvent $m_1^{\square}$. With the definition $\Gamma_1 = m_1^{\square}/m_2$, it follows from $m_2^{\S} = m_2 + m_1^{\square}$ that $m_2 = m_2^{\S}(1 + \Gamma_1)$, and on recalculating on the basis of molar mass, with $M_2 = m_2 N_L$, we have

$$M_2^{\S} = M_2(1 + \Gamma_1) \tag{9-116}$$

The parameter Γ_1 can be evaluated from the partial specific volumes of the solute $\tilde{v}_2$ or solvent $\tilde{v}_1$ and the density $\rho^{\S}$ via

$$\Gamma_1 = \frac{\tilde{v}_2 \rho^{\S} - 1}{1 - \tilde{v}_1 \rho^{\S}} \tag{9-117}$$

To a good approximation, $\tilde{v}_2$ can be determined from density measurements for the macromolecule in one-component solvents by using $\rho = \rho_1 + (1 - \tilde{v}_2 \rho_1)c$.

Equilibrium sedimentation studies are mostly used to investigate density differences in different macromolecules. The method has been used, for example, in research on the replication of ^{15}N-labeled deoxyribonucleic acids. Theoretically, it is also suitable for distinguishing between polymer blends and true copolymers. In such studies, problems usually arise from the considerable back-diffusion and the wide molar mass distribution. The gradient curves are so strongly influenced by these two effects that there is considerable overlap in the curves for substances of differing densities.

9.7.6. Preparative Ultracentrifugation

Substances of differing molar masses can be separated in sedimentation experiments. The method is the preferred method in protein and nucleic acid chemistry, since it is especially suitable for the separation of compact molecules. Three procedures can be distinguished: In the normal procedure, the particles sediment in pure solvent or in relatively dilute salt solutions. The density of the solvent or the salt solution is practically constant over the whole cell (Figure 9-15). The more rapidly moving particles tend to collect at the bottom. However, they are always contaminated with the more slowly sedimenting particles. In contrast, slower moving components cannot be isolated quantitatively.

Band ultracentrifugation is also a sedimentation velocity method, but here the particles sediment in a mixed solvent system (e.g., a salt solution). First a density gradient is produced by ultracentrifugation of the solvent system alone. Then the solute solution is added at the cell meniscus. The individual components sediment in bands which are no longer contaminated by the other components. The relatively weak density gradient is solely

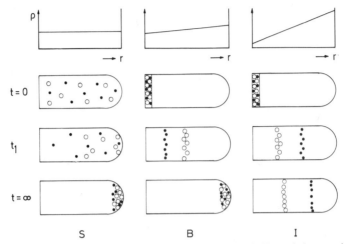

Figure 9-15. Kinds of preparative ultracentrifugation methods. (S) Normal ultracentrifugation, (B) band ultracentrifugation in a stabilized gradient, (I) isopycnic zone centrifugation; ρ is the density of the gradient-forming substance; (O) high molar mass, low density; (●) low molar mass, high density.

intended to stabilize the moving bands. As soon as the bands have formed, the sedimentation experiment is terminated. The isolated components are of high purity.

Isopycnic zone ultracentrifugation, on the other hand, deals with the preparative analog to equilibrium sedimentation in a density gradient. Here, large density gradients are used and one waits until equilibrium is attained. Both band and isopycnic zone ultracentrifugation can be carried out with very small solute concentrations. These methods have become, therefore, particularly important in the separation of biological macromolecules (biopolymers). Both methods are also described in the literature under a variety of other names.

9.8. Chromatography

Macromolecules can be fractionated according to their constitution, configuration, or molar mass by chromatographic methods. Adsorption chromatography is used rarely. Elution chromatography and gelpermeation chromatography are used much more often.

9.8.1. Elution Chromatography

The material to be separated is placed as a thin layer on an inert carrier and then eluted. Metal foil or quartz sand, for example, are suitable inert carriers. The metal foil is dipped into the macromolecular solution and then dried. The surface film is then eluted at constant temperature with solvent–precipitant mixtures of increasing solvent power. Thus, the lower molar mass fractions are removed first.

An elegant variant of this procedure is known as the Baker–Williams method. In this method, the chromatographic column is also surrounded by a thermostated heating jacket which ensures that an adequate thermal gradient is maintained. The separation efficiency is enhanced by the simultaneous concentration and temperature gradients.

9.8.2. Gel-Permeation Chromatography

In gel-permeation chromatography, the separating column consists of a solvent-swollen gel with pores of various diameters. An $\sim 0.5\%$ solution is added to the column, which is then eluted with a steady stream of solvent. In a homologous series of molecules of similar shape, molecules with the highest molar mass appear first, i.e., they have the smallest elution volume. The effect is interpreted as follows: Large molecules cannot easily, if at all, penetrate the pores of the gel material, so they have the shortest retention time (Figure 9-16). The method is known in the literature under numerous other names (gel filtration, gel chromatography, exclusion chromatography, molecular-sieve chromatography, etc.) It is an improved form of liquid–liquid chromatography.

Elution takes place under pressures of up to ~ 10 bar. The gel used must therefore not be compressed under these conditions. Cross-linked poly-(styrenes) or porous glass are usually used for organic solvent systems, and cross-linked dextrans, poly(acrylamides), or celluloses for aqueous solutions. The concentration of the eluting solutions is mostly registered automatically as a function of the volume, for example, by refractive index or by spectroscopy (Figure 9-17). The maximum in the $n_{sol} = f(V_e)$ plot is known as the elution volume. An empirical relationship exists between the elution volume V_e and the molar masses for substances of similar molecular shape and similar interaction with the solvent (Figure 9-18). For both low and high molar mass, however, elution volumes are independent of the molar mass. The exclusion limits depend on the gel material as well on the dissolved macromolecules.

The elution volumes of a homologous polymer series depend on the molar mass because the hydrodynamic volumes of macromolecules vary regularly with molecular weight. The hydrodynamic volume is proportional to the product $[\eta] M$. A "universal" calibration curve of $\log([\eta] M) = f(V_e)$ for various linear and branched polymers actually does exist for each gel material (Figure 9-19).

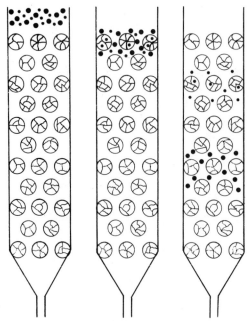

Figure 9-16. Schematic representation of the separation of molecules of different sizes on macroporous gels by gel-permeation chromatography.

The molar mass associated with the elution volume is probably represented by the following average:

$$\langle M_{\text{GPC}} \rangle = \frac{\sum_i m_i M_i^{1 + a_\eta}}{\sum_i m_i M_i^{a_\eta}} \qquad (9\text{-}118)$$

This average gives, for a Schulz–Flory distribution of degree of coupling k,

$$\langle M_{\text{GPC}} \rangle = \frac{\langle M_w \rangle (k + a_\eta + 1)}{k + 1} = \frac{\langle M_n \rangle (k + a_\eta + 1)}{k} \qquad (9\text{-}119)$$

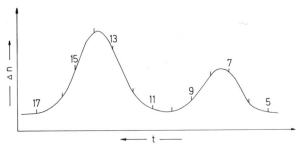

Figure 9-17. A GPC diagram. The numbers give the fraction numbers, which are proportional to the eluted volume. The refractive index difference Δn is generally measured as a function of time.

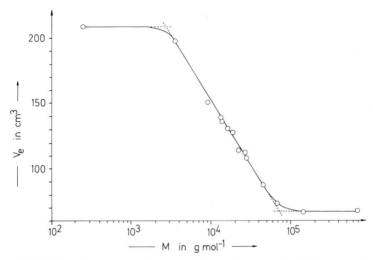

Figure 9-18. Elution volume V_e as a function of the molar mass M of dilute salt solutions of saccharose and various proteins on cross-linked dextrane (Sephadex G-75) (according to P. Andrews).

Consequently, $\langle M_{GPC} \rangle = \langle M_w \rangle$ for spheres, where $a_\eta = 0$, and $\langle M_{GPC} \rangle = \langle M_{z+1} \rangle$ for rigid rods. An average close to the weight average is obtained for coil-shaped molecules with the usual distributions when the experiments are run in theta solutions.

The width of the elution curve increases with the width of the distribution. But unimolecular substances do not give a sharp signal; an elution curve is also obtained. This is produced by what is known as the axial dispersion. According to the model described above, it signifies a distribution of residence times in the pores. The effect must also be taken into account when calculating molar mass distributions. For this, it is assumed that the total standard deviation σ_{tot} is composed of the standard deviation associated with molecular inhomogeneity σ_{mol} and the standard deviation resulting from axial dispersion σ_{ad}:

$$\sigma_{tot}^2 = \sigma_{mol}^2 + \sigma_{ad}^2 \qquad (9\text{-}120)$$

σ_{ad} is determined by reversing the flow of the liquid after development of the elution curve. The σ_{mol} can be calculated, and the true elution curve can be subsequently constructed.

9.8.3. Adsorption Chromatography

Adsorption chromatography is based on the variation in strong interactions between adsorbent and adsorbate. It is suitable for effecting a separation

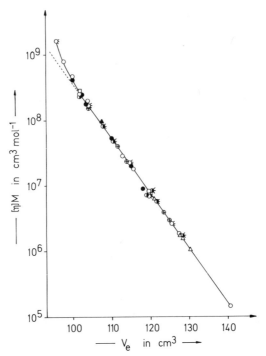

Figure 9-19. A "universal" gel-permeation chromatography calibration curve obtained from measurements on linear poly(styrene) (O), comb-branched poly(styrene) (OE), star-branched poly(styrene) (⊕), poly(methyl methacrylate) (●), poly(vinyl chloride) (△), *cis*-1,4-poly-(butadiene) (▲), poly(styrene)–poly(methyl methacrylate) block copolymer (OE), random copolymer from styrene and methyl methacrylate (◑), and ladder polymers of poly(phenyl siloxanes) (□) (according to Z. Grubisic, P. Rempp, and H. Benoit).

according to differences in constitution and configuration. In real systems, however, the adsorption–desorption equilibrium is submerged beneath a series of other effects.

Four regions can be distinguished according to the proportion of precipitant in the developer in the thin-layer chromatography of polymers (Figure 9-20). Adsorption predominates at low precipitant contents. The R_f values increase with increasing precipitant content until the desorption region is reached, when they become independent of the precipitant content. At still higher precipitant contents, a phase separation first starts, followed by precipitation of the polymer. Separation according to constitution and configuration occurs in the adsorption–desorption region, and separation according to molecular weight occurs in the precipitation region. A molecular-sieve effect from the moderately coarse-pored materials is superimposed on these effects.

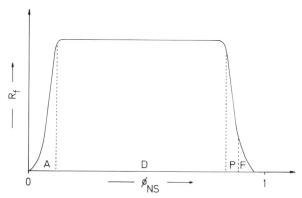

Figure 9-20. Diagrammatic representation of R_f values as a function of the volume fraction ϕ_{NS} of nonsolvent in thin-layer chromatography. A, adsorption; D, desorption; P, phase separation; F, precipitation.

9.9. Viscometry

9.9.1. Basic Principles

A relationship is shown to exist in viscometry experiments between particle size or molecular size and the viscosity of dispersions of inorganic colloids or the viscosity of macromolecular solutions. It is therefore possible to determine the molar mass from the viscosity of dilute macromolecular solutions. Since this experiment can be rapidly performed with simple equipment, it is, in practice, the most important molar mass determination method. However, the method is not an absolute one, since the viscosity depends on other molecular properties (for example, on the shape of the molecule), as well as on the molecular weight.

As early as 1906, in fact, Einstein derived a relationship between the viscosity η of solutions of unsolvated spheres, the volume fraction ϕ_2 of the spheres, and the viscosity of the pure solvent η_1:

$$\eta_{sp} \equiv \frac{\eta}{\eta_1} - 1 = 2.5\phi_2 \qquad (\phi_2 \to 0) \tag{9-121}$$

η_{sp} is called the specific viscosity and $\eta/\eta_1 = \eta_{rel}$ is called the relative viscosity. The constant 2.5, results from hydrodynamic calculations. Equation (9-121) only applies in the absence of interactions between the components of the solution, i.e., in infinitely dilute solutions. The influence of finite content can be allowed for by a series expansion in powers of the volume fraction, as was found experimentally with measurements on dispersions of glass spheres or gutta percha:

$$\eta_{sp} = 2.5\phi_2 + \alpha\phi_2^2 + \beta\phi_2^3 + \cdots, \qquad \alpha, \beta = \text{constants} \qquad (9\text{-}122)$$

Equation (9-122) can be generalized to other particles besides unsolvated spheres, e.g., to coils or rods. The volume fraction of the solute is defined by $\phi_2 \equiv V_2/V$. The volume V_2 of all the solute molecules in a solution of volume $V = V_{soln}$ (ml) is related to the hydrodynamically effective volume V_h of the individual molecule by the number N_2 of solute molecules: $V_2 = N_2 V_h$. The molar concentration $[M_2]$ of solute molecules (in mol/dm^3) is related to N_2 by $[M_2] = 10^3 N_2 / N_L V_{soln}$ and to the solute concentration c_2 (in g/cm^3) by $[M_2] = 10^3 c_2 / M_2$, where M_2 is the molar mass of the solute. If these relationships are inserted into Equation (9-121), this gives on rearrangement

$$\frac{\eta_{sp}}{c_2} = 2.5 N_L \frac{V_h}{M_2} \qquad (c_2 \to 0) \qquad (9\text{-}123)$$

where the limiting value

$$\lim_{c_2 \to 0} \frac{\eta_{sp}}{c_2} \equiv [\eta] \qquad (9\text{-}124)$$

is called the intrinsic viscosity $[\eta]$. According to Equation (9-123), $[\eta]$ depends on the molar mass M_2 and the hydrodynamic volume V_h of the solute. V_h itself is a function of the mass, shape, and density of the solute molecules. The intrinsic viscosities of material consisting of flexible macromolecules can differ by up to a factor of 5, depending on the degree of interaction between the material and its solvent (Table 9-4).

The term $[\eta]$ has the physical units of reciprocal concentration, and so those of specific volume. In the current literature the dimensions are usually quoted as cm^3/g. Older literature uses the dimensions of dl/g or $liter/g$ (the latter with the special symbol Z_η), so the numerical values are 100 or 1000

Table 9-4. *Influence of Solvent on Intrinsic Viscosity* $[\eta]$
of Poly(isobutylene) (PIB), Poly(styrene) (PS), and
Poly(methyl methacrylate) (PMMA) at 34°C

| Solvent | $[\eta]$ in cm^3/g | | |
	PIB	PS	PMMA
Cyclohexane	478	44	Nonsolvent
CCl₄	462	100	305
n-Hexane	327	Nonsolvent	Nonsolvent
Chlorobenzene	250	107	—
Toluene	247	—	403
Benzene	119	114	640
Butyl acetate	Nonsolvent	—	195

Table 9-5. Intrinsic Viscosity $[\eta]$ of Poly(styrene) and Poly(methyl methacrylate) Mixtures in CCl_4 at $25°\,C$

Mass fractions[a]		$[\eta]$ in cm^3/g	
w_{PS}	w_{PMMA}	Calculated[b]	Observed
1.0	0.0	—	120.0
0.7	0.3	90.8	102.0
0.5	0.5	76.4	84.0
0.3	0.7	58.9	65.0
0.1	0.9	41.4	44.0
0.0	1.0	—	32.7

[a]PS, Poly(styrene); PMMA, poly(methyl methacrylate).
[b]Calculations were done according to Equation (9-125).

times smaller than when the units are in cm^3/g. In solutions of low-molar-mass material, $[\eta]$ can in some cases have negative values, namely, if the solution viscosity (and thus that of the solute) is lower than the solvent viscosity.

In polymer mixtures where there is no special interaction between the polymers, e.g., in a polymer homologous series, the intrinsic viscosity is obtained as a mass average (Philippoff equation):

$$\langle\,[\eta]_w\,\rangle = \frac{\Sigma_i W_i[\eta]_i}{\Sigma_i W_i} = \frac{\Sigma_i c_i[\eta]_i}{\Sigma_i c_i} = \frac{\Sigma_i c_i[\eta]_i}{c} = \sum_i w_i[\eta]_i = [\eta] \quad (9\text{-}125)$$

Because of the attraction and repulsion forces, the conventionally determined intrinsic viscosity of mixtures of different polymers may appear higher or lower than that calculated from Equation (9-125) from the mass contributions w_i and intrinsic viscosities $[\eta]_i$ of the components. Examples of this are the values for mixtures of poly(styrene) and poly(methyl methacrylate) (Table 9-5).

9.9.2. Experimental Methods

To calculate the intrinsic viscosity $[\eta]$, the viscosities of the solvent and of solutions of different concentrations must be measured. The polymer solution concentrations must not be too high, since this makes the extrapolation of the viscosity data to infinite dilution difficult. Experience shows that solute concentrations are best chosen such that η/η_1 lies between about 1.2 and 2.0.

The upper limit of $\eta_{rel} \approx 2.0$ results from the fact that the relationship between η_{sp}/c and c becomes increasingly nonlinear with increasing concentration. The lower limit of $\eta_{rel} \approx 1.2$ arises from the fact that at low concentrations anomalies in the function $\eta_{sp}/c = f(c)$ begin to appear. These

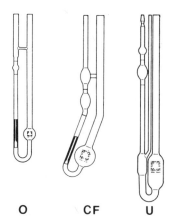

O   CF   U

Figure 9-21. Ostwald (O), Cannon–Fenske (CF), and Ubbelohde (U) capillary viscometers.

anomalies are usually considered to be apparatus dependent and to result from adsorption of macromolecules on the capillary walls.

To determine $\eta_{sp} = \eta_{rel} - 1$ for $\eta_{rel} = 1.2$ to an accuracy of $\sim\pm1\%$, the viscosity ratio must be determined to an accuracy better than $\pm0.2\%$, and the viscosities themselves to an accuracy better than $\pm0.1\%$ (compounding of errors). Capillary viscometers are particularly suitable for such determinations. The usual rotation viscometer of the Couette type at best gives an accuracy of $\pm1\%$. Falling-ball viscometers are even less exact.

The frequently used viscometer types for macromolecular chemistry are shown in Figure 9-21. In all types, the time taken for a given quantity of liquid to pass between two points is taken as an indication of the viscosity. (The pressure head under which these experiments are carried out is the pressure head of the liquid under study in the capillary. This pressure head varies during the experiment, but one can assume that an average pressure head is effective during the whole of the experiment.) As long as liquid flow is not infinitely slow, potential energy is partially lost in overcoming frictional forces during flow. Some of the potential energy is also lost through conversion to kinetic energy, which in turn is dissipated through the formation of vortices on exit from the capillary (Hagenbach). Additionally, a given amount of initial work is done on forming the parabolic shear-rate gradient (Couette). The apparent increase in viscosity caused by these two nonfrictional effects is, according to Hagenbach and Couette, taken into account by the incorporation of a correction term in the Hagen–Poiseuille equation [for derivation, see Equation (7-36)]:

$$\eta = \frac{\pi r^4 pt}{8LV} - \frac{k\rho V}{8\pi Lt} = const \times pt - const' \times \rho t^{-1} \qquad (9\text{-}126)$$

where r is the radius, p is the pressure, L is the capillary length, and V is the volume of liquid with density ρ. The constant k depends on the geometrical shape of the capillary ends. This cannot be obtained theoretically, and must be determined by calibration measurements on liquids with various viscosities.

According to Equation (9-126), the Hagenbach–Couette correction can be ignored if the viscometer capillary is sufficiently long. With commercial viscometers, the Hagenbach–Couette correction values are supplied by the manufacturer in the form of correction times. Measurement times should, preferably, never be less than 100 s; otherwise one incurs too high a percentage error. Also, the viscometer must hang vertically for every reading; otherwise the effective length of the capillary varies from one reading to another. The temperature should be constant to $\sim\pm0.01°$ C, since a temperature difference of $0.01°$ C usually means a viscosity change of $\sim0.02\%$.

Solution and solvent almost always have different densities. Therefore, the average pressure head p in capillary viscometers filled to the same height $h = h_0$ will vary in a series of measurements involving different solute concentrations. Thus, according to Equation (9-126), the following can be given for the relative viscosity in the case of a vanishingly small Hagenbach–Couette correction:

$$\eta_{rel} = \frac{\eta}{\eta_1} = \frac{const \times pt}{const \times p_1 t_1} = \frac{(h\rho)t}{(h_1\rho_1)t_1} = \frac{\rho t}{\rho_1 t_1} \qquad (9\text{-}127)$$

With relatively low-molar-mass substances, relatively high concentrations have to be used in order to obtain η_{rel} values of 1.2–2, so that in these cases particular attention should be paid to the differences in density between solution and solvent during viscosity measurements.

The capillary viscometers shown in Figure 9-21 differ in their areas of application. Because of their low price and the fact that they only require low amounts, ~3 cm^3, of liquid. Ostwald viscometers are by far the most frequently used. The amount of liquid used must be measured in very exactly and maintained constant; otherwise the pressure head will vary with the different solutions.

Suspended-level Ubbelohde viscometers are so constructed that, during a measurement, the pressure head of the suspended liquid at the capillary outlet is independent of the amount of liquid originally introduced into the viscometer. This also means that solution originally introduced into the viscometer can be diluted with solvent to provide a series of concentrations without having, as in the case of the Ostwald viscometer, to empty and clean the viscometer after every measurement. On the other hand, the amount of liquid required for a measurement is higher in the Ubbelohde than in the Ostwad viscometer. Another advantage of the Ubbelohde suspended-level viscometer is that the stress on the suspended liquid exactly compensates the surface-tension effects on the upper meniscus, and this is particularly important in the case of surface-active substances.

The Cannon–Fenske viscometer has two bulbs, and since the pressure head is different for each bulb, this allows qualitative assessment of the effect of shear stress σ_{21} on the viscosity of the solution. The shear stress is given by

$$\sigma_{21} = pr/2L \tag{9-128}$$

where p is the pressure head, and r and L are the capillary radius and length, respectively (for derivation, see Section 7.5.2). With Newtonian liquids, the viscosity, and thus also the intrinsic viscosity is independent of σ_{21}. In some cases, solutions of macromolecules of high molecular weight show non-Newtonian behavior even in dilute solutions, i.e., we have $\eta = f(\sigma_{21})$. According to theoretical studies and experimental evidence, the intrinsic viscosity measured at a certain shear stress decreases with increasing σ_{21} according to

$$[\eta]_{\sigma_{21}} = [\eta](1 - A\beta^2 \dots) \tag{9-129}$$

β is a generalized shear stress function,

$$\beta = [\eta]\eta_1 \frac{\langle M_n \rangle}{RT} \sigma_{21} \tag{9-130}$$

that takes into account the solvent viscosity η_1, the number-average molar mass $\langle M_n \rangle$, and the temperature T. In Equation (9-129), A is a constant that remains dependent on the goodness of the solvent. Poly(styrene) solutions exhibit non-Newtonian behavior when $\beta > 0.1$ (Figure 9-22), which corresponds, roughly, to a molar mass of $\langle M_n \rangle > 500\ 000$ g/mol.

Since the effects of shear stress are particularly strong in the case of rodlike macromolecules, rotation viscometers are frequently used for measurements on substances such as deoxyribonucleic acid (Figure 9-23). With a sufficiently low rotational speed and a narrow gap between rotor and stator, a linear shear-rate gradient can be produced between the rotor and stator of a rotation viscometer. With such narrow gaps between rotor and stator, the centering of rotor in the stator (or, in some viscometers, vice versa) is particularly important. In rotation viscometers of the Couette type, centering is achieved by use of a mechanical axis. A much better centering system is used in the Zimm–Crothers viscometer. This utilizes the principle that, when the viscometer contains the liquid to be measured, a suitably buoyant rotor will automatically center perfectly in the stator because of the surface tension of the liquid to be measured. The rotor contains an iron rod. The thermostated stator resides between the poles of a magnet. The magnet is connected to a motor that can be driven at a regulatable constant speed of rotation. The interaction between the external magnetic field and the magnetic field induced

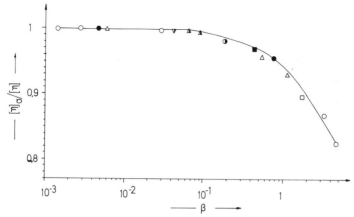

Figure 9-22. Dependence of the reduced intrinsic viscosity $[\eta]_\sigma/[\eta]$ on the reduced shear stress $\beta = ([\eta]\eta_1\langle M_\eta\rangle/RT)\sigma_{21}$ of polystyrene with molar mass $\overline{M}_w = 7.1 \times 10^6$ (O,□,△), 3.2×10^6 (O), and 1.4×10^6 (◑,◪,▲,▼) in good (O,◑,●,□,◪) and theta (△,▲,▼) solvents.

in the iron rod produces a weak torque. Shear stresses of down to 40 nN/cm³ can thus be obtained.

9.9.3. Concentration Dependence for Nonelectrolytes

The intrinsic viscosity $[\eta]$ is defined as the limiting value of the reduced viscosity η_{sp}/c $(=\eta_{sp}/c_2)$ at infinite dilution. Since measurements are made at finite concentrations, a suitable equation is required in order that values of η_{sp}/c, or related quantities, may be extrapolated to $c \to 0$. The extrapolation should be as linear as possible over the range $\eta_{rel} = 1.2\text{--}2$.

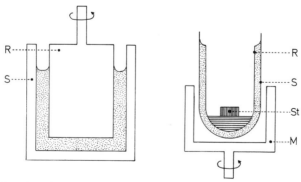

Figure 9-23. Rotation viscometers of the Couette (left) and Zimm–Crothers (right) type. R, rotor; S, stator; St, steel plate; M, magnet.

All the extrapolation formulas introduced to date are empirical. The much used expressions of Schulz and Blaschke, Huggins, and Kraemer start from the relationship

$$\frac{\eta_{sp}}{c} = \frac{[\eta]}{1 - k[\eta]c} \tag{9-131}$$

The extrapolation formula that is obtained by transforming Equation (9-131)

$$\frac{c}{\eta_{sp}} = \frac{1}{[\eta]} - kc \tag{9-132}$$

is, to date, practically unused. If Equation (9-131) is solved for c one obtains $c = \eta_{sp}/([\eta] + \eta_{sp}k[\eta])$. If this expression is inserted into the right-hand side of equation (9-131) in place of c, then the formula known as the Schulz–Blaschke equation is obtained:

$$\frac{\eta_{sp}}{c} = [\eta] + k[\eta]\eta_{sp} \tag{9-133}$$

For low values of $k[\eta]c$, the expression in the denominator in Equation (9-131) can be developed in a series: $(1 - k[\eta]c)^{-1} = 1 + k[\eta]c + \cdots$, On inserting this expression in Equation (9-131), we obtain the Huggins equation

$$\frac{\eta_{sp}}{c} = [\eta] + k[\eta]^2 c \tag{9-134}$$

If one develops $\ln \eta_{rel} = \ln(1 + \eta_{sp})$ in a Taylor series

$$\ln \eta_{rel} = \eta_{sp} - (1/2)\eta_{sp}^2 + (1/3)\eta_{sp}^3 - \cdots \tag{9-135}$$

and inserts this in Equation (9-134), the Kraemer equation is obtained:

$$\frac{\ln \eta_{rel}}{c} = [\eta] + [\eta]^2 \{k - 0.5 + [[\eta]c(0.333 - k)]\} c + \cdots \tag{9-136}$$

which is mostly written without the component $(0.333 - k)[\eta]^3 c^2$ in the shortened form

$$\frac{\ln \eta_{rel}}{c} = [\eta] + (k - 0.5)[\eta]^2 c \tag{9-137}$$

The omitted term comes from the mathematical development of Equation (9-135) in a series. Because of its magnitude, it cannot be simply ignored. Figure 9-24 shows examples of concentration dependence.

The numerical evaluation of experimental data shows that Equations (9-131)–(9-134) and (9-137) yield different values of $[\eta]$ and k. Since

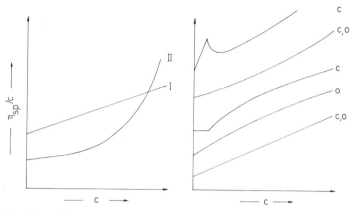

Figure 9-24. A plot of the concentration dependence of η_{sp}/c for various polymer–solvent and polymer–polymer interactions. Left diagram: nonassociating polymers; right diagram: associating polymers. (I) Good solvent, (II) poor solvent. (o) Open associaton, (c) closed association. The multiplicity of curves derives from the number of possible models as well as from the relative influences of the equilibrium association constant, the association number, the molar mass, the size and shape of the molecules and associates, and the solvent interaction.

Equations (9-134) and (9-137) are approximations of Equation (9-131), they must, *a priori*, extend over a narrower concentration range. This argument naturally assumes that Equation (9-131) really describes the concentration dependence of η_{sp}/c adequately. For wider concentration ranges, the Martin (sometimes referred to as the Bungenberg–de Jong) equation is used in the form

$$\log \frac{\eta_{sp}}{c} = \log[\eta] + kc \tag{9-138}$$

One often makes do with a viscosity measurement at a single concentration (usually 0.5%) to give what is called the inherent viscosity $\{\eta\}_c = (\ln \eta_{rel}/c)_c$ for this concentration. Fikentscher constants K are also used, particularly in the German literature, to characterize the classic polymers such as poly(styrene) and poly(vinyl chloride). [The K here is not to be confused with the K, defined by Equation (9-151), of the modified Staudinger equation.] K is evaluated from the relative viscosity at relatively high concentrations from tabular data and the equation

$$\log \eta_{rel} = \left(\frac{75k_F^2}{1 + 1.5k_F c} + k_F \right) c \tag{9-140}$$

with the definition $K = 1000k_F$. At the time it was introduced, K was considered, on the basis of limited experimental data, to be a concentration-independent constant that was related to the molar mass. However, K is concentration dependent, and at high molar masses, it becomes increasingly less sensitive to changes in the molar mass.

The molecular significance of the coefficient k has not yet been elucidated. Theoretical studies indicate that for coils k is composed of a hydrodynamic factor k_h and a thermodynamic factor $(3A_2M/[\eta])f(\alpha)$, where $f(\alpha)$ is a function of the expansion factor α:

$$k = k_h - \frac{3A_2M}{[\eta]}f(\alpha) \qquad (9\text{-}141)$$

The hydrodynamic factor k_h probably has values of between 0.5 and 0.7. In theta solvents, where $A_2 = 0$, a value of $k = 0.5$–0.7 is expected; in good solvents, since $A_2 > 0$, the value of k is <0.5–0.7. Experimentally, values of k between 0.25 and 0.35 are found for good solvents. As the molar mass increases, the second term in Equation (9-141) should, according to this equation, increase strongly initially, and subsequently increase less strongly, since, with random coils, $[\eta]$ and $f(\alpha)$ increase less than proportionally with the molar mass and A_2 decreases less than proportionally with M. The major proportion of the experimental data confirm these expectations.

9.9.4. Concentration Dependence for Polyelectrolytes

In the absence of foreign salts, η_{sp}/c increases sharply with decreasing polyelectrolyte concentration in solutions of polyelectrolytes (Figure 9-25). At increasing foreign salt concentration, this increase in η_{sp}/c becomes weaker. At low polymer concentrations, the function $\eta_{sp}/c = f(c)$ passes through a maximum, but this phenomenon is not observed with all polyelectrolyte solutions. At high polyelectrolyte concentrations, the increase of η_{sp}/c with concentration is similar to that observed with solutions of nonelectrolytes.

The effect is explained in the following manner: With decrease in polyelectrolyte concentration, the degree of ionization increases. In the case of polysalts [e.g., sodium pectinate, the sodium salt of poly(acrylic acid), etc.] the gegenions form an ion atmosphere around the chains of the polyelectrolyte macromolecule. In very dilute, foreign-salt-free solutions, the diameter of the ion atmosphere is greater than the diameter of the coiled molecule. The carboxylate ions, $-COO^-$, repel each other, increasing the chain rigidity and expanding the polymer coil, with consequent increase in the viscosity. At medium polyelectrolyte concentrations, the gegenions reside partly within and partly outside the polymer coil. At very high polyelectrolyte concentra-

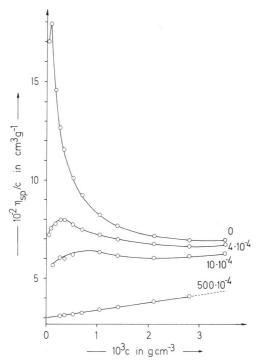

Figure 9-25. Concentration dependence of the reduced viscosity η_{sp}/c of sodium pectinate in solutions of different NaCl concentrations (in mol/liter) at 27°C (according to D. T. F. Pals and J. J. Hermans).

tions, the gegenions reside more within than outside the polymer coil. The osmotic effect that this produces causes more water to penetrate the coil and expand it. Thus at low concentrations the electrostatic effect predominates, and at high concentrations the osmotic effect predominates. The introduction of foreign salts into the polyelectrolyte solutions causes the ionic strength outside the polymer coil to be increased relative to that inside the polymer coil, and the diameter of the ion atmosphere is decreased. Both effects decrease the diameter of the polymer coil and, thus, also η_{sp}/c.

The intrinsic viscosity can be empirically determined via the Fuoss equation

$$\frac{c}{\eta_{sp}} = \frac{1}{[\eta]} + Bc^{0.5} - \cdots \qquad (9\text{-}141)$$

where the values of c/η_{sp} for $c > c_{max}$ are plotted against $c^{0.5}$ and extrapolated to $c^{0.5} \to 0$.

9.9.5. The Intrinsic Viscosity and the Molar Mass of Rigid Molecules

Unsolvated Spheres. The intrinsic viscosity $[\eta]$, according to Equation (9-123), depends on both the molar mass and the hydrodynamic volume, and the latter can also be a function of the molar mass. Unsolvated spheres present the simplest case. These spheres can be defined by their type and a density that is independent of its environment and equal to the density of the dry material. The mass m_{mol} of an individual molecule is related to its hydrodynamic volume via $m_{mol} = V_h \rho_2$, or with the molar mass M_2 via $M_2 = N_L m_{mol} = N_L V_h \rho_2$. Equation (9-123) therefore changes, for unsolvated spheres, to

$$[\eta] = 2.5/\rho_2 \qquad (9\text{-}142)$$

Solvated Spheres. To a good approximation, dispersions [e.g., poly-(styrene) latices] can be considered as unsolvated sphere systems, but the approximation no longer holds for isolated macromolecules. Certain proteins, however, are in the form of solvated spheres in aqueous solutions. With these proteins, some of the amino acid residues are in a helical conformation; the rest are in a random-coil conformation (cf. Figure 4-14). Under the influence of water, the helix and coil portions mutually align themselves via hydrophobic bonds, salt bonds, etc., to form solvated spherical particles. The mass of the hydrodynamically effective individual molecule is consequently composed of the mass of the protein component and the mass of the water of hydration, i.e., $m_h = m_2 + m_1^\square$. On average, the density of every segment of the sphere is the same. If the ratio of the masses $\Gamma = m_1^\square/m_2$ is sufficiently low, no water of hydration is exchanged with the surrounding water during the measurement. The hydrated sphere is then nondraining, and all of the water of hydration has to be considered as part of the mass and the volume of the hydrodynamically effective particle. If the hydrodynamic volume V_h in Equation (9-123) is replaced by the expression in Equation (7-6), the following is obtained:

$$[\eta] = 2.5(\tilde{v}_2 + \Gamma v_1) \qquad (9\text{-}143)$$

Thus, the intrinsic viscosity of solvated spheres depends on the partial specific volume $\tilde{v}_2$ of the solute, the specific volume v_1 of water, and the mass ratio $\Gamma = m_1^\square/m_2$ (degree of solvation) of both components in the interior of the sphere. Therefore, it is not possible to calculate the molar mass of a solvated sphere from the intrinsic viscosity alone. The intrinsic viscosities of spherical protein molecules are low, and for equal degrees of hydration are independent of the molar mass (Table 9-6). Admittedly, the proteins included in Table 9-6 are not perfectly spherical, since their coefficients of friction f are somewhat larger than those expected for a perfect sphere, f_0.

Table 9-6. Intrinsic Viscosity $[\eta]$ and Frictional
Coefficient of Approximately Spherical Proteins in Dilute
Salt Solutions at 20° C

Protein	Molar mass M_2 in g/mol	$[\eta]$ in cm^3/g	f/f_0
Ribonuclease	13 683	3.30	1.14
Myoglobin	17 000	3.1	1.11
β-Lactoglobulin	35 000	3.4	1.25
Serum albumin	65 000	3.68	1.31
Hemoglobin	68 000	3.6	1.14
Catalase	250 000	3.9	1.25

Unsolvated Rods. For unsolvated rods, the relationship between the intrinsic viscosity and the molar mass can be deduced as follows: $[\eta]$ depends on the hydrodynamic volume [cf. Equation (9-123)] and on the radius of gyration. One can write, in place of Equation (9-123), therefore,

$$[\eta] = \frac{\Phi^* \langle R_G^2 \rangle_{st}^{3/2}}{M_2} \qquad (9\text{-}144)$$

where Φ^* is a general proportionality constant.

According to Equation (4-53), the radius of gyration of rods is related to the chain end-to-end distance: $\langle L^2 \rangle = 12 \langle R_G^2 \rangle_{rod}$. For macromolecules in the all-*trans* conformation, the end-to-end distance is equal to the maximum chain length L_{max}. However, L_{max} is proportional to the degree of polymerization X [see also Equation (4-8)]. With helix-forming rodlike macromolecules, the end-to-end distance is equal to the length of the helix and is also proportional to the degree of polymerization. As a general rule, then, $L^2 = \text{const} \times X^2$ for rodlike macromolecules. With $X = M_2/M_u$, it thus follows that

$$[\eta] = \Phi^* \left(\frac{\text{const}}{12 M_u^2} \right)^{3/2} M_2^2 = K M_2^2 \qquad (9\text{-}145)$$

According to this relationship, the intrinsic viscosity of rods is proportional to the square of the molar mass. For a homologous series of rodlike molecules, i.e., a series with constant rod cross section, this functionality only applies over a limited molar mass range. At low molar masses, the rod increasingly takes on the character of a sphere, and the exponent decreases below 2. At high molar masses, a real rod is no longer inflexible, since a rod of infinite length behaves like a random coil (cf. the flexibility of thin steel wires of various lengths). Thus, at high molar masses, the exponent also falls below 2 (Figure 9-26).

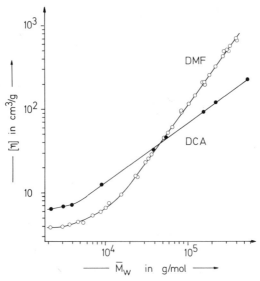

Figure 9-26. A plot according to Equation (9-151) of the molar mass dependence of the intrinsic viscosity of poly(γ-benzyl-L-glutamate) in dichloroacetic acid (DCA) and dimethyl formamide (DMF) at 25°C. Coils occur in DCA and helices occur in DMF (according to P. Rohrer and H.-G. Elias).

9.9.6. The Molar Mass and Intrinsic Viscosity of Coil-Like Molecules

Nondraining Coils. Nondraining coils are defined as those in which the solvent molecules within the coil move with the same velocity as the nearest segments of the coil itself during transport processes. The hydrodynamic volume V_h of such a coil is greater in thermodynamically good solvents than in theta solvents. This extension can be described by a coefficient of expansion α_η in a way analogous to that used in the case of the radius of gyration (see, for example, Section 4.4.5), i.e., as $V_h = (V_h)_\Theta \alpha_\eta^3$. The quantity α_η differs, however, from the expansion coefficient used for the radius of gyration, since the non-Gaussian segment distribution in the coil causes the hydrodynamic radius to vary somewhat differently with the molar mass than is the case with the radius of gyration. As a general rule, however, one can write

$$V_h = (V_h)_\Theta \, \alpha_\eta^3 = (V_h)_\Theta \, \alpha_R^q, \qquad q \neq 3 \tag{9-146}$$

According to model calculations, the factor q possesses a value of 2.43 for a spherelike coil and a value of 2.18 for an ellipsoidlike coil. Insertion of Equation (9-146) in (9-123) gives

$$[\eta] = 2.5 N_L (V_h)_\Theta \frac{\alpha_R^q}{M_2} \tag{9-147}$$

In the theta state, the hydrodynamic volume $(V_h)_\Theta$ is proportional to the third power of the radius of gyration, i.e., it is possible to write $(V_h) = \Phi' \langle R_G^2 \rangle_0^{3/2}$. Equation (9-147) then becomes

$$[\eta] = 2.5 N_L \Phi' \frac{\langle R_G^2 \rangle_0^{3/2} \alpha_R^q}{M_2} = 2.5 N_L \Phi' \left(\frac{\langle R_G^2 \rangle_0}{M_2} \right)^{3/2} M_2^{0.5} \alpha_R^q$$

$$= \Phi \left(\frac{\langle R_G^2 \rangle}{M_2} \right)^{3/2} M_2^{0.5} = \Phi \frac{(\langle R_G^2 \rangle)^{3/2}}{M_2} \tag{9-148}$$

where the relationship $\langle R_G^2 \rangle_0 = \langle R_G^2 \rangle_0 \alpha_R^2$ is used and $\Phi = 2.5 N_L \Phi' (\alpha_R / \alpha_R^3)$ is assumed.

Equation (9-148) describes the relationship between the intrinsic viscosity $[\eta]$ of nondraining coils as a function of the molar mass and the radius of gyration. Two methods can be used to obtain $[\eta]$ as a function of the molar mass alone:

1. According to Equation (4-50), $\langle R_G^2 \rangle = \text{const}' \times M_2^{1+\epsilon}$. Equation (9-148) can thus be written as

$$[\eta] = \Phi (\text{const}')^{3/2} M_2^{0.5(1+3\epsilon)} \tag{9-149}$$

or, with $K = \Phi (\text{const}')^{3/2}$ and the definition

$$a_\eta \equiv 0.5(1 + 3\epsilon) \tag{9-150}$$

also as

$$[\eta] = K M_2^{a_\eta} \tag{9-151}$$

Since ϵ rarely takes on values above 0.23 in the case of nondraining coils, a_η values up to a maximum of ~ 0.9 are obtained for nondraining coils. In the theta state, $\epsilon = 0$ and Equation (9-151) reduced to

$$[\eta]_\Theta = K_\Theta M_2^{0.5} \tag{9-152}$$

Equation (9-151) is known as the modified Staudinger equation (originally with $a_\eta = 1$) or as the Kuhn–Mark–Houwink–Sakurada equation. It was originally found empirically. K and a_η are empirical constants obtained by calibration (see also Sections 9.9.7 and 9.9.8 and Figure 9-26). In certain special cases, a_η can also be theoretically calculated (see Table 9-7).

2. According to Equation (4-51), $(\langle R_G^2 \rangle / M_2)^{3/2}$ can be expressed as

$$\left(\frac{6 \langle R_G^2 \rangle}{M_2} \right)^{3/2} = A^3 + 0.632 B M_2^{0.5} \tag{9-153}$$

Table 9-7. Theoretical Exponents a_η of the Viscosity–Molar
Mass Relationship [Equation (9-151)]

Shape	Homology	a_η
Rods	Constant diameter; height proportional to M; no rotational diffusion	2
Rods	Same as above, but with rotational diffusion	1.7
Coils	Unbranched; free draining; no excluded volume	1
Coils	Unbranched: no draining; excluded volume	0.51–0.9
Disks	Diameter proportional to M; height constant	0.5
Spheres	Constant density; unsolvated or uniformly solvated	0
Disks	Diameter constant; height proportional to M	-1
Rods	Diameter proportional to $M^{0.5}$; height constant	-1
Rods	Diameter proportional to M; height constant	-2

If Equation (9-148) is inserted into (9-153), the Burchard–Stockmayer–Fixman equation (often just known as the Stockmayer–Fixman equation) is obtained:

$$\frac{[\eta]}{M_2^{0.5}} = K_\Theta + \frac{0.632}{6^{3/2}} \, \Phi B M_2^{0.5} \qquad (9\text{-}154)$$

where it is assumed that

$$K_\Theta = \frac{\Phi}{6^{3/2}} A^3 \qquad (9\text{-}155)$$

By plotting $[\eta]/M_2^{0.5} = f(M_2^{0.5})$, the quantity K_Θ can, according to Equation (9-154), be obtained, and K_Θ contains the steric hindrance contribution in the factor A (see also Figure 9-27). Besides the steric hindrance parameter σ, A also contains, according to Equation (4-51), the bond length L between the main-chain atoms, the valence angle ϑ, and the average formula molar mass $\overline{M}_u$ of the chain units:

$$A = \left[\left(\frac{1 - \cos \vartheta}{1 + \cos \vartheta} \right) \frac{\sigma^2 l^2}{M_u} \right]^{0.5} \qquad (9\text{-}156)$$

Equation (9-156) thus allows σ to be determined from viscometric data alone. Like σ, K_Θ is only a substance-specific constant when measurements are carried out in single, apolar solvents (see also Section 4.5.1).

Both Equations (9-151) and (9-154) give the intrinsic viscosity as a function of the molar mass. It has been shown experimentally that Equation (9-151) is valid over a wider range of molar masses, temperature, and solvents than Equation (9-154). Equation (9-154) is fulfilled satisfactorily, but at higher molar mass it yields values of $[\eta]/M^{0.5}$ that are relatively too small when $T > \Theta$ (see Figure 9-27). Many other functions have been proposed to

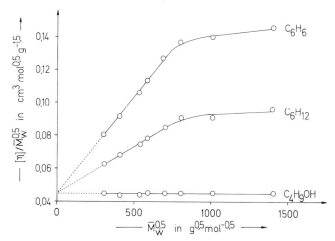

Figure 9-27. Burchard–Stockmayer–Fixman plot for poly(cyclohexyl methacrylate) in benzene and cyclohexane at 25°C and in butanol at 23°C (according to N. Hadjichristidis, M. Devaleriola, and V. Desreux).

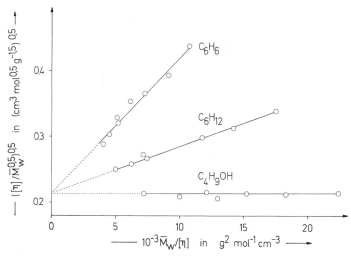

Figure 9-28. Berry plot for poly(cyclohexyl methacrylate) in benzene and cyclohexane at 25°C and in butanol at 23°C (same data as in Figure 9-27) (according to N. Hadjichristidis, M. Devaleriola, and V. Desreux).

achieve better linearity. The Berry equation has shown itself to be very useful (see Figure 9-28):

$$\left(\frac{[\eta]}{\langle M_w \rangle^{0.5}} \right)^{0.5} = K_\Theta^{0.5} + D \frac{\langle M_w \rangle}{[\eta]} \tag{9-157}$$

where D is an adjustable constant.

In Equations (9-148), (9-149), and (9-154), Φ appears everywhere as a constant. Theoretical and experimental investigations have shown that Φ does not depend on either the constitution or the configuration of the polymer or on the chemical nature of the solvent used. As a proportionality factor between the radius of gyration and the hydrodynamic volume, Φ is only related to the expansion of the coil in the relevant solvent, i.e., to the value of α or ϵ. The theoretical calculation leads to

$$\Phi = \Phi_0 (1 - 2.63\epsilon + 2.86\epsilon^2) \tag{9-158}$$

where Φ_0 is the value in the theta state ($\epsilon = 0$ or $a_\eta = 0.5$). If Φ_0 is related to the radius of gyration, then $\Phi_0 = 4.18 \times 10^{24}$ (mol macromolecule)$^{-1}$. If one uses the chain end-to-end distance, then, because $\langle R_G^2 \rangle_0 = \langle L^2 \rangle_0 / 6$, $\Phi_0 = 2.84 \times 10^{23}$ (mol macromolecule)$^{-1}$ (see also Table 9-3).

Free-Draining Coils. In the case of free-draining coils, the relative velocity of solvent movement with respect to the coil is the same both within and outside the coils. Free-draining coils may be expected in the case of relatively rigid chains in good solvents, whereas nondraining coils are to be expected in the limiting case of very flexible chains in poor solvents. All variations between these two extremes are possible. However, the concept of a partially draining coil can be described theoretically only with difficulty. In the scientific literature, two principal theories are discussed.

In the Kirkwood–Riseman theory, the perturbation of the rate of flow of the solvent by $N - 1$ chain elements is calculated for the Rth chain element and summed over all possible conformations. Suitable parameters that are used are the effective bond length b and the frictional coefficient ζ of the monomeric unit.

The Debye–Bueche theory, on the other hand, considers the partially draining coil as a sphere that is more or less permeable, within which a number of smaller beads is homogeneously distributed. The beads correspond to the monomeric units. The drag which one bead produces on the others is calculated, and this resistance is then expressed in terms of a length L, which corresponds to the distance from the surface of the sphere to where the flow rate of the solvent is reduced to $1/e$ times what it is at the surface of this sphere. The shielding ratio, or "shielding factor," ζ is given by

$$\zeta = R_s / L \tag{9-159}$$

where R_s is the sphere radius and L is the depth of shielding. ζ can also be calculated from the shielding function $F(\zeta)$ derived by Debye and Bueche:

$$F(\zeta) = 2.5 \frac{1 + (3/\zeta^2) - (3/\zeta)\cot\zeta}{1 + (10/\zeta^2)[1 + (3/\zeta^2) - (3/\zeta)\cot\zeta]} \qquad (9\text{-}160)$$

which is, in turn, related to the intrinsic viscosity:

$$[\eta] = F(\zeta)\, N_L \left(\frac{4\pi R_s^3}{3} \right) M_2^{-1} \qquad (9\text{-}161)$$

Coil macromolecules tend, however, to occur in the form of ellipsoids (cf. Section 4.4.3). With ellipsoid-shaped coils, R_s corresponds to the major rotational axis of the ellipsoid. For such coils, a simple empirical relationship between the shielding factor ζ and the quantity ϵ has been found when the coils are flexible [e.g., in the case of poly(styrene)]. The relationship is $\zeta\epsilon = 3$. The effect of the solvent on the coil expansion is described by ϵ [cf. Equation (9-158)]. Thus, Equation (9-159) becomes

$$L = R_s\epsilon/3 \qquad (9\text{-}162)$$

According to Equation (9-150), ϵ can be calculated from the exponent a_η of the viscosity–molar mass relation $[\eta] = KM^{a_\eta}$. For example, with $a_\eta = 0.8$, $\epsilon = 0.2$, and the shielding depth L is $0.067 R_s$ according to Equation (9-162). For a coil that is expanded as in this particular case, the shielding depth is only 6.7% of the major rotational axis of the ellipsoid. The solvent, therefore, only partially penetrates the coil. For the more rigid cellulose chains, the shielding depth is greater.

Molecular Dimensions from Viscosity Measurements. According to Equation (9-148), the intrinsic viscosity is related to the radius of gyration. Furthermore, in very dilute solutions, $[\eta] \approx \eta_{sp}/c$, and, consequently,

$$\frac{\eta_{sp}}{c} \approx [\eta] = \frac{\Phi\langle R_G^2\rangle^{3/2}}{M} = \frac{\Phi E}{M} \qquad (9\text{-}163)$$

where $E \equiv \langle R_G^2\rangle^{3/2}$. Correspondingly, for a polymolecular solute with i components, with $m_i = n_i M_i$, $w_i = c_i/c$, and $w_i = m_i/\Sigma_i m_i$, one has

$$\sum_i (\eta_{sp})_i = \Phi \sum_i E_i \frac{c_i}{M_i} = \Phi c \frac{\Sigma_i n_i E_i}{\Sigma_i m_i} \qquad (9\text{-}164)$$

According to Equation (9-125), the intrinsic viscosity $[\eta]$ of a polymolecular solute is the mass average of the intrinsic viscosities of the individual components:

$$[\eta] = \frac{\Sigma_i c_i [\eta]_i}{\Sigma_i c_i} \approx \frac{\Sigma_i c_i (\eta_{sp})_i/c_i}{\Sigma_i c_i} = \frac{\Sigma_i (\eta_{sp})_i}{c} \qquad (9\text{-}165)$$

Inserting Equation (9-165) into (9-164), and recalling the definition of the number-average molecular weight $M_n = \Sigma_i m_i / \Sigma_i n_i$, one obtains, after expanding with $\Sigma_i n_i / \Sigma_i n_i$.

$$[\eta] = \Phi \frac{\Sigma_i n_i E_i}{\Sigma_i m_i} = \Phi \frac{\Sigma_i n_i E_i}{\Sigma_i n_i} \frac{\Sigma_i n_i}{\Sigma_i m_i} = \Phi \frac{\overline{E}_n}{\langle M_n \rangle} = \Phi \frac{\langle R_G^2 \rangle_n^{3/2}}{\langle M_n \rangle} \quad (9\text{-}166)$$

Equation (9-166) shows that the number average of the 1.5 power of the mean square radius of gyration is obtained from viscosity measurements.

9.9.7. Calibration of the Viscosity–Molar Mass Relationship

A comparison of the expressions giving the molar mass dependence of the intrinsic viscosity shows that the relationships can be generalized for all macromolecular types in the form of the modified Staudinger equation $[\eta] = KM^{a_\eta}$ (see also Table 9-7). Both K and a_η are usually unknown. For each polymer homologous series, therefore, the modified Staudinger equation must be empirically determined. To do this, the molar masses and intrinsic

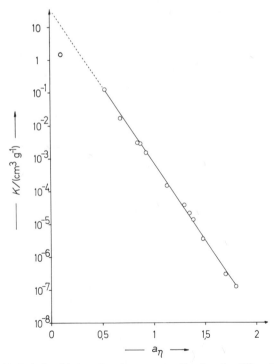

Figure 9-29. Relationship between the constants K and a_η for different polymers.

viscosities are determined for a number of samples (solvent, temperature = constant), and then $\log[\eta]$ is plotted against $\log M_2$ according to Equation (9-151) (Figure 9-26). The slope is a_η and the intercept at $\log M_2 = 0$ is equal to K. A smaller slope is obtained at low molar masses (end-group influence, deviation from random-flight statistics, etc.); this is overemphasized in Figure 9-26 because of the double logarithmic plot. With random-coil macromolecules, a_η increases with the thermodynamic goodness of the solvent. K decreases with increasing a_η (Table 9-8 and Figure 9-29), as can be described by

$$\log K = 1.507 - 4.368\, a_\eta \qquad (9\text{-}167)$$

Only measured values for ellipsoidal, compact proteins at $a_\eta = 0.1$ deviate from this relationship.

Since it is only valid to compare like averages with each other in a calibration of this kind, the question arises as to what kind of molar mass average is obtained from Equation (9-151). According to Equation (9-125), the intrinsic viscosity of a polymolecular substance is a mass average. Equation (9-151) can thus be written as

$$KM_2^{a_\eta} = [\eta] = \frac{\sum_i W_i[\eta]_i}{\sum_i W_i} = \frac{\sum_i W_i K_i M_i^{a_\eta}}{\sum_i W_i} \qquad (9\text{-}168)$$

At sufficiently high molecular weights, K_i is independent of M_2, so $K = K_i$. Solving Equation (9-151) in the form of $[\eta] = K(M_\eta)^{a_\eta}$ for $\langle M_\eta \rangle$ and inserting Equation (9-168), we obtain

$$\langle M_\eta \rangle = \left(\frac{[\eta]}{K} \right)^{1/a_\eta} = \left(\frac{\sum_i W_i M_i^{a_\eta}}{\sum_i W_i} \right)^{1/a_\eta} \qquad (9\text{-}169)$$

Table 9-8. *Constants K and a_η for Coil-Like Macromolecules*

Substance	Temperature in °C	Solvent	$10^3 K$ in cm^3/g	a_η
at-Poly(styrene)	34	Benzene	9.8	0.74
at-Poly(styrene)	34	Butanone	28.9	0.60
at-Poly(styrene)	35	Cyclohexane	78	0.50
Nylon 6,6	25	90% HCOOH	13.4	0.87
Nylon 6,6	25	m-Cresol	35.3	0.79
Nylon 6,6	25	2 m KCl in 90% HCOOH	142	0.56
Nylon 6,6	25	2.3 m KCl in 90% HCOOH	253	0.50
Cellulose tricaproate	41	Dimethyl formamide	245	0.50
Cellulose tricarbanilate	20	Acetone	4.7	0.84
Amylose tricarbanilate	20	Acetone	0.81	0.90
Poly(γ-benzyl-L-glutamate)	25	Dichloroacetic acid	2.8	0.87
Poly(γ-benzyl-L-glutamate)	25	Dimethyl formamide	0.00029	1.70

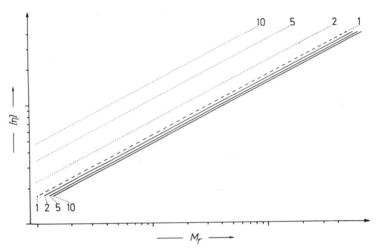

Figure 9-30. Influence of the molar mass distribution (expressed as $\langle M_w \rangle / \langle M_n \rangle$) on the molar mass ($M_r$) intrinsic viscosity ($[\eta]$) relationship at $a_\eta = 0.5. \cdots , M_r = M_n; —, M_r = M_w$.

The molar mass obtained by viscosity measurements is a viscosity average, which differs from the number average and the weight average (see also Figure 8-4). The viscosity-average molar mass is only equal to the mass-average molar mass when $a_\eta = 1$. For $a_\eta < 1, \langle M_\eta \rangle < \langle M_w \rangle$. Thus, for substances with identical molar mass distributions, if $\log[\eta]$ is plotted against $\log \langle M_w \rangle$ instead of $\log \langle M_n \rangle$, the constant K thus obtained is too low for $a_\eta < 1$ (Figure 9-30). a_η is unaffected. If $\log[\eta]$ is plotted against $\log \langle M_n \rangle$, the K obtained is, conversely, too large. If the width of the molar mass distribution, or the distribution type, varies with the molar mass, then incorrect values of K and a_η are obtained with plots of $\log[\eta]$ against either log $\langle M_w \rangle$ or log $\langle M_n \rangle$.

$\langle M_w \rangle$, $\langle M_n \rangle$, and a_η must be known before the viscosity-average $\langle M_\eta \rangle$ can be calculated for calibration. Then $\log[\eta]$ is plotted against $\log \langle M_w \rangle$, and K and a_η are determined. The viscosity average is then determined from the following approximation, which holds for distributions that are not too wide ($\langle M_w \rangle / \langle M_n \rangle < 2$):

$$\frac{\langle M_\eta \rangle}{\langle M_n \rangle} = \frac{1 - a_\eta}{2} + \frac{1 + a_\eta}{2} \frac{\langle M_w \rangle}{\langle M_n \rangle} \tag{9-170}$$

Finally, $\log[\eta]$ is plotted against $\log \langle M_n \rangle$, and K and a_η are again evaluated. This procedure is continued until K and a_η cease to change on further iteration.

9.9.8. *Influence of the Chemical Structure on the Intrinsic Viscosity*

The intrinsic viscosity is a measure of the macromolecular dimensions. Thus, for flexible macromolecules, chain skeleton parameters (bond lengths, valence angles, degree of polymerization, mass of the monomeric unit), the steric hindrance parameter σ, or a measure of the hindrance to rotation, and the expansion factor α as a measure of the interaction with the solvent, determine the magnitude of the intrinsic viscosity. In theta solvents, $\alpha = 1$. Thus, from Equations (9-148) and (9-152)

$$K_\Theta = \Phi_0 \left(\frac{\langle R_G^2 \rangle_\Theta}{M_2} \right)^{3/2} = \Phi_0 \left(\frac{\langle R_G^2 \rangle_\Theta}{X_2} \right)^{3/2} (M_u)^{-3/2} \qquad (9\text{-}171)$$

According to Equation (4-27), $\langle R_G^2 \rangle_\Theta$ is proportional to σ_2. Thus, if the hindrance parameter is independent of the type of substituents, $\log K$ plotted against $\log M_u$ would have a slope of $-3/2$. In compounds of the type $(-CH_2-CR^1R^2-)_n$, the slope is more positive (Figure 9-31). Thus, the hindrance parameter increases with the formula molar mass M_u of the

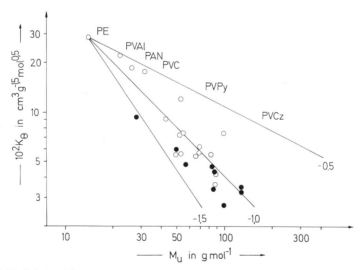

Figure 9-31. Relationship between the constant K_Θ of the molar mass relationship $[\eta] = K_\Theta M^{0.5}$ and the formula molar mass M_u of the monomeric units in polymers of the type $(-CH_2-CHR-)_n$ (**O**) or $(-CH_2-CRR'-)_n$ (**●**), PE, poly(ethylene); PVAl, poly(vinyl alcohol); PAN, poly(acrylonitrile); PVC, poly(vinyl chloride); PVPy, poly(2-vinyl pyridine), PVCz, poly(N-vinyl carbazole). The figures give the slope.

monomeric unit, i.e., σ increases with the size of the substituents R^1 and R^2. Values that are far too high are obtained, for example, with poly(vinyl N-carbazole). Figure 9-31 can also serve to estimate K values for polymers for which the viscosity–molar mass relationship is not known.

K_Θ values for copolymers cannot be obtained from a linear extrapolation of the K_Θ values for the homopolymers. For example, with poly(p-chloro-styrene) and poly(methyl methacrylate), K_Θ values of 0.050 and 0.049 cm^3/g, respectively, are found. For a copolymer with $x_{mma} = 0.484$, however, a value of $K_\Theta = 0.064$ cm^3/g is found. The copolymer coil has therefore been expanded because of the mutual repulsion of the polar groups.

Branching reduces the hydrodynamic volume relative to the mass of the coil. The intrinsic viscosity of branched macromolecules is thus lower than that of their unbranched (linear) counterparts. The effect is particularly marked in the case of long-chain branching. If the number of branch points in a polymer homologous series increases with the molar mass, then the $[\eta]$ values also decrease relative to those of linear molecules. Thus, the slope of the $\log[\eta] = f(\log M_2)$ curve continuously decreases with increasing molar mass, and such an observation can be taken as evidence of branching.

Literature

Section 9.1. Molar Mass and Molar Mass Distributions

R. U. Bonnar, M. Dimbat, and F. H. Stross, *Number Average Molecular Weights*, Interscience, New York, 1958.

P. W. Allen (ed.), *Techniques of Polymer Characterization*, Butterworths, London, 1959.

Ch'ien Jên-Yuen, *Determination of Molecular Weights of High Polymer*, Oldbourne Press, London, 1963.

S. R. Rafikov, S. Pavlova, and I. I. Tverdokhlebova, *Determination of Molecular Weights and Polydispersity of High Polymers*, Acad. Sci., USSR, Moscow 1963; Israel Program for Scientific Translation, Jerusalem, 1964.

Characterization of Macromolecular Structure, Natl. Acad. Sci. U.S.A. Pub. 1973, Washington D.C., 1968.

N. C. Billingham, *Molar Mass Measurements in Polymer Science*, Halsted Press, New York, 1977.

P-G. de Gennes, *Scaling concepts in Polymer Physics*, Cornell University Press, Ithaca, New York, 1979.

Section 9.2. Membrane Osmometry

H. Coll and F. H. Stross, Determination of molecular weights by equilibrium osmotic pressure measurements, in *Characterization of Macromolecular Structure*, Natl. Acad. Sci. U.S. Publ. 1573, Washington, D.C., 1968.

H.-G. Elias, Dynamic osmometry, *in Characterization of Macromolecular Structure*, Natl. Acad. Sci. U.S. Publ. 1573, Washington, D.C., 1968.

H. Coll, Nonequilibrium osmometry, *J. Polymer Sci. D (Macromol. Rev.)* **5**, 541 (1971).

M. P. Tombs and A. R. Peacock, *The Osmotic Pressure of Biological Macromolecules*, Clarendon Press, Oxford, 1975.

Section 9.3. Ebulliometry and Cryoscopy

R. S. Lehrle, Ebulliometry applied to polymer solutions, *Prog. High Polym.* **1**, 37 (1961).

M. Ezrin, Determination of molecular weight by ebulliometry, *in Characterization of Macromolecular Structure*, Natl. Acad. Sci. U.S. Publ. 1573, Washington, D.C., 1968.

Section 9.4. Vapor Phase Osmometry

W. Simon and C. Tomlinson, Thermoelektrische Mikrobestimmung von Molekulargewichten, *Chimia* **14**, 301 (1960).

K. Kamide and M. Sanada, Molecular weight determination by vapor pressure osmometry, *Kobunshi Kagaku (Chem. High Polym. Japan)* **24**, 751 (1967) (in Japanese).

J. van Dam, Vapor-phase osmometry, *in Characterization of Macromolecular Structure*, Natl. Acad. Sci. U.S. Publ. 1573, Washington, D.C., 1968.

D. E. Burge, Calibration of vapor phase osmometers for molecular weight measurements, *J. Appl. Polym. Sci.* **24**, 293 (1979).

Section 9.5. Light Scattering

D. MacIntyre and F. Gornick (eds.), *Light Scattering from Dilute Polymer Solutions*, Gordon and Breach, New York, 1964.

K. A. Stacey, *Light Scattering in Physical Chemistry*, Butterworths, London, 1956.

M. Kerker, *The Scattering of Light and Other Electromagnetic Radiation*, Academic Press, New York, 1969.

M. B. Huglin (ed.), *Light Scattering from Polymer Solutions*, Academic Press, London, 1972.

B. Chu, *Laser Light Scattering*, Academic Press, New York, 1974.

B. J. Berne and R. Pecora, *Dynamic Light Scattering*, Wiley, New York, 1976.

M. B. Huglin, Determination of molecular weight by light scattering, *Topics Curr. Chem.* **77**, 141 (1978).

Section 9.6. Small-Angle X-Ray Scattering and Neutron Diffraction

H. Brumberger (ed.), *Small Angle X-ray Scattering*, Gordon and Breach, New York, 1967.

T. L. Cottrell and I. C. Walker, Interaction of slow neutrons with molecules, *Q. Rev. (London)* **20**, 153 (1966).

G. Allen and C. J. Wright, Neutron scattering studies of polymers, *in Macromolecular Science* (Vol. 8 of Physical Chemistry Series 2) (C. E. H. Bawn, ed.), MTP International Review of Science, 1975, p. 223.

G. Allen and C. J. Wright, Neutron scattering studies of polymers, *Int. Rev. Sci., Phys. Chem. (2)* **8**, 223 (1975).

R. Ullman, Small angle neutron scattering of polymers, *Ann. Rev. Mater. Sci.* **10**, 261 (1980).

Section 9.7. Ultracentrifugation

T. Svedberg and K. O. Pedersen, *Die Ultrazentrifuge*, D. Steinkopff, Dresden, 1940; *The Ultracentrifuge*, Clarendon Press, Oxford, 1940.

H. K. Schachmann, *Ultracentrifugation in Biochemistry*, Academic Press, New York, 1959.

R. L. Baldwin and K. E. van Holde, Sedimentation of high polymers, *Fortschr. Hochpolym. Forschg.* **1**, 451 (1960).

H.-G. Elias, *Ultrazentrifugen-Methoden*, Beckman Instruments, Munich, 1961.

H. Fujita, *Mathematical Theory of Sedimentation Analysis*, Academic Press, New York, 1962.

J. Vinograd and J. E. Hearst, Equilibrium sedimentation of macromolecules and viruses in a density gradient, *Fortschr. Chem. Org. Naturstoffe* **20**, 372 (1962).

J. W. Williams (ed.), *Ultracentrifugal Analysis in Theory and Experiment*, Academic Press, New York, 1963.

H. Fujita, *Foundations of Ultracentrifugal Analysis*, Wiley, New York, 1975.

D. Rickwood (ed.), *Centrifugation: A Practical Approach*, Information Retrieval Ltd., London, 1978.

Section 9.8. Chromatography

G. M. Guzman, Fractionation of high polymers, *Prog. High Polym.* **1**, 113 (1961).

R. M. Screaton, Column fractionation of polymers, *in Newer Methods of Polymer Characterization* (B. Ke, ed.), Interscience, New York, 1964.

J. F. Johnson, R. S. Porter, and M. J. R. Cantow, Gel permeation chromatography with organic solvents, *Rev. Macromol. Chem.* **1**, 393 (1966).

M. J. R. Cantow (ed.), *Polymer Fractionation*, Academic Press, New York, 1967.

H. Determann, *Gelchromatographie*, Springer, Berlin, 1967.

J. F. Johnson and R. S. Porter, Gel permeation chromatography, *Prog. Polym. Sci.* **2**, 201 (1970).

K. H. Altgelt and L. Segal, *Gel Permeation Chromatography*, Marcel Dekker, New York, 1971.

N. Friis and A. Hamielec, Gel permeation chromatography—review of axial dispersion phenomena, their detection and correction, *Adv. Chromatog.* **13**, 41 (1975).

L. Fischer, *An Introduction to Gel Chromatography*, North-Holland, Amsterdam, 1969.

C. F. Simpson, *Practical High Performance Liquid Chromatography*, Heyden, London, 1976.

H. Inagaki, Polymer separation and characterization by thin layer chromatography, *Adv. Polym. Sci.* **24**, 189 (1977).

J. Cazes (ed.), Liquid chromatography of polymers and related materials, Marcel Dekker, New York, 1977.

W. W. Yan, J. J. Kirkland, and D. D. Bly, *Modern Size-Exclusion Liquid Chromatography*, Wiley, New York, 1979.

Section 9.9. Viscometry

G. Meyerhoff, Die viskosimetrische Molekulargewichtsbestimmung von Polymeren, *Fortschr. Hochpolym. Forschg.—Adv. Polym. Sci.* **3**, 59 (1961/64).

M. Kurata and W. H. Stockmayer, Intrinsic viscosities and unperturbed dimensions of long chain molecules, *Fortschr. Hochpolym. Forschg.—Adv. Polym. Sci.* **3**, 196 (1961/64).

H. van Oene, Measurement of the Viscosity of Dilute Polymer Solutions, *in Characterization of Macromolecular Structure*, Natl. Acad. Sci. U.S. Publ. 1573, Washington, D.C., 1968.

H. Yamakawa, *Modern Theory of Polymer Solutions*, Harper & Row, New York, 1971.

Other Methods

D. V. Quayle, Molecular weight determination of polymers by electron microscopy, *Br. Polym. J.* **1**, 15 (1969).

H. J. Purz, E. Schulz, Electron microscopy in polymer science. Review, *Acta Polym.* **30**, 377 (1979).

Part III
Solid State Properties

Chapter 10
Thermal Transitions

10.1. Basic Principles

10.1.1. Phenomena

Low-molar mass materials change their physical state as the temperature increases; at the melting temperature they change visibly from a crystal to a liquid, and at the boiling temperature from a liquid to a vapor. Each true phase transition is defined thermodynamically by a marked change in enthalpy or volume. However, since such changes can only be determined with difficulty, other methods are generally employed to determine the transition temperatures.

In organic chemistry melting temperatures are measured via the formation of the liquid state in the melting-point tube. This method can be used for the determination of the melting temperature because the viscosity at the melting point changes by several orders of magnitude and the viscosity of the melt is very low. The method must fail when the viscosity of the melt is so high that flow can no longer be perceived within the observation time of the experiment. This is already the case with highly fused aromatic rings such as coronene, and even more so in crystalline macromolecular substances. In the case of coronene, the melting temperature cannot be determined unambiguously with the melting-point tube. In macromolecules, the "melting point" determined in the melting-point tube is really a flow temperature, which, in certain cases, can lie well above the true melting temperature because of the high melt viscosity.

Noncrystalline materials can also show flowing in the melting-point tube. For example, radically polymerized styrene changes on heating from a brittle, rather glassy material into one which is softer and rubbery. As X-ray analysis

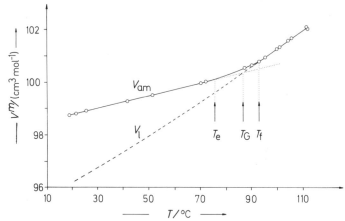

Figure 10-1. The molar volume of an atactic poly(styrene) with $\overline{M}_n = 20\ 000\ \mathrm{g/mol}$ as a function of temperature. T_e, softening point; T_G, glass transition temperature; T_f, freezing-in temperature (after G. Rehage).

shows, this poly(styrene) is amorphous, and there is no doubt that this flow temperature is *not* a melting temperature. It is more likely a softening temperature, a transition from a solid substance to a liquid. On cooling, the same effect is seen as freezing temperature. The concepts of "softening" and "freezing" temperatures are mostly defined phenomenologically, today. The freezing temperature, so defined, is the temperature at which properties first deviate from "normal" behavior observed at higher temperatures (see Figure 10-1). The "softening temperature" is similarly defined for experiments with increasing temperature. The "glass transition temperature" is defined as that temperature where the "linear" parts of the curve below the softening temperature and above the freezing temperature intersect (Figure 10-1). Consequently, the glass transition temperature lies between these two temperatures. In many cases, however, the difference between these three temperatures is conceptually and numerically indistinguishable.

As well as the glass transition and melt temperatures, additional transition or relaxation temperatures exist. Since a molecular interpretation cannot always be readily provided for these, the transition occurring at the highest temperature is called an α transition, the one occurring at the next lower temperature is called a β transition, etc.

10.1.2. Thermodynamics

Thermodynamic states are, of course, described by the Gibbs energy G or its first derivative with respect to temperature or pressure, i.e., by the enthalpy

H, the entropy *S*, and the volume *V*:

$$H = G - T\left(\frac{\partial G}{\partial T}\right)_p \tag{10-1}$$

$$S = -\left(\frac{\partial G}{\partial T}\right)_p \tag{10-2}$$

$$V = \left(\frac{\partial G}{\partial p}\right)_T \tag{10-3}$$

The second derivatives of the Gibbs energy lead correspondingly to the heat capacity C_p, the cubic expansion coefficient α, and the isothermal compressibility κ:

$$C_p = \left(\frac{\partial H}{\partial T}\right)_p = T\left(\frac{\partial S}{\partial T}\right)_p = -T\left(\frac{\partial^2 G}{\partial T^2}\right)_p \tag{10-4}$$

$$\alpha = V^{-1}\left(\frac{\partial V}{\partial T}\right)_p \tag{10-5}$$

$$\kappa = -V^{-1}\left(\frac{\partial V}{\partial p}\right)_T \tag{10-6}$$

Thermodynamic transitions can be characterized by corresponding changes in these parameters of state. First- and second-order transitions are both characterized by the thermodynamic equilibria on either side of the physical transition temperature.

First-order thermodynamic transitions show a sharp discontinuity in the *first* derivative of the Gibbs energy with temperature at the transition in question; the second derivative also has a corresponding discontinuity (Figure 10-2). The melting temperature, for example, is a typical first-order thermodynamic transition.

Conversely, second-order thermodynamic transitions, by definition, first show a sharp discontinuity for the *second* derivative of the Gibbs energy with temperature; the Gibbs energy, itself, and its first derivatives change continuously at the transition temperature (Figure 10-2). All transitions definitely known to be of second thermodynamic order are single-phase transitions such as, for example, rotational transitions in crystals and the disappearance of ferromagnetism at the Curie point.

Several aspects of a genuine second-order thermodynamic transition can be observed with what is known as the glass transition temperature (also often called the glass temperature), i.e., discontinuities in C_p, α, and κ. But the glass transition is not a genuine thermodynamic transition since there is no equilibrium between *both* sides of the glass transition point. The position of

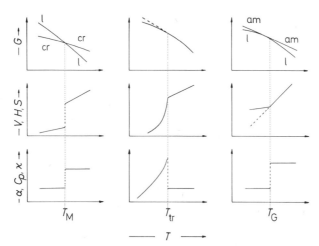

Figure 10-2. Schematic representation of various thermal transitions: the melting process as a first-order thermodynamic process, a rotational transition as a second-order thermodynamic process, and a glass transition; l, liquid; cr, crystal; am, amorphous state (after G. Rehage and W. Borchard).

the glass transition temperature depends essentially on the rate of cooling, the transition temperature lying at lower temperatures for lower rates of cooling. At very low cooling rates, no sharp discontinuity at all is obtained for the volume as a function of the temperature (Figure 10-1), that is, the glass transition temperature has disappeared.

The curve for very slow coolings cannot be obtained by direct measurement. To obtain such a curve, one can, for example, define a time $t_{1/e}$ for which the distance from the equilibrium curve is $1/e$ times the initial deviation. For the poly(styrene) data shown in Figure 10-1 ($T_G = 89°$ C), this time is 1 s at 95° C and 5 min at 89° C, but already 1 year at 77° C. The dotted curve in Figure 10-1 was consequently obtained by extrapolation of the values for a poly(styrene) solution in malonic ester. This extrapolation is possible since the cubic expansion coefficient of the polymer in the solvent is a linear function of the concentration up to a poly(styrene) concentration of 90%.

The lowering of the glass transition temperature by slower cooling indicates a kinetic cause. In genuine thermodynamic second-order transitions, the transition temperature does not depend on the rate of cooling. Another point against the glass transition temperature being a thermodynamic transition point is the fact that C_p, α, and κ are smaller below the glass transition temperature than above it. This is exactly the reverse of what would be expected of a genuine second-order thermodynamic transition (Figure 10-1).

As already mentioned in Section 5.5.1, chain-segment mobility becomes frozen in at and below the glass transition temperature. Perfect crystals do not

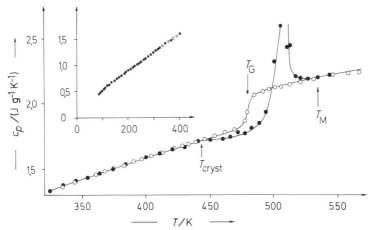

Figure 10-3. Specific heat capacity c_p at constant pressure of partially crystalline ($—\bullet—\bullet—$) and amorphous ($—\circ—\circ—$) poly[oxy-(2,6-dimethyl)-1,4-phenylene]. T_{cryst} denotes the beginning of recrystallization, T_G is the glass transition temperature, and T_M is the melting temperature (after F. R. Karasz, H. E. Bair, and J. M. O. Reilly).

have any mobile segments, since all segments are firmly set in the crystal lattice. Consequently, highly crystalline macromolecules do not show a glass transition temperature. Conversely, amorphous polymers do not have a melting temperature since a melting temperature requires a crystal lattice.

On the other hand, a partially crystalline polymer possesses both a glass transition temperature and a melting temperature (Figure 10-3). Since the chain segments still retain some mobility even below the glass-transition temperature (see also Section 10.5), crystallization can start even below the glass transition temperature (Figure 10-3) because of the practically ever-present crystallization nucleators (see Section 10.3.2). On heating, there is a constant interchange of chain segments between the crystalline and non-crystalline regions so that they are constantly being redistributed. Consequently, a melting *region* is observed instead of a melting temperature. The upper end of the melting region is defined as the melting temperature T_M of the sample since the largest and most perfect crystallites melt here. This melting temperature T_M lies lower than the thermodynamic melting temperature T_M^0 of a perfect crystal.

10.2. Special Parameters and Methods

10.2.1. Thermal Expansion

Thermal expansion depends on variations in the interatomic forces with temperature. These forces are strong for covalent bonds and weak for van der Waals forces. For example, for quartz, all atoms are three-dimensionally fixed

Table 10-1. Density ρ, Specific Heat Capacity at Constant Pressure c_p,
Linear Coefficient of Expansion β, and Thermal Conductivity λ of
Polymers, Metals, and Glass at 25°C

Material	ρ in g/cm^3	c_p in $J\,g^{-1}\,K^{-1}$	$10^5\beta$ in K^{-1}	λ in $J\,m^{-1}\,s^{-1}\,K^{-1}$
Poly(ethylene)	0.92	2.1	20	0.35
Poly(styrene)	1.05	1.3	7	0.16
Poly(vinyl chloride)	1.39	1.2	8	0.18
Poly(methyl methacrylate)	1.19	1.5	8.2	0.20
Poly(caprolactam)	1.13	1.9	8	0.29
Poly(oxymethylene)	1.42	1.5	9.5	0.23
Copper	8.9	0.39	2	350
Cast iron	7.25	0.54	1	58
Jena glass 16 III	2.6	0.78	1	0.96
Quartz (spatial average)	2.65	0.72	0.1	10.5

in space: The thermal expansion is consequently very small. On the other
hand, in liquids, intermolecular forces are dominant: The thermal expansion
is large. The main-chain atoms of organic polymers are covalently bonded in
one direction only; in the two other directions in space intermolecular forces
are operative. Thus, polymers lie between liquids and quartz (or metals) as far
as thermal expansion is concerned (Table 10-1).

Because of the great difference in the thermal expansion of polymers and
metals or glass, significant problems can arise when thermal stress is applied
to composites of these materials. The so-called dimensional stability of the
polymer is also of technological importance. Dimensionally stable polymers
must not only possess a small coefficient of thermal expansion: They should
also not exhibit recrystallization phenomena. Recrystallizations lead to
distortions because of the difference in densities between amorphous and
crystalline regions.

10.2.2. Heat Capacity

The heat capacity ("specific heat") C_p of macromolecular substances at
constant pressure is the only heat capacity readily accessible experimentally.
For theoretical considerations, however, the heat capacity at constant volume
C_V is important. According to the laws of thermodynamics, these two
quantities are related to each other via the cubic expansion coefficient α and
the isothermal compressibility κ, and so, can be calculated from each other:

$$C_p = C_V + \frac{TV\alpha^2}{\kappa} \tag{10-7}$$

The molar heat capacity of crystalline polymers at constant volume C_V^m can be theoretically calculated when the frequency spectrum is known. Atoms oscillate harmonically about their equilibrium positions in the crystalline state. In accordance with the Einstein function, each individual oscillation contributes

$$E\left(\frac{\theta}{T}\right) = \theta^2 \frac{\exp(\Theta/T)}{1 - \exp(\Theta/T)} \tag{10-8}$$

to the total heat capacity. Here $\theta = h\nu/k$ is the Einstein temperature. The molar heat capacity is simply the sum of all these contributions

$$C_V^m = R \Sigma E\left(\frac{\theta}{T}\right) \tag{10-9}$$

At very low temperatures, these lattice oscillations comprise almost all of the heat capacity. At higher temperatures, a correction for the inharmonicity of the lattice oscillations must be considered. In addition, contributions from group oscillations and rotations about main-chain bonds must also be added at higher temperatures. Finally, a contribution from defects may also be needed.

According to the law of equal distribution of energy, the maximum heat capacity is $3R$ per mole atom per monomeric unit. In actual fact, however, some degrees of freedom are always frozen in, and the molar heat capacity is consequently lower than the maximum. A value of about $1R = 8.314$ J K^{-1} mol^{-1} has been empirically found for solid polymers at room temperature per mole atom. A specific heat capacity of 1.22 J K^{-1} g^{-1} is obtained for the example in Figure 10-3 at 25°C. Thus, with a monomeric unit chemical formula of C_8H_8O, a heat capacity of 146.4 J K^{-1} mol^{-1} for one mole of monomeric unit is obtained. With 17 atoms per monomeric unit, a heat capacity of 8.61 J K^{-1} mol^{-1} per mole atom is obtained.

In fact, the heat capacities of amorphous and crystalline polymers are practically the same below the glass transition temperature (Figure 10-3). Because of the onset of new oscillations at the glass transition temperature, the heat capacity increases more or less sharply. Since such oscillations can start below the glass transition temperature, crystallizable amorphous polymers can occasionally even start to recrystallize below the glass transition temperature. The heat capacity then passes through a maximum when melting occurs. The melting temperature is then the upper end of the melting range.

10.2.3. Differential Thermal Analysis

In differential thermal analysis (DTA), a test sample and a comparison sample are heated at a constant rate. The temperature difference between the

two samples is measured. The comparison sample is so chosen so as not to
show any chemical or physical changes in the range of temperatures studied.
When, for example, the temperature reaches the melting point of the test
sample, then a fixed amount of heat (heat of fusion) must be applied until the
whole of the test sample has melted. The temperature of the test sample does
not change at the melting temperature, whereas that of the comparison sample
continues to rise. At the melting temperature, therefore, an endothermal
process is observed, i.e., a negative temperature difference between test and
comparison samples (Figure 10-4). The peak minimum is considered to be the
actual melting temperature, since the most frequently occurring crystallite size
melts here at zero heating rate. As heating continues, the test sample finally
attains the same temperature as the comparison sample again, and the
temperature difference becomes zero. This is shown by a base line running
parallel to the temperature axis. The base line, however, is mostly not at the
same level on each side of the melting temperature signal, since the test sample
usually has different heat capacities above and below the melting temperature.

Differential scanning calorimetry, DSC, is a variant of differential
thermal analysis. With DSC, the required heat for the transition is added or
removed at the transition temperature. Thus, this method is particularly
suited to quantitatively measuring heats of fusion or crystallization, as, for
example, with crystallization at a given temperature.

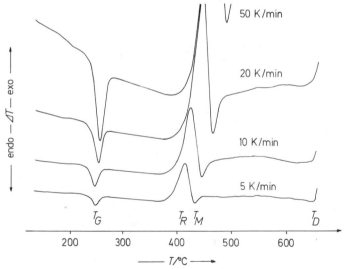

Figure 10-4. DSC diagram of a poly(oxy-2,6-dimethyl(-1,4-phenylene)). All measurements were
made with the same sample size and sensitivity, but with different rates of heating. T_G, Glass
transition temperature; T_R, onset of recrystallization; T_M, melting temperature; T_D, onset of
decomposition.

The actual differential thermal analysis is particularly suitable for routine analysis, as it can be carried out simply and quickly. On the other hand, measurements on unknown samples can often be interpreted only with difficulty and sometimes only if data from other methods are available. Quantitative information is also not readily accessible by this technique. Uncertainties in quantitative interpretation result because signal size and shape depend on the experimental conditions. For example, an endothermic peak is usually ascribed to a melting temperature, although the same signal can indicate a glass transition temperature at high rates of heating. A distinction between a melting and a glass transition temperature can be made by observing the thermal behavior of the samples under a polarizing microscope. Crystalline polymers are birefringent below the melting temperature, if the crystallite size is large enough, but not above it.

A change in slope of the DTA curve is generally interpreted as a glass transition. The glass transition temperature is usually considered to be the point of intersection of both branches of the curve. With DSC, and also sometimes with DTA, the glass transition temperature appears as a peak (Figure 10-4).

The way the results change with alteration in rate of heating and sample size should always be established. Larger sample sizes lead to a greater temperature gradient and slower temperature equilibration. This causes peak broadening and a shift to higher temperatures. Greater heating rates lead to larger peaks, since more heat per unit time is liberated (Figure 10-4).

10.2.4. Nuclear Magnetic Resonance

Atomic nuclei with uneven numbers of protons possess a magnetic moment and therefore precess about a magnetic field from an oscillating radio frequency. If the oscillating frequency of the electromagnetic field equals the precession frequency of the atomic nuclei, a resonance signal is observed whose frequency depends on the ratio of the nuclear magnetic field to the rotational torque on the nucleus and on the strength of the external, steady, magnetic field.

High-resolution nuclear magnetic resonance detects the shielding from neighboring protons in the same molecule. Such high-resolution spectroscopy methods can be used to elucidate the configuration and constitution of molecules. Much higher concentrations of, and consequently stronger interactions between, the magnetic dipoles of different nuclei exist in solids below the glass transition temperature or in melts. The magnetic dipoles of these neighboring nuclei have a distribution of orientations relative to a given nucleus. A broad signal thus results.

Chain segments become more mobile with increasing temperature. The distribution of orientations around a given nucleus thus becomes more and more random. The increased anisotropy of dipole interactions leads to a sharpening of the signals. A line width is thus a measure of the mobility of the molecules, and thence the glass transition temperature. Since the measurements are carried out at frequencies in the MHz range, the glass transition temperatures from nmr measurements are higher than those from static measurements of the temperature dependence of heat capacities or by DTA (see Section 10.5.2). Broad-line nmr also measures side-group motion, but not that from short chain segments below the glass transition temperature. The method is not very suitable for determining melting temperatures. In fact, the resonance signals continue to sharpen even well below the melting temperature of crystalline polymers, whereas the X-ray crystallinity remains constant. Therefore, there must be a certain segmental mobility within the crystal lattice even below the melting temperature.

A magnetic polarization results from the sudden imposition of a magnetic field on a sample. The magnetization generally follows an exponential function. The time constant T_1 is known as the spin–lattice relaxation time. Consequently, the nuclear magnetic resonance experiment corresponds macroscopically to a dielectric relaxation experiment. Differences exist, however, on the molecular level. The nuclear magnetization is, of course, equal to the sum of all the individual nuclear magnetic moments. The orientation of these nuclear magnets, however, is only loosely coupled with the molecular positions. Consequently, T_1 is generally much larger than the molecular relaxation time from dielectric relaxation measurements (see also Section 10.5.3).

10.2.5. Dynamic Methods

The mechanical and dielectric loss methods are based on the different mobilities of the chain segment or dipoles bound to them in the solid state and in the melt. These differences lead to an anomalous dispersion of the modulus of elasticity (see Section 11.4.4) or the relative permittivity (see Section 13.1.2) and to corresponding losses in the mechanical or the electrical alternating fields.

Part of the work performed on a sample will be converted irreversibly into random thermal motion by movement of the molecules or molecule segments. This loss passes through a maximum at the appropriate transition temperature or relaxation frequency in the associated alternating mechanical field (torsion pendulum test). A similar effect is obtained by the delayed response of the dipoles with dielectric measurements. Therefore, dielectric measurements can be made only on polar polymers. According to the frequency used, the glass-transition temperatures measured with dynamic methods lie higher than those obtained by static methods (see Section 10.5.3).

Of course, such dynamic mechanical test methods are only suitable for such probes as can support their own weight. In examining paints, lacquers, and non-self-supporting films, a glass fiber cord is impregnated with a solution of the test material and the solvent is evaporated. The impregnated cord is then subjected to periodic oscillations (torsional braid analysis).

10.2.6. Industrial Testing Methods

A whole series of empirical test methods are used in industry to determine the physical transitions. These methods are usually simultaneously influenced by various physical quantities and must therefore be standardized. The standards vary among countries.

The temperature at which a sample breaks when struck is measured in brittleness determinations. At this temperature, larger chain segments can no longer be displaced. In contrast, the mobility of much smaller chain segments is effective at the glass transition temperature, so the brittle temperature always lies higher than the glass transition temperature. The brittleness temperature does not depend solely on the mobility of larger chain segments; it also depends on the elasticity of the test sample, since the break behavior is influenced by the deformability of the sample. Thin samples are, however, more elastic than thick ones. Brittleness temperatures decrease with increasing molar mass up to a limiting value, since increased molecular length leads to increased strength (Figure 10-5).

The dimensional stability of plastics on heating is characterized by Vicat temperatures, heat distortion temperatures, or, in Germany, Martens numbers (or temperatures). The Martens number is defined as the temperature at which a standard test specimen heated at a rate of 50 K/h bends to a fixed extent under a load of 5 MPa. The heat distortion temperature is correspondingly measured with a load of 1.85 MPa on a bending rod being heated at a rate of 120 K/h. The Vicat temperature (or Vicat softening point) is the temperature at which a needle with point surface of 1 mm^2 penetrates the surface of a test sample being heated at a rate of 50 K/h to a depth of 1 mm when under a load of 10–50 N. All these "dimensional stability" tests still depend on the elasticity of the sample. The temperatures measured lie below the glass transition temperatures of amorphous polymers and below the melting temperatures of crystalline polymers.

The Vicat temperature measures the penetration of a needle into a sample under otherwise constant conditions. Thus, the method also measures elasticity and surface hardness.

The methods of determining Martens number and Vicat and heat

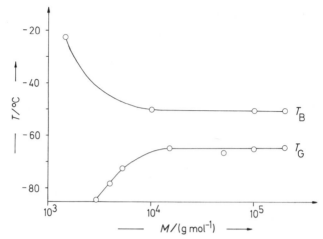

Figure 10-5. Molar mass dependence of the brittleness and the glass transition temperatures T_B and T_G of poly(isobutylene) (after A. X. Schmidt and C. A. Marlies).

distortion temperature are integral processes. Consequently, they are only suitable for studying samples with only one thermal transition temperature. Neither partially crystalline polymers nor polymer blends can be meaningfully characterized by these methods.

In laboratories where preparative work is done, "softening points" are often determined with the Kofler bar, which consists of a metal plate with a temperature gradient along it. The sample is moved with a brush from the colder to the warmer points on the metal plate. At a given point, the sample will remain stuck to the plate; the temperature associated with this point is taken as the "softening point." Since this temperature depends on both the viscosity of the sample and its adhesion to the metal surface, the softening point thus determined is very approximate. It often bears no simple relationship to the glass transition or melt temperature.

10.3. Crystallization

Many polymers crystallize when their melts or solutions are cooled. The states of order which occur depend on the constitution and configuration of the macromolecules. They also depend very much on the external conditions such as the polymer concentration, the temperature, the solvent, the method of crystallization induction, the shear stress applied to the solution or melt, etc. These internal and external parameters determine not only the formation and growth of the crystallization nuclei, but also the morphology of the crystalline structures formed.

10.3.1. Nucleation

Every crystallization is induced by crystallization nuclei. Nucleation may be homogeneous or heterogeneous. Homogeneous nuclei are formed from molecules or molecular segments of the crystallizing material itself; this kind of nucleation is therefore called spontaneous or thermal nucleation. In contrast, heterogeneous nucleation is caused by the surface of foreign bodies in the crystallizing material. These foreign bodies may be dust particles, container walls, or purposely added nucleating agents. The nucleation density varies between very wide limits; for example, from about 1 nucleus/cm^3 in the case of poly(ethylene oxide) to about 10^{12} nuclei/cm^3 in the case of poly(ethylene).

Nuclei are only stable above a certain given size. Loose aggregates of molecules or segments, called embryons, are formed and destroyed in each instant in every solution or melt. The Gibbs energy, ΔG_i, for embryon formation from i lattice units is given by the Gibbs surface energy, ΔG_σ, and the Gibbs crystallization energy, ΔG_{cryst}:

$$\Delta G_i = \Delta G_\sigma - \Delta G_{cryst} \tag{10-10}$$

or, for spherical nuclei of radius r and Gibbs surface energy of ΔG_σ^a per unit surface area as well as Gibbs crystallization energy of ΔG_{cryst}^v per volume

$$\Delta G_i(r) = 4\pi r^2 \,\Delta G_\sigma^a - (4\pi/3)r^3 \,\Delta G_{cryst}^v \tag{10-11}$$

and, correspondingly, for a nucleus of any given shape consisting of j molecules or segments

$$\Delta G_i(j) = K'j^{2/3} \,\Delta G_\sigma^a - K''j \,\Delta G_{cryst}^v \tag{10-12}$$

The surface and crystallization energies are of opposite sign, and, so, the energy of nucleus formation only becomes negative above a critical nucleus size of r_{crit} or j_{crit} (Figure 10-6). The embryons convert to stable nuclei above the critical nucleus size, and can then grow further, into, for example, spherulites.

In addition, the nucleation may be primary, secondary, or tertiary (Figure 10-7). Primary nucleation is three dimensional: a new surface is formed in each direction in space. Secondary nucleation, in contrast, is two dimensional; with tertiary nucleation, growth is in one dimension.

Homogeneous nucleation is always primary. It occurs very rarely, and appears only to have been observed with melts of poly(pivalolactone) and poly(chlorotrifluoroethylene) when they are strongly undercooled. Since the nucleation is a spontaneous process, it must be sporadic, that is, nuclei form successively and an increase in number of spherulites with time is observed. Such effects are not only observed by homogeneous nucleation, however, but

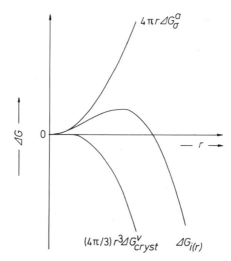

Figure 10-6. Gibbs surface energy, ΔG_σ^a, Gibbs crystallization energy, ΔG_{cryst}^v, and Gibbs nucleation energy, $\Delta G_{i(r)}$, for the formation of a spherical nucleus of radius r.

also with what is known as athermal nucleation. Partially crystalline polymers, of course, possess a wide melting range. Such samples do indeed appear molten above their conventionally defined melting temperatures, but may still contain residues of molten crystallites. These residues function as athermal crystallization nuclei and induce crystallization on cooling (Figure 10-8). Such persistent nuclei are also responsible for the "memory effect" in melts: that is, the phenomenon whereby spherulites after melting and recooling often reappear in the same positions they occupied before melting. The spherulites appear in the same positions because the viscosities of the melts are very high and athermal nuclei cannot diffuse to new positions.

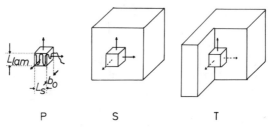

Figure 10-7. Illustration of the formation of primary P, secondary S, and tertiary T, nuclei. Growth in the --- direction produces no new surfaces, but growth in the — direction does. L_{lam}, Lamellar height; L_s, lamellar length; b_0, lamellar thickness (approximately the same size as the molecular diameter). The nucleus forms in the L_s direction, and the crystal grows in all three directions by deposition of other molecules.

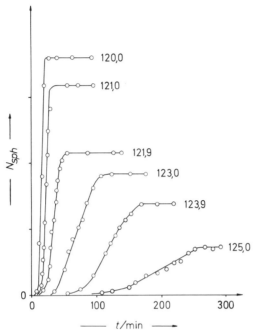

Figure 10-8. Time dependence of the number N_{sph} of spherulites formed in melts of poly-(decamethylene terephthalate) at different temperatures. N_{sph} is in arbitrary units.

Heterogeneous nucleation occurs when the crystallizing molecules or segments can wet the nucleus surface or deposit in cracks or holes on the nucleus. If the interaction between the nuclei and the crystallizing material is strong, then the number of nuclei or the spherulites formed from them is constant from the start of the crystallization and all spherulites are of the same size.

In heterogeneous nucleation, a chainlike molecule deposits onto the existing nucleus surface with chain folding (Figure 10-7). The Gibbs energy of secondary nucleation is given by the surface energies of the fold surface σ_f and of the side surfaces σ_s, as well as the crystallization energy per unit volume ΔG_{cryst}^v:

$$\Delta G_i = 2L_s b_0 \sigma_f + 2L_{lam} b_0 \sigma_s - b_0 L_s L_{lam} \Delta G_{cryst}^v \qquad (10\text{-}13)$$

Only two of the four possible side surfaces need to be considered. since, in the b_0 direction, no new surface is created. Differentiating Equation (10-13) with respect to L_s and equating to zero, we obtain

$$(L_{lam})_{theor} = 2\sigma_f / \Delta G_{cryst}^v \qquad (10\text{-}14)$$

for the theoretical critical lamellar height. The critical lamellar thickness or side-surface length is correspondingly given by differentiation with respect to L_{lam}:

$$(L_s)_{theor} = 2\sigma_s / \Delta G^v_{cryst} \qquad (10\text{-}15)$$

The Gibbs energy of crystallization, however, depends on the crystallization temperature:

$$\Delta G^v_{cryst} = (\Delta H_M)_u - T_{cryst}(\Delta S_M)_u \qquad (10\text{-}16)$$

The melting point, on the other hand, is given by

$$T^0_M = \frac{(\Delta H_M)_u}{(\Delta S_M)_u} = \frac{\Delta H^0_M}{\Delta S^0_M} \qquad (10\text{-}17)$$

Inserting Equations (10-16) and (10-17) into (10-14) leads to

$$(L_{lam})_{theor} = \frac{2\sigma_f T^0_M}{(\Delta H_M)_u (T^0_M - T_{cryst})} \qquad (10\text{-}18)$$

Thus, the critical theoretical lamellar height decreases with increasing supercooling $(T^0_M - T_{cryst})$. This has been observed for actual lamellar heights (Figure 10-9).

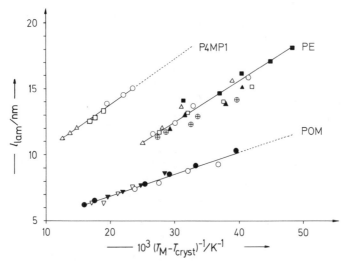

Figure 10-9. Dependence of the lamellar height L_{lam} on the reciprocal undercooling for poly(4-methyl pentene-1), poly(ethylene), and poly(oxymethylene) for experiments carried out in different solvents (after A. Nakajima and F. Hameda).

10.3.2. Nucleators

Homogeneous and athermal nuclei are formed sporadically. As a result, crystallization proceeds only slowly, and the final properties of the polymer are only achieved after a relatively long time, which, for example, leads to large cycle times in injection moulding. In addition, first formed nuclei grow to large spherulites, which influence the mechanical properties unfavorably.

For this reason, attempts are made to control crystallization through the use of solid nuclei. In theory, nuclei of the same polymer can be added to externally produce homogeneous nucleation. As an example of this droplet technique, the sample is ground so small that there are many small particles and the probability that a foreign body resides in a given particle decreases to near zero.

A method of internal homogeneous nucleation utilizes the tendency of chainlike molecules to chain-fold. Stiff macromolecules do not fold easily. If flexible segments are polymerized into the macromolecule, these segments will preferentially reside in the fold surface, and homogeneous nucleation is facilitated by chain folding.

Special external nucleating agents, however, are used in industry. Alkali, alkaline earth, aluminum, and titanium salts of organic carboxylic and sulfonic and phosphonic acids are suitable for poly(olefines). Flavanthrone, copper phthalocyanines, or other planar aromatic ring systems may be used. Quartz, graphite, titanium dioxide, carbon black, and alkali halogenides are added to polyamides. The nucleating mechanism of these nucleators obviously does not only depend on their wettability by the polymer melt. It appears that all effective nucleating agents have flat furrows on their surfaces. These furrows force the adsorbed polymer molecules to adopt stretched conformations, and these are the precursors of crystallization by chain folding.

It is not only the crystallization rate, however, which is influenced by nucleating agents, but also the morphology. Isotactic poly(propylene) crystallizes monoclinically in the presence of *p-t*-butyl benzoic acid and pseudohexagonally when the quinacridone dyestuff, Permanent Red E3B, is added.

10.3.3. Crystal Growth

Immediately below the melting point, the crystallization rate is very small, since already formed nuclei can disassociate. At a temperature T_{ch} about 50 K below the glass transition temperature, the mobility of the chain segments and molecules, however, is virtually zero. Consequently, crystallization generally occurs only between the melting temperature and the glass

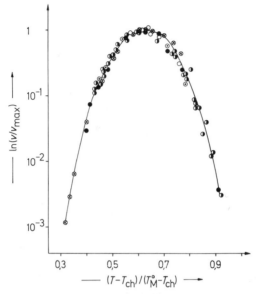

Figure 10-10. Plot of the natural logarithm of the reduced growth rate of spherulites of different polymers as a function of a reduced temperature. T_{ch} is the characteristic temperature, about 50 K below the glass transition temperature, at which all molecular motion ceases. T_M^0 is the thermodynamic melting point (after A. Gandica and J. H. Magill).

transition temperature, and the crystallization rate passes through a maximum (Figure 10-10). A universal curve, independent of the polymer type, is obtained when $\ln(v/v_{max})$ is plotted against $(T - T_{ch})/(T_M^0 - T_{ch})$. The maximum occurs at about a value of 0.63. If T_{ch} is expressed in terms of $T_G - 50$, and considering that the values of T_M^0/T_G lie between 2 and 1.5, then it can be shown that the maximum rate of crystallization occurs at about $0.8{-}0.87 T_M^0$.

The time dependence of the overall (primary) crystallization is described by the Avrami equation, which was originally derived for the crystallization of metals. The crystallinity is expressed as the volume fraction ϕ of the crystalline material in the total sample. For the derivation, it is assumed that each nucleus leads to an entity (e.g., a rod, disk, or sphere). After an infinitely long time the whole sample is filled with these shapes. The crystallinity of the sample is then ϕ_∞. This, however, is the crystallinity of a single entity which is assumed not to change during crystallization. At the time t, the fraction of the sample volume filled by these entities is ϕ/ϕ_∞. With randomly distributed nuclei the probability p that a point does not lie in any one entity is proportional to this fraction, i.e.,

$$p = 1 - \frac{\phi}{\phi_\infty} \tag{10-19}$$

The probability p_i that a point does not lie in a given entity with the volume V_i is

$$p_i = 1 - \frac{V_i}{V} \tag{10-20}$$

The probability that a point lies outside all the entities is equal to the product of all the individual probabilities:

$$p = p_1 p_2 \ldots p_n = \prod_{i=1}^{n} \left(1 - \frac{V_i}{V}\right) \tag{10-21}$$

or

$$\ln p = \sum_{i=1}^{n} \ln \left(1 - \frac{V_i}{V}\right) \tag{10-22}$$

Should the volume of every entity be very much smaller than the total volume ($V_i \ll V$), then the logarithmic expression can be developed into a series $\ln(1 - x) = -x - x^2/2 - \cdots$, whereby terms in x^2 and higher can be neglected. We have

$$\ln p = -\sum_{i=1}^{n} \frac{V_i}{V} = -V^{-1} \sum_{i=1}^{n} V_i \tag{10-23}$$

The mean volume $\overline{V}_i$ of a single entity is given by $\overline{V}_i = (\sum_{i=1}^{n} V_i)/N$, where N is the number of entities. The concentration ν of entities per unit volume is $\nu = N/V$. Accordingly, when the antilogarithm has been evaluated, Equation (10-23) becomes $p = \exp(-\nu \overline{V}_i)$, and Equation (10-19) becomes

$$\phi = \phi_\infty [1 - \exp(-\nu \overline{V}_i)] \tag{10-24}$$

If all the entities are formed simultaneously, then the density of nuclei is constant ($\nu = k_0$). Furthermore the growing entities all have the same volume $\overline{V}_i$, which naturally increases with time. For rodlike entities with constant cross section A, this increase results entirely from the increase with time of the length L:

$$\overline{V}_i = AL = Ak_1 t \qquad \text{(rods)} \tag{10-25}$$

With disk-shaped entities, the thickness d remains constant, and the radius grows with time; thus, for average volume,

$$\overline{V}_i = \pi dr^2 = \pi d(k_2 t)^2 \qquad \text{(disks)} \tag{10-26}$$

With spheres, the radius also increases proportionately with time, giving

$$\overline{V}_i = (4/3)\pi (k_3 t)^3 \qquad \text{(spheres)} \tag{10-27}$$

The nuclei are not formed all at once, but appear randomly with time in sporadic nucleation. Therefore ν is a time-dependent entity, increasing with time. When it is assumed that the nuclei appear randomly in both time and space, the mathematics becomes simpler. The final result is not altered by this double assumption because when the treatment concerns the whole volume, including that already filled with growing forms, the fictitious nuclei within the already crystallized regions do not affect the volume fraction of space available for growth. With sporadic nucleation, it is possible that the concentration ν of nuclei increases linearly with time t:

$$\nu = kt \tag{10-28}$$

The mean volume $\overline{V}_i$ of a nucleus is then given by reasoning that every nucleus has the same probability of being formed in the same interval of time. For the time interval from $t - \tau$ to $t - \tau + d\tau$ this interval is $d\tau/t$. Equations (10-25)–(10-27) then become

$$\overline{V}_i = k_1 A \int_0^t (t - \tau) \frac{d\tau}{t} = 0.5 k_1 A t \qquad \text{(rods)} \tag{10-29}$$

$$\overline{V}_i = \pi d k_2^2 \int_0^t (t - \tau)^2 \frac{d\tau}{t} = \frac{1}{3} \pi d k_2^2 t^2 \qquad \text{(disks)} \tag{10-30}$$

$$\overline{V}_i = \frac{4}{3} \pi k_3^3 \int_0^t (t - \tau)^3 \frac{d\tau}{t} = \frac{1}{3} \pi k_3^3 t^3 \qquad \text{(spheres)} \tag{10-31}$$

If these expressions for ν and $\overline{V}_i$ are inserted into Equation (10-24), one obtains equations of the general type

$$\phi = \phi_\infty [1 - \exp(-zt^n)] \tag{10-32}$$

Equation (10-32) is known as the Avrami equation. The physical significances of the constants z and n are given in Table 10-2.

A double logarithmic form of Equation (10-32) is used to evaluate crystallization kinetic data:

$$\ln[-\ln(1 - \phi\phi_\infty^{-1})] = \ln z + n \ln t \tag{10-33}$$

The constant n can be found from a plot of the left-hand side of Equation (10-33) against $\ln t$. For example, when poly(chlorotrifluoroethylene) is crystallized, values for n of 1 or 2 are found according to crystallization conditions. A value of $n = 3$ is obtained for poly(hexamethylene adipamide). Poly(ethylene terephthalate) gives values of between 2 and 4 according to the crystallization temperature. Extra difficulties in the interpretation of values of n can be seen from the fact that nonintegral numbers and values of $n = 6$ are

Table 10-2. Constants z and n of the Avrami Equation

Shape	z		n	
	Instantaneous	Sporadic	Instantaneous	Sporadic
Rod	$k_0 k_1 A$	$0.5 k k_1 A$	1	2
Disk	$k_0 k_2^2 \pi d$	$\frac{1}{3} k k_2^2 \pi d$	2	3
Sphere	$\frac{4}{3} k_0 k_3^3 \pi$	$\frac{4}{3} k k_3^3 \pi$	3	4

also known. Different methods can yield different values of n when the methods measure different aspects of the crystallization. For example, dilatometry generally measures the growth of spherulites, whereas calorimetry also measures the growth of lamellae in spherulites. The Avrami equation, however, is, of course, suitable in all cases up to the point where growth forms impinge on each other.

The rate of crystallization depends on the constitution and the configuration of the polymer, and varies between wide limits (Table 10-3). Polymers of symmetrical structure mostly crystallize rapidly; conversely, polymers with hindering substituents and main-chain groups crystallize slowly. Since the crystallization rate depends on nucleation as well as on crystal growth, athermal crystallization is generally faster than spontaneous crystallization. Poly(ethylene terephthalate), for example, can be obtained almost completely amorphous by rapidly supercooling from the melt, but the rapidly crystallizing poly(ethylene) has never been produced in the almost completely amorphous state, even when the melt is cooled directly with liquid nitrogen. It is also for this reason that PET crystallizes much more slowly than PE.

Table 10-3. Linear Crystallization Rates of Various
Polymers from the Melt When Supercooled to
Approximately 30° C below the Melting Point

Polymer	Crystallization rate in μm/min
Poly(ethylene)	5000
Poly(hexamethylene adipamide)	1200
Poly(oxymethylene)	400
Poly(caprolactam)	150
Poly(trifluorochloroethylene)	30
it-Poly(propylene)	20
Poly(ethylene terephthalate)	10
it-Poly(styrene)	0.25
Poly(vinyl chloride)	0.01

10.3.4. Morphology

It has as yet not been completely established how the microstructure of crystalline polymers or the crystallization conditions relate to the morphology. The radius of gyration of partially deuterated poly(ethylenes) does not change on passing from the melt to the partially crystalline state, according to neutron scattering measurements. In both cases, the radius of gyration varies with the square root of the molar mass. It is also independent of the long period, which characterizes the crystal thickness in the main-chain direction. The long period, of course, can be varied by annealing. An explanation for this radius of gyration phenomenon and the classification of macroscopic states of order remains outstanding.

In general, melt crystallization leads to spherulites which form from the melt, and have a fibrillar internal structure (see Figure 5-24) resulting from fractionation crystallization during spherulitic growth. Polymer fractions which are more strongly branched and/or are of lower molar mass have lower melting temperatures than the more linear and higher molar mass fractions; they thus require a stronger undercooling in order to crystallize. These more poorly crystallizing fractions are excluded from the zones of crystal growth and accumulate between these zones. They impede crystallization between the zones, which leads to preferential growth in the crystal growth zones, giving a fibrillar structure.

Spherulites can be considered as special forms of dendritic growth. Dendrites are always produced when crystals grow in the direction of a strong temperature gradient in supercooled melts (see also Section 5.4.4). The liquid phase occurring between the dendrites often solidifies in a microcrystalline form. In addition, the nucleation can occur from the melt surface. Trans-crystallization is the name given to this process, which is strongly influenced by diffusional processes.

Strongly surface-deficient lamellalike single crystals (see Section 5.4.2) are produced by single-crystal growth in very dilute polymer solutions. Preferential one-dimensional segment growth along the lamellar side surfaces occurs. Two-dimensional growth can also occur along the sides or fold surfaces, thus forming steps or spirals. The lamellar height (also called the fold length) for a given crystallization temperature is practically independent of the molar mass when the degree of polymerization is sufficiently large. The lamellar height varies, however, with the solvent when different polymer macroconformations are stable in different solvents. For example, amylose tricarbanilates crystallize from a dioxane/ethanol mixture in the form of folded chains, whereas from pyridine/ethanol they grow in the form of folded helices. For very high degrees of polymerization and moderate polymer concentrations, aggregate nuclei, among others, are formed since the

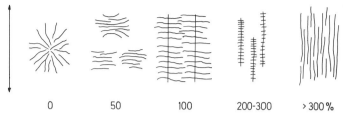

0	50	100	200-300	> 300%

Figure 10-11. Influence of shear on the crystallization of natural rubber. ↔, Shear stress direction (after E. H. Andrews).

viscosities are much higher and the simpler folded structures aggregate. At still higher concentrations, fibrillar, or even network structures are formed.

Special morphologies are produced by crystallizing under flow conditions. Such flow conditions are observed in nature during the formation of cellulose and natural silk fibers, during blood coagulation as well as during the mechanical denaturing of proteins. Flow-induced crystallization is utilized technologically in flash-spinning, and in the production of high modulus fibers and certain synthetic papers.

Spherulites are at first deformed during crystallization of polymer melts under shear stress. Finally, "shish-kebab" type structures are formed by epitaxial growth and main-chain orientation in the direction of stress (Figure 10-11). At higher shear stresses, the crystallization is obviously induced by a continuous series or row of crystal nuclei arranged along the direction of flow. The "row structure" of oriented lamellae or chains is produced in this way. Simultaneously, the Avrami exponent n changes from the value of about 3–4 for spherulitic growth to 1–2 for row crystallization. Shish-kebab-type fiber structures are also formed when strong (turbulent) shear stresses are applied to dilute crystallizing polymer solutions. Poly(ethylene) fibers up to 2000 m long with about 40% of the theoretical tensile strength can be produced in this way for growth rates up to 160 cm/min.

10.4. Melting

10.4.1. Melting Processes

The corners and edges of crystals are always so disordered that the melting process can here be induced by introducing sufficient energy. In contrast to crystallization, therefore nucleating agents are not required. However, the melting process is influenced by the rate of heating (Figure 10-12). With solution-grown single cyrstals, the melting temperature first

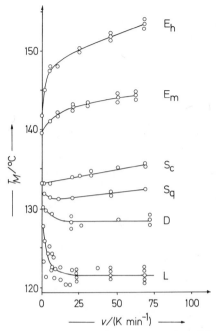

Figure 10-12. Melting behavior of a poly(ethylene) as a function of the morphology and the rate of heating. L, Lamellar single crystals from solution; *D*, dendrites obtained by shock cooling solutions; Sq spherulites obtained by shock cooling melts under normal pressures; Sc, spherulites produced by crystallization under normal pressures; E, extended chain crystals obtained by crystallization from high molar mass E_h and low molar mass E_m poly(ethylene) melts under high pressures (after B. Wunderlich).

decreases with increasing rate of heating and then remains constant. Reorganization of the crystals obviously occurs at lower heating rates. On the other hand, an increase in the melting temperature with heating rate is observed for polymers crystallized from the melt, since, in this case, superheating of crystals is increased. Superheating is more marked for more perfect crystals, that is, it is greater for extended-chain crystals than it is for spherulites.

The melting temperature is defined as that temperature at which the crystalline layer is in thermodynamic equilibrium with the melt. It must of necessity depend on the lamellar thickness *before* the onset of the melting process. Each monomeric unit contributes an enthalpy of fusion $(\Delta H_M)_u$ to the observed enthalpy of fusion ΔH_M. The enthalpy of fusion is also lowered by an amount equal to the surface enthalpy ΔH_f of both sides of the lamellae. Thus, for a lamella of N_m monomeric units

$$\Delta H_M = N_u (\Delta H_M)_u - 2 \Delta H_f \qquad (10\text{-}34)$$

The melting temperature observed for such a lamella is

$$T_M = \frac{\Delta H_M}{\Delta S_M} \qquad (10\text{-}35)$$

whereas the melting temperature of an infinitely thick lamella has the value

$$T_M^0 = \frac{N_u(\Delta H_M)_u}{N_u(\Delta S_M)_u} = \frac{(\Delta H_M)_u}{(\Delta S_M)_u} \qquad (10\text{-}36)$$

Inserting Equations (10-34)–(10-36) into each other leads to

$$T_M = T_M^0 - \frac{2\,\Delta H_f}{(\Delta S_M)_u}\frac{1}{N_u} = T_M^0\left[1 - \frac{2\,\Delta H_f}{(\Delta H_M)_u}\frac{1}{N_u}\right] \qquad (10\text{-}37)$$

Consequently, the thermodynamic melting temperature T_M^0 can be obtained by extrapolation from a plot of the melting temperature T_M against the number of monomeric units per lamella, that is, the lamellar height (Figure 10-13).

The number of monomeric units per chain section in the lamella is related to the observed lamellar thickness L_{obs} via the crystallographic monomeric unit length L_u:

$$L_{obs} = N_u L_u \qquad (10\text{-}38)$$

The observed lamellar thickness is, however, always greater than the lamellar

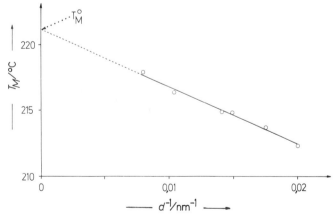

Figure 10-13. The melting temperature dependence of the reciprocal lamellar thickness $1/d$ of lamellae for poly(trifluorochloroethylene). The lamellar thickness was measured as the inter-lamellar distance by small-angle X-ray analysis, and thus contains both the crystalline component and the amorphous surface layer (after J. D. Hoffman from data of P. H. Geil and J. J. Weeks).

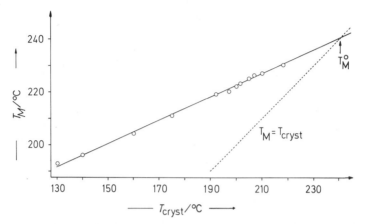

Figure 10-14. Influence of the crystallization temperature T_{cryst} on the melting point of it-poly(styrene) (after N. Overbergh, H. Bergmans, and G. Smets).

thickness calculated from crystal growth theory (L_{lam})$_{theor}$ by a factor of γ:

$$L_{obs} = \gamma (L_{lam})_{theor} \tag{10-39}$$

Insertion of Equations (10-17), (10-18), (10-38), and (10-39) into Equation (10-37) leads to (with $\sigma_f = L_u \Delta H_f$)

$$T_M = T_M^0(1 - \gamma^{-1}) + \gamma^{-1} T_{cryst} \tag{10-40}$$

Consequently, a plot of the melting temperature T_M of the crystals against their crystallization temperature should give a straight line (Figure 10-14). The intersection of this line with the line $T_M = T_{cryst}$ gives the thermodynamic melting point T_M^0. The γ value for many polymers is found to be about 2.

10.4.2. Melting Temperature and Molar Mass

Most of the melting temperatures in the literature do not apply to perfect crystals of infinitely long chains (T_M^{00}) or even to perfect crystals of finite chains (T_M^0). They are more often the melting temperatures of imperfect crystals from polymers of finite molar mass, T_M. The melting temperatures, T_M^0, are obtained through the following considerations:

The enthalpy of fusion of polymers, (ΔH_M)$_X$, depends on the enthalpy of fusion per monomeric unit, (ΔH_M)$_u$, the degree of polymerization X, the enthalpy of fusion of the end groups, (ΔH_M)$_E$, and the change in heat capacity on passing from crystal to melt:

$$(\Delta H_M)_X = X(\Delta H_M)_u + 2(\Delta H_M)_E + X \Delta C_p(T_M^0 - T_M) \tag{10-41}$$

The entropy of fusion also contains an entropy of mixing contribution, ΔS_{mix}, resulting from mixing of monomeric units with end groups. The end groups are, of course, excluded from the crystalline zone but may mix with the monomeric units in the melt:

$$(\Delta S_M)_X = X(\Delta S_M)_u + 2(\Delta S_M)_E + \Delta S_{\text{mix}} + X \Delta C_p \ln (T_M^0 / T_M)$$

$$(10\text{-}42)$$

with

$$\Delta S_{\text{mix}} = R \ln \Omega = \ln \{X! / [2!(X - 2)!]\}$$
$$= R \ln [X(X - 1)/2] \approx 2R \ln X - R \ln 2$$

$$(10\text{-}43)$$

Inserting Equations (10-41)–(10-43) into the expressions for the melting temperature gives

$$(\Delta H_M)_X = T_M(\Delta S_M)_X \qquad (10\text{-}44)$$

$$(\Delta H_M)_u = T_M^0(\Delta S_M)_u \qquad (10\text{-}45)$$

which rearranges to

$$T_M = T_M^0 - [2 RT_M T_M^0 / (\Delta H_M)_u][(\ln X)/ X] + f(X) \qquad (10\text{-}46)$$

where

$$f(X) = \frac{2[(\Delta H_M)_E - T_M(\Delta S_M)_E] + RT_M \ln 2 + \Delta C_p[(T_M^0 - T_M) - X \ln (T_M^0/ T_M)]}{X(\Delta H_M)_u}$$

The expression $f(X)$ can be neglected with respect to the other terms on the right-hand side of Equation (10-46). Thus, since $T_M^0 \approx T_M$, a straight line of ordinate intercept equal to the melting temperature T_M^0, and slope equal to the enthalpy of fusion per monomeric unit is obtained when the melting temperature is plotted against $(\ln X)/ X$ (Figure 10-15). The melting temperatures obtained in this way agree well with those obtained from lamellar heights or crystallization temperatures.

10.4.3. *Melting Temperature and Constitution*

The magnitudes of the enthalpies and entropies of fusion determine the temperature of melting (Table 10-4). The entropy of fusion can be further separated into the contribution from the conformational change on melting, ΔS_c, and the contribution resulting from the change in volume, $(\alpha/\beta) \Delta V_M$, where α is the cubic expression coefficient and β is the compressibility

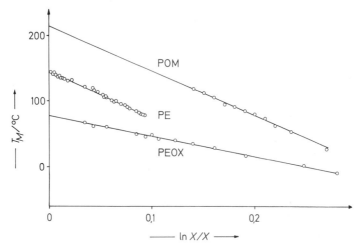

Figure 10-15. Melting temperature T_M as a function of the degree of polymerization parameter, $(\ln X)/X$, for poly(oxymethylenes), poly(ethylenes), and poly(oxyethylenes) (according to data from N. Hay).

coefficient in the neighborhood of the melting temperature;

$$\Delta H_M = T_M^0 \Delta S_M = T_M^0 (\Delta S_c + (\alpha/\beta)\Delta V_M) \qquad (10\text{-}47)$$

Chain units gain in conformational degrees of freedom on melting. In the ideal cases of three conformers (e.g., T, G^+, G^-) of the same conformational energy, the conformational entropy would increase by $R \ln 3 = 9.13$ J mol^{-1} K^{-1} on melting. But the conformational energies are not all equally large, and, so, a somewhat smaller conformational entropy change is obtained. In addition, the volume almost always increases at the temperature of heating (see Table 10-4). Consequently, the entropy of fusion is lower than the conformational entropy, according to Equation (10-47). The conformational entropy, however, contributes at least 75% to the entropy of fusion. In fact, most entropies of fusion per chain unit lie between 6 and 8 J mol^{-1} K^{-1}. The values for 1,4-*trans*-poly(butadiene) and 1,4-*cis*-poly(isoprene) are significantly lower. But, according to broad-line nuclear magnetic resonance measurements, these polymers already exhibit high segmental mobility below the temperature of melting, and consequently, the conformational entropy gain on melting is less.

The enthalpies of melting of polymers usually lie between 2 and 3 kJ/mol chain unit. Lower values are expected for higher chain mobilities below the melting temperature, and higher values are expected if intramolecular hydrogen bonding is present.

It is often supposed that the melt temperatures are predominantly influenced by cohesive energies. Of course, cohesive energies are a measure of

Table 10-4. Enthalpies of fusion, ΔH_M, entropies of fusion, ΔS_M, and specific volume changes on melting, Δv_M, of polymers. N = Number of "Free" Chain Units per Repeating Unit (after B. Wunderlich).

Repeat unit of the polymer	N	ΔH_M in kJ mol^{-1}		ΔS_M in J mol^{-1} K^{-1}		T_M in °C	Δv_M in cm^3 g^{-1}
		per repeat unit	per chain unit	per repeat unit	per chain unit		
CF$_2$	1	3.42	3.42	5.69	5.7	327	0.065
CH$_2$	1	4.11	4.11	9.91	9.9	142	0.173
CH$_2$—CH(CH$_3$) (it)	2	6.94	3.47	15.1	7.6	187	0.112
CH$_2$—CH(C$_2$H$_5$) (it)	2	7.01	3.51	17.0	8.5	139	0.112
CH$_2$—CH(C$_3$H$_7$) (it)	2	6.31	3.16	15.6	7.8	131	0.093
CH$_2$—CH(C$_6$H$_5$) (it)	2	10.0	5.00	19.4	9.7	242	0.061
CH$_2$CH=CHCH$_2$ (cis)	3	9.20	3.07	32	10.7	14	0.121
CH$_2$CH=CHCH$_2$ (trans)	3	3.61	1.20	8.7	2.9	142	0.157
CH$_2$CCH$_3$=CHCH$_2$ (cis)	3	4.36	1.45	14.4	4.8	30	0.108
CH$_2$CCH$_3$=CHCH$_2$ (trans)	3	12.9	4.3	36.4	12.1	81	0.153
O—CH$_2$	2	9.79	4.90	21.4	10.7	184	0.085
O—(CH$_2$)$_2$	3	8.66	4.33	25.3	8.4	69	0.081
O—(CH$_2$)$_5$	6	14.4	2.4	43.7	7.3	56	0.116
OOC—CH$_2$	3	11.1	3.7	22	7.3	231	0.078
OOC—(CH$_2$)$_2$	4	9.08	2.27	25.5	6.4	83	0.041
OOC—(CH$_2$)$_5$	7	16.2	2.31	48.1	6.9	64	0.076
OOC(CH$_2$)$_4$COO(CH$_2$)$_2$	10	21.0	2.10	62.2	6.2	64	0.092
OOC(CH$_2$)$_6$COO(CH$_2$)$_2$	12	26.6	2.22	76.6	6.4	74	0.115
OOC(CH$_2$)$_8$COO(CH$_2$)$_2$	14	32.0	2.29	89.8	6.4	83	0.132
OOCC$_6$H$_5$COO(CH$_2$)$_2$	7	26.9	3.84	48.6	6.9	280	0.088
NH(CH$_2$)$_5$CO	7	26.0	3.71	48.8	7.0	260	0.077
NH(CH$_2$)$_4$NHCO(CH$_2$)$_6$CO	14	67.9	4.85	123	6.8	279	0.111
Cellulose tributyrate	(5)	12.6	2.52	26.2	5.24	208	0.111

the intermolecular forces acting on the liquid/gaseous phase transition, but melting is considered to be a solid/liquid transition. The cohesive energies for both kinds of transition are not necessarily comparable. Infrared measurements on polyamide melts, for example, have shown that the majority of hydrogen bonds remain above the melting temperature. Thus, the cohesive energy must be relatively unimportant.

If the melting temperatures were primarily determined by cohesive energy, then they should increase with increasing numbers of groups of high cohesive energy per monomeric unit. The cohesive energy of a methylene group is 2.85 kJ mol^{-1}, of an ester group, 12.1 kJ mol^{-1}, and of an amide group, 35.6 kJ mol^{-1}. Consequently, the melting temperatures of aliphatic polyamides and polyesters should increase with decreasing methylene group content. But the reverse behavior is observed for polyesters (Figure 10-16). On the other hand, ester groups have a lower rotational potential energy barrier than methylene and amide groups (Table 4-3). Thus, the flexibility of the individual molecule, and not the intermolecular interaction between chains, is the primary factor in determining the magnitude of the melting temperature.

The flexibility of a molecule depends on the constitution and configuration of the chains and the conformations produced by these. Given the same conformation, the larger the distance between the chain atoms and the greater their valence angles, the greater will be the flexibility. It is higher when the steric hindrance to rotation is lower. Poly(ethylene) ($T_M^0 = 144°C$), with its

Table 10-5. Comparison of Cohesive Energies, Melting Temperatures, and Glass Transition Temperatures of Polymers

Chain units		Polymeric examples			
Group	Cohesive energy in kJ/mol group	Monomeric unit	Mean cohesive energy in kJ/mol group	T_M in °C	T_G in °C
—CH$_2$—	2.85	—CH$_2$—	2.85	144	−80
—CF$_2$—	3.18	—CF$_2$—	3.18	327	127
—O—	4.19	—CH$_2$—O—	3.52	188	−85
		—CH$_2$—CH$_2$—O—	3.31	67	−67
—C(CH$_3$)$_2$—	8.00	—CH$_2$—C(CH$_3$)$_2$—	5.40	44	−73
—CCl$_2$—	13.0	—CH$_2$—CCl$_2$—	7.91	198	−17
—CH(C$_6$H$_5$)—	18.0	—CH$_2$—CH(C$_6$H$_5$)—	10.4	250[a]	100
—CHOH—	21.4	—CH$_2$—CHOH—	12.1	265[b]	85
—COO—	12.1	—(CH$_2$)$_5$—COO—	4.4	55	
—CONH—	35.6	—(CH$_2$)$_5$—CONH—	8.3	228	45

[a] Isotactic.
[b] Probably syndiotactic.

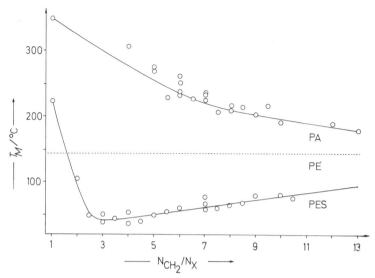

Figure 10-16. Dependence of the melting temperature T_M of aliphatic polyamides (PA) and aliphatic polyesters (PES) with X = amide or ester groups as a function of the group content. Result for polyethylene is given by the lashed line.

relatively high rotational potential energy barrier for the CH_2/CH_2 bond rotation, thus has a higher melting temperature than the ether-oxygen-containing poly(tetrahydrofuran) $+CH_2-CH_2-CH_2-CH_2-O+_n$, with $T_M \sim 35°C$ (see also Section 4.2). Rigid groups (phenylene rings, etc.) raise the melting temperature.

According to conformational sequence, helices are more tightly or more loosely constructed. The helices of poly(oxymethylene), with the conformational sequence GG, for example, have a much smaller diameter than poly(ethylene oxide) helices, which possess the TTG conformation and have roughly the same number of main-chain atoms per unit length of chain along the chain direction of the helix. The poly(oxymethylene) sequences are therefore more rigid. Consequently, the melting temperature of poly(oxymethylene) is higher than that of poly(oxyethylene).

The tendency of the helix to become more rigid because of closer packing can also be affected by substituents. Increased substitution in direct proximity to the main-chain of helix-forming macromolecules widens the helix and lowers the melting temperature. The melting point of it-poly(butene-1) is therefore lower than that of it-poly(propylene). Because of tighter intermolecular packing, poly(3-methylbutene-1) possesses a higher melting temperature than poly(butene-1):

$$CH_3$$
$$|$$
$$CH_3 \qquad CH_2-CH_3 \qquad CH-CH_3 \qquad CH_3-C-CH_3$$
$$| \qquad\qquad | \qquad\qquad | \qquad\qquad |$$
$$+CH_2-CH)_{\overline{n}} \qquad +CH_2-CH)_{\overline{n}} \qquad +CH_2-CH)_{\overline{n}} \qquad +CH_2-CH)_{\overline{n}}$$

it-poly(propylene)   it-poly(butene-1)   it-poly(3-methylbutene-1)   it-poly(3.3'-dimethyl-butene-1)

3_1 helix 3_1 helix 4_1 helix

$T_M = 186°$ C $T_M = 136°$ C $T_M = 304°$ C   ?

(monoclinic)   (rhombic)   (monoclinic) $T_M = 320°$ C

In the series of poly(α-olefins), with linear aliphatic substituents, the substituent bonded directly to the main-chain backbone is always a CH group. Thus the chain conformation is retained. The longer side chains decrease the efficiency with which the chains can pack together, however, and so the melting temperature falls (Figure 10-17). Only when the side chains are very long does their ordering give additional order, so that the melting temperature again rises as the number of carbon atoms increases (the so-called side chain crystallization).

10.4.4. Copolymers

The different base units in a copolymer can be isomorphous. If, in addition, they occur randomly along the chain, then the melting temperatures of such copolymers increase   regularly with the mole fraction of the

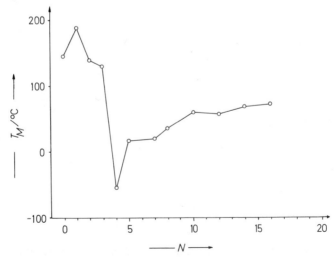

Figure 10-17. Melting temperature T_M of isotactic poly(α-olefins), $+CH_2-CHR)_n$ as a function of the number of methylene groups N in the unit R = $+CH_2)_N$ H.

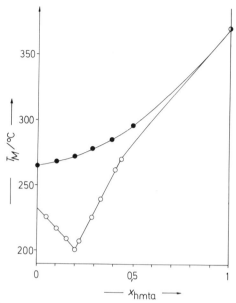

Figure 10-18. Melting temperature T_M of copolymers of hexamethylene terephthalamide (HMTA) and hexamethylene adipamide (●) or hexamethylene sebacamide (○) (after O. B. Edgar and R. Hill).

comonomer that melts at the highest temperature. An example of this is the copolymer of hexamethylene terephthalamide and hexamethylene adipamide (Figure 10-18). With nonisomorphous base units, on the other hand, the lengths of the crystallite regions in the solid polymer are decreased as the proportion of the second comonomer is increased. The melting temperatures fall, reaching a minimum at a specific copolymer composition, as shown in Figure 10-18 for the copolymers of hexamethylene terephthalamide and hexamethylene sebacamide.

10.5. Glass Transitions

10.5.1. Free Volumes

The viscosity at the freezing temperature was found to be about 10^{12} Pa s and independent of the substance in the earliest measurements of glass transitions. An "isoviscous" process was considered to characterize this freezing-in process. Today, the freezing-in temperature is considered to be

Table 10-6. Cubic or Volume Expansion Coefficients, α_l and α_{am}, of Polymers

Polymer	T_G in K	$10^4\alpha_l$ in K^{-1}	$10^4\alpha_{am}$ in K^{-1}	Equation (10-51)	Table 5-7	$(\alpha_l-\alpha_{am})T_G$
				f_{exp}		
PE	193	7.97	2.87	0.104	—	0.098
PIB	200	5.79	1.86	0.082	0.125	0.079
PS	373	5.65	2.09	0.144	0.127	0.133
PVAC	300	6.53	2.26	0.137	0.14	0.128
PMMA	378	5.28	2.16	0.128	0.13	0.118

that temperature at which all substances have the same free volume (see Section 5.6.1). Which free volume is involved is still a matter of dispute.

To a first approximation, volumes change linearly with temperature (see Figure 10-1). Consequently, with the definition of the cubic expansion coefficient α, the following is obtained for the liquid state and the amorphous state:

$$(V_l^0)_T = (V_l^0)_0 + (V_l^0)_T\alpha_l T \tag{10-48}$$

$$(V_{am}^0)_T = (V_{am}^0)_0 + (V_{am}^0)_T\alpha_{am} T \tag{10-49}$$

The volumes of liquid and amorphous material are approximately equal at the glass transition temperature. Consequently, $(V_l^0)_G = (V_{am}^0)_G$. Equating equations (10-48) and (10-49) and reintroducing Equation (10-49) we obtain

$$\left[\frac{(V_{am}^0)_0 - (V_l^0)_0}{(V_{am}^0)_0}\right](1 - \alpha_{am} T_G) = (\alpha_l - \alpha_{am}) T_G \tag{10-50}$$

The volumes of liquid and crystal must be equal at 0 K. The term in square brackets in Equation (10-50) must give the free-volume fraction [see Equation (5-11)]. We obtain

$$f_{exp} = (\alpha_l - \alpha_{am}) T_G/(1 - \alpha_{am} T_G) \approx (\alpha_l - \alpha_{am}) T_G \tag{10-51}$$

The free-volume fraction f_{exp} calculated from Equation (10-51) agrees very well with the values given in Table 5-7 (Table 10-6). According to the empirical Boyer–Simha rule, the product $(\alpha_l - \alpha_{am}) T_G$ generally has a value of about 0.11 for a large number of polymers. Partially crystalline polymers and polymers with relaxation mechanisms lying below the glass transition temperature are exceptions to this rule.

10.5.2. Molecular Interpretation

Coupled chain segment movements must be initiated or frozen in at the glass transition temperature. This can be deduced from the behavior of

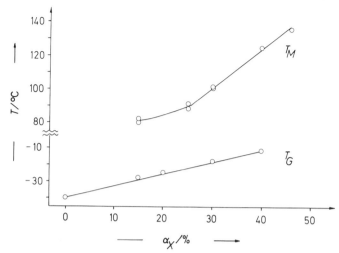

Figure 10-19. Dependence of the melting temperature T_M and glass transition temperature T_G of a 90% syndiotactic 1,2-poly(butadiene) on the X-ray crystallinity, α_x.

chemically or physically cross-linked networks. Cross-linking, of course, limits chain segment mobility when the distance between cross-link points is smaller than the segment length of those segments responsible for the mobility. It has been found for cross-linking copolymerizations that the mobility of segments of 30–50 chain links is responsible for the glass transition temperature. Physical cross-linking produces the same effect: the glass transition temperature of partially crystalline polymers often increases with the degree of crystallization (Figure 10-19), for example with poly(vinyl chloride), poly(ethylene oxide) and poly(ethylene terephthalate), but not with it-poly(propylene), or poly(chlorotrifluoroethylene).

Since both the glass transition temperature and the melting temperature depend on the mobility of segments or molecules, a relationship between these two parameters should exist. If the cumulative frequencies of T_G / T_M ratios are plotted against these ratios, then a smooth curve is obtained for more than 70 homopolymers (Figure 10-20). Deviations are only found for low T_G / T_M ratios; and these belong to unsubstituted polymers such as poly(ethylene), poly(oxymethylene), poly(oxyethylene), etc. The median of the curve is independent of the constitution of the polymers, and corresponds to the empirical Beaman–Boyer rule:

$$T_G \approx (2/3) T_M \qquad (10\text{-}52)$$

The glass transition temperature is influenced by intra- or intermolecular cooperative segment mobilities. Since similar transition temperatures are observed in the amorphous state and in solution, these temperatures must result from intramolecular effects. The good agreement between the glass

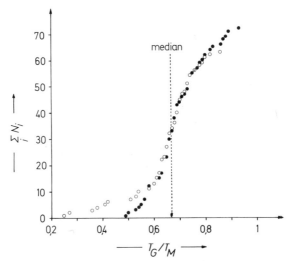

Figure 10-20. Cumulative frequency of occurrence of a certain T_G / T_M ratio as a function of this ratio for (○) "symmetrical" polymers such as $+CH_2CR_2+$, and for (●) "unsymmetrical" polymers such as $+CH_2CHR+$ (after D. W. van Krevelen).

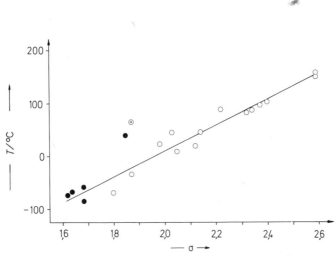

Figure 10-21. The relationship between the glass transition temperature T_G and the steric hindrance parameter σ for carbon–carbon chains (○), carbon–oxygen chains (●), and carbon–nitrogen chains (⊙).

transition temperature and the hindrance parameter for rotation about main-chain atoms is in accord with this conclusion (Figure 10-21). Thus, the glass transition temperature depends primarily on the flexibility of the individual chain, and only secondarily on the forces acting between chains.

10.5.3. Static and Dynamic Glass Transition Temperatures

Chain segments move with a definite frequency above the glass transition temperature. Consequently, the frequency of the measurement method used or the deformation time of the sample must determine the numerical value of the glass transition temperature observed. Thus, the methods of measurement are classified as static or dynamic according to the speed of the measurement.

The static methods include determinations of heat capacities (including differential thermal analysis), volume change, and, as a consequence of the Lorentz–Lorenz volume–refractive index relationship, the change in re-fractive index as a function of temperature. Dynamic methods are represented by techniques such as broad-line nuclear magnetic resonance, mechanical loss, and dielectric-loss measurements.

Static and dynamic glass transition temperatures can be interconverted. The probability p of segmental mobility increases as the free-volume fraction f_{WLF} increases (see also Section 5.6.1). For $f_{WLF} = 0$, of necessity, $p = 0$. For $f_{WLF} \to \infty$, it follows that $p = 1$. The functionality is consequently

$$p = \exp\left(\frac{-B}{f_{WLF}}\right), \qquad B = \text{const} \tag{10-53}$$

The extent of the deformation depends on the time t. To a good approximation, it can be assumed that $pt = \text{const}$, and Equation (10-53) becomes

$$\log pt = -\frac{B}{f_{WLF}} \log e + \log t = \log (\text{const}) \tag{10-54}$$

For the differences in the logarithms of the times t_2 and t_1, therefore, we have, with the corresponding fractions of free volumes,

$$\log t_2 - \log t_1 = \Delta (\log t) = B(\log e)\left[\frac{1}{(f_{WLF})_2} - \frac{1}{(f_{WLF})_1}\right] \tag{10-55}$$

A change in the time scale consequently corresponds to a change in the free volume (a smaller time corresponds to a larger volume). On the other hand, the free-volume fraction must increase with increasing temperature. This increase is linear in the vicinity of the glass transition temperature

$$(f_{WLF})_2 = (f_{WLF})_1 + (\alpha_l - \alpha_{am}) (T_2 - T_1) \tag{10-56}$$

where α_1 and α_{am} are the expansion coefficients of the liquid and amorphous state of the polymer. The free-volume fractions of these two states, however, are of the equal magnitude at the glass-transition temperature. With $T_1 = T_G$ and any desired temperature $T_2 = T$, equations (10-55) and (10-56) give

$$\Delta (\log t) = \frac{[B(\log e)/(f_{WFL})_G] \, (T - T_G)}{[(f_{WFL})_G/(\alpha_l - \alpha_{am})] + (T - T_G)} \tag{10-57}$$

or, solved for T, and with $\Delta (\log t) = - \log a_t$,

$$T = T_G + \frac{[(f_{WFL})_G/(\alpha_1 - \alpha_{am})] \log a_t}{(B \log e)/(f_{WLF})_G - \log a_t} \tag{10-58}$$

It has been found empirically that $(f_{WLF})_G \approx 0.025$ (see Section 5.6.1). B can, to a good approximation, be made equal to unity. Since $\alpha_l - \alpha_{am}$ is about $4.8 \times 10^{-4} \, K^{-1}$ for many materials (see also Table 10-6), Equation (10-58) can be given as

$$T = T_G + \frac{51.6 \log a_t}{17.4 - \log a_t} \tag{10-59}$$

Equation (10-59) or (10-58) is known as the William–Landels–Ferry (WLF) equation. It applies to all relaxation processes, and therefore also for the temperature dependence of the viscosity (see Section 7.6.4). Its validity is limited to a temperature range from T_G to about $T_G + 100 \, K$. Outside this temperature range the expansion coefficient α_1 varies, not linearly, but with the square root of temperature.

The WLF equation enables static glass transition temperatures T_G and various dynamic glass transition temperatures T to be interconverted. To do this, the deformation times of the various individual methods must be known

Table 10-7. *Deformation Times or Periods (Reciprocal Effective Frequencies) of Various Methods and the Glass Transition Temperature Observed with Poly(methyl methacrylate)*

Method	Deformation period in s	Glass transition temperature in °C
Thermal expansion	10^4	110
Penetrometry	10^2	120
Mechanical loss[a]	10^3—10^7	—
Rebound elasticity	10^{-5}	160
Dielectric loss[a]	10^4—10^{11}	—
Broad-line NMR	10^{-4}—10^{-5}	—
NMR spin-lattice relaxation	10^{-7}—10^{-8}	—

[a]The frequency can be altered with this method.

(Table 10-7). The shift factor a_t for the calculation is obtained as the difference between the logarithms of the deformation times.

According to the method used, then, one and the same material can exhibit a quite different mechanical behavior. Poly(methyl methacrylate), according to Table 10-7, is a glass at 140°C with respect to measurement of the rebound elasticity of spheres, but a rubbery elastic body with respect to penetrometric measurements. Static and dynamic glass transition temperatures thus also have direct practical significance. In the case of the static glass transition temperature, the body changes from the brittle to the tough state under slow stresses such as drawing, bending, etc., whereas the dynamic glass transition temperature is important over shorter periods of stress (shocks, jolts).

10.5.4. Constitutional Influences

Glass transition temperatures increase with increasing molar mass in a polymer homologous series, and become practically constant only above degrees of polymerization of about 100–600. The dependence on the degree of polymerization appears to separate into three regions with quite distinct transitions between them (Figure 10-22). The influence of the end groups and the packing densities presumably reproduces this behavior.

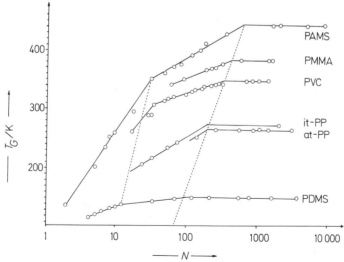

Figure 10-22. The glass transition temperature as a function of the number N of chain units for poly(α-methyl styrene), poly(methyl methacrylate), poly(vinyl chloride), iso- and atactic poly(propylene), and poly(dimethyl siloxane) (J. M. G. Cowie).

Differences in packing density are also responsible for the effect of number and kind of branch points on the glass transition temperature. Branched polymers always have higher glass transition temperatures than their unbranched counterparts, and randomly branched polymers have higher glass transition temperatures than the same polymers with star-shaped branching and the same degree of branching (Figure 10-23).

Different side chain lengths also affect packing. Increasing alkyl residue lengths in poly(alkyl methacrylates) decrease the possibilities for dense packing of the chains, and both the glass transition temperatures and brittleness temperatures decrease. If the alkyl residues become sufficiently long, they reinforce the positions taken up by each other and thereby reduce the flexibility of the main chain. The consequence is that the brittleness temperature then begins to increase (Figure 10-24).

Very little research has been done on the relations between glass transition temperatures and tacticity. Atactic and isotactic poly(styrenes) almost always have the same glass transition temperatures, and this is also the case for at- and it-poly(methacrylate). The glass transition temperature of it-poly(methyl methacrylate) (42° C), on the other hand, is distinctly lower than that of the atactic product (103° C).

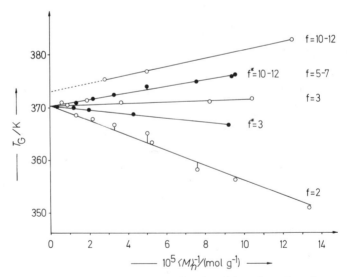

Figure 10-23. Glass transition temperature as a function of the reciprocal number average molar mass for linear, $f = 2$, and star-shaped branched poly(styrenes) with $f^* = 3$ or $f^* = 10-12$ arms, as well as for poly(styrenes) cross-linked with divinyl benzene with mean degrees of cross-linking per chain of $f = 3$, 5–7, or 10–12 (according to data from F. Rietsch, D. Daveloose and D. Froelich).

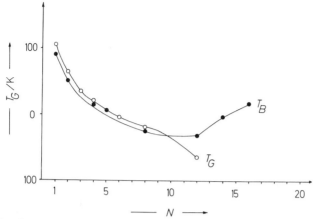

Figure 10-24. Glass transition temperatures T_G or brittleness temperatures T_B for poly(alkyl methacrylates) with the monomeric unit $+CH_2-C(CH_3)(COO(CH_2)_N H)+$.

The glass transition temperature of copolymers depends on the mass functions w_A and w_B of A and B monomeric units, the probabilities p of occurrence of AA, AB, and BB diads, and on their corresponding glass transition temperatures. Considering $(T_G)_{BA} = (T_G)_{AB}$, the following can be used empirically:

$$\frac{1}{T_G} = \frac{w_A p_{AA}}{(T_G)_{AA}} = \frac{w_A p_{AB} + w_B p_{BA}}{(T_G)_{AB}} + \frac{w_B p_{BB}}{(T_G)_{BB}} \tag{10-60}$$

Depending on the composition, the glass transition temperatures of co-polymers can thus increase or decrease or even pass through a maximum or minimum (Figure 10-25). If the glass transition temperature of a polymer is decreased by copolymerizing with a second monomer, then this is referred to as "internal plasticization." For example, butyl methacrylate is such an internal plasticizer for poly(styrene). Analogously, plasticization obtained by mixing in another substance is referred to as "external plasticization," or, simply, plasticization.

10.6. Other Transitions and Relaxations

Polymers may possess a series of other transition and relaxation temperatures besides the glass transition and melting temperatures. Such temperatures may lie either above or below the glass transition temperature. Like the glass transition temperature, these temperatures also depend on the frequency of the method of measurement.

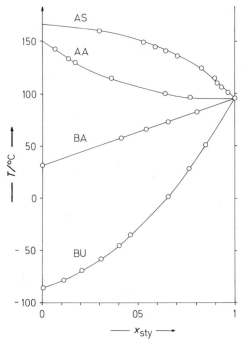

Figure 10-25. Glass transition temperatures T_G of free-radically-polymerized copolymers of styrene and acrylic acid (AS), acrylamide (AA), t-butyl acrylate (BA), and butadiene (BU) as a function of the mole fraction x_{sty} of styrene monomeric units (after K. H. Illers).

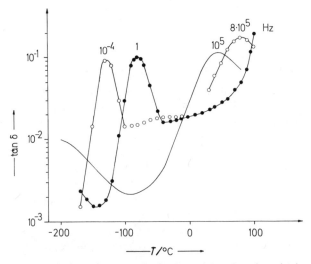

Figure 10-26. Mechanical loss factor tan δ of poly(cyclohexyl methacrylate) as a function of temperature for various frequencies (after J. Heijboer).

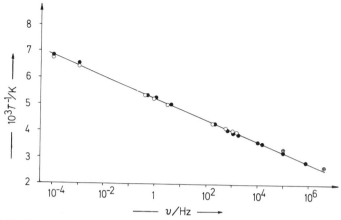

Figure 10-27. Temperature dependence of the loss maxima of poly(cyclohexyl methacrylate) (●), poly(cyclohexyl acrylate) (○), and cyclohexanol (⊕) (after J. Heijboer).

A typical example of these measurements is the mechanical loss factor (for definition, see Section 11). Here a loss maximum for poly(cyclohexyl methacrylate) is observed at $-125°C$ when the frequency is 10^{-4} Hz (Figure 10-26). The maximum is shifted to higher temperatures when the frequency is increased. In addition, the reciprocal loss temperature depends linearly on the logarithm of the frequency (Figure 10-27). Studies on different chemical compounds show that this loss maximum is specific to the cyclohexyl group. The values for both poly(cyclohexyl methacrylate) and poly(cyclohexyl

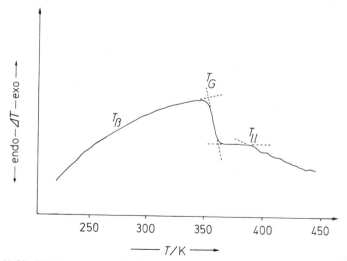

Figure 10-28. DSC thermogram of a low-molar-mass poly(styrene) with a β transition T_β, a glass transition T_G, and a liquid/liquid transition T_{ll}, temperature (after R. F. Boyer).

acrylate) as well as for cyclohexanol can, of course, be plotted on the same curve, but the values for, for example, poly(phenyl acrylate) do not fall on the same plot. Thus, the loss maxima must result from the chair/boat transformation of the cyclohexyl ring.

The chair/boat conversion of the cyclohexyl ring is one of the few examples whereby a molecular interpretation can be given to an observed β transition. However, such transitions are very frequent below the glass transition temperature. They can often be recognized as a kink in the curve of DSC diagrams (Figure 10-28). In many cases, these β transitions appear to be caused by the coupled movement of very short main-chain segments.

A liquid/liquid transition is observed above the glass transition temperature with amorphous polymers (Figure 10-28). The molar mass dependence of this transition temperature proceeds parallel to that of the glass transition temperature. A simple relationship independent of the constitution appears to exist between the glass transition temperature and the liquid/liquid transition temperature for infinitely high molar mass. This relationship is $T_{ll} = 1.2\,T_G$ (Figure 10-29).

10.7. Thermal Conductivity

The usual polymers do not conduct electricity. Consequently, heat cannot likewise be transported by electrons in these polymers; heat must

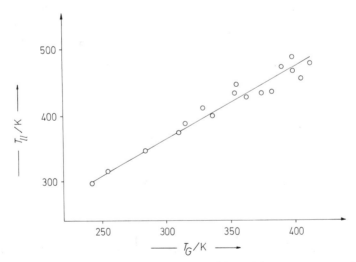

Figure 10-29. Relationship between the liquid/liquid transition temperature and the glass transition temperature for various polymers (after R. F. Boyer).

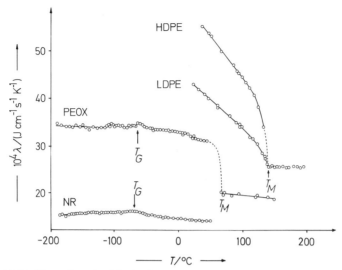

Figure 10-30. Thermal conductivity λ of natural rubber (NR), poly(oxyethylene) (PEOX), and poly(ethylene) (PE) of various densities as a function of temperature. T_G is the glass transition temperature, T_M is the melting temperature. (From data of various authors in the compilation of W. Knappe.)

mostly be transported by elastic waves (phonons in the corpuscular picture). The distance at which the intensity of the elastic waves has decreased by $1/e$ is known as the free path length of the phonons. This free path is comparatively independent of temperature for glasses, amorphous polymers, and liquids and is about 0.7 nm. From this, it can be concluded that the weak decrease in thermal conductivity observed for amorphous polymers below the glass transition temperature is essentially due to the decrease in the heat capacity with temperature (Figure 10-30).

At still lower temperatures, a plateau is reached at 5–15 K. A slower decrease with temperature is then observed until, below 0.5 K, the thermal conductivity becomes proportional to the square of the temperature.

Above temperatures of 150 K, heat is essentially transported by intermolecular collisions. A decrease in the thermal conductivity above the glass transition temperature can be expected because of the increasingly loose arrangement of the molecules. The thermal conductivity above and below the glass transition temperature does not differ very much because the molecular packing above and below this temperature is also not very different. The thermal conductivity exhibits only a weak maximum at the glass transition temperature.

On the other hand, the packing density of crystalline polymers changes drastically at the melting temperature. The decrease becomes stronger with increasing crystallinity of the polymer involved. The decrease begins well below the temperatures at which the onset of melting is first observed by other methods.

Literature

10.1. Basic Principles

B. Wunderlich and H. Bauer, Heat capacities of linear high polymers, *Adv. Polym. Sci.* **7**, 151 (1970).

A. D. Jenkins, *Polymer Science*, a materials science handbook, 2 vols., North-Holland, Amsterdam, 1972.

G. Patfoort, *Polymers: An Introduction to Their Physical, Mechanical and Rheological Behavior*, Story-Scientia, Ghent 1974.

G. M. Bartenev and Yu. V. Zenlenev (eds.), *Relaxation Phenomena in Polymers*, Halsted, New York, 1974.

A. M. North, Relaxations in polymers, *Int. Rev. Sci., Phys. Chem.* (2) **8**, 1 (1975).

D. J. Meier (ed.), Molecular basis of relaxations and transitions of polymers (Midland Macromolecular Monographs, Vol. 4), Gordon and Breach, New York, 1978.

10.2. Methods of Measurement

M. Dole, Calorimetric studies and transitions in solid high polymers, *Fortsch. Hochpolym. Forschg.* **2**, 221 (1960).

W. J. Smothers and Y. Chiang, *Handbook of Differential Thermal Analysis*, Chem. Publ. Co., New York, 1966.

D. Schultze, *Differentialthermoanalyse*, Verlag Chemie, Weinheim, 1969.

R. C. MacKenzie (ed.), *Differential Thermal Analysis*, Vol. 1 (1970), Vol. 2 (1972), Academic Press, New York.

J. K. Gillham, Torsional braid analysis—A semimicro thermomechanical approach to polymer characterization, *Crit. Rev. Macromol. Sci.* **1**, 83 (1972).

A. M. Hassan, Application of wide-line NMR to polymers, *Crit. Rev. Macromol. Sci.* **1**, 399 (1972).

V. J. McBrierty, N.M.R. of solid polymers: a review, *Polymer (London)* **15**, 503 (1974).

T. Murayama, *Dynamic Mechanical Analysis of Polymeric Material*, Elsevier, Amsterdam, 1978.

P. Tormälä, Spin label and probe studies of polymeric solids and melts, *J. Macromol. Sci.-Rev. Macromol Chem.* **C17**, 297 (1979).

W. Wrasidlo, Thermal analysis of polymers, *Adv. Polym. Sci.* **13**, 1 (1974).

J. Chiu, Dynamic thermal analysis of polymers, an overview, *J. Macromol. Sci. (Chem.)* **A8**, 1 (1974).

10.3. Crystallization

L. Mandelkern, *Crystallization of Polymers*, McGraw-Hill, New York, 1964.

A. Sharples, *Introduction to Polymer Crystallization*, Arnold, London, 1966.

B. Wunderlich, Crystallization during polymerization, *Adv. Polym. Sci.* **5**, 568 (1968).

J. N. Hay, Application of the modified Avrami equation to polymer crystallization kinetics, *Brit. Polym. J.* **3**, 74 (1971).

S. K. Bhateja and K. D. Pae, The effects of hydrostatic pressure on the compressibility, crystallization, and melting of polymers, *J. Macromol. Sci. (Revs.)* **C13**, 77 (1975).

J. P. Mercier and R. Legras (eds.), Recent advances in the fields of crystallization and fusion of polymers, *Polym. Symp.* **59** (1977).

R. L. Miller (ed.), Flow induced crystallization of polymers (Midland Macromolecular Monographs, Vol. 6), Gordon and Breach, New York, 1979.

K. A. Mauritz, E. Baer, A. J. Hopfinger, The Epitaxial Crystallization of Macromolecules, *Macromol. Rev.* **13**, 1 (1978).

D. W. van Krevelen, Crystallinity of polymers and the means to influence the crystallization process, *Chimia* **32**, 279 (1978).

G. S. Ross and L. J. Frolen, Nucleation and Crystallization (of Polymers), *Methods Exp. Phys.* **16B**, 339 (1980).

10.4. Melting

H. G. Zachmann, Das Kristallisations- und Schmelzverhalten hochpolymerer Stoffe, *Fortschr. Hochpolym. Forschug.—Adv. Polymer Sci.* **3**, 581 (1961/64).

B. Wunderlich, *Macromolecular Physics*, Vol. 3, *Crystal Melting*, Academic Press, New York, 1980.

10.5. The Glass Transition

R. F. Boyer, The relation of transition temperatures to chemical structure in high polymers, *Rubber Chem. Technol.* **36**, 1303 (1963).

A. J. Kovacs, Transition vitreuse dans les polymères amorphes. Etude phénoménologique, *Fortschr. Hochpolym. Forschg.—Adv. Polym. Sci.* **3**, 394 (1961/64).

M. C. Shen and A. Eisenberg, Glass transitions in polymers, *Rubber Chem. Technol.* **43**, 95 (1970).

N. W. Johnston, Sequence distribution—glass transition effects, *J. Macromol. Sci. (Revs. Macromol. Chem.)* **C14**, 215–250 (1976).

M. Goldstein and R. Simha (ed.), The glass transition and the nature of the glassy state, *Ann. N. Y. Acad. Sci.* **279**, 1–246 (976).

R. F. Boyer, Transitions and relaxations, in *Encycl. Polym. Sci. Technol., Suppl. Vol.* **2**, 745–839 (1977).

Y. Lipatov, The Iso-Volume State and Glass Transitions in Amorphous Polymers: New Development of the Theory, *Adv. Polym. Sci.* **26**, 63 (1978).

10.6. Other Transitions

A. Hiltner and E. Baer, Relaxation processes at cryogenic temperatures, *Crit. Rev. Macromol. Sci.* **1**, 215 (1972).

10.7. Thermal Conductivity

D. R. Anderson, Thermal conductivity of polymers, *Chem. Revs.* **66**, 677 (1966).

W. Knappe, Wärmeleitung in Polymeren, *Adv. Polym. Sci.* **7**, 477 (1971).

D. Hands, K. Lane, and R. P. Sheldon, Thermal conductivities of amorphous polymers, *J. Polym. Sci. (Symp.)* **42,** 717, (1973).

C. L. Choy, Thermal conductivity of polymers, *Polymer* **18,** 984–1004 (1977).

D. Hands, The thermal transport properties of polymers, *Rubber Chem. Technol.* **50,** 480–522. (1977).

D. W. Phillips and R. A. Pethrick, Ultrasonic studies of solid polymers, *J. Macromol. Sci.—Rev. Macromol. Chem.* **C16,** 1–22 (1977/78).

Chapter 11

Mechanical Properties

11.1. Phenomena

Macromolecular materials react quite differently to mechanical stresses. Beakers of conventional poly(styrene) are very brittle, and a short, quick blow will break them. In contrast, beakers of nylon 6 are very tough. Weakly cross-linked natural rubber expands on stretching by several hundred percent; after being released, it returns to what is practically its original form. When plasticine is deformed, on the other hand, it completely retains its new shape.

The reaction of a material to stress often seems related intuitively to the aggregate state. Accordingly, low-molar-mass substances can be solid, liquid, or gaseous. With low-molar-mass substances, a classification according to the aggregate state is usually also a classification according to the states of order. A classification according to the three classic aggregate states proves to be too restrictive, however, in the case of macromolecular substances.

Low-molar-mass materials are termed *solid* when they show a high order and a high resistance toward deformation. Iron and common salt are solid in this sense. The order is brought about by their high crystallinity. A small stress displaces the atoms from their rest position and the atomic distances increase. The atoms again take up their original rest positions when the stress is released. In order for the deformation to be reversible, extension should not exceed amounts of $\sim 1\%-2\%$. This type of body is called *ideal elastic* or *energy elastic*.

Wood and glass, however, are also solids in the usual sense of the word. Both materials exhibit a high resistance to deformation at room temperature, but are not X-ray crystalline. On the other hand, there are "liquid crystals" which do show an order that can be detected optically, but exhibit little resistance to deformation.

True liquids, in contrast, do not show an extensive order. Even with very slight stresses applied for a short time, they deform so completely that they very quickly adopt the form of the surrounding container. Low-molar-mass liquids thus behave in a purely viscous way under normal conditions. When stress is applied, the molecules are displaced irreversibly in relation to one another. In high-molar-mass substances above the glass transition temperatures, flow can be produced relatively easily. Deformations are much more difficult below the glass transition temperature of amorphous polymers. For this reason, and because of their lack of order, amorphous substances below their glass-transition temperatures are often termed *supercooled liquids*.

According to rheology (the science of flow), viscous flow and energy elasticity are only two extreme forms of the possible types of behavior of matter. It is appropriate to consider the entropy-elastic (or rubber elastic), viscoelastic, and plastic bodies as other special cases.

Plastic bodies only show irreversible deformation above a given shear stress (see Section 7.5.1).

Entropy-elastic or highly elastic bodies, such as, for example, weakly cross-linked elastomers, contrary to the energy-elastic materials, can be extended reversibly by very large amounts (see Section 11.3.1). With the exception of large elongations of several hundred percent, entropy-elastic bodies therefore behave like solid, low-molar-mass substances under stress. On the other hand, like liquids, elastic bodies show a limited dimensional stability, but a high volume stability. The coefficients of expansion and contraction are smaller, however, than those of liquids. Weakly cross-linked rubbers, for example, are almost purely entropy-elastic bodies. Entropy elasticity is therefore often termed *rubber* or *elastomeric elasticity*. The cause of the behavior is a displacement of the molecular segments out of their rest position such that they take up new conformations. After relaxation, the most probable distribution of conformations reappears. The elasticity thus depends on a change in entropy. Since, in this kind of material, the molecular chains are fixed in relation to one another by cross-linking, extensive slipping of the molecular chains past one another, i.e., viscous flow, cannot occur.

Linear macromolecular substances also show entropy-elastic behavior to a certain extent. One can imagine, therefore, that the molecular chains are partly entangled. The entanglements cannot free themselves during short periods of stress. Thus, the entanglements behave here as cross-links, and the body shows an entropy-elastic behavior. During long periods of stress, the chains disentangle; the substance flows. Materials showing comparable entropy-elastic and viscous behavior are called *viscoelastic*.

The scientific classification of materials according to their flow behavior corresponds, in a limited sense, to the classification according to their commercial application. A distinction is made here between thermoplasts, fibers, elastomers, and thermosets. This classification naturally only applies at the application or processing temperature under consideration (see Chapter 33).

11.2. Energy Elasticity

11.2.1. Basic Parameters

An ideal-elastic or energy-elastic body deforms under the influence of a force by a definite amount which does not depend on the duration of the force. For comparison purposes, reference is made, not to the force, but to the force per unit surface area, i.e., the stress. The deformation may be a stretching, shearing, turning, compression, or bending (see Table 11-1).

Hooke's law describes energy-elastic bodies under stress/strain. This law relates the tensile stress (stress), $\sigma_{11} = F/A_0$, to the deformation (strain) $\epsilon = \Delta l/l_0$ by

$$\sigma_{11} = E\epsilon, \qquad E = (F/A_0)/(\Delta l/l_0) \qquad (11\text{-}1)$$

The proportionality constant E in stress/strain measurements is known as the modulus of elasticity, the E modulus, or Young's modulus. The E modulus is always related to the cross section of the sample *before* stretching. The reciprocal of the modulus of elasticity is known as the tensile compliance.

Similar relationships exist for other kinds of deformation of energy

Table 11-1. Classification of Moduli According to the Deformation

Force	Deformation	Modulus
Tensile stress	Tensile strain	Modulus of elasticity
Shear stress (tangential)	Shear strain	Shear modulus, torsional modulus
Torque (rotational shear stress)	Torsional strain	Torsional modulus
Compressive stress	Compressive strain	Modulus of elasticity (if pressure on all sides, modulus of compression)
Flexural stress	Flexural strain	Modulus of elasticity (as average of the tension and the compression)

elastic bodies. If a compression occurs on all sides, then the proportionality constant between the pressure p and the compression $-\Delta V / V_0$ is the modulus of compression K:

$$p = K \frac{-\Delta V}{V_0} \tag{11-2}$$

The shear stress σ_{21} and the deformation γ are related via the shear modulus G:

$$\sigma_{21} = G\gamma \tag{11-3}$$

The shear modulus is often known as the rigidity. The reciprocal is the shear compliance. The reciprocal of the compression modulus, on the other hand, is called the compressibility.

The moduli are related to each other by the Poisson number, or Poisson ratio, which is the ratio of the relative lateral contraction to the axial extension or axial strain:

$$\mu = (\Delta d / d_0) / (\Delta l / l_0) \tag{11-4}$$

According to calculations for simple bodies, μ can only vary between 0.5 and 0. The upper limiting value is reached for deformations at constant volume, i.e., with liquids. The lower limiting value occurs for deformations without lateral contraction, e.g., with energy-elastic solid bodies.

For geometrically simple bodies, the relationship between the Poisson ratio and the various moduli is given by

$$E = 2G(1 + \mu) = 3K(1 - 2\mu) \tag{11-5}$$

In this way, the Poisson ratio can be calculated from any two moduli. However, the identity required by Equation (11-5) is not always experimentally found for polymers, since such measurements are often carried out under different loading periods, and, consequently, the viscoelastic contributions become marked. On the basis of theoretical limiting Poisson ratios, however, it can be predicted that the shear modulus must always lie between $1/3$ and $1/2$ of the E modulus.

Theoretical considerations require that the E modulus should vary linearly with the pressure p. From the same derivations, the pressure compliance should be given by the Poisson ratio:

$$E_p = E_0 + 2(5 - 4\mu_0)(1 - \mu_0)p \tag{11-6}$$

where E_p and E_0 are the E moduli at pressures p and 1 MPa, and μ_0 is the Poisson ratio at 1 MPa. Thus, the modulus ratio E_p/E_0 at constant pressure should vary linearly with the reciprocal of the E modulus, E_0, for various polymers having approximately the same Poisson ratio (Figure 11-1).

Table 11-2. Poisson Ratios, Moduli of Elasticity, E, Shear Moduli, G, and Compression Moduli, K, of Various Polymers

Polymer	ν exp.	G/GPa exp.	K/GPa exp.	E, GPa		
				Exp.	From G	From K
Natural rubber	0.50	0.00035	2	0.0011	0.0011	—
Poly(ethylene) low density	0.49	0.070	3.3	0.20	0.21	0.20
Polyamide 6.6	0.44	0.70	5.1	1.9	2.0	1.8
Epoxy resin	0.40	0.90	6.4	2.5	2.5	3.8
Poly(methyl methacrylate)	0.40	1.1	5.1	3.2	3.1	3.0
Poly(styrene)	0.38	1.15	5.5	3.4	3.2	4.0

11.2.2. Theoretical Moduli of Elasticity

The modulus of elasticity of polymers can be calculated theoretically for the chain direction. For example, assume a polymer chain in an all-*trans* conformation with N bonds of length b and bond angle τ. The maximum length of this chain is

$$L_{\max} = Nb \sin(0.5\tau) = Nb \cos\theta \tag{11-7}$$

where $\theta = (180 - \tau)/2$, and, so, is the complimentary bond angle. A change in polymer chain length of ΔL can be caused by a bond length deformation of Δb or by a bond angle expansion of $\Delta\theta$. The deformation, Δb, is given by the force required for this deformation, $F \cos\theta$, and the force constant, k_b, obtainable from infrared measurements, and is $\Delta b = (F \cos\theta)/k_b$. A corresponding force of $(1/2)Fb \sin\theta$ acts on each complimentary angle, and so, with the force constant, k_τ, for the bond angle deformation, the expansion of the bond angle is given by $\Delta\theta = -(1/4)Fb \sin\theta/k_\tau$. Together with the

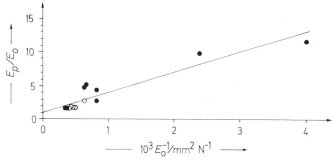

Figure 11-1. Dependence of the modulus ratio E_p/E_0 of various crystalline (●) and amorphous (O) polymers on the modulus of elasticity E_0 at a pressure of 689 MPa. (After H.-G. Elias, S. K. Bhateja, and K. D. Pae.)

modulus of elasticity definition, Equation (11-1), the following is obtained for molecular parameters

$$E = \frac{b \cos\theta}{A_m} \left(\frac{\cos^2\theta}{k_b} + \frac{b^2 \sin^2\theta}{4k_\tau} \right)^{-1} \tag{11-8}$$

where A_m is the cross-sectional area of the polymer chain.

Thus, the theoretical modulus of elasticity depends on the two force constants for the bond length deformation and bond angle expansion, as well as on bond length, bond angle, and molecular cross-sectional area. Thus, polymers with about the same cross-sectional area can have very different moduli of elasticity (Table 11-3).

Theoretical moduli of elasticity calculated in this way are largely confirmed experimentally by what are known as lattice moduli, or crystal moduli. The lattice moduli are determined by following the Bragg angle of selected reflexes by X-ray crystallography as a function of applied stress. The apparent lattice modulus so obtained:

$$E_{cryst}^{app} = \frac{\text{applied force/cross-sectional area}}{\text{measured crystal elongation}} \tag{11-9}$$

is identical to the true lattice modulus E_c only when the stress distribution is homogeneous. This condition is more likely to be fulfilled at lower temperatures because of reduced chain mobility. Thus, the apparent lattice moduli decrease with increasing temperatures, exhibiting plateau regions at high and low temperatures (Figure 11-2).

The low-temperature lattice moduli obtained in this way are more or less in agreement with theoretically calculated moduli of elasticity. They are also independent of the density crystallinity α_{cryst}, which indicates that, in fact, the stress distribution is homogeneous.

11.2.3. Real Moduli of Elasticity

The moduli of elasticity determined by stress/strain measurements are generally much lower than the lattice moduli of the same polymers (Table 11-3). The difference is to be found in the effects of entropy elasticity and viscoelasticity. Since the majority of the polymer chains in such polymer samples do not lie in the stress direction, deformation can also occur by conformational changes. In addition, polymer chains may irreversibly slide past each other. Consequently, E moduli obtained from stress/strain measurements do not provide a measure of the energy elasticity. Such E moduli are no more than proportionality constants in the Hooke's law equation. The proportionality limit for polymers is about 0.1%–0.2% of the

Table 11-3. Theoretical and experimental moduli of elasticity of polymers of different degrees of crystallinity, α_{cryst}.

Polymer	$10^{16} A_m{}^a$ in cm^2	E/GPa Theory	E/GPa Lattice	E/GPa Tensile stress	α_{cryst}/%
Poly(ethylene)	18.3	186–347	250	200	98
Poly(ethylene)			250	15	84
Poly(ethylene)			250	2.4	64
Poly(ethylene)			250	0.7	52
Poly(tetrafluoroethylene), Mod. II	27.5	163	156		
Poly(oxymethylene), trigonal	17.2	48	54	23	
orthorhombic	18.2	220	105		
Poly(oxyethylene)	21.5	9	10		
Polyamide 6,6	17.6	196		5	
Poly(p-benzamide)	19.8	203		77	
Poly(p-phenylene terephthalamide)	20.3			132	
Poly(ethylene terephthalate)	20.0	122–146	140	13	
Cellulose I	32.3	129	130	12^b–105^c	
Cellulose II			90		
Poly(cis-acetylene derivate)^d		45		45	100
it-Poly(propylene)	34.3	50	35	6	
it-Poly(styrene)	69.8		12		
st-Poly(vinyl fluoride), β-mod.	21.1	212	181	21	
st-Poly(vinyl chloride)	21.4	230			
st-Poly(vinyl alcohol)	21.6		255	25	
st-Poly(acrylonitrile)	30.8	236			
st-Poly(methyl methacrylate)		63			
Poly(isobutylene)	41.2	84			
Graphite		1020		420^e	
Steel		270		210	
E-Glass				69–138	

$^a A_m$ = molecular cross-sectional area.
^{b} Rayon.
^{c} Flax.
^{d} Polymers from $C_6H_5NHCOOCH_2$—$C\equiv C$—$C\equiv C$—$CH_2OOCNHC_6H_5$.
^{e} Modmor I.

strain produced. Above this so-called proportionality limit, the relationship between stress and strain can be quite different (Section 11.5). For this reason, the modulus of elasticity of polymers is usually measured over a strain of 0.2% and over a time of 100 s. Moduli measured over higher strains or longer times are lower.

The influence of the degree of crystallinity on the modulus of elasticity can in many cases be described with the aid of the two-phase model in the same way as for blends or fiber-reinforced plastics (see Chapter 35). The partially crystalline polymer is considered as a composite material consisting of the

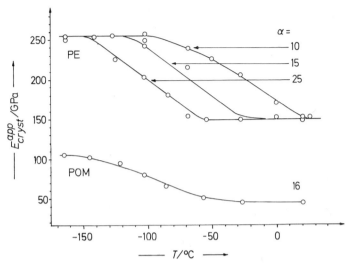

Figure 11-2. Temperature dependence of the apparent lattice moduli E_{cryst}^{app} on the draw temperature and the draw ratio α for a poly(ethylene), PE, and a poly(oxymethylene), POM. (From data of B. Brew, J. Clements, G. R. Davies, R. Jakeways, and I. M. Ward.)

crystalline and amorphous phases. The E modulus of real polymers is given by the ideal E moduli of the crystalline and amorphous phases according to the additivity rules corresponding to the different kinds of orientation of the crystallites. In some cases, a continuity factor which appears to a first approximation to be the aspect ratio for the crystallites must also be used.

The E moduli often depend on their environment. Water acts as a plasticizer in many cases, and, so, reduces the modulus of elasticity by increasing chain mobility. The time-dependent diffusion of water into the polymer thus leads to time-dependent E moduli. For example, a polyamide had an E modulus of 2.75 GPa in the dry state, an E modulus of 1.7 GPa in humid air, and, after four months in the air, an E modulus of 0.86 GPa.

Table 11-4. Moduli of Elasticity of Various
Materials at Room Temperature

Material	$E/$GPa
Vulcanized rubber	0.001–0.01
Crystalized rubber	0.1
Unoriented, partially crystalline polymers	0.1–10
Fibers, reinforced plastics	10–100
Inorganic glasses	100–1000
Crystals	1000–10000

E moduli vary according to their material state by one or more orders of magnitude (see also Section 11.4). However, guidance values may be given for each state of matter, and these values are largely independent of the chemical structure of the polymers (Table 11-4). For the "normal" unoriented and partially crystalline polymers, *E* moduli of about 0.1–10 GPa are obtained (Table 11-3). These values are too low for load-bearing polymer applications, but may be strongly reinforced by careful drawing or stretching and/or by adding suitable fillers (see also Section 35).

11.3. Entropy Elasticity

11.3.1. Phenomena

Weakly cross-linked polymers are rubberlike above their glass transition temperatures. Such rubbers simultaneously exhibit characteristic properties of solids, liquids, and gases. Like solids, they have dimensional stability and behave as Hookean bodies for small deformations. On the other hand, they possess similar expansion coefficients and moduli of elasticity as liquids. Just as the pressures of compressed gases increase with increasing temperature, the stresses for rubbers also increase (Figure 11-3). In contrast, the stresses increase with decreasing temperature below the glass transition temperature. The gaslike behavior is characteristic for what are known as entropy-

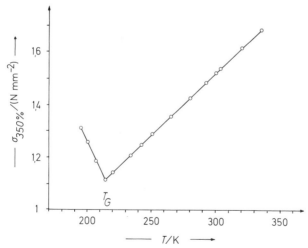

Figure 11-3. Tensile stress $\sigma_{350\%}$ at 350% elongation as a function of temperature for weakly cross-linked natural rubber. (After data from K. H. Meyer and C. Ferri.)

elastic bodies. If these bodies are deformed, the segments are displaced from their equilibrium positions into entropically unfavorable states. Because of the cross-links, however, the segments cannot slide past each other; consequently, the body is not irreversibly deformed. On stress release, the segments return from an ordered state into disordered positions: the entropy increases. Thus, this "rubber elasticity" can be described in various ways. On a thermodynamic basis, the deformation produces an entropy reduction. On the molecular level, a change in conformation is forced by the deformation. In the language of mechanics, one is dealing with the occurrence of a normal stress. This stress is so named because it occurs perpendicular or normal to the strain direction.

Entropy-elastic bodies differ very characteristically from energy-elastic bodies:

1. Energy-elastic bodies exhibit large moduli of elasticity for small deformations of about 0.1%. In contrast, entropy-elastic bodies possess low moduli of elasticity and high reversible deformabilities of several hundred percent.
2. Energy-elastic bodies cool under strain, whereas entropy-elastic bodies become warmer.
3. Energy-elastic bodies such as steel and rubber under strains of less than about 10% expand on heating, but strongly stretched rubber contracts.

11.3.2. Phenomenological Thermodynamics

The changes in the states of entropy-elastic bodies described in the previous section can be expressed quantitatively by phenomenological thermodynamics, starting with one of the fundamental equations in thermodynamics. The relationship of interest here relates the pressure p with the internal energy U, the volume V, and the thermodynamic temperature T (see textbooks of chemical thermodynamics):

$$p = -\left(\frac{\partial U}{\partial V}\right)_T + \left(\frac{\partial p}{\partial T}\right)_V T \qquad (11\text{-}10)$$

Instead of the change in volume dV, the change in length dl on application of a stretching force F is considered. F and p have opposite signs. Equation (11-10) thus becomes

$$F = \left(\frac{\partial U}{\partial l}\right)_T + \left(\frac{\partial F}{\partial T}\right)_l T \qquad (11\text{-}11)$$

If the second law of thermodynamics $A = U - TS$ is differentiated with respect to the length l, this gives

$$\left(\frac{\partial A}{\partial l}\right)_T = \left(\frac{\partial U}{\partial l}\right)_T - T\left(\frac{\partial S}{\partial l}\right)_T \qquad (11\text{-}12)$$

By setting the two expressions for $(\partial U/\partial l)_T$ in Equations (11-11) and (11-12) equal, one finds

$$\left(\frac{\partial A}{\partial l}\right)_T + T\left(\frac{\partial S}{\partial l}\right)_T = F - T\left(\frac{\partial F}{\partial T}\right)_l \qquad (11\text{-}13)$$

In thermodynamics, furthermore, the following is quite generally valid:

$$\left(\frac{\partial S}{\partial V}\right)_T = \left(\frac{\partial p}{\partial T}\right)_V \qquad (11\text{-}14)$$

Since the force F is proportional to the pressure p, and the length l to the volume V, it is possible to write, in analogy to Equation (11-14),

$$\left(\frac{\partial S}{\partial l}\right)_T = -\left(\frac{\partial F}{\partial T}\right)_l \qquad (11\text{-}15)$$

If Equation (11-15) is inserted into (11-13), one obtains

$$\left(\frac{\partial A}{\partial l}\right)_T - T\left(\frac{\partial F}{\partial T}\right)_l = F - T\left(\frac{\partial F}{\partial T}\right)_l \qquad (11\text{-}16)$$

The second term on both sides of Equation (11-16) is identical. What is called the equation of the thermal state for entropy-elastic bodies is therefore

$$\left(\frac{\partial A}{\partial T}\right)_l = F \qquad (11\text{-}17)$$

It has been found experimentally that the force F is directly proportional to the temperature T in the case of weakly cross-linked natural rubber extended by less than 300%. From this it follows that $F = const \times T$ or $\partial F/\partial T = const$, or

$$\frac{F}{T} = \frac{\partial F}{\partial T} \qquad (11\text{-}18)$$

If Equation (11-18) is inserted into Equation (11-11), the result is

$$\left(\frac{\partial U}{\partial l}\right)_T = 0 \qquad (11\text{-}19)$$

Thus, the internal energy U of an entropy-elastic body does not alter during elongation. An entropy-elastic body differs fundamentally from an energy-elastic one in this respect. The change of force F on heating can be obtained by combining Equations (11-18) and (11-15):

$$dS = -\frac{F}{T}\,dl \qquad (11\text{-}20)$$

$T\,dS$ and F are positive. Therefore, the change in length dl must be negative; entropy-elastic bodies have a negative thermal coefficient of expansion.

The total differential of the change in length of an entropy-elastic body is

$$dl = \left(\frac{\partial l}{\partial F}\right)_T dF + \left(\frac{\partial l}{\partial T}\right)_F dT \qquad (11\text{-}21)$$

On heating at constant length, $dl = 0$. Equation (11-21) then becomes

$$\left(\frac{\partial F}{\partial T}\right)_l = -\frac{(\partial l/\partial T)_F}{(\partial l/\partial F)_T} \qquad (11\text{-}22)$$

According to Equation (11-20), the thermal coefficient of expansion $\gamma = (1/l)(\partial l/\partial T)$ of an entropy-elastic body is negative. The denominator $(\partial l/\partial F)_T$ in Equation (11-22) corresponds to the increase in length when the stress is increased and is positive, so that $(\partial F/\partial T)_l$ must be positive in Equation (11-22). When an entropy-elastic body is heated, the stress increases.

In contrast to ideal entropy-elastic bodies, real entropy-elastic bodies have an energy-elastic component. The force F_e resulting from this component is given for a uniaxial deformation by

$$F_e = (\partial U/\partial l)_{T,V} = T(\partial F/\partial T)_{V,l} \qquad (11\text{-}23)$$

Thus, the energy-elastic component F_e/F can in principle be obtained by force/temperature measurements at constant volume. Since such measurements are experimentally difficult to carry out, measurements are mostly made at constant length, and evaluated with an equation, not given here, applicable to this case.

As expected from theory, the energy-elastic component F_e/F so obtained is independent of the strain ratio α over quite a considerable range (Figure 11-4). In this range, the data are also independent of the measurement method, the degree of cross-linking, the cross-linking conditions, the kind of deformation (extension, compression, boring), the nature of the swelling solvent, and the degree of dilution. The F_e/F values decrease for higher strain ratios because the natural rubber crystallizes under stress. The sharp increase of the F_e/F values at low strain ratios is presumably due to intermolecular interaction, which becomes noticeable in this region.

Energy elastic components may be either positive or negative (Table 11-5). With polymers having a *trans* conformation as lowest energy conformation, a transition from *gauche* to *trans* will therefore lead to an energy

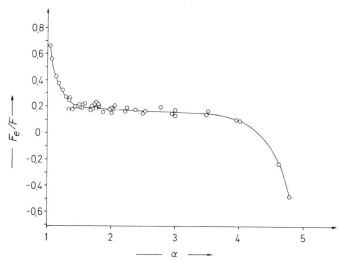

Figure 11-4. Dependence of the energy-elastic component F_e/F on the draw ratio α for natural rubber according to data from various authors (M. Shen, D. A. McQuarrie, and J. L. Jackson).

decrease. The energy-elastic component is then negative, as has been found for cross-linked poly(ethylene). In contrast, the energy-elastic component for poly(dimethyl siloxane) cross-linked networks is positive; in this case, a *trans* conformation must be transformed by extension into a *gauche* conformation.

11.3.3. Statistical Thermodynamics of Ideal Networks

Phenomenological thermodynamics describes changes in energies, temperatures, volumes, etc. Unless additional assumptions are made, however, it cannot give any information about the molecular phenomena that lie behind these processes. Statistical thermodynamics attempts to obtain such information through the use of probability functions.

When a weakly cross-linked material is stretched, the molecular segments

Table 11-5. Energy-Elastic Component of Various Cross-Linked Polymers

Polymer	F_e/F	Polymer	F_e/F
Poly(ethylene)	−0.42	Poly(isobutylene)	−0.06
Poly(oxyethylene)	0.08	Poly(dimethyl siloxane)	0.19
Poly(oxytetramethylene)	−0.47	*cis*-1,4-Poly(butadiene)	0.12
Poly(vinyl alcohol)	0.42	*trans*-1,4-Poly(butadiene)	−0.25
Poly(styrene)	0.16	*cis*-1,4-Poly(isoprene)	0.17
Poly(ethylene-co-propylene)	−0.43	*trans*-1,4-Poly(isoprene)	−0.09

between two cross-linking points adopt a more unfavorable (less random) position. The ends of the segments move away from one another.

With one end of the chain at the origin of the coordinate system, the probability $\Omega_i(x, y, z)\,dx\,dy\,dz$ of finding the other end point in a volume element $dx\,dy\,dz$ it then [see also Equation (A4-37) for the corresponding expression for *one* direction]

$$\Omega_i(x,y,z)dx\,dy\,dz = (B/\pi^{0.5})^3 \exp[-B^2(x_i^2 + y_i^2 + z_i^2)]dx\,dy\,dz \qquad (11\text{-}24)$$

with

$$B^2 = \frac{3/2}{Nb^2} \qquad (11\text{-}25)$$

N is the number of segments in the chain and b is the segment length. Inserting Equation (11-24) into the Boltzmann equation $s_i = k \ln \Omega_i$, we obtain the following for the entropy of a chain:

$$s_i = k \ln[const \times (B/\pi^{0.5})^3 \, dx\,dy\,dz] - kB^2(x_i^2 + y_i^2 + z_i^2) \qquad (11\text{-}26)$$

The constant serves to maintain dimensionality, and is, therefore, purely a factor of convenience (in the subsequent mathematical treatment it is canceled out). It is then assumed, in fact, that the coordinates of each individual segment change in the same proportion as the external coordinates of the test body. Thus if the external coordinates are extended by an elongation ratio α, the coordinate in the x direction of the i segment should in this affine deformation be α times as large after elongation as the initial unextended value $(x_{i,0})$, etc.:

$$x_i = \alpha_x x_{i,0}, \qquad y_i = \alpha_y y_{i,0}, \qquad z_i = \alpha_z z_{i,0} \qquad (11\text{-}27)$$

Equation (11-27), together with Equation (11-26) for the change in entropy of a segment becomes, for extension,

$$\Delta s_i = s_i - s_{i,0} = -kB^2[(\alpha_x^2 x_{i,0}^2 + \alpha_y^2 y_{i,0}^2 + \alpha_z^2 z_{i,0}^2) - (x_{i,0}^2 + y_{i,0}^2 + z_{i,0}^2)]$$

$$(11\text{-}28)$$

The total change in entropy must be additive. With chains of equal length it then follows that

$$\Delta S = \sum_i \Delta s_i = -kB^2\left[(\alpha_x^2 - 1)\sum_i x_{i,0}^2 + (\alpha_y^2 - 1)\sum_i y_{i,0}^2 + (\alpha_z^2 - 1)\sum_i z_{i,0}^2\right]$$

$$(11\text{-}29)$$

For the end-to-end chain distance [see Equation (4-24)]

$$\langle L^2 \rangle_{00} = Nb^2 = \langle x_0^2 \rangle + \langle y_0^2 \rangle + \langle z_0^2 \rangle \qquad (11\text{-}30)$$

With the number N_i of chains, we have

$$\sum_i x_{i,0}^2 = N_i \langle x_0^2 \rangle, \qquad \sum_i y_{i,0}^2 = N_i \langle y_0^2 \rangle, \qquad \sum_i z_{i,0}^2 = N_i \langle z_0^2 \rangle \qquad (11\text{-}31)$$

and for an isotropic material

$$\langle x_0^2 \rangle = \langle y_0^2 \rangle = \langle z_0^2 \rangle \qquad (11\text{-}32)$$

It therefore follows from Equations (11-29)–(11-31) and (11-25) that

$$\Delta S = -0.5 k N_i (\alpha_x^2 + \alpha_y^2 + \alpha_z^2 - 3) \qquad (11\text{-}33)$$

According to Equation (11-33), the entropy change of a rubber on stretching is not due to a particular chemical structure, but simply to the number of chains and hence the number of cross-linking points linking the chains.

Equation (11-33) describes a special case, namely, that of elongation without volume change ($\alpha_x \alpha_y \alpha_z = 1$). Various equations have been proposed for the general case, and they can all be formulated as

$$\Delta S = -A[(\alpha_x^2 + \alpha_y^2 + \alpha_z^2 - 3) - D] \qquad (11\text{-}34)$$

where A and D have different significances according to the assumptions made in different theoretical treatments (for example, D is often taken as $D = \ln \alpha^3$).

According to Equation (11-20), the elongation force F can be expressed by

$$F = -T \left(\frac{\partial S}{\partial l} \right)_{T,V} = -T \left(\frac{\partial \Delta S}{\partial l} \right)_{T,V} \qquad (11\text{-}35)$$

ΔS can be found from Equation (11-33). If, on stretching, the chain is elongated in one direction ($\alpha_x = \alpha$) and simultaneously contracted in the two other directions ($\alpha_y = \alpha_z = 1/\alpha_x^{0.5}$), then Equation (11-33) becomes

$$\Delta S = -0.5 k N_i \left(\alpha^2 + \frac{2}{\alpha} - 3 \right) \qquad (11\text{-}36)$$

and, with $\alpha = l/l_0$, we find, after differentiation,

$$\left(\frac{\partial \Delta S}{\partial l} \right)_{T,V} = -\frac{k N_i (\alpha - \alpha^{-2})}{l_0} \qquad (11\text{-}37)$$

or, inserted into Equation (11-35),

$$F = \frac{k T N_i (\alpha - \alpha^{-2})}{l_0} \qquad (11\text{-}38)$$

If both sides are divided by the original cross-sectional area $A_0 = V_0/l_0$, then, with the definition of the tensile stress $\sigma_{ii} \equiv F/A_0$, the gas constant $R = k N_L$,

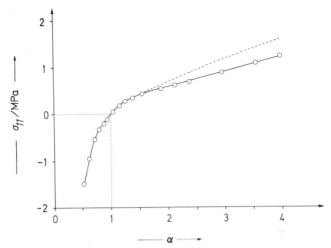

Figure 11-5. Relationship between the stress σ_{11} and the draw ratio $\alpha = L/L_0$ for cross-linked natural rubber: $-O-O-$, Experimental; ---, calculated according to Equation (11-39). Measurements by elongation (at $\alpha > 1$) or compression ($\alpha < 1$). (After L. R. G. Treloar.)

and the molar concentration $[M_i] = N_i/V_0 N_L$ of the network chains, Equation (11-38) becomes

$$\sigma_{ii} = kT\frac{N_i}{V_0}(\alpha - \alpha^{-2}) = RT[M_i](\alpha - \alpha^{-2}) \tag{11-39}$$

Experiment and theory agree well with regard to the relationship between σ_{ii} and α in the case of compression and small elongations (Figure 11-5). Deviations occur at large elongations, which could be due to beginning rubber crystallization, to a non-Gaussian distribution of cross-linking sites, or to time effects.

11.3.4. Real Networks

Experimentally found deviations from Equation (11-39) are mostly explained in terms of the empirical Mooney–Rivlin equation (Figure 11-6):

$$[F^*] = \frac{\sigma_{ii}}{\alpha - \alpha^{-2}} = 2C_1 + 2C_2\alpha^{-1} \tag{11-40}$$

The constant $2C_1$ is often identified with the corresponding expression in Equation (11-39), that is, with

$$\frac{\sigma_{ii}}{\alpha - \alpha^{-2}} = RT[M_i] = kT(N_i/V_0) = 2C_1 \tag{11-41}$$

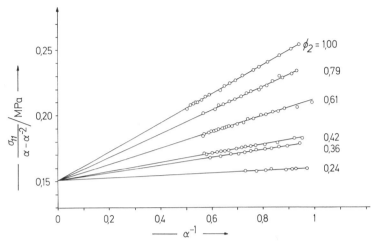

Figure 11-6. Reduced stress $\alpha_{11}/(\alpha - \alpha^{-2})$ of cross-linked natural rubber as a function of the reciprocal strain ratio at 45° C in the unswollen state ($\phi_2 = 1$) and in the swollen state in decane at various swell ratios. (After J. E. Mark.)

The constant $2C_2$ decreases according to Figure 11-6, with decreasing volume fraction ϕ_2 of the polymer in swollen cross-linked networks, that is, with increasing degree of swelling. The constant is also influenced by the cross-linking conditions, such as, for example, chain orientation during cross-linking, presence of solvent, degree of cross-linking, etc. Many indications suggest that the expression $2C_2$ derives from entanglements between chain segments. The ratio C_2/C_1, of course, decreases with the cross-sectional area of the polymer molecule (Figure 11-7), which, in turn, is related to the stiffness of the polymer chains and their tendency to entangle.

11.3.5. Sheared Networks

Shearing can be treated in the same way as drawing. When sheared, the sample is elongated in the x direction and correspondingly contracted in the y direction. The coordinates in the z direction remain constant. Thus, $\alpha = \alpha_x$, $\alpha = 1/\alpha_y$, and $\alpha_z = 1$. Equation (11-33) therefore becomes

$$\Delta S = -0.5kN_i(\alpha^2 + \alpha^{-2} - 2) \qquad (11\text{-}42)$$

and correspondingly, for the change in entropy ΔS_V which is related to the unit volume V_0, with $[M_i] = N_i/V_0 N_L$, $R = kN_L$, and $\Delta S_V = \Delta S/V_0$,

$$\Delta S_V = -0.5R[M_i](\alpha^2 + \alpha^{-2} - 2) \qquad (11\text{-}43)$$

The shear strain γ produced by a shearing load is given by $\gamma = \alpha - \alpha^{-1}$. From

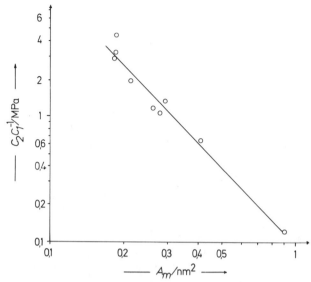

Figure 11-7. Ratio C_2/C_1 of the Mooney–Rivlin constants of different elastomers as a function of the chain cross section for $2C_1 = 0.2$ MPa. (After R. F. Boyer and R. L. Miller.)

Equation (11-42), therefore, since $\gamma^2 = (\alpha - \alpha^{-1})^2 = \alpha^2 + \alpha^{-2} - 2$,

$$\Delta S_V = -0.5R[M_i]\gamma^2 \qquad (11\text{-}44)$$

The relationship between shear stress σ_{21} and shear strain γ is given, in analogy to Equation (11-35), by

$$\sigma_{21} = -T\frac{\partial \Delta S_V}{\partial \gamma} \qquad (11\text{-}45)$$

After differentiating Equation (11-44), we obtain from Equation (11-45)

$$\sigma_{21} = RT\gamma[M_i] = G\gamma \qquad (11\text{-}46)$$

According to Equation (11-46), the shear stress σ_{21} is directly proportional to the shear strain γ. The rubber, therefore, deforms on shearing according to Hooke's law with the shear modulus G [see Equation (11-3)], but is non-Hookean during elongation [see Equation (11-39)].

11.3.6. Entanglements

Non-cross-linked chain molecules, as well as chemically weakly cross-linked materials, show entropy-elastic behavior under given stresses. It is also conceivable that the chains of long, non-cross-linked, flexible macromolecules can become entangled or interpenetrate each other to a certain extent. When deformation by drawing is rapid, the entanglements act as cross-links. The parts of the chain adopt more random positions in an attempt to return to the original random positions. A normal stress is created. Both the normal stress and the shear stress can be measured in a cone and plate viscometer. During rotation there is shearing. It is possible to calculate the shear stress from the rotational moment (torque) applied to the rotor. The cone and plate are pushed away from one another, however, by the normal stress developed by the sample when sheared. In order to prevent this, it is necessary to apply a force, which is then proportional to the normal stress. The normal stress can be much greater than the shear stress (Figure 11-8).

Shear and normal stresses cannot be measured individually in a capillary viscometer. If material with entropy-elastic components is pressed through a nozzle, the macromolecules are deformed. If this time of loading is short and the load not too great, however, the segments cannot slip away from one another, because of entanglements. At the nozzle opening, the material again has more space at its disposal. The material will therefore expand when it leaves the nozzle. The effect is known as the Barus or memory effect in melts and the Weissenberg effect in solutions. In extrusion it is called swelling and in the blowing of hollow samples it is called swelling behavior. Of course, the effect is also time dependent, since, as the time increases, the chains can slip away from one another more readily.

The Barus effect can be studied with the aid of the Bagley plot. Solving

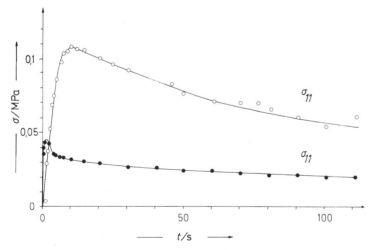

Figure 11-8. Shear stress σ_{21} and normal stress σ_{11} of a poly(ethylene) at 150°C and a shear gradient of 8.8 s^{-1} as a function of time in second. Measurements with a cone-and-plate viscometer. (After BASF). Read σ_{21} instead of σ_{11} for ●—●—●.

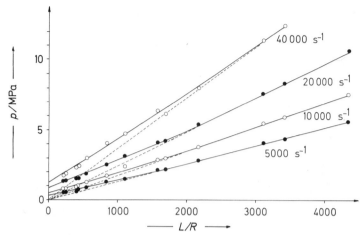

Figure 11-9. Pressure p as a function of the capillary geometry L/R of a 6% solution of poly(isobutylene) ($\langle M_w \rangle = 6 \times 10^6$, $\langle M_n \rangle = 0.55 \times 10^6$ g/mol) in toluene for various shear rates at room temperature. L is the length and R is the radius of the capillary. (After J. Klein and H. Fusser.)

Equation (7-32) for the pressure p and replacing the shear stress σ_{21} for non-Newtonian liquids by $\sigma_{21} = \eta_{app} \, \dot{\gamma}$,

$$p = 2\sigma_{21} \frac{L}{R} = 2\eta_{app} \, \dot{\gamma} \, \frac{L}{R} \qquad (11\text{-}47)$$

In the Bagley plot, the pressure p is plotted against the nozzle geometry L/R at a constant shear rate $\dot{\gamma}$. With Newtonian liquids, according to Equation (7-32), for $L/R = 0, p$ will also be 0; the slope of the straight lines is given by $2\sigma_{21}$. Behavior of this type is also found for concentrated polymer solutions at high values of L/R (Figure 11-9). Obviously, one is then working in the second Newtonian viscosity region, η_∞. At low L/R values, however, the function $p = f(L/R)$ for $\dot{\gamma} =$ const deviates from this straight line, and tends toward a new linear relationship. This straight line occurring at low L/R values does not intersect the p axis at $p = 0$, but at a finite value p_0 (Figures 11-9 and 11-10). Equation (11-47) thus becomes

$$p = p_0 + \text{const}' \times \frac{L}{R} \qquad \dot{\gamma} = \text{const} \qquad (11\text{-}48)$$

Usually p_0 is identified with the loss of pressure caused by the elastically accumulated energy of the flowing liquid and by the formation of a stationary flow profile at both ends of the capillary. Since the loss in pressure disappears at large L/R values, it may be assumed that the elastic deformation can equilibrate in time in very long capillaries (disentangling). This is also

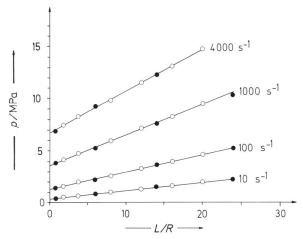

Figure 11-10. Bagley plot of a high-impact poly(styrene) at 189° C at shear gradients of 10, 100, 1000, and 4000 s^{-1} from measurements with capillaries of $R = 1$ mm in diameter (●) or $R = 0.6$ mm (○) and different lengths L. (After BASF.)

confirmed by the fact that the p_0 values are greater, the larger the shear rates, i.e., the shorter the retention times. In addition, according to measurements on poly(ethylene) melts, the extrusion swelling falls off at very high L/R values. According to the general theory, however, the extrusion swelling is a measure of the elastic energy accumulated in the system.

Apart from these theoretical considerations, however, p_0 has a direct practical meaning in the extrusion and spinning of plastics. That is, the greater the nozzle ratio L/R, the higher the pressure that has to be applied. In practice, therefore, as small a nozzle length as possible is used.

11.4. Viscoelasticity

11.4.1. Basic Principles

In the preceding discussion about the energy- and entropy-elastic behavior of matter, it was tacitly assumed that the body returns immediately and completely to the original state when the load is removed. In actual fact, this process always takes a certain time in macromolecular substances. In addition, not all bodies return completely to the original position; in some cases they are partially irreversibly deformed.

In these bodies, then, there must be simultaneous cooperation between time-independent elastic and time-dependent viscous properties. Many

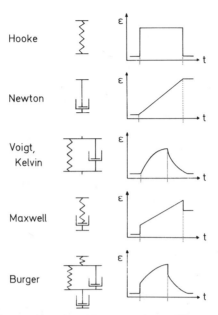

Figure 11-11. Deformation ϵ as a function of time, according to different models. The sample was loaded up to the time t_1 and the load was released at the time t_2 (represented by dotted lines).

phenomena can be described by different combinations of the basic equations for elastic and viscous behavior.

The two extreme cases of mechanical behavior can be reproduced very well by mechanical models. A compressed Hookean spring can serve as a model for the energy-elastic body under load (Figure 11-11). On releasing the load, the compressed spring immediately returns to its original position. The relationship between the shear stress $(\sigma_{21})_e = \sigma_e$, the shear modulus G_e, and the elastic deformation γ_e is given by Hooke's law [Equation (11-1)]:

$$(\sigma_{21})_e = \sigma_e = G_e \gamma_e \tag{11-49}$$

Differentiation with respect to time yields

$$\frac{\mathrm{d}(\sigma_{21})_e}{\mathrm{d}t} = G_e \frac{d\gamma_e}{dt} \tag{11-50}$$

The model for rate effects is a dash pot containing a piston in a viscous Newtonian liquid. On removing the load, a certain time passes before the piston returns to its original position. This time dependence is already included in the Newton's law equation, Equation (7-28). For greater clarity, and to conform with Equation (11-50), the shear stress σ_{21} in Newton's law will be written as $(\sigma_{21})_\eta = \sigma_\eta$ and the shear rate dv/dy as the deformation rate $d\gamma_\eta/dt$:

$$\sigma_\eta = \eta \frac{d\gamma_\eta}{dt} \tag{11-51}$$

Integration leads to

$$\gamma_\eta = \frac{\sigma_\eta}{\eta} t \tag{11-52}$$

Maxwell bodies are obtained if Hookean and Newtonian bodies are connected in series (Figure 11-11). The Kelvin or Voigt model, on the other hand, contains Hookean and Newtonian bodies in a parallel arrangement (Figure 11-11). The Maxwell body is a model for relaxation phenomena and the Kelvin body is a model for retardation processes.

11.4.2. Relaxation Processes

The mountains melted before the Lord.

Judges 5,5

A relaxation is defined in mechanics as the decrease in stress with constant deformation. That is, a stress must be applied during the deformation of a viscous liquid. When the deformation ceases, the stress will fall off (relax) as the molecules or molecular segments return to their rest state. The Maxwell model obviously describes this behavior very well. During rapid deformation, the spring will very quickly elongate since a viscous liquid responds only slowly to a rapidly applied stress. If deformation is kept constant, then, because of the relaxation of the spring to its equilibrium stress position, the piston subsequently begins to move slowly through the viscous liquid (see also Figure 11-11). If the stress is suddenly removed, the spring contracts immediately but the piston remains in the elongated state. Although originally intended only for energy-elastic bodies, the spring model can also be used to describe the shearing of entropy-elastic bodies, as exemplified by Equation (11-46).

The rates of deformation $d\gamma/dt$ are additive in these processes. By combining the expressions for the rates of deformation according to Hooke's law [Equation (11-50)] and according to Newton's law [Equation (11-51)], we obtain the following for the total rate of deformation:

$$\frac{d\gamma}{dt} = \frac{d\gamma_e}{dt} + \frac{d\gamma_\eta}{dt} = G_e^{-1} \frac{d\sigma_e}{dt} + \frac{\sigma_\eta}{\eta} \tag{11-53}$$

The indices e and η can be neglected, since the stress is the same in both elements. When the deformation is constant, $d\gamma/dt = 0$, and Equation (11-53) becomes

$$G^{-1} \frac{d\sigma}{dt} = -\frac{\sigma}{\eta}$$

(11-54)

or, when integrated,

$$\sigma = \sigma_0 \exp\left(\frac{-Gt}{\eta}\right) = \sigma_0 \exp\left(\frac{-t}{t_{rel}}\right)$$

(11-55)

$t_{rel} = \eta/G$ is the relaxation time. It indicates the time interval required for the stress to fall to a value $1/e$ times the original value. The ratio t_{rel}/t is known as the Deborah number. The Deborah number is zero for liquids, unity for polymers at the glass transition temperature, and infinite for ideal energy-elastic solid bodies.

With real polymers, however, there exists not only one, but a whole spectrum of relaxation times. In an ideal rubber, for example, all cross-linking points are separated by equal distances. With short periods of stress, the resulting stresses are compensated by the largely "free" rotation about the chain bonds within short relaxation periods of $\sim 10^{-5}$ s. With long stress periods, the cross-linking points can also shift in relation to one another. The long relaxation times characteristic of this process hinder the viscous flow of the material during short periods of stress. Between these two relaxation times there is a region in which the modulus of elasticity remains almost constant. With real rubbers, however, the distances between the cross-linking points are not equal, but vary over a wide range. Thus a whole spectrum of relaxation times is to be expected. This spectrum can be exemplified through the use of a model with a number of Maxwell bodies in a parallel arrangement.

11.4.3. Retardation Processes

Retardation is defined as the increase in deformation with time under constant stress. Retardation processes in a material can be recognized as "creep" or "strain softening." Since the phenomenon is also observed in apparently solid materials at room temperature, it is also called "cold flow." When the load is removed, a change in the deformation is often observed. In some cases, the sample readopts its original dimensions. Creep is therefore better described as retarded elasticity than as viscous flow. Although creep can be modeled by a series of coupled Maxwell elements, the corresponding mathematical equations are difficult to solve and a special model with parallel spring and dash-pot elements (Kelvin or Voigt element) is preferred. Since creep is a deformation at constant stress, it is only necessary to add the two expressions for the stress in Hookean bodies [Equation (11-49)] and in Newtonian fluids [Equation (11-51)]:

$$\sigma = \sigma_e + \sigma_\eta = G_e \gamma_e + \eta \frac{d\gamma_\eta}{dt} \tag{11-56}$$

From this it follows through integration that (index K characterizes the Kelvin element)

$$\gamma_K = \frac{\sigma}{G}\left[1 - \exp\left(\frac{-Gt}{\eta}\right)\right] = \gamma_\infty\left[1 - \exp\left(\frac{-t}{t_{ret}}\right)\right] \tag{11-57}$$

Here, the indices e and η are again omitted because the strain is the same in both elements. In Equation (11-57), γ_∞ is a constant and t_{ret} the retardation time. As a rule, there is usually a whole spectrum of retardation times in the retardation processes also. Retardation and relaxation time distributions are similar, but they are not identical since they pertain to different models of the deformation behavior.

11.4.4. Combined Processes

Macromolecular materials usually possess entropy elasticity together with viscous and energy-elastic components. Such behavior was only partly comprehensible by use of the models discussed up to now. It can be described very satisfactorily, however, by a four-parameter model in which a Hooke body, a Kelvin body, and a Newton body are combined (see the lowest figure in Figure 11-11). With this model, the deformation must again be added, i.e., with Equations (11-49), (11-52), and (11-57),

$$\gamma = \gamma_e + \gamma_K + \gamma_\eta = \frac{\sigma}{G} + \gamma_\infty\left[1 - \exp\left(\frac{-t}{t_r}\right)\right] + \frac{\sigma}{\eta}t \tag{11-58}$$

or by introducing the compliance $C = \gamma/\sigma$,

$$C = C_0 + C_\infty\left[1 - \exp\left(\frac{-t}{t_r}\right)\right] + \frac{t}{\eta} \tag{11-59}$$

According to this equation, the mechanical behavior observed depends very much on the ratio of the test time to the orientation time t_r. With $t \gg t_r$, the exponential term in Equation (11-59) makes almost no contribution to the total deformation. Conversely, the damping contribution as described by this term in the total deformation will become more noticeable the nearer the test and orientation times t and t_r approach each other (see Figure 11-12).

Similar reasoning applies to the temperature dependence. At low

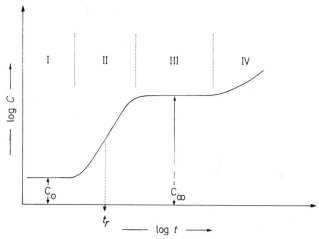

Figure 11-12. Schematic representation of the compliance C as a function of time. t_r is the orientation time. I, Glassy state; II, viscoelastic state; III, entropy-elastic state; IV, viscous flow.

temperatures, $t_r = \eta/G$ tends toward infinity as the viscosity η becomes very large; only a Hookean elasticity is observed. At high temperatures, on the other hand, the third term (viscous flow) predominates. In between lies a range of temperatures at which the test and orientation times are comparable. A damping will then be observed at these temperatures.

11.4.5. Dynamic Loading

With dynamic measurements, the sample is subjected to a periodic stress. In the simplest case, the stress is applied sinusoidally to an ideal energy-elastic body. The applied stress $\sigma(t)$ then alters with time t and the angular frequency ω according to

$$\sigma(t) = \sigma_0 \sin \omega t \tag{11-60}$$

where σ_0 is the amplitude. In ideal energy-elastic bodies, the deformation instantly follows the applied stress, and consequently,

$$\gamma(t) = \gamma_0 \sin \omega t \tag{11-61}$$

Polymers are not ideal energy elastic, but viscoelastic. In such cases, the deformation lags behind the applied stress. With ideal viscoelastic bodies, the resulting phase angle ϑ in the corresponding vector diagram can be assumed constant, such that

$$\gamma(t) = \gamma_0 \sin (\omega t - \vartheta) \tag{11-62}$$

The stress vector can be similarly considered to be the sum of two components. One component, $\sigma' = \sigma_0 \cos \vartheta$, is in phase with the deformation, the other component, on the other hand, $\sigma'' = \sigma_0 \sin \vartheta$, is not. A modulus can be assigned to each of these two components. The real modulus, or storage modulus, G', measures the rigidity and resistance to deformation of the sample. It is given by

$$G' = \sigma'/\gamma_0 = (\sigma_0/\gamma_0) \cos \vartheta = G * \cos \vartheta \qquad (11\text{-}63)$$

The imaginary, or loss modulus, G'', on the other hand, gives the loss of useful mechanical energy through dissipation as heat. G'' is given by

$$G'' = \sigma''/\gamma_0 = G * \sin \vartheta \qquad (11\text{-}64)$$

The same expressions can be derived after introduction of complex variables. Equations (11-60) and (11-62) become in this formalism

$$\sigma * = \sigma_0 \exp (i\omega t) \qquad (11\text{-}65)$$

$$\gamma * = \gamma_0 \exp i(\omega t - \vartheta) \qquad (11\text{-}66)$$

and the complex modulus $G *$ is given as

$$G * = G' + iG'' \qquad (11\text{-}67)$$

or, from Equations (11-63) and (11-64) as

$$G* = [(G')^2 + (G'')^2]^{1/2} \qquad (11\text{-}68)$$

The ratio of the imaginary to the real modulus is called the loss factor:

$$\Delta = G''/G' = \tan \vartheta \qquad (11\text{-}69)$$

Instead of following the deformation $\gamma(t)$ produced by a given stress $\sigma(t)$,

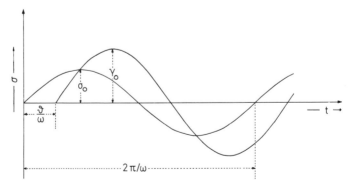

Figure 11-13. Schematic representation of the tensile stress σ as a function of the time t with dynamic (sinusoidal) loading (see text).

the sample can, of course, be subject to a given deformation, and the resulting stress can be measured. The complex compliance $J^* = 1/G^*$ is obtained in this case, and the storage and loss compliances are correspondingly given by

$$G' = J'/[(J')^2 + (J'')^2] \tag{11-70}$$

$$G'' = J''/[(J')^2 + (J'')^2] \tag{11-71}$$

Dynamic mechanical measurements can mostly be divided into two groups. The reaction of a sample to a once applied light torque can be measured with the torsion pendulum. The sample oscillates freely, whereby the amplitude decreases steadily with each cycle for viscoelastic materials. The ratio of two successive amplitudes is constant for ideal viscoelastic materials. This procedure yields shear moduli. The torsion pendulum allows measurements to be relatively easily made; the disadvantage is that the frequency is not an independent variable with this method.

Another group of procedures subjects the sample to continuous forced oscillations. The resulting stresses and deformations can be measured independently of each other with the most commonly used instrument of this group, the rheovibron. Since the rheovibron applies a tensile stress, the moduli obtained are tensile moduli and not shear moduli, as is the case with the torsion pendulum.

11.5. Deformation Processes

11.5.1. Tensile Tests

In a typical tensile strength test, a standard rod of the sample is stretched in a tensile strength machine and elongated at a constant rate, while the stress σ_{11} and strain ratio $\alpha = L/L_0$ (or time) are recorded. The stress is given relative to the original cross-sectional area A_0 of the sample. The elongation $\epsilon = (L - L_0)/L_0$ is often given instead of the strain ratio α. When, for example, the sample has been stretched to 2.5 times its original length, it is usually said to be 150% elongated.

A few typical stress/strain diagrams are shown in Figure 11-14. From the diagram it is seen that elastomers exhibit an increasingly strong stress with increasing strain until they rupture at the point IV. Typical thermoplasts, on the other hand, behave differently. At low stresses or strains, Hooke's law is followed in the region from the origin to point I. Point I is consequently called the *proportionality limit* or *elastic limit*. The latter name is incorrect because of the entropy elasticity remaining above point I. By definition, the proportionality limit is reached when the sample shows a remaining strain of

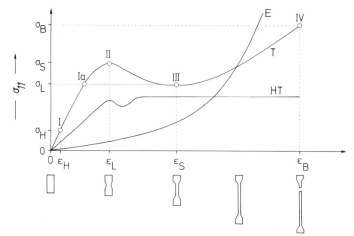

Figure 11-14. Schematic representation of the tensile stress σ_{11} as a function of strain ϵ at constant temperature for an elastomer E, a partially crystalline thermoplast T, and a hard-elastic thermoplast HT. The ductile region is Ia-II-III. The necking effect shown below the diagram is typical of normal thermoplasts, but does not occur with elastomers or hard-elastic thermpolasts. The diagram is not drawn to scale; for example, elastomers show a much larger elongation at break than do thermoplasts.

0.1% after removal of the stress. On the other hand, the *technical elastic limit* is defined as an extension of 0.2%.

Point II is the maximum of the stress/strain curve and is characterized by the upper yield stress σ_S and the *upper flow limit* ϵ_S. Correspondingly, the lower yield stress σ_L and *lower flow limit* ϵ_L occur at point III. The region between the points Ia, II, and III is what is known as the ductile region. And, finally, the *tenacity at break* or the *tensile strength*, σ_B, and the *elongation at break* or *ultimate elongation*, ϵ_B, occur at the point IV. A parting break occurs at point IV. The decrease in tensile stress with increasing strain between II and III is called stress softening. The increasing stress with increasing strain between III and IV is called stress hardening. Stress softening, however, is only nominal, since it is not seen when the stress is given relative to the actual cross section of the sample during elongation.

The cross section of the sample, of course, is reduced on elongation. Consequently, the actual true stress σ_{11}' is greater than the nominal or engineering stress σ_{11}:

$$\sigma_{11}' = F/A = (F/A_0)(L/L_0) = \sigma_{11}(L/L_0), \qquad V = \text{const} \quad (11\text{-}72)$$

The true strain (Hencky strain) ϵ' is also correspondingly different from the nominal or Cauchy or engineering strain ϵ:

$$\epsilon' = \int_{L_0}^{L} dL/L = \ln (L/L_0) = \ln (A_0/A) \quad (11\text{-}73)$$

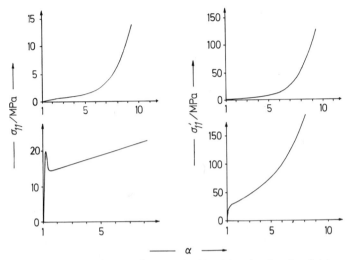

Figure 11-15. Stress/strain diagram of a natural rubber (above) and an it-poly(styrene) (below) at room temperature. Left: experimental curves; right: true stress curves. Numerical values were not given for the lower left figure in the original work. (After P. I. Vincent.)

Nominal and true stress/strain diagrams differ characteristically (Figure 11-15).

Still further differences are observed for stress/strain diagrams of what are known as hard elastic or springy polymers. These polymeric states should exhibit a large energy-elastic component which is attributed to a special network structure (see Figure 38-10). However, electron microscopic studies do not provide any evidence for the proposed network structure.

11.5.2. Necking

Practically all polymers exhibit a necking effect on stretching within a specific temperature region (Figure 11-14). The effect can be recognized by the formation of a neck during elongation due to the occurrence of a constriction after the upper flow limit is reached. The cross section of this constriction continues to decrease up to the lower flow limit. With continued elongation, the cross section of this constriction remains practically constant, but the length of the constricted portion continues to increase at the expense of adjacent parts of the sample. The flow zone moves along the test sample, and a kind of neck is formed.

The difference between the test and brittle temperatures for the sample is decisive for the occurrence of necking (Figure 11-16). If the temperature of the sample is sufficiently low, every body behaves as if it is brittle, and no stress softening or necking is observed. The light stretchability achieved by stress

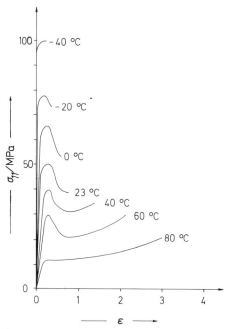

Figure 11-16. Stress/strain curves for a poly(vinyl chloride) at temperatures between −40° C and +80° C. (After R. Nitsche and E. Salewski.) The samples appear to be brittle at −40° C, to be ductile (tough) from −20 to 23° C, to show cold flow from 40 to 60° C, and are rubberlike at 80° C.

softening was first observed at room temperature in the absence of additional heating and so was called cold flow. A stress which does not increase with strain, but decreases, or at least remains constant, is characteristic for cold flow.

The temperature can rise locally to 50° C above the surrounding temperature during neck formation. This causes a decrease in the viscosity, which, of course, leads to increased flow. The effect, however, is also observed under isothermal conditions, and so must be primarily caused by some other factor. There are, of course, locally microscopically small differences in cross section, which, for the same applied force, cause the stress to be greater at the smaller cross sections. A localized neck is formed and is stabilized by molecular orientation and/or heat released on stretching. The released heat causes a localized decrease in viscosity, etc.

11.5.3. Elongation Processes

Amorphous polymers assume their unperturbed dimensions in the solid state (see Section 5.6.2). Even the dimensions of partially crystalline polymers are not significantly different from their unperturbed dimensions in solution. According to Equation (4-27), the chain end-to-end distance of an unperturbed coil is given by the number N of bonds, the bond length b, the bond angle τ, and the hindrance parameter σ.

The maximum attainable chain end-to-end distance on stretching is that when the whole molecule lies in a *trans* conformation. This maximum achievable length is given by Equation (4-8). Consequently, the maximum elongation ratio α_{max} is

$$\alpha_{max} = L_{max}/\langle L^2 \rangle_0^{1/2} = \frac{N^{1/2}(1 - \cos\tau)^{1/2}\sin(0.5\tau)}{\sigma(1 + \cos\tau)^{1/2}} \qquad (11\text{-}74)$$

Thus, a poly(ethylene) of molar mass 140,000 g/mol, that is, with $N = 10^4$, bond angle between carbon bonds of $\tau = 112°$, and hindrance parameter of $\sigma = 1.63$, has a maximum elongation of 75 times its original length. This elongation ratio, however, is seldom attained. What is known as the "natural" draw ratio is obtained under normal conditions, and this ratio is given by the termination of necking or the onset of strain hardening. The natural draw ratio is about 1.5–2.5 for amorphous polymers, 4–5 for low-crystallinity polymers, and 5–10 for high-crystallinity polymers. Methods producing higher elongation ratios are called ultradrawing methods.

The crystallinity of partially crystalline polymers during elongation can increase, decrease, or remain constant (Figure 11-17). Chain orientation, on the other hand, increases continuously during elongation. With quenched samples, i.e., those with low degrees of crystallinity, the available crystallites with their molecular axes are obviously first oriented in the stress direction.

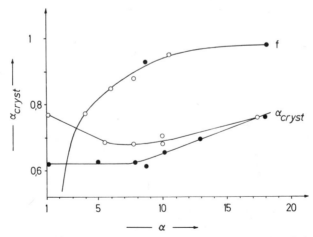

Figure 11-17. Change in the crystallinity α_{cryst} and the degree of orientation f with draw ratio α on stretching a poly(ethylene) at 60°C. The poly(ethylene) was either quenched (●) or slowly cooled (○) from the melt before drawing. (After W. Glenz and A. Peterlin.)

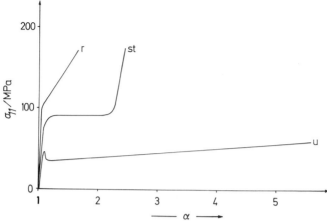

Figure 11-18. Stress/strain plot of a poly(ethylene terephthalate) after different pretreatments; u, undrawn; st, biaxially oriented; r, drawn and annealed. (After R. A. Hudson.)

Then, the amorphous regions crystallize. Annealed polymers, however, are more highly crystalline. In this case, the crystallites must be partially dissolved before the molecular axes can be oriented.

The stress/strain diagrams of stretched polymers differ significantly from those of unstretched polymers (Figure 11-18). The absence of an upper flow limit, that is, the absence of cold flow, is especially noticeable. Of course, orientation of chain segments and crystallites hinders viscoelastic and viscous flow.

Freezing in of orientations is made use of with what are called heat-shrink films. Heat-shrink films are stretched or drawn biaxially, and these films are used to solidly encapsulate mold components. The wrapped goods are then warmed, whereby the film expands slightly below the glass transition temperature (for amorphous polymers) or the melting temperature (for crystalline polymers). At the transition temperature, however, the molecular segments become mobile: the molecules tend to assume their unperturbed dimensions and the film shrinks in upon itself.

11.5.4. Hardness

The hardness of a body is understood to be the mechanical resistance of this body to its penetration by another. A generally valid definition of hardness suitable for all materials does not exist; neither are there generally applicable test methods. The various hardness scales, however, can be roughly related to each other (Table 11-6).

In Brinell's method, a small steel sphere is pressed into the test body with

Table 11-6. Comparison of Various Hardness Scales

| | | | | Rockwell | | Shore | | |
| | | | | | | | | |
Material	Mohs	Vickers	Brinell	M	$\alpha \approx R$	D	C	$A \approx$ IRHD
	10	2500	1000	500				
	9	1650	785	450				
	8	1030	600	400				
	7	600	440	350				
	6	325	310	300				
	5	155	200	250				
	4	63	120	200				
	3	20	65	150				
Hard plastics	2	4	25	100				
			16	80				
			12	70	100	90		
			10	65	97	86		
			9	63	96	83		
			8	60	93	80		
			7	57	90	77		
			6	54	88	74		
	1		5	50	85	70		
			4	45		65	95	
Soft plastics			3	40	50	60	93	98
			2	32		55	89	96
			1.5	28		50	80	94
			1	23		42	70	90
Elastomers			0.8	20		38	65	88
			0.6	17		35	57	85
			0.5	15		30	50	80
						25	43	75
						20	36	70
						15	27	60
						12	21	50
						10	18	40
						8	15	30
						6.5	11	20
						4	8	10

a certain force. The depth of penetration, i.e., the retained plastic deformation, is measured. The measurement is only carried out, therefore, after the load is removed. The Brinell hardness test is especially suitable for hardness tests on metals, where the measurements are made above the flow limit in the plastic region.

Rockwell's hardness test works in a similar way to Brinell's hardness test, i.e., it uses the depth of penetration. Contrary to Brinell's method, however, it measures the penetration of a sphere while still under a load, and then measures the remaining elastic deformation. For this reason, the Rockwell method always gives lower degrees of hardness than the Brinell method. In addition, the degrees of hardness according to the Rockwell method are not measured as mechanical stress, but in scale numbers of 0–120. Steel balls are used with soft materials, and diamond points with hard ones. The Vicker's hardness test uses a diamond pyramid. A modified Rockwell method is used for plastics. It should be noted that, with the Rockwell hardness thus determined, the plastic deformation contribution increases only gradually, because of creep. With metals, on the other hand, the deformation is always plastic, and therefore, also independent of time. Plastics, therefore, exhibit a relatively high Rockwell hardness compared to metals.

The so-called Shore hardnesses are measured differently for metals and plastics. With hard materials (metals), a scleroscope is used to measure the rebound of a small steel ball. This Shore hardness is thus measured by a dynamic method, which yields the rebound hardness (the "impact elasticity" of the rubber industry). Soft plastics, on the other hand, are tested with a Shore durometer. This measures the resistance to the penetration of the point of a cone through the contraction of a calibrated spring. The durometer thus works according to a static method, and yields the true Shore hardness as understood by the rubber industry. Like the Rockwell hardness, the Shore hardness is given in scale divisions.

The pendulum hardness is used to test painted steel surfaces. For this method, a duroscope is used. Here, a small hammer similar to a pendulum is made to fall on the sample. There are also many other standard test methods for pendulum hardness tests.

In the Mohs hardness test, the resistance of the sample to scratching is tested. The Mohs hardness scale is divided into ten degrees of hardness. These are fixed arbitrarily (e.g., talc = 1, Iceland spar = 3, quartz = 7, diamond = 10). A similar hardness scale is based on the scratching power of pencils of different hardness.

In all methods of testing hardness, the thickness of the material and the type of substrate are very important because the elasticity is usually also measured. In addition, it should be noted that hardness tests always measure the hardness of the surface, and not that of the material within the sample. The surface of a sample can be plasticized, for example, by water vapor from the air. If a plastic that can be crystallized is injected into a cold mold, then in some cases the surface is less crystalline than the interior, etc.

11.5.5. Friction and Wear

In a certain sense, wear tests can also be included among the hardness tests. Wear, or abrasion, is affected partly by the hardness and partly by the frictional properties of the test sample.

On rolling a hard sphere over a soft material, the friction results almost entirely from energy loss through deformation of the soft material. Thus, the friction depends on the viscoelastic properties of the material that the sphere is rolled over. If the deformation only results from pressure (and not from shear stress), then the frictional coefficient is given as

$$\mu = \beta \, (F/E)^a r^{a-1} \tag{11-75}$$

where F is the force acting, E is the modulus of elasticity, and r the radius. The coefficient β is related to the loss modulus. The exponent assumes a value of $1/3$ for a cylinder and $1/4$ for a sphere.

On sliding hard bodies along other hard bodies, the frictional coefficient $\mu = F_{21}/F_{11}$ is given by the ratio of tangentially active forces to perpendicularly active forces. The frictional coefficient determined in this way lies between 0.15 and 0.5. It is independent of the chemical nature of the bodies, which suggests that it is influenced by surface roughness.

11.6. Fracture

11.6.1. Concepts and Methods

A polymer can fracture in many different ways, according to the type, conditions, and stress. Many polymers fracture or break almost immediately as stress is applied. With others, no change can be seen even after days and months. The break can be "clean" or rough. The elongation at break can be less than 1% or more than several thousand percent.

In the limiting case, two types of breaking processes are possible, namely, a brittle fracture and a tough or ductile break. With a brittle fracture, the material fractures perpendicular to the direction of stress with little if any flowing processes. With a tough break, on the other hand, tearing takes place in the direction of the pressure stress through shearing processes, and by positional transitions in crystalline regions. Consequently, a body is defined as being brittle when the extension at break is less than 20%.

The fracture behavior of brittle substances is frequently tested by flexural strength tests (Figure 11-19). In a flexural strength test, the body is loaded slowly with a continually increasing force. For this, the test sample is either supported at two points or else clamped at one point. The flexural strength is an indication of the capacity of a body to change its form. Soft bodies can bend so much that the sample slips from the support.

To test the impact strength, the sample is quickly bent under stress up to the breaking point. The impact can be a pendulum or a flexural or a tensile impact. To test the notched-bar impact strength, the sample is first notched

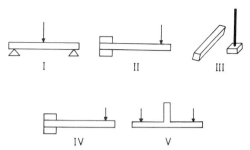

Figure 11-19. Schematic representation of different types of flexural tests. I, Flexural test with sample supported at two points; II, flexural text, sample clamped at one point; III, pendulum impact test to determine the flexural impact strength; IV, flexural impact test; V, tensile impact test.

distinctly, and the subsequent tensile strength, so to speak, is measured. The ratio of the work done to the cross-sectional width is termed the *impact strength*.

11.6.2. Theory of Fracture

In principle, the force necessary to cause a brittle break $F = E/L$ can be calculated from the energy E required to separate chemical and physical bonds by an interbond-partner distance L. To break extended-chain poly-(ethylene) crystals perpendicular to the chain direction (i.e., breaking covalent bonds), a force of about 20 000 MPa is necessary, whereas to cause a break parallel to the chain direction (i.e., working only against dispersion forces), only 200 MPa is required. Experimentally, however, a maximum tensile strength of 20 MPa is observed (the so-called crystal paradox). Consequently, the break must occur at inhomogeneities, since these lead to an inhomogeneous distribution of the tensile stress onto "disruption points" and thus lead to stress concentrations.

The break behavior of energy-elastic and entropy-elastic bodies is different. According to the break theory of Ingles for energy-elastic bodies, there is a relationship between the critical break stress $(\sigma_{11})_{\text{crit}}$, the stress operating at the top of a crack σ_{11}, the geometry of the crack, and the modulus of elasticity. In the simplest case of a crack of length L with a round tip of radius R, we have

$$(\sigma_{11})_{\text{crit}} = \sigma_{11} \left(\frac{R}{4L} \right)^{0.5} \tag{11-76}$$

The Ingles theory offers a very good description of the break behavior of silicate glass, since silicate glasses are practically solely energy elastic and the crack propagation energies are of the same order of magnitude as the surface energies.

The break behavior of any desired elastic body is described by the Griffith theory. According to Griffith, a crack in an elastic body only propagates further when the elastically stored energy just exceeds the energy required to break chemical bonds. Combination of this with the Ingles concept leads, for long cracks, to

$$(\sigma_{11})_{\text{crit}} = \left(\frac{2E\gamma}{\pi L}\right)^{0.5} \qquad (11\text{-}77)$$

where E is the modulus of elasticity and γ is the break surface energy, that is, the energy required to form a new, crack-free surface. The predicted dependence of the critical tensile stress on the reciprocal square root of the crack length has actually been observed (Figure 11-20). For small crack lengths, measured data deviate from the Griffith theory and concentrate at a finite value of $(\sigma_{11})_{\text{crit}}$ for $L = 0$. The crack length at which the deviating behavior is first observed results from crazing, and not from "naturally occurring" cracks.

Crazes occur perpendicular to the stress direction shortly before a destructive break. They may be up to 100 μm long and up to 10 μm wide. Crazes are not hairline cracks, that is, they are not totally void between the break surfaces, in contrast to what are known as white breaks.

Because certain materials whiten with crack formation on flexing, these

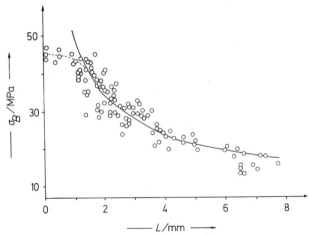

Figure 11-20. Dependence of the critical tensile stress σ_B on the length of cracks artificially introduced into poly(styrene) rods of cross sections between 0.3 × 0.5 and 2.8 × 0.5 cm^2 at elongation rates between 0.05 and 0.5 cm/min. The solid line gives the functionality predicted by the Griffith theory. (After J. P. Berry.)

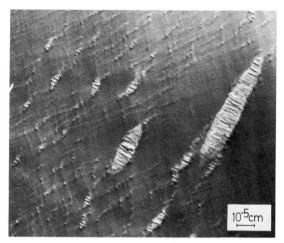

Figure 11-21. Crazing by a 25% elongated poly(styrene) of molar mass 97 000 g/mol. (After S. Wellinghoff and E. Baer.)|

cracks are called white breaks. White breaks contain irregularly arranged large voids. On the other hand, the spaces between the break surfaces in crazes contain bundles of molecules partially oriented in the stress direction and anchored in the rest of the sample (Figure 11-21). Consequently, in contrast to genuine breaks, crazes possess a structural and mechanical continuity.

According to electron spin resonance data, free radicals are produced at chain ends even before a macroscopic break occurs. The free radical concentration depends only on the extension, and not on the tensile stress. Concentrations of 10^{14}–10^{17} free radicals/cm^2 are generally observed. Since free radical concentrations of only about 10^{13} free radicals/cm^2 occur on the surface, free radicals must form in the test sample interior, that is, from the breaking of polymer chains. In addition, chemical decomposition products are produced by a ductile break, but not by a brittle fracture.

In such break cases, break occurs generally in the amorphous regions, since the amorphous phase is set under stress on extension. Thus, break occurs at interlamellar bonds and at the surfaces of spherulites (Figure 11-22). Oriented samples generally exhibit a greater tensile strength in the orientation direction than do nonoriented samples (Figure 11-23). For each polymer a threshold tensile strength and molecular weight value exist above which the tensile strength either does not increase or increases very slowly.

The tensile strength of polymers at first increases strongly, and then weakly with increasing molar mass (Figure 11-23). Below a certain critical molar mass, the tensile strength is practically zero. The transition from sharp to weak molar mass dependence is probably due to entanglements. At very

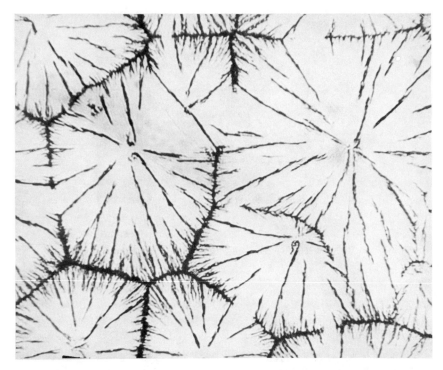

Figure 11-22. The onset of break at the amorphous positions in it-poly(propylene) spherulites, that is, between spherulites and radially within the spherulites. (After H. D. Keith and F. J. Padden, Jr.).

high molar masses, the tensile strength is practically independent of the molar mass.

11.6.3. Stress Cracking

The term *stress cracking* is used with metals and plastics when the material is damaged by the simultaneous action of chemical agents and mechanical forces. With plastics, a crazing is usually observed on the surface of the material under such conditions. It is also described, therefore, as stress corrosion cracking or crazing, or—since chemical reactions mostly play a minor role or none at all—stress crazing. Stress cracking plays an important role in bottles, tubing, cables, etc., that come into contact with chemical reagents.

The extent of the stress corrosion cracking varies according to the

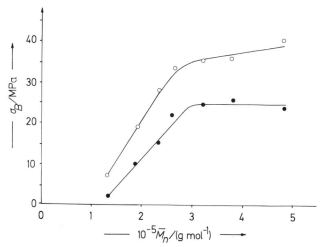

Figue 11-23. Tensile stress at break (tensile strength) σ_B of at-poly(styrenes) of narrow molar mass distribution and various number-average molar masses $\overline{M}_n$. Measurements were made at 23°C and 50% relative humidity. Processing was by compression molding (●) or injection molding (O). Injection-molded samples are oriented. (After H. W. McCormick, F. M. Brower, and L. Kin.)

surrounding medium. The effects are usually small with nonwetting media. Here, stress corrosion cracking proceeds in three phases. In the case of stress corrosion cracking below the breaking point, weak points grow into visible hairline cracks or crazes perpendicular to the direction of stress. The crazes subsequently deepen and develop into cracks up to a limiting value, when the material is again strengthened. With wetting media, on the other hand, there is no limiting value to the development of the cracks.

The cause of stress cracking is not yet fully understood. It is established that there is still amorphous material in the cracks. This material can be deformed by cold flow. Of course, the extent of the cold flow is also determined by the diffusion and degree of swelling of the surrounding medium into the material. Wetting substances can build up a swelling pressure at the weak points.

The susceptibility of a material to stress cracking falls with increasing molecular weight and rises in the same material with increasing density. Stress cracking only occurs at the surface. Polymer softeners plasticize the surface, causing the stresses to compensate one another, so that the susceptibility to stress cracking falls. The mobility of the chain elements also ensures that no stress cracking occurs above the glass transition (if amorphous) or melt temperature (if crystalline) of a material. The susceptibility to stress cracking is lowered by cross-linking.

11.6.4. Fatigue

In some cases, materials do not suffer damage immediately after a given stress is applied, but only after a certain time. Here, a distinction must be made between the fatigue limit and the fatigue strength. The fatigue limit is understood to be the stress under which a material suffers no damage even after an infinitely long time. Fatigue strength is understood to be the stress that destroys or damages the material after a given time.

The stress here can be either static or oscillating. In static tests (creep strength tests) the sample is, for example, subjected to a specific force, and the time up to the break is then measured. The same test is subsequently carried out with different forces. If the stress applied is a tensile stress, this is called the tensile creep strength. A static stress under compression would correspondingly be referred to as the compression creep strength. In order to determine the creep strength, the quantities that are proportional to the force (for example, tensile stress) are usually plotted directly or as logarithms against the logarithm of the time (Figure 11-24). The creep strengths can vary considerably according to the polymer. At a load of 39 MPa, for example, normal poly(styrene) samples have creep strengths of 0.01–10 h, whereas those of high-impact-strength poly(styrenes) are up to 10^4 h. Partially crystalline polymers in addition often also show a kink in the creep strength plot. In these cases, a viscous break occurs at shorter times and a brittle break is seen for longer times.

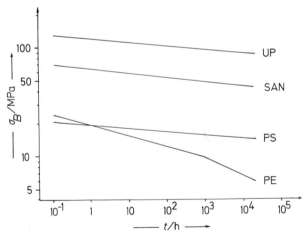

Figure 11-24. Fatigue strength under tensile stress (tensile strength) σ_B as a function of time for a glass-fiber-reinforced unsaturated polyester (UP), a high-impact-strength poly(styrene) (SAN), poly(styrene) (PS), and poly(ethylene) (PE) (after BASF).

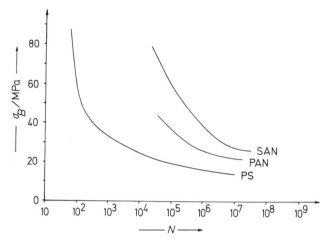

Figure 11-25. Wöhler curve for the alternating flexural stress applied to a high-impact-strength poly(styrene)(SAN), a humid polyamide(PA), and a poly(styrene)(PS). The flexural stress σ_B is measured as a function of the number N of alternations in the load. Read PA instead of PAN.

Isochronic stress/strain diagrams are more informative than creep strength diagrams. They are obtained by holding samples under a constant load for various lengths of time. Finally, a stress/strain diagram is obtained for each load time.

With alternating stresses, a distinction is made between those with alternating load and those with alternating torque. With the former, the fatigue strength under alternating flexing load is measured, and with the latter, fatigue strength under alternating torque. In analogy to the determination of creep strength, the quantities that are proportional to the force are again plotted against the logarithm of the number of changes in alternating load or torque to determine the "oscillation fatigue strength" (Figure 11-25). Curves of this type are called Wöhler curves. Here, again, large differences are found between normal and high-impact-strength poly(styrenes). With flexural stresses of 40 MPa, normal poly(styrenes) break after only 300 changes of load, but high-impact poly(styrenes) only after one million.

Literature

General

A. V. Tobolsky, *Properties and Structures of Polymers*, Wiley, New York, 1960.
H. Oberst, *Elastische und viskose Eigenschaften von Werkstoffen*, Beuth-Vertrieb, Berlin, 1963.
L. E. Nielsen, Crosslinking—effect on physical properties of polymers, *Rev. Macromol. Chem.* **4**, 69 (1970).

I. M. Ward, *Mechanical Properties of Solid Polymers*, Wiley–Interscience, London, 1971.

A. Peterlin, Mechanical properties of polymeric solids, *Am. Rev. Mater. Sci.* **2**, 349 (1972).

D. W. van Krevelen, *Properties of Polymers—Correlation with Chemical Structures*, Elsevier—North-Holland, Amsterdam, 1972.

J. R. Martin, J. F. Johnson, and A. R. Cooper, Mechanical properties of polymers: the influence of molecular weight and molecular weight distribution, *J. Macromol. Sci. C (Rev. Macromol. Chem.)* **8**, 57 (1972).

E. A. Meinecke and R. C. Clark, *The Mechanical Properties of Polymeric Foams*, Technomic Publ. Co., Westport, Connecticut, 1972.

L. Nielsen, Mechanical Properties of Polymers and Composites, Marcel Dekker, New York, Vol. 1, 1974; Vol. 2, 1976.

R. G. C. Arridge, *Mechanics of Polymers*, Clarendon Press, Oxford, 1975.

R. M. Jones, *Mechanics of Composite Materials*, Scripta Book Co., Washington, D.C., 1975.

K. D. Pae and S. K. Bhateja, The Effects of Hydrostatic Pressure on the Mechanical Behaviour of Polymers, *J. Macromol. Sci. (Rev.)* **C13**, 1 1975.

D. W. van Krevelen, *Properties of Polymers—Correlation with Chemical Structure*, second ed., Elsevier, Amsterdam, 1976.

A. Malmeisters, V. Tamusz, and G. Teters, *Mechanik der Polymerwerkstoffe*, Akademie Verlag, Berlin, 1978.

A. Casale and R. S. Porter, *Polymer Stress Reactions*, Vol. 1, Academic Press, New York, 1978.

Section 11.2. Energy Elasticity

I. Sakurada and K. Kaji, Relation between the polymer conformation and the elastic modulus of the crystalline region of polymers, *J. Polym. Sci.* **C31**, 57 (1970).

Section 11.3. Entropy Elasticity

L. R. G. Treloar, *Physics of Rubber Elasticity*, Oxford University Press, Oxford, 1958.

M. Shen, W. F. Hall, and R. E. De Wames, Molecular theories of rubber-like elasticity and polymer viscoelasticity, *J. Macromol. Sci. C (Rev. Macromol. Chem.)* **2**, 183 (1968).

K. J. Smith, Jr., and R. J. Gaylord, Rubber elasticity, *ACS Polymer Div. Polymer Preprints* **14**, 708 (1973).

J. E. Mark, Thermoelastic properties of rubber-like networks and their thermodynamic and molecular interpretation, *Rubber Chem. Technol.* **46**, 593 (1973).

L. R. G. Treloar, The elasticity and related properties of rubber, *Rep. Prog. Phys.* **36**, 755 (1973); *Rubber Chem. Technol.* **47**, 625 (1974).

J. E. Mark, The constants $2C_1$ and $2C_2$ in phenomenological elasticity theory and their dependence on experimental variables, *Rubber Chem. Technol.* **48**, 495 (1975).

L. R. G. Treloar, *The Physics of Rubber Elasticity*, third ed., Clarendon Press, Oxford, 1975.

J. E. Mark, Thermoelastic results on rubberlike networks and their bearing on the foundations of elasticity theory, *Macromol. Revs.* **11**, 135 (1976).

S. Kawabata and H. Kawai, Strain energy density function of rubber vulcanizates from biaxial extension, *Adv. Polym. Sci.* **24**, 89 (1977).

Section 11.4. Viscoelasticity

N. G. McCrum, B. E. Read, and G. Williams, *Anelastic and Dielectric Effects in Polymeric Solids*, Wiley, London, 1967.

R. M. Christensen, *Theory of Viscoelasticity: An Introduction*, Academic Press, New York. 1970.

J. J. Aklonis, W. J. MacKnight, and M. Shen, *Introduction to Polymer Viscoelasticity*, Wiley–Interscience, New York, 1972.

D. W. Hadley and I. M. Ward, Anisotropic and nonlinear viscoelastic behaviour in solid polymers, *Rep. Prog. Phys.* **38**(10) 1143 (1975).

R. F. Boyer, Mechanical motions in amorphous and semicrystalline polymers, *Polymer* **17**, 996 (1976).

T. Murayama, *Dynamic Mechanical Analysis of Polymeric Material*, Elsevier, Amsterdam, 1978.

J. D. Ferry, *Viscoelastic Properties of Polymers*, third ed., Wiley, New York, 1980.

K. Murakami and K. Ono, *Chemorheology of Polymers*, Elsevier, Amsterdam, 1979.

J. G. Williams, *Stress Analysis of Polymers*, second ed., Wiley, Chichester, 1980.

Section 11.5. Deformation Processes

A. J. Durelli, E. A. Phillips, and C. H. Tsao, *Introduction to the Theoretical and Experimental Analysis of Stress and Strain*, McGraw-Hill, New York, 1958.

J. W. Dally and W. F. Riley, *Experimental Stress Analysis*, McGraw-Hill, New York, 1965.

O. H. Varga, *Stress–Strain Behavior of Elastic Materials*, Wiley–Interscience, New York, 1966.

A. Peterlin, *Plastic Deformation of Polymers*, Marcel Dekker, New York, 1971.

J. G. Williams, *Stress Analysis of Polymers*, Longmans, Harlow, Essex, U.K., 1973.

A. R. Payne, Physics and physical testing of polymers, *Prog. High Polym.* **2**, 1 (1968).

H. J. Orthmann and H. J. Mair, *Die Prüfung thermoplastischer Kunststoffe*, Hanser, Munich, 1971.

G. C. Ives, J. A. Mead, and M. M. Riley, *Handbook of Plastic Test Methods*, Iliffe, London, 1971.

S. Turner, *Mechanical Testing of Plastics*, Butterworths, London, 1973.

J. K. Gillham, Torsional braid analysis, *Crit. Rev. Macromol. Sci.* **1**, 83 (1972).

K. Sato, The hardness of coating films, *Prog. Org. Coatings* **8**, 1 (1980).

B. Carlowitz, *Tabellarische Uebersicht über die Prüfung von Kunststoffen*, fourth ed., Umschau-Verlag, Frankfurt am Main, 1972.

J. J. Bikerman, Sliding friction of polymers, *J. Macromol. Sci. (Rev.)* **C11**, 1 (1974).

L. H. Lee (ed.), *Advances in Polymer Friction and Wear*, 2 vols., Plenum, London, 1975.

Section 11.6. Fracture

R. W. Hertzberg and J. A. Manson, *Fatigue of Engineering Plastics*, Academic Press, New York, 1980.

E. H. Andrews, *Fracture in Polymers*, Oliver and Boyd, Edinburgh, 1968.

H. H. Kausch and J. Becht, Elektronenspinresonanz, eine molekulare Sonde bei der mechanischen Beanspruchung von Thermoplasten, *Kolloid-Z. Z. Polym.* **250**, 1048 (1972).

E. H. Andrews, Fracture of polymers, in *Macromolecular Science* (Vol. 8 of Physical Chemistry Series 1), C. E. H. Bawn (ed.), MTP International Review of Science, 1972.

H. Liebowitz (ed.), *Fracture*, Vol. 7, *Fracture of Nonmetals and Composites*, Academic Press, New York, 1972.

G. H. Estes, S. L. Cooper, and A. V. Tobolsky, Block copolymers and related heterophase elastomers, *J. Macromol. Sci. C (Rev. Macromol. Chem.)* **4**, 313 (1970).

P. F. Bruins (ed.), *Polyblends and Composites* (Appl. Polymer Symposia) Vol. 15, Interscience, New York, 1970.

S. Rabinowitz and P. Beardmore, Craze formation and fracture in glassy polymers, *Crit. Rev. Macromol. Sci.* **1**, 1 (1972).

R. P. Kambour, A review of crazing and fracture in thermoplastics, *J. Polym. Sci. D* **7**, 1 (1973).

J. A. Manson and R. W. Hertzberg, Fatigue failure in polymers, *Crit. Rev. Macromol. Sci.* **1**, 433 (1973).

H. H. Kausch, *Polymer Fracture*, Springer, New York, 1978.

E. H. Andrews and P. E. Reed, Molecular fracture in polymers, *Adv. Polym. Sci.* **27**, 1 (1978).

L. G. E. Struik, *Physical Aging in Amorphous Polymers and Other Materials*, Elsevier, Amsterdam, 1978.

E. H. Andrews (ed.), *Developments in Polymer Fracture*, Appl. Sci. Publ., Barking, Essex, 1979.

Chapter 12

Interfacial Phenomena

12.1. Spreading

Insoluble molecules spread on liquid surfaces which are also known as the hypophases. This spreading behavior corresponds to that of a two-dimensional gas if the coverage is small. Analogous to the ideal gas equation, the following is valid for the relationship between the surface pressure, given as $\gamma_0 - \gamma$, the difference between the surface tensions of the covering and the covered surface, and the surface area per molecule of spreading material, for a surface area of A per molecule:

$$(\gamma_0 - \gamma)A = kT \qquad \text{(for } A \to \infty) \tag{12-1}$$

In principle, then, the molar mass of the spreading material can be determined with Equation (12-1). The surface pressure and the specific surface area are measured in a Langmuir trough. Here, a fixed quantity of material spreads out over a given surface, which is separated on the one side by an easily moved float. The pressure exerted on this float by a given surface area of a given quantity of material is then the surface pressure. These measurements are not simple to carry out, since only low pressures are found with low quantities of material, and the surface of the hypophase must be meticulously clean. Therefore the method has not become a routine method for determining molar masses.

The plot of $\gamma_0 - \gamma = f(A)$ yields interesting conclusions concerning molecules of different shape or conformation. The more rigid poly(vinyl benzoate) molecules, for example, lead to a collapse of the surface at low values of A, but the more flexible poly(vinyl acetate) molecules do not (Figure 12-1). Isotactic and syndiotactic molecules also show different behavior

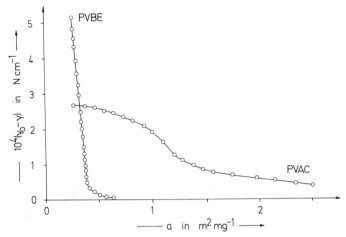

Figure 12-1. Dependence of the spreading pressure ($\gamma_0 - \gamma$) on the specific surface area a in poly(vinyl acetate) (PVAC) and poly(vinyl benzoate) (PVBE). (After N. Berendjick.)

during spreading studies, but the curves are difficult to interpret quantitatively as a function of molecular quantities. In addition, there is the fact that the apparent molar mass calculated at finite concentrations using Equation (12-1) can also depend in come cases on the chemical nature of the hypophase. Such behavior is found during spreading studies with proteins, and indicates association–dissociation phenomena in the proteins.

12.2. Surface Tension of Liquid Polymers

Liquids have a surface tension with respect to the gaseous phase and an interfacial tension with respect to other liquids. Such interfacial and surface tensions may be measured by a whole series of different methods in the case of low-molar-mass liquids. But polymer liquids have very high viscosities, and so, only a few methods of measurement are suitable. The capillary method and the wire ring method are not suitable since the measured surface tension depends on the speed at which the test is carried out. All of what are known as static methods are suitable, i.e., the suspended drop method and the Wilhelmy plate method.

The Wilhelmy-plate method consists in partially immersing a plate in a wetting liquid. The surface tension of the liquid γ_{lv} acts downward on the plate. When the plate is wetted and its lower edge just resides on the liquid surface, the force acting on the plate is $\gamma_{lv} l_{per}$, where l_{per} is the perimeter of the plate. By measuring the restraining force on the plate in air and in contact with the liquid surface the surface tension can be calculated. The method is only used to measure the surface tension, but is not used to measure the interfacial tension between two polymer liquids since it is difficult to attain a contact angle ϑ of zero.

The shape of a suspended drop depends on the surface tension as well as on the gravitational force. The drop is photographed and the diameter at various positions is measured. A consistent shape factor can be evaluated when hydrodynamic equilibrium is reached.

The *surface tension* of a liquid polymer depends on its end groups, its molar mass, and on the temperature. A theoretical derivation of the molar mass dependence based on free volume theory leads to

$$\gamma_{lv}^{-1/4} = (\gamma_{lv}^{\infty})^{-1/4} + K_s \langle M \rangle_n^{-1} \tag{12-2}$$

and an empirical molar mass dependence is given by

$$\gamma_{lv} = \gamma_{lv}^{\infty} - K_e \langle M \rangle_n^{-2/3} \tag{12-3}$$

The slope constant K_e is influenced by the chemical nature of the end groups, as shown in Figure 12-2. γ_{lv}^{∞} is independent of the molar mass and the nature of the end groups. Typical surface tensions of polymer liquids of finite molar mass are given in Table 12-1. The surface tensions do not vary very much with temperature.

The interfacial tensions between two liquid polymers are generally small. They increase in size with stronger polarity differences between the two liquids (Table 12-1).

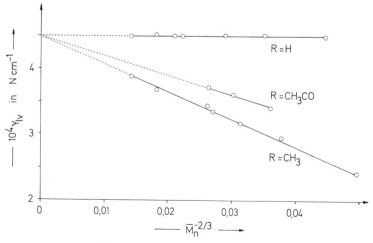

Figure 12.2 Dependence of the surface tension γ_{lv} of poly(oxyethylenes), $RO(CH_2CH_2O)_n R$, on the number-average molar mass $\langle M_n \rangle$ at 24°C. (Data from various authors.)

Table 12.1. Interfacial Tension γ_{ll} between Two Liquid Polymers and Surface Tensions γ_{lv} of the Pure Polymers at 150°C

Polymer[a]	$10^5\gamma_{lv}$ in N/cm	$10^5\gamma_{ll}$ in N/cm							
		PDMS	it-PP	PBMA	PVAC	PE	PS	PMMA	PEO
PDMS	13.6	0	3.0	3.8	7.4	5.4	6.0	—	9.8
it-PP	22.1	3.0	0	—	—	1.1	5.1	—	—
PBMA	23.5	3.8	—	0	2.8	5.2	—	1.8	—
PVAC	27.9	7.4	—	2.8	0	11.0	3.7	—	—
PE	28.1	5.4	1.1	5.2	11.0	0	5.7	9.5	9.5
PS	30.8	6.0	5.1	—	3.7	5.7	0	1.6	—
PMMA	31.2	—	—	1.8	—	9.5	1.6	0	—
PEOX	33.0	9.8	—	—	—	9.5	—	—	0

[a]See Table VII-6 of the Appendix for polymer identification.

12.3. Interfacial Tension of Solid Polymers

12.3.1. Basic Principles

A drop of liquid forms a certain contact angle ϑ on a solid, smooth surface. The value of the contact angle is vectorially determined with the aid of the Young equation from the three interfacial tensions liquid/vapor (γ_{lv}), solid/liquid (γ_{sl}), and solid/vapor (γ_{sv}), whereby γ_{sv} can be separated into the equilibrium pressure π_e and the surface energy γ_s^0 of the solid (Fig. 12-3):

$$\gamma_{sl} + \gamma_{lv} \cos \vartheta = \gamma_{sv} = \gamma_s^0 + \pi_e \qquad (12\text{-}4)$$

The parameter π_e gives the spreading pressure of saturated solvent vapor on the solid polymer surface at equilibrium. It tends to a value of zero as the contact angle approaches a value of zero. But π_e can assume considerable magnitude for finite contact angle values, i.e., 14×10^{-5} N/cm for water on

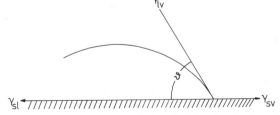

Figure 12-3. Definition of contact angle ϑ and the interfacial surface tensions γ_{sl} (solid/liquid), γ_{lv} (liquid/vapor), and γ_{sv} (solid/vapor).

poly(ethylene). In vacuum, π_e becomes zero. At a contact angle $\vartheta = 0°$, there is complete spreading of the liquid on the surface, whereas at an angle of $\vartheta = 180°$ no spreading occurs. Real systems have contact angles of between $0°$ and $180°$. Since the contact angle determines the spreadability, its cosine is a direct measure of the wettability of the surface.

Real surfaces are rough, not smooth. The ratio true surface/geometric surface is defined as the roughness r, and can only be equal to or greater than 1. Freshly cleaved mica has r values close to 1, polished surfaces have r values between 1.5 and 2.

As a result of this roughness, an experimental average value ϑ_{exp} is measured instead of the theoretical contact angle ϑ. The roughness of the surface will tend to enlarge the liquid–polymer contact area. The opposed effects of cohesion and adhesion affect the response to an enlarged interface. In the case of liquids that spread poorly ($\vartheta > 90°$), cohesion predominates. The enlargement of the surface due to roughness is then counterbalanced by an increase in the contact angle ($\vartheta_{exp} > \vartheta$). With liquids that spread well ($\vartheta < 90°$), adhesion predominates. Therefore the liquid can cover a greater surface area on the roughened surface than on a smooth one, and the contact angle decreases ($\vartheta_{exp} < \vartheta$). The roughness can thus also be given as $r = \cos\vartheta_{exp}/\cos\vartheta$. The true contact angle ϑ can then be calculated from the roughness r, the true surface area, and the experimentally observed contact angle ϑ_{exp}.

12.3.2. Surface Energy and Critical Surface Tension

The surface energy γ_s of a solid is an important material constant. Since it cannot be measured directly, various methods of estimating it have been tried.

Instead of determining the interfacial tension of a solid polymer with respect to a low-molar-mass liquid and then calculating the value of γ_{sv} for the solid polymer from the surface tension, γ_{lv}, of this liquid and the measureable contact angle ϑ, the interfacial tensions of the molten polymer with respect to this liquid can be measured at various temperatures and the calculated γ_{sl} values can be extrapolated back to a temperature at which the polymer is solid. The method is not above suspicion, since the extrapolation may have to be carried out over a considerable temperature range and the surface structures of the molten and solid polymer are not necessarily identical. Such influences are apparent, for example, in the contact angles of molten polymers on solid polymers: the contact angle of molten poly(butyl methacrylate) on solid poly(vinyl acetate) is equal to zero, but the contact angle of molten poly(vinyl acetate) on poly(butyl methacrylate) is 42°.

In what is known as the Zisman procedure, the contact angle, ϑ, of vari-

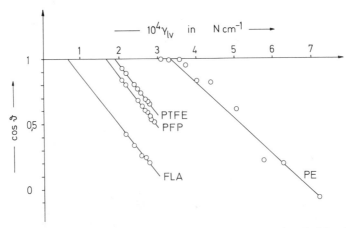

Figure 12-4. Determination of the cosine of the contact angle ϑ on the interfacial surface tension γ_{lv} of the liquid used on different substrates in contact with air (20° C). PE, poly(ethylene); PTFE, poly(tetrafluorethylene); PFP, poly(hexafluoropropylene); FLA, perfluorolauric acid (mono-molecular layer on platinum). (After R. C. Bowers and W. A. Zisman.)

ous solvents on a polymer is determined at constant temperature, and then, the surface tension γ_{lv} of these solvents are plotted against cos ϑ (Figure 12-4):

$$\gamma_{lv} = \gamma_{crit} - a(1 - \cos \vartheta) \qquad (12\text{-}5)$$

At a value of cos $\vartheta = 1$, the limiting value of the surface tension corresponds to complete wetting of the substrate and is therefore termed the critical surface tension γ_{crit} of the substrate. This relationship between cost ϑ and γ_{lv} applies, in the case of a given substrate, not only for a homologous series of liquids, but also quite well for liquids that are very different from one another. An example of this is found in the measurements on poly(ethylene) at 20° C with such different liquids as benzene ($\gamma_{lv} = 28.9 \times 10^{-5}$ N/cm), 1, 1, 2, 2-tetrachloroethane (36.0), formamide (58.2), and water (72.0) (Figure 12-4). Thus, the critical surface tension γ_{crit} of the polymer appears to be almost constant for a given substrate.

Thus, the critical surface tension γ_{crit} appears to be a material constant. However, it is neither identical to the surface energy γ_s^0 nor to the interfacial tension γ_{sv} of the polymer, as can be seen on comparing Equations (12-4) and (12-5). The theoretical significance of the critical surface tension is consequently a matter of dispute. On the other hand, the values of γ_{crit} and γ_{sv} do not differ much from each other (Table 11-2).

The critical surface tension of all known solid polymers is lower than the surface tension of water at 72×10^{-5} N/cm (Table 12-2). All polymers are therefore relatively poorly wetted by water. The critical surface tension of polymers containing fluorine is particularly low, and they are poorly wetted

Table 12-2. *Critical Surface Tensions*
γ_{crit} *of Clean, Smooth Polymers*
and Metals at 20°C

Surface covered with:	$10^5 \gamma_{crit}$ in N/cm	$10^5 \gamma_{sv}$ in N/cm
—CF	6	—
PHFP[a]	16.2	—
PTFE	18.5	14.0
—CH₃	22	—
PDMS	23	—
PVDF	25	30.3
PVF	28	36.7
PE	33	33.1
PS	34	42.0
PVAL	37	—
PVC	39	41.5
PVDC	40	45
PET	43	41.3
Copper	44	—
Wool	45	—
Aluminum	45	—
Iron	46	—
PA 66	46	43.2
Sodium silicate	47	—
UF	61	—
Wool, chlorinated	68	—
Quartz	78	—
Titanium dioxide (anatase)	91	—
Tin (II) oxide	111	—

[a]Poly(hexafluoropropylene).

by oils and fats as well as by water. Oils, fats, and glycerol esters possess surface tensions of $\sim(20–30) \times 10^{-5}$ N/cm. Commerical use is made of this phenomenon, for example, in coating frying pans with poly(tetrafluoroethylene) to make them nonsticking.

12.4. Adsorption of Polymers

The adsorption of high-molar-mass compounds onto solid interfaces differs characteristically from that of low-molar-mass compounds. To a first approximation, low-molar-mass compounds are more or less spherical; each molecule has only one contact point with the surface and the number of contacts per unit surface area determines the coverage. Thus, it is sufficient to

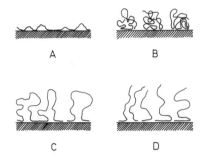

Figure 12-5. Modeled conception of the adsorption of coil-like macromolecules on solid surfaces. A, Two-dimensional coil; B, three dimensional coil; C, loops; D, bristles.

measure the adsorption isotherms and their temperature dependence in the case of low-molar-mass substances.

Coil-like macromolecules, however, may form a great many contacts with a substrate, and the shape of the adsorbed coil molecule may be very different according to polymer–substrate, polymer–solvent, polymer–polymer, and substrate–solvent interactions (Figure 12-5). The number and order of adsorbed segments leads to a definite macroconformation, and a definite macroconformation, in turn, determines the thickness of and polymer concentration in the adsorbed layer.

Layer thickness and polymer concentration is obtained directly by ellipsometry, that is, the change in elliptically polarized light after reflection from a surface covered by an adsorbed layer. The number of adsorbed segments is accessible via infrared spectroscopic studies as well as via calorimetric adsorption enthalpy measurements.

According to the concentration and molar mass of the polymer, the adsorption equilibrium may be reached in minutes or hours. During approach to equilibrium, the adsorbed quantity of polymer and adsorbed layer thickness increase to constant final values, and the polymer concentration in the adsorbed layer first decreases and then remains constant (Figure 12-6). The adsorbed layer thickness and the adsorbed mass per unit surface area is greater and the polymer concentration in the adsorbed layer is lower when the initial solution concentration is higher. At low initial concentrations (B in Figure 12-5), the macromolecules are first adsorbed three dimensionally, and then spread more and more "two dimensionally" over the surface with increasing time (A in Figure 12-5). At higher initial concentrations, the polymer molecules compete more and more for substrate surface: the adjustment time becomes longer, layer thicknesses are greater, and the polymer concentrations in the adsorbed layer are lower. In addition, the layer thickness increases with the square root of the molar mass at high initial concentrations: polymer molecules are adsorbed as unperturbed coils in the case of adsorption from the theta system, poly(styrene)/cyclohexane/36° C.

Adsorption is stronger when adsorption occurs for a polymer in a poorer solvent. Poly(styrene) does not adsorb at all on chromium from the good

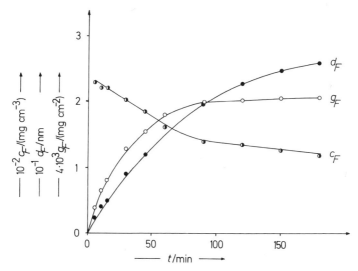

Figure 12-6. Time dependence of the layer thickness d_F of adsorbed mass per unit surface area g_F and of the polymer concentration c_F in the adsorbed layer for the adsorption of poly(styrene) of $\langle M_w \rangle = 176\ 000$ g/mol from a solution of 5 mg/cm^3 in cyclohexane onto a chromium surface. (From data of E. Killmann, J. Eisenlauer, and M. Korn.)

solvent, dioxane. With increasing polymer polarity, on the other hand, increasingly more contacts are formed with the substrate and the adsorbed layer is flatter and more compact. The adsorbed layer of poly(ethylene oxide) on chromium is only 2 nm thick but is so compact that the refractive index of the adsorbed layer is identical to that of the crystallized polymer.

Adsorption of polymers from poor solvents is marked by only meager interactions between the polymer segments. Completely different relationships, however, are apparent for the adsorption from polymer melts. In these cases, segment/segment interactions are strong and the polymer molecules are probably first adsorbed in the form of loops (C in Figure 12-5). At equilibrium, however, bristles should occur (D in Figure 12-5), since, in this case, more conformative degrees of freedom occur for the same number of polymer segments/surface contacts than is the case for loops.

Literature

12.0. General Reviews

J. F. Danielli, K. G. A. Parthurst, and A. C. Biddiford, *Surface Phenomena in Chemistry and Biology*, Pergamon Press, New York, 1958.
I. R. Miller and D. Bach, Biopolymers at interfaces, *Surf. Colloid Sci.* **6**, 185 (1973).

L.-H. Lee (ed.), *Characterization of Metal and Polymer Surfaces*, Vol. 2, *Polymer Surfaces*, Academic Press, New York, 1977.
D. T. Clark, W. J. Feast (eds.), *Polymer Surfaces*. Wiley, New York, 1979.

Section 12.1. Spreading

W. D. Harkins, *The Physical Chemistry of Surface Films*, Reinhold, New York, 1952.
D. J. Crisp, Surface films of polymers, in *Surface Phenomena in Chemistry and Biology*, J. F. Danielli, K. G. A. Pankhurst, and A. C. Riddifort (eds), Pergamon Press, New York, 1958.
F. H. Müller, Monomolekulare Schichten, in *Struktur und physikalisches Verhalten der Kunststoffe*, Vol. 1, R. Nitzsche and K. A. Wolf (eds.), Springer, Berlin, 1962.

Section 12.2. Surface Tension of Liquid Polymers

G. L. Gaines, Jr., Surface and interfacial tension of polymer liquids, *Polym. Eng. Sci.* **12**, 1 (1972).
S. Wu, Interfacial and surface tension of polymers, *J. Macromol. Sci. C* (*Rev. Macromol. Chem.*) **10**, 1 (1974).

Section 12.3. Interfacial Tension of Solid Polymers

W. A. Zisman, Relation of the equilibrium contact angle to liquid and solid constitution, *Adv. Chem. Ser.* **43**, American Chemical Society, Washington, D.C., 1964.
Yu. S. Lipatov and A. E. Feinerman, Surface tension and surface free energy of polymers, *Adv. Colloid Interface Sci.* **11**, 195 (1979).

Section 12.4. Adsorption of Polymers

F. Patat, E. Killmann, and C. Schliebener, Die Adsorption von Makromolekülen aus Lösungen, *Fortschr. Hochpolym.-Forschg.* **3**, 332 (1961/64).
Yu. S. Lipatov and L. M. Sergeeva, *Adsorption of Polymers* (in Russian), Naukova Dumka, Kiev, 1972; Wiley, New York, 1974.
S. G. Ash, Polymer adsorption of the solid liquid interface, in *Colloid Science*, Vol. 1 (D. H. Everett (ed.), Chemical Society, London, 1973.
I. R. Miller and D. Bach, *Biopolymers at interfaces*, *Surface Colloid Sci.* **6**, 185 (1973).
L. E. Smith and R. S. Stromberg, *Polymers at Liquid-Solid Interfaces*, Loughborough, 1975.
E. Killmann, J. Eisenlauer, and M. Korn, The adsorption of macromolecules on solid/liquid interfaces, *Polym. Symp.* **61**, 413 (1978).
T. Sato and R. Ruch, *Stabilization of Colloidal Dispersions by Polymer Adsorption*, Marcel Dekker, New York, 1980.

Chapter 13

Electrical Properties

Matter can be classified according to its "specific" electrical conductivity σ into electrical insulators ($\sigma = 10^{-22}$–10^{-12} Ω^{-1} cm^{-1}), semiconductors ($\sigma = 10^{-12}$–10^{3} Ω^{-1} cm^{-1}), and conductors ($\sigma > 10^{3}$ Ω^{-1} cm^{-1}). Super conductors have a specific electrical conductivity of about 10^{20} Ω^{-1} cm^{-1}. The electrical conductivity is the reciprocal of the electrical resistance. Since electrical resistance is measured in ohms, the unit of conductivity is often written as mho instead of ohm^{-1} in American scientific literature.

Macromolecules with certain constitutional characteristics possess semi-conductor properties (Section 13.2). The majority of the commercially used polymers, however, are insulators (Section 13.1). A consequence of their limited conductivity is that these polymers readily become electrostatically charged (Section 13.1.5). Specific conductivities are, for example, $\sim 10^{-17}$ Ω^{-1} cm^{-1} for poly(ethylene), 10^{-16} Ω^{-1} cm^{-1} for poly(styrene), and 10^{-12} Ω^{-1} cm^{-1} for polyamides (containing water?).

13.1. Dielectric Properties

When an electrical field is applied, the groups, atoms, or electrons of the insulator molecules are polarized. With stronger fields, electrons are displaced, giving rise to ions. With even stronger fields, the conductivity of the ions finally becomes so great that the material no longer shows any electrical resistance: It discharges. Electrical conduction need not only take place in the interior, it can also occur on the surface.

13.1.1. Polarizability

If a static electric field is applied to an nonconductor, then the electrons tend to be displaced relative to the atomic nuclei (electron polarizability). The corresponding displacement of atomic nuclei is called atomic polarizability. The electric moment μ_i thus induced is directly proportional to the electric field E_i, i.e., for the displacement polarizability (electronic and atomic polarizability)

$$\mu_i = \alpha E_i \qquad (13\text{-}1)$$

Here α is the polarizability of the atom, group, or molecule. The greater α, the more energy will be adsorbed by the material.

Molecules with polar groups possess a permanent dipole moment μ_p. In these molecules, a static electric field produces an orientation polarizability, in addition to induced atomic or electron polarization; i.e., the most probable rest position for the permanent dipole lies preferentially in the direction of the field. Molecules with permanent dipoles thus often store more electrical energy than those with induced dipoles.

In general, polarizability, α, is difficult to determine experimentally. However, the ratio of the capacity of a condenser in a vacuum to that in the medium under consideration, i.e., the relative permittivity (earlier, dielectric constant), of the medium, can be measured. At low frequencies, the relative permittivity of electrical nonconductors is almost independent of the frequency. At high frequencies, the relative permittivity depends on the frequency, since the permanent dipoles are no longer able to establish a preferred orientation, because of rapid alteration of the field.

13.1.2. Behavior in an Alternating Electric Field

If a dielectric material is suddenly placed in an electric field, the permanent molecular dipoles in the dielectricum will attempt to orientate. The orientation occurs by a random process, that is, via diffusion or jumps. The applied electrical field, of course, influences the mean orientation more than the reorientation of the individual molecular dipoles. Since molecular dipolar reorientation is coupled with reorientation of molecules or molecular groups, the time required for macroscopic reorientation corresponds to that for the reorientation of the molecules or groups.

In an alternating electric field, the dipoles of the dielectric medium attempt to align themselves in the direction of the field. The more rapidly the direction of the alternating field is changing, the less easily they are able to achieve this. The more the adjustment of the dipoles lags behind the applied alternating field, the greater is the electrical energy consumed in this effect (power loss). The available output power or voltage is thereby decreased, since power is lost by conversion into thermal energy.

The power loss depends on the phase difference between the alternating current produced by an applied alternating voltage. When the material behaves as a perfect dielectric, the phase difference between the alternating potential and the amplitude of the current is 90° and the power loss is zero. If current and voltage are in phase, then all of the electrical energy is converted into heat and the power output is zero. The ratio of power loss N_v to power output N_b is called the dielectric *dissipation factor*, tan δ:

$$\frac{N_v}{N_b} = \frac{UI \cos (90 - \vartheta)}{UI \sin (90 - \vartheta)} \equiv \tan \delta \qquad (13\text{-}2)$$

The real power output and the power loss can also be given in terms of the real ϵ' and imaginary ϵ'' dielectric constants (relative permittivities), respectively:

$$\epsilon = \epsilon' - i\epsilon'' \qquad (13\text{-}3)$$

when ϵ is the (complex) dielectric constant. The dissipation factor is

$$\tan \delta = \frac{\epsilon''}{\epsilon'} = \frac{\epsilon \sin \delta}{\epsilon \cos \delta} \qquad (13\text{-}4)$$

ϵ' and ϵ'' may depend on the frequency ν. The function $\epsilon' = f(\nu)$ corresponds to an energy storage and the function $\epsilon'' = g(\nu)$ to an energy dissipation (Figure 13-1). The loss of energy per second, i.e., the power loss, is given as

$$N_v = E^2 2\pi\nu\epsilon \tan \delta \qquad (13\text{-}5)$$

where E is the amplitude, ν the frequency of the alternating field, ϵ the relative permittivity (dielectric constant), and tan δ the dissipation factor. The term ϵ' tan δ is called the *loss factor*, and is not the same as the dissipation factor. Materials with a high ϵ' tan δ are suitable for high-frequency-field heating, i.e., they can be welded in a high-frequency field. These materials are not suitable, on the other hand, as insulating materials for high-frequency conductors. Nonpolar plastics such as poly(ethylene), poly(styrene), poly(isobutylene), etc., have low relative permittivities ($\sim$2–3) and dissipation factors (tan δ = 10^{-4} to 8×10^{-4}). As insulating materials they are of considerable importance in high-frequency-field technology. Polar materials such as poly(vinyl chloride), by contrast, have an ϵ' tan δ that is at least 100 times greater than the corresponding value of poly(styrene) or poly(ethylene). Therefore PVC can be welded extremely well using high-frequency fields.

The glass transition temperature and other relaxation temperatures can be determined by investigating the behavior of polar macromolecules in an alternating electric field (see also Section 11.4.5). If the frequencies are low and the sample is above the glass transition temperature, then the dipoles

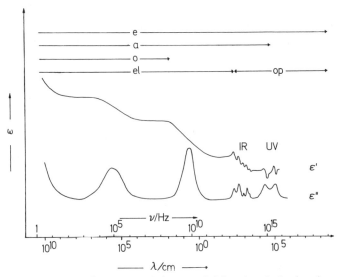

Figure 13-1. Dependence of the real relative permittivity ϵ' and the imaginary relative permittivity ϵ'' on the frequency or the wavelength (schematic); e, electron polarization; a, atomic polarization; o, orientation polarization; el, electrical region; op, optical region; ir, infrared region; uv, ultraviolet region.

oscillate in phase with the alternating field. At high frequencies and below the glass transition temperature this is no longer possible. The behavior of a polar macromolecule in an alternating field also depends on whether the dipoles are in the chain or in pendant side groups. In the case of poly(oxymethylene), $(-CH_2-O-)_n$, the dipoles are in the chain. They are only capable of orienting themselves, therefore, when the segment mobility is high, i.e., only above the glass transition temperature. In materials such as poly(vinyl ethers), $(-CH_2-CHOR-)_n$, on the other hand, the dipoles are in pendant side groups. The orientation of these flexible groups, therefore, can take place either via segmental motion in the main chain or else by side-group movement. Thus, two areas of dissipation are observed here: at low frequencies, that due to segmental mobility, and at high frequencies, that arising from orientation of the side groups. If the experiment is carried out below the glass transition temperature, then, of course, only the effect due to mobility of the side groups will be observed.

13.1.3. Dielectric Field Strength

Heat is developed within the polymer by the imaginary component of the dielectric constant. If the field is allowed to be effective for a very long time, then, because of the poor heat conductivity of the material, the heat produced may not be dissipated and the material may become hot. The imaginary component results from out-of-phase orientation of polar groups in the polymer or from conduction arising from impurities. These impurities must be of an ionic nature, since the conductivity of the polymer depends very much on the temperature. On the other hand, the electronic conductivity varies much less with temperature. Because of the strong dependence on temperature of the ionic conductivity, the accumulation of heat leads to increasingly better conductivity until, finally, a threshold (thermal breakdown) is reached. Thin films show a higher dielectric field strength than thick films, because of the better heat dissipation.

13.1.4. Tracking

Tracking is defined as a leakage current which flows across the surface of a material that is a good insulator in the dry, clean state, and occurs between charged particles consisting of conductive impurities. Since the surface resistance is two orders of magnitude lower than the specific resistance and is usually difficult to measure, the discharge resistance is tested under standard conditions. For this, a "standard impurity" is used, namely, a salt solution with an added detergent. This test solution is allowed to drip uniformly between electrodes at a standard potential difference and is separated by a standard distance on the surface of the test sample. The number of drops required before a discharge occurs is a measure of the discharge resistance. Initially, the surface becomes contaminated with conducting impurities, and when the contamination becomes sufficient, sparking occurs, which can cause decomposition or carbonization. Tracking then occurs.

Tracking does not occur in a material that does not carbonize under these conditions. If, for example, sparking produces monomer from the polymer by depolymerization, then the vaporizing monomer molecules prevent contaminating salt films from being deposited. Sparking occurs between gaps in the conducting film; an example of this is poly(methyl methacrylate). The same effect is shown when volatile degradation products are formed under the influence of sparking or arcing as, for example, in the case of poly(ethylene) or polyamides. Poly(N-vinylcarbazole), on the other hand, does not form volatile degradation products, and, therefore, has only a poor discharge resistance despite its good insulating properties.

13.1.5. Electrostatic Charging

Static electricity, or what is known as electrostatic charging, can be defined as an excess or deficiency of electrons on an insulated or nonearthed surface. Such charging can occur through friction or contact of surfaces with ionized air or by contact of two surfaces with their subsequent separation. Materials become electrostatically charged when the specific electrical conductivity is lower than about $10^{-8}\ \Omega^{-1}\ cm^{-1}$ and the relative air humidity is less than about 70%.

If two plastics are rubbed together or a plastic is rubbed against metal, different degrees of charging will be observed, depending on the materials rubbed together and the duration of rubbing. Poly(oxymethylene) rubbed against polyamide 6, for example, produces a charge of $+360$ V/cm on just one rubbing, a value of 1400 V/cm with 10 rubbings, and finally a limiting value of 3000 V/cm. If an ABS polymer with added antistatic agent is rubbed against poly(acrylonitrile), the extreme value is 120 V/cm. The same ABS polymer, however, has a limiting value of -170 V/cm when rubbed against polyamide 6. The test methods are standardized.

The origin of this behavior is the transfer of electrical charges. Regions with both positive and negative charges are thus produced on the surface of the material.

However, the charges are not homogeneously distributed. For example, "islets" of positive charges may exist in a "sea" of negative charges, and vice versa, as is seen on dusting the surface with differently charged dyestuffs. One charge type generally predominates. The sign of the overall charge depends on the position of the rubbing partner in the triboelectric series of nonmetallic materials (Table 13-1).

Because of the poor surface conductivity of most macromolecular materials, the charge which is produced can only flow away slowly. The half-lives for such dissipation of the charge are usually different for positively and negatively charged substances (Table 13-2). The half-lives, which are often high, frequently have unpleasant effects in industry and in the household, for example, in the charging of godet rolls during spinning processes or in the accumulation of dust on household articles made from plastics.

Table 13-1. Triboelectric Series of
Nonmetallic Materials

Material	Charge density in 10^{-6} C/g
Melamine resin	-14.70
Phenolic resin	-13.90
Graphite	-9.13
Epoxy resin	-2.13
Silicone	-0.18
Poly(styrene)	$+0.37$
Poly(tetrafluoroethylene)	$+3.41$
Poly(trifluorochloroethylene)	$+8.22$

Table 13-2. Half-Lives for Charge Loss
from Charged Materials

Material	Half-lives in s	
	Positively charged	Negatively charged
Cellophane	0.30	0.30
Wool	2.50	1.55
Cotton	3.60	4.80
Poly(acrylonitrile)	670	690
Polyamide 66	940	720
Poly(vinyl alcohol)	8500	3800

To a first approximation, the discharging time depends directly on the electrical resistance R and the capacity C:

$$t = k \ R \ C = RQ/U \tag{13-6}$$

It is difficult to influence the charge Q and voltage U and, consequently, the capacity C of the plastics. The discharging time, so, can only be reduced by lowering the discharging resistance. The resistance to discharging, in turn, depends on the surface resistivity and the resistance to conduction as well as on the resistance of the surrounding air. The lowest of these three resistances provides the lowest time constant, and, so determines the magnitude of the charging.

The electrostatic charging effect can be prevented by various methods. One group of methods "grounds" the charges, e.g., by neutralizing them with ionized air (as in the textile industry) or by encasing rubber pipes in metal sheaths (as at gas stations). Alternatively, the materials can be protected externally or internally with antistatic agents. If, for example, up to 30% carbon black is worked into a copolymer of ethylene and vinylidene chloride, the material still retains practically all the good properties of the plastic, but this additive increases the specific conductivity up to about $10^{-2} \ \Omega^{-1}$ cm. The material no longer becomes electrostatically charged. In the case of an external antistatic treatment, materials that utilize the humidity of the air are applied to the surface. Contrary to internal addition, this form of treatment does not affect the specific conductivity, but does change the surface resistance. Of course, external antistatic treatments are not permanent, and retreatments are required from time to time. The electrostatic charge can also be lessened if friction is lowered, e.g., by adding lubricants or by coating with poly(tetrafluoroethylene).

Conversely, the effects of the electrostatic charge can also be used commercially, namely, in electrostatic paint spraying and in the flocking of fabrics to produce velvety surfaces.

13.1.6. Electrets

Electrets are dielectric bodies that can retain an electric field for a certain time after it has been applied. They are only formed by polymers with poor electrical conductivity, for example, poly(styrene), poly(methyl methacrylate), poly(propylene), polyamides, or carnauba wax.

There are two procedures for the manufacture of electrets. In the first method, the polymer is heated to temperatures above the glass transition temperature and then an electric field is applied (e.g., 25 kV/cm) and the polymer is allowed to form a glass while still under the influence of the field. An optimum working temperature seems to be at $\sim 37°$C above the glass transition temperature T_G. In the second method, the polymer is allowed to glassify while flowing under pressure. Here, the optimum temperature appears to be at $(T_G + 57)°$C. When the electric field is removed, the bodies are positively charged on one side and negatively charged on the other. The difference in the charge diminishes only slowly, in a process that can extend over months.

As yet, the principles of electret formation are not fully understood. It is likely that both volume and surface polarizations can occur. A volume polarization is obtained with fields below ~ 10 kV/cm. That is, if an electret is parted parallel to the charged surface, two new capacitors result. With fields of more than ~ 10 kV/cm, an ionization, electronic failure, or breakdown due to the field takes place, giving a surface polarization. The polarizations at different field strengths also support this interpretation. At low field strengths, polarization opposes the electric field, which may be due to a charge migration by, for example, ionic impurities. At temperatures above the glass transition temperature, the ion separation should occur readily, and then at $T < T_G$ the ion positions become frozen in. At high field strengths, air is ionized, and the surfaces of the electrets are polarized in the same sense as used in the case of electrodes.

13.2. Electronic Conductivity

13.2.1. Basic Principles

The electronic conductivity σ of a material is determined by the number N of charge carriers per volume V as well as by their charge e and mobility μ:

$$\sigma = (N/V)e\,\mu = (N_0/V)e\,\mu\,\exp(-E^{\ddagger}/kT) \qquad (13\text{-}7)$$

The concentration of electrons increases with increasing temperature.

Since the activation energy $E^{\ddagger}$ is positive, the electronic conductivity of semiconductors increases with temperature. In contrast, the conductivity of metals decreases with increasing temperature.

The electrical properties of macromolecular semiconductors are generally characterized by the conductivity, the activation energy of the conductivity, the free radical concentration, and the thermal electromotive force (thermal emf). Since the polymers are usually in the form of amorphous powders, they are compressed into tablets. The contacts are either metal electrodes pressed into the surface or conductive pastes. The samples may not have any ionic conductivity (due to impurities) or surface conductivity and must be free of moisture, otherwise the conductivities will be too high.

To determine the thermal emf, the sample is placed between two plates at different temperatures. The thermal voltage that occurs with a difference in temperature of $1°C$ is called the Seebeck coefficient. The Seebeck coefficient is positive when the hotter pole is positive. A positive Seebeck coefficient originates from an excess of electron defects (*p*-type conductivity) and a negative coefficient from an excess of conducting electrons (*n*-type conductivity). The concentration of free radicals, measured using electron-spin resonance, need not be identical, of course, with the concentration of conducting electrons.

The specific conductivities of organic semiconductors extend into the range of those of metalloids or metals. The concentrations of charge carriers are also, at 10^9–10^{21} particles/cm^3, almost as high in some cases as those of metals (10^{21}–10^{22} particles/cm^3). The mobility of the charge carriers, on the other hand, is 10^{-6}–10^2 cm^2 V^{-1} s^{-1}, in general, considerably lower than that of metals and inorganic semiconductors (10–10^6 cm^2 V^{-1} s^{-1}). It is therefore doubtful whether the simple Brillouin zone (band) model used for inorganic semiconductors can be used with organic semiconductors. Solids with closely packed atoms and good long-range order, as, for example, with metallic semiconductors, form different energy levels of valence and conductivity bands which are well separated from each other. The energy level difference is about double the activation energy defined in Equation (13-7). With suitable excitation, electrons may leave the valence band and enter the conductivity band. Defects are produced in the valence band, which may also act as charge carriers. Electrons and defects may move freely within the bands; they have a high mobility.

Organic molecules are not so close packed as the atoms of metals or metalloids. In contrast, they are quite far apart from each other, and, in addition, are only held together by weak van der Waals forces. Thus, the electronic interaction between organic molecules is only small. Charges can only be transported by thermally activated jumping from defect to defect, and the mobility of the charge carriers is low.

Table 13-3. *Specific Conductivity of Polymers and Low-Molecular-Weight Compounds*[a]

| Material | | Temp. in °C | Specific conductivity in Ω^{-1} cm^{-1} | Activation energy in eV |
Name	Chemical constitution			
Cellulose, dry	—	25	10^{-18}	?
Gelatine, dry	—	130	2×10^{-14}	3.1
Tobacco mosaic virus	—	130	9×10^{-14}	2.9
Deoxyribonucleic acid	—	130	2×10^{-12}	2.4
Corene		15	6×10^{-18}	0.85
Ovalene		15	4×10^{-16}	0.55
Circumanthracene		15	2×10^{-13}	?

		T_p^a		
Graphite	—	25	10^5	0.025
Violanthrene	(structure)	15	5×10^{-15}	0.43
Violanthrone	(structure)	15	4×10^{-11}	0.39
Poly(methylene)	$-(CH_2)_n-$	25	$<10^{-17}$	?
Poly(vinylene)	$-(CH=CH)_n-$	25	$<10^{-8}$	?
Poly(acetylene)	$-(C≡C)_n-$	25^b	$<10^{-8}$	0.83
		25^c	$<10^{-4}$	?
Poly(phenylene)	(structure)	25	10^{-11}	0.94
Poly(p-divinylbenzene)^d	(structure) $-(CH_2-CH)_n-$	25	10^{-15}	?
			10^{-12}	?
			10^{-6}	?
			10^2	?
Poly(carbazene)	$-(N=CR)_n-$	25	$\sim 10^{-5}$	~ 0.2
Poly(azasulfene)	$-(NS)_n-$	25	~ 8	~ 0.02

$^a T_p$ is the pyrolysis temperature: 1 eV = 1.6021×10^{-19} J.
^{b}Amorphous.
^{c}Crystalline.
^{d}Oxidized and pyrolized at 500, 600, 700, and 1000° C, respectively, for the four values listed for specific conductivity.

13.2.2. Influence of Chemical Structure

The specific conductivity of a polymer depends on two factors: charge carrier transport within the individual molecule and transport from molecule to molecule.

For good intramolecular electron transport to occur, the molecule must have an extensive delocalized π-electron system. The specific conductivity thus increases with conjugated ring system size. This is the case, for example, in the series coronene, ovalene, circumanthracene, graphite (Table 13-3). For equal ring systems size, the specific conductivity increases with the extensiveness of the delocalized π-electron system, as can be seen with a comparison of violanthrene and violanthrone.

Of course, inter- and intramolecular electron transport cannot be rigorously separated from each other. For example, intermolecular electron passage is facilitated by molecular chains being in a high state of order. Thus, the specific conductivity of crystalline poly(acetylene) is four orders of magnitude higher than for the amorphous polymer. The electronic conductivity of amorphous polymers is promoted by molecular cross-linking.

For example, poly(p-phenylene), polymerized by converting p-dichlorobenzene with sodium, has practically linear polymer chains and only a relatively low specific conductivity of $10^{-11}\ \Omega^{-1}\ cm^{-1}$; presumably because the chains are not completely planar. In contrast, the polymerization of benzene with Friedel–Crafts catalysts produces cross-linked or branched poly-(phenylenes) with specific conductivities of $0.1\ \Omega^{-1}\ cm^{-1}$ and activation energies of $0.025\ eV$. Still higher specific conductivities of over $5\ \Omega^{-1}\ cm^{-1}$ are obtained for the conversion products of hexachlorobenzene with sodium. Crosslinked systems with conjugated double bonds and graphitelike structures are also produced by the oxidation and pyrolysis of poly(p-divinyl benzenes), whereby specific conductivities up to $100\ \Omega^{-1}\ cm^{-1}$ can occur.

Cross-linked polymers produced in this way cannot be worked easily. In this respect, charge transfer complexes from polymeric donors and acceptors are more advantageous. The poly(2-vinyl pyridine) and iodine complex, for example, has a specific conductivity of $10^{-3}\ \Omega^{-1}\ cm^{-1}$. It is used as a cathode in Li/I_2 batteries for implantable heart pacemakers. This solid state battery has a higher energy density than the best lead accumulators and a lifetime of about ten years.

Similar conductivities are also shown by free radical ions. For example, systems from vinyl pyridine containing copolymers with tetracyano-p-quinodimethane belong to this group. In contrast to cross-linked semiconductor polymers, these systems are soluble and can be cast into films. But they decomposed slowly in air and lose thereby their conductivity.

13.2.3. Photoconductivity

Light can produce radical ions, and thus, conductivity, in certain suitable systems. The effect finds application in what is known as Xerography. A metal cylinder is coated with a photoconducting material and negatively charged in darkness by a corona discharge in this reproduction process. The object to be reproduced is projected onto the photoconductor, and the brigher areas discharge. A black, positively charged developer encapsulated in resin is sprayed onto the latent picture so produced, and a negatively charged sheet of paper is passed over the cylinder. The copy is then heated, the resin sinters together, and the picture is fixed.

At first, diarsenic triselenide was used as the photoconducting material. Poly(N-vinyl carbazole) is now used. This polymer absorbs ultraviolet light, producing an exciton, which ionizes in an electric field. Poly(vinyl carbazole) behaves as an insulator in visible light, but can, however, be sensitized with certain electron donors to form a charge transfer complex.

Literature

Section 13.1. Dielectric Properties

N. G. McCrum, B. E. Read, and G. Williams, *Anelastic and Dielectric Effects in Polymeric Solids*, Wiley, London, 1967.

E. Fukada, Piezoelectric dispersion in polymers, *Prog. Polym. Sci. Jpn.* **2**, 329 (1971).

M. E. Baird, *Electrical Properties of Polymeric Materials*, Plastics Institute, London, 1973.

A. D. Moore, *Electrostatics and Its Applications*, Wiley, New York, 1973.

Dechema Monography, Vol. 72, *Elektrostatische Aufladung*, Verlag Chemie, Weinheim, 1974.

M. W. Williams, The dependence of triboelectric charging of polymers on their chemical compositions, *J. Macromol. Sci.—Rev. Macromol. Chem.* **C14**, 251 (1976).

P. Hedvig, *Dielectric Spectroscopy of Polymers*, Halsted, New York, 1975.

A. R. Blythe, *Electrical Properties of Polymers*, Cambridge University Press, Cambridge, 1979.

E. Fredericq and C. Houssier, *Electric Dichroism and Electric Birefringence*, Clarendon Press, Oxford, 1973.

G. M. Sessler (ed.), *Electrets* (Topics in Applied Physics Vol. *33*), Springer, Heidelberg, 1980.

13.2. Electronic Conductivity

J. E. Katon, ed., *Organic Semiconducting Polymers*, M. Dekker, New York, 1968.

W. L. McCubbin, Conduction processes in polymers, *J. Polym. Sci.* **C30**, 181 (1970).

R. H. Norman, *Conductive Rubbers and Plastics*, Elsevier, Amsterdam, 1970.

H. Meier, Zum Mechanismus der organischen Photoleiter, *Chimia* **27**, 263 (1973).

Ya. M. Paushkin, T. P. Vishnyakova, A. F. Lunin, and S. A. Nizova, *Organic Polymeric Semiconductors*, Wiley, New York, 1974.

E. P. Goodings, Polymeric conductors and superconductors, *Endeavour* **34**, 123 (1975).

E. P. Goodings, Conductivity and superconductivity in polymers, *Chem. Soc. Rev.* **5**, 95 (1976).

Y. Wada and R. Hayakawa, Piezoelectricity and pyroelectricity of polymers, *Jpn. J. Appl. Phys.* **15**, 2041 (1976).

J. M. Pearson, Photoconductive polymers, *Pure Appl. Chem.* **49**, 463 (1977).

M. Stolka and D. M. Pai, Polymers with photoconductive properties, *Adv. Polym. Sci.* **29**, 1 (1978).

R. G. Kepler, Piezoelectricity, pyroelectricity, and ferroelectricity in organic materials, *Ann. Rev. Phys. Chem.* **29**, 497 (1979).

W. E. Hatfield (ed.), *Molecular Metals*, Plenum, New York, 1980.

J. Mort, Conductive polymers, *Science* **208**, 819 (1980).

Chapter 14
Optical Properties

The appearance of a material depends on its optical properties, which, in turn, depend on the interaction with the electromagnetic field of the incident light. Two principal groups of optical properties can be distinguished: those resulting from molecular property averages and those caused by local deviations from these averages. Refraction, absorption, and diffraction phenomena belong to the first group; scattering effects belong to the second group. Conversely, the appearance of a material can also be related to its geometrical properties or its color characteristics. The former influences such properties as gloss, haze, transparency, and opacity; the latter influences parameters such as shade, saturation, and strength of color.

14.1. Light Refraction

If a ray of light is incident on a transparent body at an angle α with respect to the normal to its surface, it passes inside the body and is found to form a different angle β with respect to that normal (Figure 14-1): The light is "refracted." The refractive index n is a numerical measure of the refraction and depends on the angle of incidence α and the angle of refraction β:

$$n = \frac{\sin \alpha}{\sin \beta} = \frac{\sin \alpha'}{\sin \beta'} \tag{14-1}$$

The refractive index varies with the wavelength of the incident light. The Abbé number ν is given as a measure of this "dispersion." ν is obtained from three refractive index measurements at the wavelengths 656.3, 589.3, and 486.1 nm:

$$\nu = \frac{n_{589} - 1}{n_{486} - n_{656}} \tag{14-2}$$

The capacity to separate the colors of white light increases as ν decreases.

The refractive index n of a material depends according to Lorenz–Lorentz relationship, on the polarizability P of all the molecules residing in a uniform field:

$$\frac{n^2 - 1}{n^2 - 2} = \frac{4}{3}\pi P = \frac{4}{3}\pi N\alpha = \frac{4}{3}\pi N\frac{\mu}{E} \tag{14-3}$$

The polarizability P is given by the number N of molecules per unit volume and the polarizability α of the isolated molecule. The polarizability α in turn, depends on the dipole moment μ induced by an electric field of strength E. Consequently, both α and n increase with increasing number and mobility of electrons in the molecule. Thus, carbon has a much higher polarizability than hydrogen. Since for this reason the hydrogen contribution to the polarizability can, to a first approximation, be ignored, most carbon–carbon chain polymers have about the same refractive index (1.5). Deviations from this "normal value" only occur when there are large side groups [e.g., poly(N-vinyl carbazole)] or if highly polar groups are present (i.e., fluorine-containing polymers) (see also Figure 14-2). Further, on the basis of the molecular structure, one can estimate that the refractive indices of all organic polymers should lie within a range of only 1.33–1.73.

14.2. Light Interference and Color

14.2.1. Basic Principle

Some of the light incident on a homogeneous, transparent body is reflected from the surface (external reflection) and some passes inside, where it is reflected at an interior boundary of the body (internal reflection). According to Fresnel, the relationship between the intensity of the reflected light I_r, and the intensity of the incident light I_0 involves the angle of incidence α and the angle of refraction β (for angle definitions, see Figure 14-1):

$$R = \frac{I_r}{I_0} = \frac{1}{2}\frac{\sin^2(\alpha - \beta)}{\sin^2(\alpha + \beta)} + \frac{\tan^2(\alpha - \beta)}{\tan^2(\alpha + \beta)} \tag{14-4}$$

The reflectivity R is small for small α and begins to increase sharply for high α (Figure 14-3).

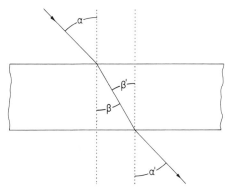

Figure 14-1. Definition of the angle of incidence α and angle of refraction β; for light incident on a homogeneous plate with plane parallel sides, $\beta = \beta'$, and if the two sides are in contact with the same medium, $\alpha = \alpha'$, i.e., the ray emerges parallel to its initial direction but displaced a distance that depends on the refractive index n.

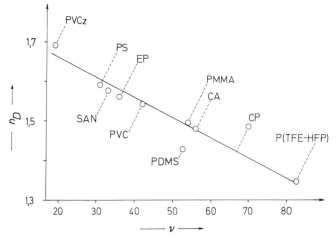

Figure 14-2. The relationship between the refractive index n_D for the sodium D line (589.3 nm) and the Abbé dispersion ν for various polymers [see Equation (14-12)]. PVCz, poly(N-vinyl carbazole); P(TFE-HFP), copolymer of tetrafluoroethylene and hexafluoropropylene. For other abbreviations, see Table VII-6, Appendix.

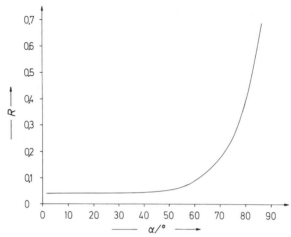

Figure 14-3. The reflectivity R as a function of the angle of incidence α for a material with $n = 1.5$.

14.2.2. Iridescent Colors

Iridescent colors can arise from the interference of light incident on films consisting of many layers. At each interface between layers, only a small portion of the incident light is reflected. The rest (neglecting absorption) passes through to the next interface, where it is again partly reflected and partly transmitted. If all the layers are of equal thickness, then for particular wavelengths (which depend on the optical densities and thickness of the films) the light reflected at the interfaces may be in phase, so that constructive interference occurs and the reflected light will have high intensity. If two polymers of refractive indices n_1 and n_2 make up alternate layers of thickness d_1 and d_2, then the wavelength λ_m of the mth-order reflections for perpendicularly incident light is given by

$$\lambda_m = \frac{2}{m}(n_1 d_1 + n_2 d_2) \tag{14-5}$$

The relative intensities of the individual wavelengths depend on the optical density fractions of the two polymers, i.e.,

$$f_1 = \frac{n_1 d_1}{n_1 d_1 + n_2 d_2} \tag{14-6}$$

When the optical densities are equal ($f_1 = f_2 = 0.5$), the reflections will be suppressed for even-number orders and they will be of maximum intensity for odd orders. If $f_1 = 0.33$, on the other hand, third-order reflections are suppressed. The first-order reflections are still strong, and the second-, fourth-, etc., order reflections have less than maximum intensity. When the first-order reflections is $\lambda_1 = 1$ μm for $f_1 = 0.50$, there is a strong reflection for $(1.5/3)$ μm $= 0.5$ μm, no reflection for $(1.5/4)$ μm $= 0.375$ μm, etc. Such a film would reflect in the near-infrared (1.5 μm) and in the blue–green (0.5 μm).

Bandwidths are larger for variable layer thickness. In some circumstances, the whole visible spectrum is reflected when there is a suitable number of layers of two polymers with the right choice of refractive indices and the right choice of the various layer thicknesses. Such films have a metallic appearance.

14.2.3. Light Transmission and Reflection

Total reflection occurs when the incident light is reflected without loss. This is especially important in the case of total internal reflection, since this principle is used in what is called fiber optics.

Total internal reflection only occurs above a quite specific minimum (critical) internal angle of incidence. The relationship $\sin \alpha \geqq 1/n_1$ is valid for a material of refractive index n_1 in air. Consequently, $\alpha_{\text{crit}} = 42°$ for $n_1 = 1.5$. The light is totally reflected on the interior interface and in a suitable array will be transmitted in a zigzag path through the system (Figure 14-4).

When the light transmitter is surrounded by air, the optically effective external surface is free. Surface scratches and dust deposits lead to light scattering and, consequently, loss of light intensity. Therefore, a smooth housing consisting of a transparent material of lower refractive index n_2 is used. The refractive index difference $n_1 - n_2$ should be as large as possible since it determines the light entry angle $2\alpha_0$ via

$$n_0 \sin \alpha_0 = (n_1^2 - n_2^2)^{0.5} \tag{14-7}$$

The entry angle $2\alpha_0$ is the maximum entry angle for the transmission of light through the light transmitter in a surrounding medium of refractive index n_0 (see Figure 14-4). For example, a core of poly(methyl methacrylate) and covering of partially fluorinated polymers, and a core of high-purity soda glass in a covering of poly(tetrafluoroethylene-co-hexafluoropropylene), have been introduced technologically for the transmission of, respectively, visible light and ultraviolet light.

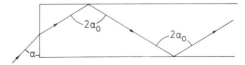

Figure 14-4. The principle of light transmission inside a body. $2\alpha_0$ is the entrance angle.

With flexible bundles of light transmitters, light can, for example, be transmitted "around corners," and one can even "look around corners." Light transmitters are used in medicine to illuminate or observe internal organs, in industry for rear lights on autos, and for the postmarking of postage stamps on letters and parcels, etc.

14.2.4. Transparency

When light is incident perpendicularly on an optically homogeneous sample, since the angles $\alpha = 0$ and $\beta = 0$, the Fresnel equation (14-4) reduces to

$$R_0 = \frac{(n - 1)^2}{(n + 1)^2} \qquad (14\text{-}8)$$

The internal transmittance τ_i (transmittivity, transparency) is thus

$$\tau_i = 1 - R_0 \qquad (14\text{-}9)$$

The refractive index is about $n \approx 1.5$ for most polymers. Consequently, the transparency can be a maximum of 96%, with at least 4% of the light being reflected at the polymer–air interface.

This ideal transparency is only rarely achieved, since the light is always absorbed and/or scattered to some extent. The most transparent polymer, poly(methyl methacrylate), has a maximum transparency of 92% (Figure 14-5) and this only in the range of about 430–1110 nm. On each side of this range, the transparency decreases because of absorption. Polymers generally absorb infrared radiation. Exceptions are the halogenated poly(ethylenes).

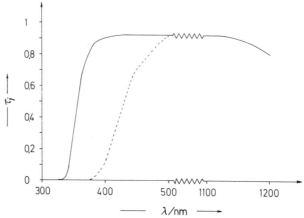

Figure 14-5. Internal transmittance as a function of wavelength for poly(methyl methacrylate). The maximum possible transparency of 96% is almost achieved in the region 430–1115 nm. The broken line indicates the shift obtained with a uv absorber.

There is an industrial distinction between transparent and translucent materials. Transparent bodies transmit up to more than 90%; they are still mostly clear even for large thicknesses. Translucent bodies have light transmittances of less than 90% and are only clear for small thicknesses. They are also called contact clear, since the material is indeed turbid but appears clear as a packaging material when in contact with the contents.

The "hiding power" of a paint can, to a first approximation, also be estimated with the Fresnel equation. In this case, the refractive indices, n_1, for the pigment, and n_2, for the polymer of pigmented paints, are to be considered:

$$R_0 = (n_1 - n_2)^2 / (n_1 + n_2)^2 \qquad (14\text{-}10)$$

Thus, the hiding power of a paint increases with increasing refractive index difference. For this reason, rutile (a TiO_2 modification with $n_D = 2.73$) is almost exclusively used when a white pigment is required. Other white pigments only have hiding powers of 78 (anatase, another TiO_2 modification), 39 (ZnS), 18 (lithopone, a mixture of 28% ZnS and 72% $BaSO_4$) and 14% (ZnO) relative to rutile. Microporous fillers are more advantageous than compact fillers, since the polymer binder cannot penetrate the pores, and, so, air remains as an inclusion. The refractive index difference between air and filler or binder brings an additional hiding power. Small variations in polymer composition can produce large differences in hiding power because of the exponential increase in hiding power with refractive index difference.

Hiding power is not only influenced by the reflection: it is also influenced by light scattering. Light incident on a particle will be scattered in all directions (see Section 9.5). The scattering intensity increases with the particle size. Back-scattering decreases with particle size, and a large back-scattering is desirable for good hiding power. Consequently, the hiding power as a function of particle size passes through a maximum. The scattering intensity increases with increasing pigment concentration. If the pigment concentration is too high, the same light ray will be scattered many times. Multiple scattering lowers the relative scattering intensity, and the hiding power decreases. This loss in hiding power becomes marked when the interparticle distance becomes less than three times the particle diameter.

14.2.5. Gloss

Gloss is defined as the ratio of the reflection of the sample to that of a standard. In the paint industry, for example, the standard is a sample of refractive index $n_D = 1.567$. Consequently, the gloss as a ratio of two reflections depends, according to Equation (14-4), on the refractive indices of the test sample and the standard, as well as on the angles of incidence and refraction of the light (Figure 14-6). The gloss of the polymer increases with increasing refractive index.

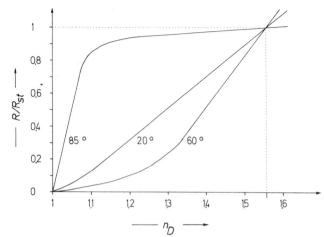

Figure 14-6. Gloss (R/R_{st}) as a funtion of refractive index n_D of the sample for various angles of incidence. A body with $n_D = 1.567$ is chosen as standard.

The theoretically maximum possible gloss calculated in this way is only rarely achieved in practice. The surfaces are always a little uneven. Also, optical inhomogeneities below the surface, that is, in the medium itself, cause marked light scattering. The relative fraction of light scattered from the surface and from the medium depends on the angle of incident light. In general the two contributions can be separated by measuring the scattering in air and then, after immersion of the test sample, in a liquid having the same refractive index as the test sample. By subtraction, the light scattering fraction coming from the surface of the test sample can be obtained.

Glistening is a gloss phenomenon at preferential places. It is produced by an intensified, directional light reflection and/or a color and light intensity contrast between glistening positions and their surroundings. Glistening occurs in films and shaped structures because of added metal pigments, and in threads and fibers because of their triangular or trilobal structure.

14.3. Light Scattering

14.3.1. Phenomena

All considerations so far discussed are only valid for optically homogeneous systems. In optically inhomogeneous systems, the medium acts in two ways on an electromagnetic wave traversing it. On the one hand, the amplitude and the phase of the wave are altered, that is, the wavefront is distorted. In optical terms, there is a loss in "resolving power"; in the plastics industry terminology, there is a loss in "clarity".

On the other hand, the electromagnetic wave loses some of its energy by scattering on encountering an inhomogeneity (see Section 9.4). The contrast lost because of forward scattering is called "haze." The combination of contrast loss by forward and backward scattering makes a sample "milky."

The internal transmittance of optically inhomogeneous materials is given by the reflected R, scattered f_{sc}, and absorbed f_{abs} fractions:

$$\tau_i = 1 - R - f_{sc} - f_{abs} \qquad (14\text{-}11)$$

The reflected fraction can be eliminated by immersing the sample in a liquid of the same refractive index. In this case, the changes in both scattering and absorption are proportional to the thickness of the sample L. The proportionality coefficient is, consequently, given by the absorption coefficient K and the scattering coefficient, the so-called turbidity S:

$$\tau_i = 1 - (K + S)L \qquad (14\text{-}12)$$

Equation (14-12) assumes a once-only scattering. For the general case, one starts from an infinitely thin sample, and integration gives

$$\tau_i = \exp\left[-(K + S)L\right] \qquad (14\text{-}13)$$

Previously, the sum $K + S$ was known as the extinction coefficient.

The Kubelka–Munk theory relates the extinction coefficient to the reflection. In the simplest case, it is assumed that light is only scattered in two directions: in the incident and in the backward direction for an incident ray normal to the surface of the test sample. Also, both incident light and emitted light are diffuse. According to Kubelka and Munk, then,

$$\frac{K}{S} = \frac{(1 - R_\infty)^2}{2R_\infty} \qquad (14\text{-}14)$$

where R_∞ is the reflection for complete hiding power. Theoretically, the reflected fraction for incomplete hiding power by films of finite thickness is given by

$$R = \frac{1 - R_{sub}\,(a - b\,\cotanh\,bSL)}{a - R_{sub} + b\,\cotanh\,bSL} \qquad (14\text{-}15)$$

where

$$a = (S + K)/S \qquad \text{and} \qquad b = (a^2 - 1)^{1/2}$$

14.3.2. Opacity

A light-scattering body appears opaque when there is a variation in refractive index or differences in the orientation of anisotropic volume elements, or both.

Local variations in the refractive index only lead to opacity when different structures are present whose dimensions are greater than the wavelength of incident light. On the other hand, these structures may not be too large, since an infinitely large single crystal does not scatter light.

Consequently, the clarity of a material can be considerably increased by decrease in the structure size. On the other hand, approach of the refractive index values of the two phases to each other leads to only a small increase in the clarity. If the refractive index of PVC is greater than that of the disperse phase in a PVC/ABS mixture, the material appears milky yellow in reflected light. It is milky blue in the reverse case.

Effects resulting from variations in the refractive index can be distinguished from those resulting from variation in orientation of anisotropic volume elements by the use of polarized light. The horizontally polarized scattering observed with incident vertically polarized light (H_v scattering) originates from anisotropy of the scattering elements. The V_v scattering depends on the anisotropy of the scattering elements as well as on the differences in refractive index.

An ordered spherulite is spherically symmetric. Consequently, the scattering in the interior of an H_v sample (see also Figure 5-25) should be zero. A finite scattering intensity in the center thus indicates disorder. The size of the spherulite can be determined from the angle at which maximum scattering occurs.

Lamellar structures with order over regions of dimensions greater than the wavelength of incident light are less optically heterogenous than spherulitic structures. Thus, they scatter less light; they are more transparent. Consequently, poly(ethylene) films drawn and oriented under certain conditions are clear, although the samples are crystalline and even have superstructures of dimensions greater than the wavelength of incident light.

14.4. Color

14.4.1. Introduction

Color is an impression of the senses. It occurs only in the presence of light. What are known as the six psychological basic colors are classified as bright colors (red, green, yellow, blue) and neutral colors (white, black). The sense of color can be produced in different ways:

With *self-illuminating colors*, color mixing is additive. An example is a mixture of colored light produced by a prism.

Non-self-illuminating colors lead to subtractive color mixtures. Such mixing occurs by absorption or by scattering. In this case, a distinction between transmission and remission must be made. If, for example, a neutral color light source is placed behind a transparent colored body, then a certain wavelength region of the light is selectively absorbed. The observer then sees the remaining wavelength region as a transmitted color.

If, on the other hand, light of neutral color falls on a body at a certain angle, then a surface color due to absorption as well as to scattering is observed because of the selective return radiation. White pigments predominantly reemit by scattering. Colored pigments absorb light, whereby inorganic pigments scatter relatively strongly, while organic pigments scatter only weakly. The color impression received is composed of both color re-emission and gloss together and the proportions of these two components vary with the angle of observation. Ideal mat surfaces reflect equally well in all directions, but ideal glossy surfaces reflect preferentially in the direction of observation. All the colors can be classified in various color systems. The color systems can, in turn, be subdivided into two classes: those that are based on collections of physical color samples and those that are not. The first group can be further subclassified into (a) random collections that do not allow a given intermediate color to be derived, (b) subjective systems without regular guidelines, and (c) systems based on set principles. The best-known system in the last group is the Munsell system (Section 14.4.2).

The most important of the systems not based on collections of physical color samples is what is known as the CIE system. This system was developed by the *C*ommission *I*nternational de l'*E*clairage (International Commission on Illumination).

14.4.2. Munsell System

The Munsell system arranges color samples according to a definite systematic process. The differences between the individual colors have been revised many times over the years; the improved system is known as the Munsell Renotation System. Two different MR systems are used, one for glossy, the other for mat surfaces.

A color is described by three different parameters in the MR system: chroma, hue, and value. The *hue* (nuance, shade) is classified in the same way as the spectral colors: it may, for example, be blue, blue-green, yellow, yellow-red, red-yellow, etc. The *value* (brightness) gives the degree of darkness of a given color shade; for example, a reddish color may be bright pink or dark red. The *chroma* (color strength, brilliance, saturation) describes the grey concentration of the sample; a color is more saturated when the white component is smaller. Neutral colors are described by the chroma alone, bright colors are, in contrast, described by chroma, hue, and value.

One page is reserved for each hue in the Munsell color book. Colors of the same hue are arranged two dimensionally on this page according to value and

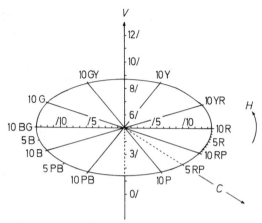

Figure 14-7. Munsell coordinate system with hue (H), value (V), and chroma (C).

chroma. The color quality (hue, value, chroma) is then described by a combination of letters and numbers. The hue is specified by ten letters which give the five main colors (*R*ed, *Y*ellow, *G*reen, *B*lue, *P*urple) and the five intermediate combinations (for example, GY = Green-Yellow). A further subdivision of the colors is made with the numbers 1–10, which are written before the letters. Value and chroma, on the other hand, are described only with numbers, which follow the letters and are separated from them by a sloped line or stroke. Thus, a certain greenish yellow color is, for example, fully described as 5 GY 2/6 according to the Munsell terminology.

The Munsell color system is sometimes also shown graphically (Figure 14-7). The colors are then arranged anticlockwise in a circle, that is, in the direction Y-GY-G-BG-B-PB-P-RP-R-YR. The value is written along an axis from top to bottom which passes through the center of the circle. The chroma is given as the distance from the center of the circle.

The color quality of the colorant is influenced by a whole series of factors. Thus, the same colorant may have quite different Munsell classifications according to its surroundings (Table 14-1).

Table 14-1. Influence of the Environment (Surface Structure,
Gloss, Concentration, etc.,) on the Color Quality of a
Green Phthalocyanine Pigment

Material	Hue	Value/Chroma
Pigment powder	4.4 BG	4.3/6.3
Printing ink, solid	4.9 G	6.6/12.6
Printing ink, 50%	5.7 G	8.4/3.2
Acrylic lacquer, 25%	6.0 G	5.8/10.2
Acrylic lacquer, 1%	9.7 G	8.6/3.7

14.4.3. CIE System

In contrast to the Munsell system, the CIE system does not use color samples. The CIE system is based on the Grassmann laws, which state that a color valency is completely described by the sum of the vector products of three color values and their color value proportions:

$$F = xX + yY + zZ \qquad (14\text{-}16)$$

The color valency is the color sensation registered by the eye, which depends on light source and observer. Consequently, what is known as a standard observer is defined in the CIE system, and this standard observer has a quite definite eye sensitivity. In addition, three standard light sources A, B, and C are introduced, and these correspond to incandescence (at a temperature of 2854 K for black bodies), the midday sun, and the daylight on an overcast day. The three color values can be more or less freely chosen. But for various reasons, three monochromatic basic colors of wavelengths 700.0 nm (red), 546.1 nm (green), and 435.8 nm (blue) are used as what are known as basic colors or primary valencies. In principle, the test color can be reproduced by mixing the three basic colors. Certain colors are characterized by two positive and one negative color proportion instead of by three positive color proportions. In this case, a third basic color is added to the test color, and not to the comparison or standard color. Characterization by negative numbers, however, can be avoided if three hypothetical (or virtual) basic colors are recalculated and introduced. These virtual colors, or standard valencies, are then used in Equation (14-16).

Thus, a color is characterized by three numbers, X, Y, and Z, and, consequently, can be shown as a point in a three-dimensional coordinate system. Since each kind of color may be defined by its color proportion, or what is known as the standard color value proportion (trichromatic coefficient), and, by definition, the sum of all the trichromatic coefficients is unity, then, the color can be shown as a point in a two-dimensional diagram. This diagram is also known as the chromaticity diagram or color table (Figure 14-8). The line joining the points at 400 and 700 nm is called the purple line. Thus, all real colors lie within the horseshoe structure in Figure 14-8.

The CIE system, however, is not a metric sensitivity system, that is, color separations are not equally large for equally large sensitivity differences. The human eye has a maximum sensitivity at 555 nm, and often sees a certain color displaced from rather than at the absorption spectrum maximum. The CIE system and the Munsell system are not directly interconvertible for this reason. The "shade" of the Munsell system, however, corresponds to a ray coming from the neutral color point for the wavelength of the same color shade in the CIE system. The "chroma" of the Munsell system can be assigned to the spectral color proportions of the CIE system which are given as

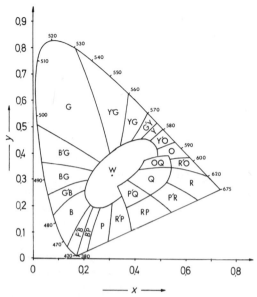

Figure 14-8. CIE chromaticity diagram for the trichromatic coefficients y and x. W, white point or neutral color point. The other letters give the positions of the commonly known colors: R, red; R'O, reddish orange; O, orange; Y'O, yellowish orange; Y, yellow; G'Y, greenish yellow; YG, yellow-green; Y'G, yellowish green; G, green; B'G, bluish green; BG, blue-green; G'B, greenish blue; B, blue; P'B, purplish blue; B'P, bluish purple; P, purple; R'P, purplish red; OQ, orange pink; Q, pink; P'Q, purplish pink. The numbers around what is called the spectral color course correspond to the wavelengths of light in nanometers.

coincentric lines about the neutral color point. Finally, the "value" of the Munsell system corresponds to the brightness density of the CIE system.

Literature

14.0. General Reviews

R. S. Hunter, *The Measurement of Appearance*, Wiley, New York, 1975.

14.2. Light diffraction; 14.3. Light Scattering

N. S. Capany, *Fiber Optics*, Academic Press, New York, 1967.

G. Kortüm, *Reflexionsspektroskopie*, Springer, Berlin, 1969.

T. Alfrey, Jr., E. F. Gurnee, and W. J. Schrenk, Physical optics of iridescent multilayered plastic films, *Polym. Eng. Sci.* **9**, 400 (1969).

U. Zorll, New aspects of gloss of paint films and its measurement, *Progr. Org. Coatings* **1**, 113 (1972).

G. Ross and A. W. Birley, Optical properties of polymeric materials and their measurement, *J. Phys. D* **6**, 795 (1973).

H. Dislich, Plastics as Optical Materials, *Angew. Chem. Int. Ed. Eng.* **18**, 49 (1979).

14.4. Color

R. M. Evans, *An Introduction to Color*, Wiley, New York, 1948.

Munsell Book of Color, Munsell Color Co., Baltimore, pocket edition (mat) 1929–1960, cabinet edition (renotations, glossy), 1958.

G. Wyszecki, *Farbsysteme*, Musterschmidt, Berlin, 1960.

R. W. Burnham, R. M. Hanes, and C. J. Bertleson, *Color: A Guide to Basic Facts and Concepts*, Wiley, New York, 1963.

W. D. Wright, *The Measurement of Color*, third ed., Hilger and Watts, London, 1964.

W. Schultze, *Farbenlehre und Farbenmessung*, seventh ed., Springer, Berlin, 1966.

D. B. Judd, *Color in Business, Science and Industry*, third ed. Wiley, New York, 1975.

Index

Prefixes of chemical compounds such as 2-, o-, p-, D-, N-, α-, it-, etc. were disregarded in the alphabetical arrangement. Page numbers in parentheses indicate numerical data for chemical compounds without extensive description. Pages 1–507 will be found in Volume 1, pages 509–end in Volume 2.